Mechanisms in Classical Conditioning
A Computational Approach

What mechanisms are involved in enabling us to generate predictions of what will happen in the near future? Although we use associative mechanisms as the basis to predict future events, such as using cues from our surrounding environment, timing, attentional and configural mechanisms are also needed to improve this function. Timing mechanisms allow us to determine when those events will take place. Attentional mechanisms ensure that we keep track of cues that are present when unexpected events occur and that we disregard cues present when everything happens according to our expectations. Configural mechanisms make it possible to combine separate cues into one signal that predicts an event different from that predicted individually by separate cues. Written for graduates and researchers in neuroscience, computer science, biomedical engineering and psychology, the author presents neural network models that incorporate these mechanisms and shows, through computer simulations, how they explain the multiple properties of associative learning.

DR. SCHMAJUK has been an Associate Professor of Biomedical Engineering in Buenos Aires (Argentina), an Assistant Professor of Psychology at Northwestern University, and is presently a Professor of Psychology and Neuroscience at Duke University. Here he has developed several neural network models of classical conditioning, operant conditioning, animal communication, creativity, spatial learning, cognitive mapping and prepulse inhibition. Previous books by this author: *Animal Learning and Cognition. A Neural Network Approach* (1997), *Occasion Setting. Associative Learning and Cognition in Animals* (1998), *Latent Inhibition and its Neural Substrates* (2002).

Mechanisms in Classical Conditioning

A Computational Approach

NESTOR SCHMAJUK

CAMBRIDGE UNIVERSITY PRESS
Cambridge, New York, Melbourne, Madrid, Cape Town, Singapore,
São Paulo, Delhi, Dubai, Tokyo

Cambridge University Press
The Edinburgh Building, Cambridge CB2 8RU, UK

Published in the United States of America by Cambridge University Press, New York

www.cambridge.org
Information on this title: www.cambridge.org/9780521887809

© Cambridge University Press 2010

This publication is in copyright. Subject to statutory exception
and to the provisions of relevant collective licensing agreements,
no reproduction of any part may take place without the written
permission of Cambridge University Press.

First published 2010

Printed in the United Kingdom at the University Press, Cambridge

A catalogue record for this publication is available from the British Library

Library of Congress Cataloguing in Publication data
Schmajuk, Nestor A.
  Mechanisms in classical conditioning : a computational approach / Nestor Schmajuk.
    p.   cm.
  ISBN 978-0-521-88780-9 Hardback
  1. Conditioned response–Computer simulation.   2. Neural networks
(Computer science)   I. Title.
  BF319.S357 2010
  153.1'526–dc22   2009036649

ISBN 978-0-521-88780-9 Hardback

Cambridge University Press has no responsibility for the persistence or
accuracy of URLs for external or third-party internet websites referred to in
this publication, and does not guarantee that any content on such websites is,
or will remain, accurate or appropriate.

Every effort has been made in preparing this publication to provide accurate
and up-to-date information which is in accord with accepted standards and
practice at the time of publication. Although case histories are drawn
from actual cases, every effort has been made to disguise the identities of
the individuals involved. Nevertheless, the authors, editors and publishers
can make no warranties that the information contained herein is totally
free from error, not least because clinical standards are constantly
changing through research and regulation. The authors, editors and
publishers therefore disclaim all liability for direct or consequential
damages resulting from the use of material contained in this publication.
Readers are strongly advised to pay careful attention to information provided
by the manufacturer of any drugs or equipment that they plan to use.

To my wife, Mabel, and my children Mariana, Gabriela, and Hans.

Contents

Preface ix
Acknowledgments xi
Abbreviations xv

Part I Introduction 1

1 Classical conditioning: data and theories 3

Part II Attentional and associative mechanisms 19

2 An attentional–associative model of conditioning 21

3 Simple and compound conditioning 34

4 The neurobiology of fear conditioning 57

5 Latent inhibition 64

6 The neurobiology of latent inhibition 93

7 Creativity 119

8 Overshadowing and blocking 137

9 Extinction 165

10 The neurobiology of extinction 234

Part III Configural mechanisms 243

11 A configural model of conditioning 245

12 Occasion setting 257

13 The neurobiology of occasion setting 316

Part IV Attentional, associative, configural and timing mechanisms 357

14 Configuration and timing: timing and occasion setting 359

15 Attention and configuration: extinction cues 392

16 Attention, association and configuration: causal learning and inferential reasoning 403

Part V Conclusion: mechanisms of classical conditioning 415

17 Conclusion: mechanisms of classical conditioning 417

References 425
Author Index 460
Subject Index 468

Preface

This book extends the application of the neural network models described in my previous book on "Animal learning and cognition: A neural network approach" to a whole range of important classical conditioning paradigms, including recovery from overshadowing, recovery from blocking, backward blocking and recovery from backward blocking; extinction, and occasion setting, as well as the neurophysiology of some of those phenomena.

In the last decades, models of conditioning have shown increasing completeness and precision. This book describes a number of computational mechanisms (associations, attention, configuration, and timing) that first seemed necessary to explain a small number of conditioning results and then proved able to account for a large part of the extensive body of conditioning data. These computational mechanisms are implemented by artificial neural networks, which can be mapped onto different brain structures. Therefore, the approach permits to establish clear brain-behavior relationships.

The book is organized as follows. Part I presents major classical conditioning data and describes several theories proposed to explain them. Part II presents a neural network theory, which includes attentional and associative mechanisms, and applies it to the description of conditioning, latent inhibition, overshadowing and blocking, extinction, and creative processes. In addition, it examines the neurobiological bases of latent inhibition and extinction. Part III describes another neural network, which includes configural mechanisms, and applies it to the description of occasion setting. In addition, it examines the neurobiological bases of occasion setting. Finally, Part IV shows how the combination of attentional, associative, configural and timing mechanisms applies to timing in occasion setting, extinction cues, causal learning and inferential reasoning.

In addition to summarizing recent progress in conditioning theory, the material in the book suggests directions for future efforts. The ability of the models both to replicate and explain behavior opens the door to the design of novel experiments, as well as to the conception of more effective therapeutic methods to treat some psychological disorders and the development of adaptive robots.

Acknowledgments

This book reflects the work during the years I have spent so far at Duke University. I want to thank my collaborators during this time, including Ying Wan Lam, Jeffrey Gray, Catalin Buhusi, Jef Lamoureux, Peter Holland, Landon Cox, Beth Christiansen, Jose Larrauri, Kevin LaBar, Gunes Kutlu, Diana Aziz and Margaret Bates. Many of them were my colleagues at the time of our collaborations; others were students who later became colleagues.

I thank the many people who gave me feedback at different stages of this project. The list is long and includes Andy Barto, Mark Bouton, Cody Brooks, Gonzalo de la Casa, Jennifer Higa, John Moore, Santiago Pellegrini, Oskar Pineño, Juan Manuel Rosas, John Staddon, Mabel Tyberg, Lise Wallach and Michael Wallach. I am also grateful to my editor, Martin Griffiths, for his help during the preparation of the book.

This book includes material from the following papers:

1. Schmajuk, N.A., Lam, Y.W. & Gray, J.A. (1996). Latent inhibition: a neural network approach. *Journal of Experimental Psychology: Animal Behavior Processes*, **22**, 321–349. (With kind permission from the American Psychological Association)
2. Buhusi, C. & Schmajuk, N.A. (1996). Attention, configuration, and hippocampal function. *Hippocampus*, **6**, 621–642. (With kind permission from John Wiley & Sons, Inc.)
3. Schmajuk, N.A. & Buhusi, C. (1997). Occasion setting, stimulus configuration, and the hippocampus. *Behavioral Neuroscience*, **111**, 235–258. (With kind permission from the American Psychological Association)
4. Schmajuk, N.A., Lamoureux, J. & Holland, P.C. (1998). Occasion setting and stimulus configuration: a neural network approach. *Psychological Review*, **105**, 3–32. (With kind permission from the American Psychological Association)

5. Schmajuk, N. A. & Buhusi, C. V. (1997). Spatial and temporal cognitive mapping: a neural network approach. *Trends in Cognitive Sciences*, **1**, 109–114. (With kind permission from Elsevier)
6. Buhusi, C. V., Gray, J. A. & Schmajuk, N. A. (1998). The perplexing effects of hippocampal lesions on latent inhibition: a neural network solution. *Behavioral Neuroscience*, **112**, 316–351. (With kind permission from the American Psychological Association)
7. Schmajuk, N. A., Buhusi, C. V. & Gray, J. A. (1998). The pharmacology of latent inhibition: a neural network approach. *Behavioural Pharmacology*, **9**, 711–730. (With kind permission from Lippincott, Williams & Wilkins)
8. Buhusi, C. V. & Schmajuk, N. A. (1999). Timing in simple conditioning and occasion setting: a neural network approach. *Behavioral Processes*, **45**, 33–57. (With kind permission from Elsevier)
9. Schmajuk, N. A., Cox, L. & Gray, J. A. (2001). Nucleus accumbens, entorhinal cortex and latent inhibition: a neural network approach. *Behavioral Brain Research*, **118**, 123–141. (With kind permission from Elsevier)
10. Schmajuk, N. A., Christiansen, B. A. & Cox, L. (2000). Haloperidol reinstates latent inhibition impaired by hippocampal lesions: data and theory. *Behavioral Neuroscience*, **114**, 659–670. (With kind permission from the American Psychological Association)
11. Schmajuk, N. A. (2001). Hippocampal dysfunction in schizophrenia. *Hippocampus*, **11**, 599–613. (With kind permission from John Wiley & Sons, Inc.)
12. Schmajuk, N. A., Gray, J. A. & Larrauri, J. A. (2005). A pre-clinical study showing how dopaminergic drugs administered during pre-exposure can impair or facilitate latent inhibition. *Psychopharmacology*, **177**, 272–279. (With kind permission of Springer Science and Business Media)
13. Schmajuk, N. A. (2005). Brain-behaviour relationships in latent inhibition: a computational model. *Neuroscience and Biobehavioral Reviews*, **29**, 1001–1020. (With kind permission from Elsevier)
14. Schmajuk, N. A. & Larrauri, J. A. (2006). Experimental challenges to theories of classical conditioning: application of a computational model of storage and retrieval. *Journal of Experimental Psychology: Animal Behavior Processes*, **32**, 1–20. (With kind permission from the American Psychological Association)
15. Schmajuk, N. A., Larrauri, J. A. & LaBar, K. (2006). Reinstatement of conditioned fear: an attentional-associative model. *Behavioral Brain Research*, **177**, 242–253. (With kind permission from Elsevier)
16. Larrauri, J. A. & Schmajuk, N. A. (2008). Attentional, associative, and configural mechanisms in extinction. *Psychological Review*, **115**, 640–676. (With kind permission from the American Psychological Association)

17. Schmajuk N. A. & Larrauri J. A. (2008). Associative models can describe both causal learning and conditioning. *Behavioural Processes*, **77**, 443–445. (With kind permission from Elsevier)
18. Schmajuk, N. A., Aziz, D. R. & Bates, M. J. B. (2009). Attentional-associative interactions in creativity. *Creativity Research Journal*, **21**, 92–103. (With kind permission from Taylor & Francis Group)
19. Schmajuk, N. (2008). Attentional and error-correcting associative mechanisms in classical conditioning. *Journal of Experimental Psychology: Animal Behavior Processes*, in press. (With kind permission from the American Psychological Association)
20. Schmajuk, N. (2010). From latent inhibition to retrospective revaluation: an attentional–associative model. In *Latent Inhibition*, eds. I. Weiner & B. Lubow. New York: Cambridge University Press. (With kind permission from Cambridge University Press)
21. Schmajuk, N. A. & Kutlu, G. M. (2009). The computational nature of associative learning. *Behavioral Brain Science*, **32**, 223–224. (With kind permission from Cambridge University Press)
22. Dunsmoor, J., and Schmajuk, N. A. (2009). Interpreting patterns of brain activation in human fear conditioning with an attentional–associative learning model. *Behavioral Neuroscience*, **123**, 851–855.

Abbreviations

ACQ	acquisition
Amyg	amygdala
an_j	output of the hidden units
as_i	output of the input units
BB	backward blocking
B_{CS}	aggregate prediction of CS
B_{CX}	aggregate prediction of CX
$\bar{B}_k$	average aggregate prediction of event k
BS	between-subject
BT	buffer threshold
B_{US}	aggregate prediction of the US
BW	between-subject with interspersed water presentations
CA	cornu ammonis
CAT	conditioned attention theory
CAQ	creative achievement questionnaire
CC	classical conditioning
CER	conditioned emotional response
CL	cortical lesion
CN	central nucleus *or* configural stimulus
CPS	creative personality scale
CR	conditioned response (strength)
CS	conditioned stimulus (intensity)
CSg	intensity of common elements (generalized) in different contexts
CXc	intensity of contextual cues in the training cage
CXg	intensity of contextual cues shared by all contexts
CXh	intensity of contextual cues in the home cage
DA	dopaminergic
dlPFC	dorsolateral prefrontal cortex

EC	entorhinal cortex *or* extinction cue
EH	error signal for hidden units
EO	error signal for output units
fMRI	functional magnetic resonance imaging
FN	feature negative
FP	feature positive
FTI	feature-target interval
GABA	γ-aminobutyric acid
HAL	haloperidol
HFL	hippocampus formation lesion
HP	hippocampus proper
HPL	hippocampus proper lesion
ISI	interstimulus interval
ITI	intertrial interval
LI	latent inhibition
$\bar{\lambda}_k$	average observed value of event k
LN	lateral nucleus
LTM	long-term memory
MGB	medial geniculate body
NAC	nucleus accumbens
NCX	neocortex
NM	nictitating membrane
NMDA	*N*-methyl D-aspartate
NMR	nictitating membrane response
Novelty'	total novelty normalized between 0 and 1
NP	negative patterning
OR	orienting response (strength)
PFC	prefrontal cortex
PIN	posterior intralaminar nucleus
PP	positive patterning
PRE	preresponse
PREE	partial reinforcement extinction effect
PUP	paired–unpaired–paired
RAT	remote associates test
RF_E	reinforcement excitatory
RF_I	reinforcement inhibitory
RRP	rest–rest–paired
RW	model Rescorla–Wagner model
SAL	saline

SCR	skin conductance response
SD model	Schmajuk–DiCarlo model
SL	sham lesion
SLG model	Schmajuk–Lam–Gray model
SLH model	Schmajuk–Lamoureux–Holland model
SOP model	Wagner's Sometimes Opponent Process Standard Operating Procedures model
STM	short-term memory
SUB	subiculum
τ_{CS}	trace of the CS
τ_{CX}	trace of the CX
THAL	thalamus
TTX	tetrodotoxin
t.u.	time unit
US	unconditioned stimulus
UUP	unpaired–unpaired–paired
$V_{CS,CX}$	association between CS and CX
$V_{CS,US}$	association between CS and US
$V_{CS1,CS2}$	association between CS_1 and CS_2
VEH	vehicle
VH	CS–CN association
VN	CN–US association
VP	ventral pallidum
VS	CS–US association
VTA	ventral tegmental area
WS	within-subject
X_{CS}	representation of the CS
X_{CX}	representation of the CX
z_{CS}	attention to the CS
z_{CX}	attention to the CX

Part I INTRODUCTION

1

Classical conditioning: data and theories

During classical (or Pavlovian) conditioning, human and animal subjects change the magnitude and timing of their conditioned response (CR), as a result of the contingency between the conditioned stimulus (CS) and the unconditioned stimulus (US).

In this chapter we briefly describe results of a number of classical conditioning paradigms that are discussed in detail in different chapters of the book (see Schmajuk, 2008a, 2008b). Then we introduce different types of learning theories. Finally, we present a number of computational models of classical conditioning.

Classical conditioning data

A. *Excitatory conditioning*

1. Acquisition. After a number of CS–US pairings, the CS elicits a conditioned response (CR) that increases in magnitude and frequency.
2. Partial reinforcement. The US follows the CS only on some trials, and might lead to a lower conditioning asymptote.
3. Generalization. A CS_2 elicits a CR when it shares some characteristics with a CS_1 that has been paired with the US.
4. US- and CS-specific CR. The nature of the CR is determined not only by the US, but also by the CS.

B. *Inhibitory conditioning*

1. Conditioned inhibition. Stimulus CS_2 acquires inhibitory conditioning with CS_1 reinforced trials interspersed with, or followed by, CS_1–CS_2 nonreinforced trials.

2. Extinction of conditioned inhibition. Inhibitory conditioning is extinguished by CS_2-US presentations, but not by presentations of CS_2 alone.
3. Differential conditioning. Stimulus CS_2 acquires inhibitory conditioning with CS_1 reinforced trials interspersed with CS_2 nonreinforced trials.
4. Contingency. A CS becomes inhibitory when the probability that the US will occur in the presence of the CS, p(US/CS), is smaller than the probability that the US will occur in the absence of the CS (p[US/noCS]).

C. *Preexposure effects*

1. Latent inhibition (LI). Preexposure to a CS followed by CS–US pairings retard the generation of the CR.
2. Context preexposure. Preexposure to a context facilitates the acquisition of fear conditioning.
3. US-preexposure effect. Presentation of the US in a training context, prior to CS–US pairings, retards production of the CR.
4. Learned irrelevance. Random exposure to the CS and the US retards conditioning even more than combined latent inhibition and US preexposure.

D. *Compound conditioning*

1. Relative validity. Conditioning to X is weaker when training consists of reinforced XA trials alternated with XB nonreinforced trials, than when training consists of XA trials alternated with XB trials each type reinforced half of the time.
2. Blocking. Conditioning to CS_1-CS_2 following conditioning to CS_1 results in a weaker conditioning to CS_2 than that attained with CS_1-CS_2-US pairings.
3. Unblocking by increasing the US. Increasing the US during CS_1-CS_2 conditioning increases responding to the blocked CS_2.
4. Unblocking by decreasing the US. Responding to CS_2 can be increased by decreasing the US during CS_1-CS_2 conditioning.
5. Overshadowing. Conditioning to CS_1-CS_2 results in a weaker conditioning to CS_2 than that attained with CS_2-US pairings.
6. Potentiation. Conditioning to CS_1-CS_2 results in a stronger conditioning to CS_2 than that attained with CS_2-US pairings.
7. Backward blocking. Conditioning to CS_1 following conditioning to CS_1-CS_2 results in a weaker response to CS_2 than that attained with CS_1-CS_2-US pairings.

8. Overexpectation. Reinforced CS_1–CS_2 presentations following independent reinforced CS_1 and CS_2 presentations, result in a decrement in their initial associative strength.
9. Superconditioning. Reinforced CS_1–CS_2 presentations following inhibitory conditioning of CS_1 increase CS_2 excitatory strength compared with the case when it is trained in the absence of CS_1.

E. *Recovery from compound conditioning*

1. Recovery from latent inhibition. Presentation of the US in the context of preexposure and conditioning results in renewed responding to the preexposed CS.
2. Recovery from overshadowing. Extinction of the CS_1 results in increased responding to the overshadowed CS_2.
3. Recovery from forward blocking. Extinction of the blocker CS_1 results in increased responding to the blocked CS_2.
4. Recovery from backward blocking. Extinction of the blocker CS_1 results in increased responding to the blocked CS_2.

F. *Extinction*

1. Extinction. When CS–US pairings are followed by presentations of the CS alone, or by unpaired CS and US presentations, the CR decreases.
2. External disinhibition. Presenting a novel stimulus immediately before a previously extinguished CS might produce renewed responding.
3. Spontaneous recovery. Presentation of the CS after some time after the subject stopped responding might yield renewed responding.
4. Renewal. Presentation of the CS in a novel context might yield renewed responding.
5. Reinstatement. Presentation of the US in the context of extinction and testing might yield renewed responding.
6. Reacquisition. CS–US presentations following extinction might result in faster or slower reacquisition.
7. Partial reinforcement extinction effect (PREE). Extinction is slower following partial than following continuous reinforcement.

G. *Nonlinear combinations of multiple stimuli*

1. Positive patterning. Reinforced CS_1–CS_2 presentations intermixed with nonreinforced CS_1 and CS_2 presentations result in stronger responding to CS_1–CS_2 than to the sum of the individual responses to CS_1 and CS_2.

2. Negative patterning. Nonreinforced CS_1–CS_2 presentations intermixed with reinforced CS_1 and CS_2 presentations result in weaker responding to CS_1–CS_2 than to the sum of the individual responses to CS_1 and CS_2.

H. *Occasion setting*

1. Simultaneous feature-positive discrimination. Reinforced simultaneous CS_1–CS_2 presentations, alternated with nonreinforced presentations of CS_2, result in stronger responding to CS_1–CS_2 than to CS_2 alone. In this case, CS_1 gains a strong excitatory association with the US.
2. Serial feature-positive discrimination. Reinforced successive CS_1–CS_2 presentations, alternated with nonreinforced presentations of CS_2, result in stronger responding to CS_1–CS_2 than to CS_2 alone. In this case, CS_1 acts as an occasion setter.
3. Simultaneous feature-negative discrimination. Nonreinforced simultaneous CS_1–CS_2 presentations, alternated with reinforced presentations of CS_2, result in weaker responding to CS_1–CS_2 than to CS_2 alone. In this case, CS_1 gains a strong inhibitory association with the US.
4. Serial feature-negative discrimination. Nonreinforced successive CS_1–CS_2 presentations, alternated with reinforced presentations of CS_2, result in weaker responding to CS_1–CS_2 than to CS_2 alone. In this case, CS_1 acts as an occasion setter.

I. *Temporal properties*

1. Interstimulus interval (ISI) effects. Conditioning is negligible with short ISIs, increases dramatically at an optimal ISI, and gradually decreases with increasing ISIs.
2. Intertrial interval (ITI) effects. Conditioning to the CS increases with longer ITIs.
3. Timing of the CR. The CR peak tends to be located around the end of the ISI.
4. Temporal specificity of blocking. Blocking is observed when the blocked CS is paired in the same temporal relationship with the US as the blocking CS.
5. Temporal specificity of occasion setting. A serial feature-positive discrimination is best when the feature–target interval during testing matches the training interval.

J. *Combination of multiple conditioning events*

1. Sensory preconditioning. When CS_1–CS_2 pairings are followed by CS_1–US pairings, presentation of CS_2 generates a CR.

2. Second-order conditioning. When CS_1–US pairings are followed by CS_1–CS_2 pairings, presentation of CS_2 generates a CR.

Learning theories

Some classical conditioning theories stress the importance of mechanisms that act at the time of the presentation of the CS and the US. These theories assume that the association between events CS_i and CS_k, $V_{CSi, CSk}$, represents the *prediction* that the CS_i will be followed by CS_k (Dickinson, 1980). Neural network, or connectionist theories frequently assume that the association between CS_i and CS_k is represented by the efficacy of the synapses, $V_{CSi, CSk}$, that connect a presynaptic neural population excited by CS_i with a postsynaptic neural population that is excited by CS_k (event k might be another CS, or the US). When CS_k is the US, this second population controls the generation of the conditioned response (CR).

Following Hebb's (1949) ideas, changes in synaptic strength, $V_{CSi, CSk}$, might be described by $\Delta V_{CSi, CSk} = f(CS_i)f(CS_k)$, where $f(CS_i)$ represents the presynaptic activity, and $f(CS_k)$ the postsynaptic activity. Different $f(CS_i)$ and $f(CS_k)$ functions have been proposed. Learning rules for $V_{CSi, CSk}$ either assume variations in the effectiveness of CS_i, $f(CS_i)$, the US, $f(CS_k)$ (Dickinson & Mackintosh, 1978), or both.

Variations in the effectiveness of the CS during learning

Attentional theories assume that the effectiveness of CS_i to form CS_i–US associations (associability) depends on the magnitude of the "internal representation" of CS_i. In neural network terms, attention may be interpreted as the modulation of the CS representation that activates the presynaptic neuronal population involved in associative learning. Attentional theories include Makintosh's (1975), Grossberg's (1975) and Pearce and Hall's (1980) theories.

Variations in the effectiveness of the US during learning

A popular rule, proposed independently in psychological (Rescorla & Wagner, 1972) and neural network (Widrow & Hoff, 1960) domains, has been termed the "delta" rule. The delta rule describes changes in the synaptic connections between the two neural populations by way of minimizing the squared value of the difference between the output of the population controlling the CR generation, and the US. According to the "simple" delta rule, CS_i–US associations are changed until the difference between the US intensity and the "aggregate prediction" of the US computed upon all CSs present at a given moment, (US $- \Sigma_j V_{CSj,US} CS_j$), is zero. The term (US $- \Sigma_j V_{CSj,US} CS_j$) can be interpreted as the effectiveness of the US to become associated with the CS.

Schmajuk and DiCarlo (1992) introduced a model (the SD model) that, by employing a "generalized" delta rule (also known as backpropagation, see Rumelhart, Hinton & Williams, 1986) to train a layer of hidden units that *configure* simple CSs, is able to solve exclusive-or problems, and hence, negative patterning.

Variations in the effectiveness of both the CS and the US during learning

In order to account for a wider range of classical conditioning paradigms, some theories have combined variations in the effectiveness of both the CS and the US. For example, Frey and Sears (1978) proposed a model of classical conditioning that assumed variations in the effectiveness of both the CS and the US: $f(CS_i)$ is modulated by $V_{i,US}$. Wagner (1978) suggested that CS_i–US associations are determined by (a) $f(US) = (US - \Sigma_j V_{j,US} CS_j)$ as in the Rescorla–Wagner model; and (b) $f(CS_i) = (CS_i - V_{i,CX} CX)$, where CX represents the context, and $V_{i,CX}$ the strength of the CX–CS_i association. Other theories that incorporate changes in the effectiveness of both the CS and the US include Wagner's (1981) sometimes opponent process (SOP) theory, Schmajuk, Lam and Gray's (1996) attentional–associative theory, Le Pelley's (2004) hybrid model, and Harris's (2006) elementary model.

Performance theories

Some classical conditioning theories stress the importance of mechanisms that act during performance to control the generation of the CR. Examples of this approach are Miller's comparator hypothesis (e.g. Miller & Schachtman, 1985), Wagner's (1981) SOP model, Schmajuk, Lam and Gray's (1996) attentional–associative model, and Harris's (2006) elementary model.

CS–US and CS–CS associations and decision processes during performance

The comparator hypothesis (Miller & Schachtman, 1985; Miller & Matzel, 1988; Denniston *et al.*, 2001; Stout & Miller, 2007) suggests that the magnitude of the CR is determined by a comparator that uses the CS–CS and CS–US associations of the CSs present at a given time as inputs.

CS–CS associations and inference generation during performance

During classical conditioning, animals learn to expect (predict) that a CS is followed by another CS, or by the US. Tolman (1932) proposed that multiple expectancies (predictions) can be integrated into larger units, through a reasoning process called inference. One simple example of inference formation is sensory preconditioning (see Bower & Hilgard, 1981, page 330). As summarized above, sensory preconditioning consists of a first phase in which two conditioned stimuli, CS_1 and CS_2, are paired together in the absence of the US. In a second phase, CS_1 is paired with the US. Finally, when CS_2 is presented alone, it generates a CR: the animal has inferred that CS_2 predicts the US. Tolman hypothesized that a large number of expectancies can be combined into a cognitive map (see Chapters 2, 7 and 16).

Dickinson (1980) suggested that knowledge can be represented in declarative or procedural form. Whereas in the declarative form knowledge is represented as a description of the relationships between events (knowing that), in the procedural form, knowledge is represented as the prescription of what should be done in a given situation (knowing how). Examples of declarative knowledge are classical CS–CS associations (CS_1 precedes CS_2) or CS–US (CS_2 precedes the US) associations. An example of procedural knowledge is the operant S–R association (if S is present, then do R). Dickinson indicated that declarative, but not procedural, knowledge can be integrated through inference rules.

By including CS–CS associations, some models of classical conditioning are able to generate inferences and, therefore, to describe sensory preconditioning. For instance, Gelperin, Hopfield and Tank (1985; Gelperin, 1986) proposed an autoassociative recurrent network capable of describing stimulus–stimulus associations during classical conditioning. The network can simulate first- and second-order conditioning, extinction, sensory preconditioning and blocking in the terrestrial slug, *Limax maximus*. Schmajuk (1987) proposed a dual memory architecture that incorporates an autoassociative nonrecurrent network capable of cognitive mapping in classical conditioning. The network separately computes CS–CS and CS–US predictions, and combines them to generate new expectancies. For instance, if CS(A) predicts (is associated to) CS(B), and CS(B) predicts the US, the network *infers* that CS(A) also predicts the US. Schmajuk defined first-order predictions as the prediction of the US by CS(B), and higher-order predictions as the predictions involving a chain of two or more predictions. The network describes complex classical conditioning paradigms such as sensory preconditioning, second-order conditioning, compound conditioning and serial-compound conditioning. Similarly, the models presented by Schmajuk and Moore (1988, 1989) and Schmajuk, Lam and Gray (1996) are able to describe sensory preconditioning and second-order conditioning.

Computational models of classical conditioning

As suggested by Hinzman (1991), the inherent unreliability of verbal intuitive reasoning for relating hypotheses and experimental results favors theories that provide precise quantitative descriptions. Furthermore, only formal models can be simultaneously examined at different levels. At the behavioral level, simulated behavioral results are compared with experimental data describing behavior. At the neuroanatomical level, interconnections among neural elements in the model are compared with neuroanatomical data, and the model performance is compared with animal performance after lesioning. At the computational level, simulated activity of the neural elements of the model is compared with the activity of single neuron or neural population activity. At the neurophysiological level, model performance is compared with animal performance after inducing long-term changes (e.g. lesions) or short-term changes (e.g. drug infusions) in different brain areas.

Below, we review in detail some computational models of classical conditioning that have been applied to a number of the conditioning paradigms described before. Some other models (e.g. Brandon, Vogel & Wagner, 2000; Kruschke, 2001; Pearce, 1994) are described later in the book, when they are relevant to the experimental results being discussed.

The Rescorla–Wagner (1972) model

In their classic article, Rescorla and Wagner (1972) indicated that the impetus for their new theoretical model was not new data which clearly disconfirmed existing theories, but rather the accumulation of a pattern of data which appeared to invite a more integrated account. The salient pattern of data the authors referred to was a set of observations involving Pavlovian conditioning with compound CSs. The central notion of the theory was that organisms only learn when the actual value of the US differs from its expected value. By proposing the novel principle that this expected value of the US is computed as a linear combination of the associative strength of all active CSs, the effect of reinforcement or nonreinforcement on the associative strength of a CS depends upon the existing associative strength, not only of that CS, but also of other CSs concurrently present.

Rescorla and Wagner (1972) proposed to formalize the basic idea of their theory by modifying Hull's (1943) account of the growth of habit-strength (stimulus–response associations), as described by Bush and Mosteller's (1955) linear operator. In the Rescorla–Wagner (RW) model, variations in the strength of the CS–US association, $V_{i,US}$, are given by $\Delta V_{i,US} = \alpha_i \beta_{US} (\lambda_{US} - B_{US})$, where α_i represents the salience of CS_i, β_{US} represents the learning rate parameter corresponding to the US, and B_{US} is the linear combination of the prediction of the US by all active CSs. B_{US} is given by $B_{US} = \Sigma_j V_{j,US}$. By this equation, CSs compete to gain association with the US. The conditioned response (CR) was assumed to be proportional to B_{US}.

As observed by Sutton and Barto (1981), a rule similar to the RW equation had been described in the neural network field by Widrow and Hoff (1960). This rule, termed the "delta" rule (Rumelhart, Hinton & Williams, 1986), describes changes in the CS–US associations by way of minimizing the squared value of the difference between the predicted and observed values of the US. Most interestingly, as indicated by Duda and Hart (1973), the rule is able to solve simultaneous systems of equations; a power that provides a different perspective of the processes that take place during classical conditioning.

The RW model correctly described many Pavlovian conditioning phenomena such as acquisition and extinction of conditioned excitation, partial reinforcement, conditioned inhibition, overshadowing, blocking, unblocking by increasing the US strength, overprediction, generalization, US–preexposure effect and contingency effects. The success of the model in making specific correct predictions, inaugurated the modern era of experimental psychology.

In spite of its significant achievements, the RW model was unable to describe several aspects of classical conditioning including (a) the effects of temporal parameters, such as stimulus duration, interstimulus intervals (ISI) or intertrial intervals (ITI); (b) Pavlovian paradigms whose solution require a nonlinear combination of the prediction of the US by all active CSs, such as negative patterning; (c) conditioned inhibition not being extinguished by presentations of the inhibitory CS alone; (d) latent inhibition; (e) backward blocking; and (f) the recovery from blocking and overshadowing.

The Van Hamme and Wasserman (1994) version of the Rescorla and Wagner (1972) model

Van Hamme and Wasserman (1994) offered a modified version of the Rescorla and Wagner (1972) model that is able to explain some of the results mentioned above. Van Hamme and Wasserman (1994) proposed that the association of a CS with the US decreases when the CS is absent ($\Delta V_{i,US} < 0$), instead of staying constant, as in the original model ($\Delta V_{i,US} = 0$ because $\alpha_i = 0$). The model can explain the effects of extinction of the companion CS overshadowing and blocking.

Dickinson and Burke (1996) observed that the Van Hamme and Wasserman (1994) rule did not specify when an absent CS was allowed to decrease its association with the US. Following previous suggestions (Chapman, 1991; Markman, 1989; Tassoni, 1995), they indicated that the expectation of an absent CS, via its (within-compound) association with a present CS, could serve that purpose.

The Van Hamme and Wasserman (1994) version of the Rescorla–Wagner (1972) rule is able to describe that (a) extinction of the blocking CS results in the recovery of the response to the blocked CS (Blaisdell et al., 1999); (b) extinction of the overshadowing CS results in the recovery of the response to the overshadowed CS (Matzel et al., 1985); but cannot explain (c) extinction of the context following latent inhibition (LI) results in the recovery of the response of the CS (Grahame et al., 1994); or (d) LI and overshadowing counteracting each other (Blaisdell et al., 1998) (see Chapter 3). In addition, the Van Hamme and Wasserman (1994) version explains backward blocking assuming a nonzero value of the CS on the trials in which it is absent.

Grossberg's (1975) attentional network

Grossberg (1975) suggested that a CS activates neural populations whose activity constitutes a sensory representation, or short-term memory (STM), of the CS. A US activates neural populations of the drive representation of the US. Sensory representations compete among themselves for a limited-capacity STM activation that is reflected in a long-term memory (LTM) storage. The pairing of a CS with a US causes a long-term association of the sensory representation of the CS with the drive representation of the US (conditioned reinforcement learning). In addition, the pairing of the CS with the US causes a second long-term association of the drive representation of the US with the sensory representation of CS ("incentive motivation" learning). (Although incentive motivation has mainly an appetitive connotation, in the Grossberg model the term applies to both appetitive, and aversive USs.) Incentive motivation associations reflect the association of the US with a CS representation, and mediate the enhancement of the sensory representation of the CS according to the strength of this association.

In Grossberg's (1975; Schmajuk & DiCarlo, 1989) architecture, CS_i activates a representation, X_{i1}. Changes in X_{i1} are given by $d(X_{i1})/dt = -K_1 X_{i1} + K_2(K_3 - X_{i1}) I_{i1} - K_4 X_{i1} J_{i1}$, where $-K_1 X_{i1}$ represents the passive decay of X_{i1}, K_2 represents the rate of increase of X_{i1}, constant K_3 is the maximum possible value of X_{i1}, I_{i1} the total excitatory input and J_{i1} represents the total inhibitory input. Schmajuk and DiCarlo (1989) proposed that the total excitatory input, I_{i1}, is given by $I_{i1} = XCS_i + K_5 X_{i1}^m + K_6 X_{i2} XCS_i$, where $K_5 X_{i1}^m$ represents a positive feedback from X_{i1} to itself, and $K_6 X_{i2} CS_i$ represents a signal from X_{i2} to X_{i1} that is active only if CS_i is present. XCS_i is a short-term memory of CS_i given by $XCS_i = -K_7 XCS_i + K_8(K_9 - XCS_i)CS_i$. The output of the XCS_i node is a sigmoid given by $f(XCS_i) = XCS_i^n/(\beta^n + XCS_i^n)$. Total inhibitory input J_{i1} is given by $J_{j1} = \Sigma_{j \neq i} X_{j1}$ for $X_{j1} >$ threshold.

Activity in the drive representation node is given by $dY/dt = -K_{10}Y + K_{11}(\Sigma_i X_{i1} V_i + US) + K_{11}(+ US)$, where $-K_{10}Y$ represents the passive decay of drive representation activity, and $\Sigma_i X_{i1} V_i$ is the sum of sensory representations multiplied by their associations with the US. Simultaneous activation of the drive representation and CS sensory representations causes X_{i1} to become associated with the output of the drive representation. Changes in V_i are given by $dV_i/dt = -K_{14} V_i X_{i1} + K_{15}(K_{16} - V_i) Y X_{i1}$, where $-K_{14} V_i X_{i1}$ is the active decay in V_i when

X_{i1} is active, $K_{15}(K_{16} - V_i)YX_{i1}$ is the increment in V_i when $K_{15}Y$ and X_{i1} are active together, and K_{16} is the upper limit of V_i. After X_{i1} becomes associated with the drive representation, Y, it becomes a secondary reinforcer for other CSs.

The association between Y and X_{i1}, Z_i, changes according to $dZ_i/dt = -K_{17}Z_iX_{i1} + K_{18}(K_{19} - Z_i)YX_{i1}$. Conditioning of the Y–X_{i1} pathway increases X_{i1} sensory representation by incentive motivation. A sensory cue with large V_i and Z_i can augment the activity of its sensory representation. Sensory representations compete among themselves for a limited short-term memory capacity, which is implemented by letting activity X_{i1} be excited by CS_i, and inhibited by the sum of all other STM activities, J_{i1}. Through this competition, CSs with strong V_i and Z_i inhibit CSs with weak V_i and Z_i.

The output of the system (i.e. the CR and UR) is sigmoid, given by $R_3(Y) = Y^2/(Y^2 + \beta^2)$. Schmajuk and DiCarlo (1991b) showed that the model describes (a) acquisition of delay and trace conditioning, (b) extinction of delay and trace conditioning, (c) acquisition–extinction series of delay conditioning, (d) latent inhibition, (e) blocking, (f) overshadowing, (g) discrimination acquisition, (h) discrimination reversal, and (i) secondary reinforcement. Latent inhibition is the result of an initial value of Z_i that decreases during CS_i preexposure. Overshadowing is the consequence of the competition between X_{11} and X_{21} to enter short-term memory. Because V_1 and Z_1 increase during the first phase of blocking, X_{11} is larger than X_{21} when CS_1 and CS_2 are presented together, thereby limiting the value of V_2. Therefore, $CR_1 = X_1V_1$ will be larger than $CR_2 = X_2V_2$.

Mackintosh's (1975) theory

Mackintosh's (1975) attentional theory suggests that CS_i associability, $f(CS_i)$, increases whenever CS_i is the best predictor of the outcome of a trial, and decreases otherwise. Mackintosh's model describes latent inhibition, overshadowing and blocking. Moore and Stickney (1980) and Schmajuk and Moore (1989) generalized Mackintosh's approach to include CS–CS associations, and assumed that $f(CS_i)$ increases when CS_i is the best predictor of other CSs and the US, but decreases otherwise.

Pearce and Hall's (1980) theory

In contrast to Mackintosh's (1975) view, Pearce and Hall (1980) suggested that $f(CS_i)$ increases when CS_i is a poor predictor of the US, $f(CS_i) = |US - \Sigma_j V_{j,US} CS_j|$, where $\Sigma_j V_{j,US} CS_j$ represents the "aggregate prediction" of the US computed upon all CSs present on a given trial. In addition to latent inhibition, blocking and overshadowing, the Pearce and Hall model correctly predicts that latent inhibition might be obtained after training with a weak US.

Wagner's (1981) SOP model

In the sometimes opponent process (SOP) model, a stimulus representation can be in one of three states, A_1 (high activation), A_2 (low activation) or I (inactive). Acquisition is explained in terms of the excitatory association formed between a CS and a US when their representations are both in the A_1 state. After training, presentation of the CS by itself activates a representation of the US (initially in the I state) in the A_2 state. Because activation of the US representation in the A_2 state limits the US representation in the A_1 state, increments in the CS–US association will be smaller as conditioning progresses. Extinction is the result of the inhibitory association formed between a CS and a US formed when the CS representation is in the A_1 state and the US representation is in the A_2 state, that is, the US is not present, but is evoked by the excitatory CS–US association. Similarly, conditioned inhibition is the result of the inhibitory association formed between a CS and a US formed when the CS representation is in the A_1 state and the US representation is in the A_2 state, that is, the US is not present, but is evoked by an excitatory CS. Latent inhibition is the result of the CS being represented in A_2 when predicted by the context, thereby limiting the CS A_1 representation, and retarding its association with the US. Blocking and overshadowing are the consequence of the US representation being activated in the A_2 state, and limiting the A_1 activation by the US, and the growth of the association of the US with CS_1 and CS_2. In addition, SOP is able to predict when the CR will resemble, and when it will differ from the UR.

The Dickinson and Burke (1996) version of Wagner's (1981) SOP model

Dickinson and Burke (1996) proposed a revised version of Wagner's (1981) SOP theory. Whereas Wagner (1981) had suggested that if the representations of two stimuli are in the A_2 state no learning occurs, Dickinson and Burke (1996) postulated that in this situation an excitatory association is formed. This association is weaker, however, than that formed when both stimuli are in the A_1 state. In addition, whereas Wagner (1981) had suggested that if the CS is represented in the A_2 state and the US in the A_1 state no learning occurs, Dickinson and Burke (1996) postulated that in this situation an inhibitory association is formed between the CS and the US.

The Dickinson and Burke (1996) modified SOP model can describe (a) extinction of the context following LI results in the recovery of the response of the CS, (b) extinction of the blocking CS results in the recovery of the response to the blocked CS, (c) extinction of the overshadowing CS results in the recovery of the response to the overshadowed CS, but not (d) LI and overshadowing counteract each other.

According to Dickinson and Burke, presentation of the context alone, after conditioning, activates both the representations of CS and US in the A_2 state, thereby increasing the CS–US association, and producing recovery from LI. Extinction of CS_2 activates the A_2 representations of both CS_1 and the US. Because associations between two stimuli in the A_2 state increase, the CS_1–US association will increase and CS_1 will recover from overshadowing. Extinction of the blocker CS activates the A_2 representations of both CS_2 and the US, thereby increasing the CS_2–US association, and CS_2 will recover from blocking. Backward blocking is explained in the following terms. Presentation of blocker CS_1 with the US during the second phase of backward blocking makes the representation of CS_2 and US present in A_2. Presentation of the US will excite the inactive elements of its representation into A_1. This results in incongruence for some of the elements of CS_2 and the US. According to the model, the net increment in the association between CS_2 and the US will be smaller in the experimental group (CS_1–CS_2–US, CS_1–US) than in the control group (CS_1–CS_2–US, CS_3–US).

The Miller and Schachtman (1985) comparator hypothesis

According to the comparator hypothesis (Miller & Schachtman, 1985; Miller & Matzel, 1988; Stout & Miller, 2007), during testing, the CS generates two representations of the unconditioned stimulus (US): a direct one through its own CS–US association, and an indirect one through CS–Comparator CS and Comparator CS–US associations. The Comparator CS (i.e. another CS, the context, or both) is the one with which the CS was trained. When the strength of the direct representation is greater than the indirect one, the potential for excitatory responding is larger than that for inhibitory responding. When the strength of the indirect representation is greater than the direct one, the potential for inhibitory responding is larger than that for excitatory responding.

According to Miller and Schachtman's (1985) comparator hypothesis, LI is the consequence of a strong CS–CX association, acquired during preexposure, which makes the indirect representation of the US stronger than the direct one, therefore attenuating responding to the CS. According to Grahame *et al.* (1994), extinction of the training context following training results in decreased CX–US associations and the attenuation of LI, because it reduces the indirect representation of the US. According to the comparator hypothesis, overshadowing of CS_1 is the consequence of strong CS_1–CS_2 and CS_2–US associations, which increase the indirect representation of the US, thereby attenuating responding to CS_1. Extinction of CS_2 following training results in decreased CS_2–US associations and the attenuation of overshadowing, because of the reduced indirect representation of the US. According to the comparator hypothesis, blocking of CS_2 is the consequence of a strong CS_1–US association, acquired during both phases of the experiment, which makes the indirect representation of the US (CS_2–CS_1 and CS_1–US) stronger than the direct one (CS_2–US), therefore attenuating responding to CS_2. Extinction of CS_1 following training results in decreased CS_1–US associations and the attenuation of blocking, because of the reduced indirect representation of the US. Like forward blocking, backward blocking is the consequence of a strong A–*Outcome* association, acquired during both phases of the experiment, which makes the indirect representation of the *Outcome* (through X–A and A–*Outcome*) stronger than the direct one (X–*Outcome*), therefore attenuating responding to X (see Chapter 8).

The Blaisdell et al. (1998) extended comparator hypothesis

Because the original hypothesis was unable to describe data showing that a blocked CS is incapable of blocking another CS (Williams, 1996) and that LI and overshadowing counteract each other (Blaisdell *et al.*, 1998), Denniston *et al.* (2001) introduced the "extended comparator hypothesis." According to this hypothesis, responding is still determined by the CS–US association compared with the CS–Comparator CS_1 association combined with the Comparator CS_1–US associations. But, in addition, both CS–Comparator CS_1 and Comparator CS_1–US associations are the result of additional comparisons: (a) the strength of the CS–Comparator CS_1 link is determined by the CS–Comparator CS_1 association compared with a CS–Comparator CS_2 association combined with the Comparator CS_2–Comparator CS_1 association, and (b) the strength of the Comparator CS_1–US link is determined by the Comparator CS_1–US association compared with a Comparator CS_1–Comparator CS_3 association combined with Comparator CS_3–US association.

In addition to correctly describing the above-mentioned data, the comparator hypothesis has been successfully applied to overshadowing (Matzel, Schachtman & Miller, 1985), the US–preexposure effect (Matzel, Brown & Miller, 1987), conditioned inhibition (Kasprow, Schachtman & Miller, 1987), and backward blocking (Miller, Hallam & Grahame, 1990). In spite of all its successes, Denniston *et al.* (2001) pointed out that the extended comparator hypothesis cannot account for (a) spontaneous recovery, external desinhibition and renewal; (b) trial order; and (c) mediated extinction and sensory preconditioning.

Le Pelley's (2004) hybrid model

According to Le Pelley (2004), CS_A–US associations, V_A, increase according to $\Delta V_A = \alpha_A \sigma_A \beta_E (1 - V_A + \overline{V}_A) |\lambda_{US} - (\Sigma V_{CS} - \Sigma \overline{V}_{CS})|$, when $\lambda_{US} - (\Sigma V_{CS} - \Sigma \overline{V}_{CS}) > 0$; whereas

CS_A–US antiassociations (see Schmajuk and Moore, 1985), $\overline{V}_A$, increase according to $\Delta \overline{V}_A = \alpha_A \sigma_A \beta_I (1 - \overline{V}_A + V_A) |\lambda_{US} - (\Sigma_{CS} V_{CS} - \Sigma \overline{V}_{CS})|$, when $\lambda_{US} - (\Sigma_{CS} V_{CS} - \Sigma_{CS} \overline{V}_{CS}) < 0$. The net associative strength of CS_A is given by $V_A^{NET} = V_A - \overline{V}_A$. In these equations, β_E is the learning-rate parameter for excitatory learning, β_I is the learning-rate parameter for inhibitory learning, λ_{US} represents the US strength, ΣV_{CS} is the sum of CS–US associations of all present CSs, and $\Sigma \overline{V}_{CS}$ is the sum of all CS–US antiassociations of all present CSs.

Changes in "attentional" associability α_A are defined as in the extended Mackintosh model: $\Delta \alpha_A = - \theta_E (|\lambda_{US} - (V_A - \overline{V}_A)| - |\lambda_{US} - (V_X - \Sigma \overline{V}_X)|)$, when $R > 0$, and $\Delta \alpha_A = - \theta_E (||R| - |-V_A + \overline{V}_A|| - ||R| - |-V_X + \overline{V}_X||)$, when, when $R < 0$ ($R = |\lambda_{US} - (\Sigma V_{CS} - \Sigma \overline{V}_{CS})|$). "Salience" associability σ_A on trial n is defined as in the Pearce–Hall model, $\sigma_A(n) = \gamma (|\lambda_{US} - \Sigma_{CS} V_{CS}^{NET}| + (1 - \gamma) \sigma_A(n - 1)$, where $\gamma < 1$ and $n - 1$ refers to the trial previous to the present trial.

According to Le Pelley (2004, page 235), changes in attentional associability in his hybrid model explain learned irrelevance, attentional decrements during blocking and overtraining reversal effect. Changes in salience associability explain latent inhibition, negative transfer and better learning by poor predictors of events. The separable error term explains the greater associative change by an inhibitory (or neutral CS) when reinforced, or not reinforced together with an excitatory CS. The total error term explains overshadowing, blocking, conditioned inhibition, overexpectation and supernormal conditioning.

Harris's (2006) model

Harris (2006) proposed a model in which (a) a CS activates a number of elements that correspond to its different microfeatures – each element is activated in proportion to the salience of its corresponding feature. Changes in activity X_Y of element Y are given by $\Delta X_Y = X_Y - \Sigma_i X_i V_{iY}$ when $X_Y > 0$, and by $\Delta X_Y = 0 - \Sigma_i X_i V_{iY}$ when $X_Y = 0$; (b) activated elements compete to enter a fixed-capacity attention buffer as a function of their activation level X (an idea similar to that proposed by Grossberg, 1975; see also Schmajuk and DiCarlo, 1991b); (c) the buffer threshold (BT) is defined by the activity of the element n (X_n), for which $\Sigma_i^n X_i \leq$ buffer capacity; (d) if an element enters the buffer, its activity is increased in proportion to the increase in activation that allowed it to enter the buffer; (e) excitatory conditioning occurs when US elements are active in the attention buffer, and inhibitory conditioning when US elements are active outside the attention buffer. Because US elements are more active in the buffer than outside the buffer, excitatory conditioning occurs at a faster rate than inhibitory conditioning. The location of the activation of CS elements does not determine the sign of the association; (f) the association between element X, X_X (e.g. of a CS) and element Y, X_Y (of a US) increases according to $\Delta V_{XY} = 2X_X \beta_Y \Delta X_Y$ when $\Delta X_Y \geq BT$, and decreases according to $\Delta V_{XY} = X_X \beta_Y(-\Delta X_Y)$ when $\Delta X_Y < BT$. β_Y is a learning rate parameter; and (g) responding to CS(A) is the sum of the products of the activation weights of all elements and their associative strength, $R(A) = \Sigma_i X A_i V A_i$.

According to the model, latent inhibition is the result of associations being formed among the elements of a CS, and between the elements of the CS and those of the context. As the activation of the CS elements is reduced, fewer enter the attention buffer, and conditioning is retarded in the context of pre-exposure. In addition to describing latent inhibition as an acquisition deficit, because the model describes latent inhibition as a performance deficit, it expects latent inhibition to decrease when the CS is tested in a different context. The model explains negative patterning, because more elements enter the buffer when *A* or *B* are presented separately on *A+* or *B+* trials, than when they are presented together on *AB−* trials. On *AB−* trials, inhibitory associations are formed between the stronger elements of each CS (referred to as *A* and *B*) and the weaker elements of the other CS (referred to as *a* and *b*), because *A* and *B* are in the attention buffer, and a and b are outside the buffer. (This is similar to one possible solution found by the SD model presented in Chapter 11, in which inhibitory *A–B* and *B–A* associations are formed in the hidden units.) The model explains overshadowing in terms of the competition of the elements of each CS to enter the attentional buffer. The model also explains why, during the second phase of blocking, the blocking CS gains less association with the US than the neutral CS (see Chapter 3).

Summary

Attempts to find a simple principle to describe the many properties of classical conditioning appeared to be unsuccessful. Instead, theories needed to incorporate more than one mechanism to explain the experimental data. In the following chapters, we describe several of the proposed mechanisms, show how they address some important experimental results, and how they can be combined to describe most of the known properties of the conditioning.

Part II ATTENTIONAL AND ASSOCIATIVE MECHANISMS

2

An attentional–associative model of conditioning

Schmajuk, Lam and Gray (SLG, 1996; Schmajuk & Larrauri, 2006) proposed a neural network model of classical conditioning. The SLG model shares properties with other associative models described in Chapter 1, including equations that portray behavior on a moment-to-moment basis (Grossberg, 1975; Wagner, 1981), the attentional control of the formation of CS–US associations (Pearce & Hall, 1980), the competition among CSs to become associated with the US (Rescorla & Wagner, 1972) or other CSs (Schmajuk & Moore, 1988) and the combination of attention and competition (Wagner, 1979).

Important properties of the SLG model include that (a) it describes not only CS–US associations, as in the above-mentioned models, but also the formation and *combination* of CS–CS and CS–US associations; (b) attention to the CS is controlled by the CS–US associations (as in the Pearce & Hall, 1980, model), by context–CS (CX–CS) associations (as in the Wagner, 1979, model), and by CS–CS associations (as in the Schmajuk & Moore, 1988, model; a property that explains why latent inhibition can become context independent – see Chapter 5); and (c) retrieval of CS–US and CS–CS associations is controlled by the magnitude of the attention to the CS (as in Wagner's, 1981, SOP model).

Figure 2.1 shows a block diagram of the SLG network. The diagram includes (a) a short-term memory and feedback system, (b) an attention system, (c) an association system, and (d) a Novelty' system. The attentional mechanism is regulated by the novelty (defined below) of the US, the CSs and the context (CX); the associative network is controlled by a real-time competitive rule, and output from the associative network is sent back to the input through a feedback loop. The attentional mechanism was designed to explain the retardation of conditioned responding following CS preexposure (latent inhibition). The associative network was designed to describe changes

22 Attentional and associative mechanisms

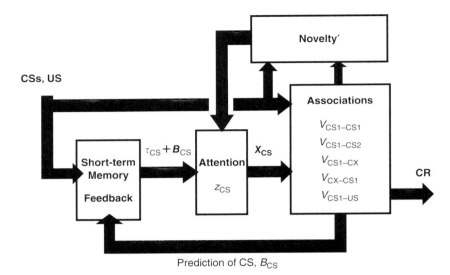

Figure 2.1 Block diagram of the Schmajuk–Lam–Gray (SLG, 1996) network. CS = conditioned stimulus; US = unconditioned stimulus; τ_{CS} = short-term memory trace of the CS; B_{CS} = prediction of the CS; z_{CS} = attentional memory; X_{CS} = internal representation of the CS; $V_{CS1-CS1}$, $V_{CS1-CS2}$, ..., V_{CS1-US} = associations CS_1–CS_1, CS_1–CS_2, ..., CS_1–US; CR = conditioned response.

in CS–US associations compatible with overshadowing, blocking and conditioned inhibition (including its extinction properties). The feedback loop was included to describe sensory preconditioning and second-order conditioning through the combination of CS–CS and CS–US associations. The resulting model proved capable of correctly describing many experimental results for which it was not specifically designed, that is, those results that are "emergent properties" of the model.

Figure 2.2 shows a simplified diagram of the model that illustrates the different mechanisms involved in the generation of a CR when a given CS is presented. Table 2.1 lists the variables used in the model and the symbols that denote them.

Short-term memory and feedback

In order to allow a CS to establish associations with other CSs or the US even when separated by a temporal gap (trace conditioning), a CS activates a short-term memory trace, τ_{CS} (see Hull, 1943; Grossberg, 1975). Changes in τ_{CS} are given by

$$d\tau_{CS}/dt = K_1(\lambda_{CS} - \tau_{CS}), \qquad [2.1]$$

Table 2.1. *Symbols and meaning of the variables in the model*

Symbol	Meaning
λ_{CS}	Intensity of the conditioned stimulus
λ_{US}	Strength of the unconditioned stimulus
τ_{CS}	Trace of the CS (or CX)
B_{CS}	Aggregate prediction of CS (or CX)
B_{US}	Aggregate prediction of the US
z_{CS}	Attention to the CS (or CX)
Novelty'	Total novelty normalized between 0 and 1
X_{CS}	Representation of the CS (or CX)
$V_{CS,CX}$	Association between CS and CX
$V_{CS,US}$	Association between CS and US
OR	Strength of the orienting response
CR	Strength of the conditioned response

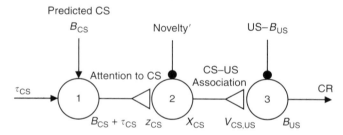

Figure 2.2 Simplified circuit of the Schmajuk-Lam-Gray (1996) model. CS: conditioned stimulus; US: unconditioned stimulus; τ_{CS}: trace of CS; z_{CS}: attention to CS; X_{CS}: internal representation of the CS; $V_{CS,US}$: X_{CS}-US association; B_{US}: predicted US; CR: conditioned response; Novelty': detected novelty. Triangles: variable connections (associations) between nodes that modulate the activation of the node. Arrows: inputs that control the output of the node. Solid circles: inputs that modify connections z_{CS} and V_{CS-US} without affecting outputs X_{CS} and CR.

where K_1 is the rate of increase and decay of τ_{CS}, and λ_{CS} the intensity of the CS. By Equation [2.1], τ_{CS} increases over time from zero to a maximum when the CS is present, and then gradually decays back to its initial value when the CS is absent.

The output of Node 1 is given by $(\tau_{CS} + K_3 B_{CS})$, where B_{CS} is the prediction of the CS by (a) itself, through CS_i-CS_i associations; (b) other CSs, through CS_i-CS_j associations; and (c) the CX, through CX-CS_i associations. Therefore, output

24   Attentional and associative mechanisms

$\tau_{CS} + B_{CS}$ is active either in the presence of a CS, or when an absent CS is predicted by other CSs or the CX. As suggested by Konorski (1967), whereas τ_{CS} can be regarded as a perceptual input, B_{CS} can be considered an "imagined" input. As shown in Chapters 3 and 16, B_{CS} allows the network to describe sensory preconditioning (Brogden, 1939) and second-order conditioning (Pavlov, 1927).

Attention

Sokolov (1960) proposed that the strength of the orienting response (OR) might be an index of the amount of processing afforded to a given stimulus, and that this amount of processing is proportional to the novelty of the stimulus. Gray (1971) suggested that attention to environmental stimuli is increased in response to novelty. In order to increase attention to the CSs when novelty is sensed, the output of Node 1, ($\tau_{CS} + B_{CS}$), becomes associated with the normalized value of total novelty detected in the environment, Novelty' (defined below).

Attentional memory, z_{CS}, reflects the association between ($\tau_{CS} + K_3 B_{CS}$) and Novelty'. The change in z_{CS} over one time unit is given by

$$dz_{CS}/dt = (\tau_{CS} + K_3 B_{CS})(K_{12} K_5 \text{Novelty}'(1 - z_{CS}) - K_6(1 + z_{CS})), \qquad [2.2]$$

where K_{12} reflects the presence of dopaminergic drugs (see Chapter 6), K_5 is the rate of increase of z_{CS}, K_6 is the rate of decay of z_{CS}, and Novelty' is given by Equation [2.8]. Equation [2.2] has some similarities with a variant of the Pearce and Hall (1980) attentional rule proposed by Pearce *et al.* (1983), which assumes that attention is an always positive, running average of the absolute value of the difference between the actual and predicted US.

The synaptic weight (represented by the triangle) connecting Node 1 to Node 2 reflects the positive value of attention z_{CS}, which can vary between 1 and −1. The initial value of z_{CS} is zero. When Novelty' is *larger* than a certain value, z_{CS} *gradually* increases; when Novelty' is *smaller* than another value, z_{CS} decreases. Notice that Novelty' need not change for z_{CS} to increase or decrease. Because attention z_{CS} is controlled by Novelty', changes in z_{CS} always lag behind changes in Novelty'. Therefore, z_{CS} reflects the history of exposure of the CS to previous values of Novelty'.

Importantly, because the output of Node 1 is proportional to $\tau_{CS} + B_{CS}$, either when the CS is present ($\tau_{CS} > 0$), or absent ($\tau_{CS} = 0$) but predicted by other CSs through $B_{CS}(B_{CS} > 0)$, Novelty' can modify z_{CS}. This property of z_{CS} is important in the paradigms described, such as those involving recovery from latent inhibition (Chapter 5), overshadowing and blocking (Chapter 8).

The output of Node 2 is the attention-modulated representation of the CS, X_{CS}. Importantly, the more negative z_{CS} becomes, the longer it will take to become positive again and have an effect on X_{CS}. Therefore, whereas positive values of z_{CS} can be interpreted as a measure of the *attention* directed to CS, negative values of z_{CS} can be interpreted as a measure of the *inattention* to CS (see Lubow, 1989, page 192). Representation X_{CS}, is given by

$$X_{CS} = K_2(\tau_{CS} + K_3 B_{CS})(K_4 + z_{CS}). \qquad [2.3]$$

By Equation [2.3], X_{CS} is active either when (a) CS is present and τ_{CS} is greater than zero, or (b) when CS is predicted by other CSs and B_{CS} is greater than zero. Increasing values of z_{CS} increase the magnitude of X_{CS}. We assume that when $z_{CS} < 0$, then $X_{CS} = K_2(\tau_{CS} + K_3 B_{CS})K_4$. This means that when z_{CS} becomes negative, input $(\tau_{CS} + K_3 B_{CS})$ activates X_{CS} only through an unmodifiable connection K_4 (not shown in Figure 2.2). This connection ensures that Node 1 and Node 2 are always minimally connected.

Changes in z_{CS} and X_{CS} are used to explain latent inhibition (Chapter 5). According to the model, latent inhibition is the consequence of the decreased X_{CS} that results from a decreased attention (small positive z_{CS}) or the inattention (negative z_{CS}) to the preexposed CS. The magnitude of the effect depends on the time needed to increase X_{CS} by increasing a positive z_{CS} or by reversing inattention (negative z_{CS}) into attention (positive z_{CS}) during conditioning.

One interesting characteristic of the model is that, for a given value of Novelty′, attention to the CX, z_{CX}, tends to increase or decrease more than attention to the CS, z_{CS}. This is so because the CX, which is treated as a tonic CS that lasts for the duration of the entire intertrial interval (ITI), is active for a longer time than a CS. Then, when Novelty′ is relatively small, z_{CX} will reach lower negative values than z_{CS} and, therefore, attention to the CX will take longer to recover than attention to the CS. This property is important in spontaneous recovery (Chapter 9), a phenomenon that the model describes as a transient response to an extinguished CS produced because attention to the excitatory CS increases earlier than attention to the inhibitory CX.

Conscious and unconscious conditioning

Gray, Buhusi and Schmajuk (1997) showed that the SLG model describes automatic (or unconscious) and controlled (or conscious) processing (Schneider & Shiffrin, 1977; Pearce & Hall, 1980). In the framework of the model, a CS might be processed in controlled or conscious mode when Novelty′, z_{CS}, and X_{CS} are large, and in automatic or non-conscious mode when Novelty′, z_{CS}, and X_{CS} are small. According to the model, in the case of latent inhibition (see Chapter 5) a preexposed CS with a small X_{CS} remains unconscious. Such an "unconscious" CS can still control behavior and generate an automatic response if presented in combination with strong CS–US association.

Novelty

The novelty of event k (a CS, CX or the US) is computed as the absolute value of the difference between the average observed value of that event, and the average of the sum of all predictions of that event by all active CSs and CXs:

$$\text{Novelty}_k = |\bar{\lambda}_k - \bar{B}_k|, \qquad [2.4]$$

where $\bar{\lambda}_k$ is the average observed value of event k, and $\bar{B}_k$ is the average aggregate prediction of event k.

Changes in the average observed value of event k is given by

$$d\bar{\lambda}_k/dt = (1 - \bar{\lambda}_k)\lambda_k - K_8\bar{\lambda}_k, \qquad [2.5]$$

where K_8 is the rate of decay of $\bar{\lambda}_k$.

Changes in the average aggregate prediction of event k is given by

$$d\bar{B}_k/dt = (1 - \bar{B}_k)B_k - K_8\bar{B}_k, \qquad [2.6]$$

where K_8 is the rate decay of $\bar{B}_k$.

Total novelty, Novelty, at a given time is given by the sum of the novelty of all stimuli present or predicted at a given time. Novelty is given by

$$\text{Novelty} = \Sigma_k |\bar{\lambda}_k - \bar{B}_k|, \qquad [2.7]$$

where k includes all CSs and the US.

We assume that a given CS can be predicted by other CSs, the CX or itself. Therefore, either repeated presentations of that CS in a given context, or simply repeated presentations of CS, lead to a decrease in its novelty. Whereas CS–CS associations decrease CS novelty in a context-independent manner, CS_j–CS_k or CX–CS_k associations decrease CS novelty in a context-dependent way. Because decrements in novelty are responsible for LI, CS–CS associations are responsible for context-nonspecific LI, whereas CS_j–CS_k or CX–CS_k associations are responsible for context-specific latent inhibition (see Good & Honey, 1993).

The normalized value of Novelty, Novelty', is given by

$$\text{Novelty}' = \text{Novelty}^2/(K_9^2 + \text{Novelty}^2). \qquad [2.8]$$

Novelty' is used by the attentional system to define the value of z_{CS} (see Equation [2.2]). Following Pearce and Hall's (1980) view, we assume that the orienting response, OR, is directly proportional to Novelty'.

An interesting emergent property of the model is that it always tends to maintain a minimum level of Novelty'. When Novelty' increases, attention to the environmental stimuli increases, associations are formed, the stimuli are predicted and Novelty' decreases. When Novelty' decreases too much, attention decreases, existing associations are not activated, predictions decrease and

Novelty′ increases again. In sum, Novelty′ might decrease, but it rarely becomes zero (see Figure 9.1 in Chapter 9).

Changes in CS–US associations

Sokolov (1960) suggested that animals build an internal model of their environment. Whenever there is a mismatch between predicted and actual environmental events, the internal model is modified. When there is coincidence between the observed and the predicted stimulus, the animal may respond without changing its neural model of the world. In the network, environmental regularities are stored in the associative system as CS–CS and CS–US associations in a recurrent autoassociative network (Kohonen, 1977). In the diagram shown in Figure 2.1, environmental regularities are stored in the associative system as (a) associations of each X_{CS} with its corresponding CS, $V_{CS1-CS1}$; (b) associations between X_{CS} with other CSs, $V_{CS1-CS2}$; (c) associations of X_{CS} with the context (CX), V_{CS1-CX}; (d) associations of X_{CX} of the CX with the CS, V_{CX-CS1}; and (e) associations of X_{CS} with the US, V_{CS1-US}. Presentation of a CS or a CX will increase the activity of the representations of their associated CS, CX or US if the associations are excitatory. If the associations are inhibitory, those presentations decrease the activity of the representations of their associated CS, CX or US.

Figure 2.2 shows that Node 3 receives input from Node 2, X_{CS}, as well as from the US. The synaptic weight connecting Node 2 to Node 3, V_{CS-US}, reflects the (excitatory or inhibitory) association of X_{CS} with the US. Changes in this association, $V_{CS,US}$, are given by

$$dV_{CS,US}/dt = K_7 X_{CS}(\lambda_{US} - B_{US})(1 - |V_{CS,US}|), \quad [2.9a]$$

where X_{CS} is the internal representation of the CS, λ_{US} is the intensity of the US, and B_{US} is the aggregate prediction of the US by all Xs active at a given time (See Equation [2.3]). As in the Rescorla–Wagner (RW) (1972) model, changes in $V_{CS,US}$ are proportional to a common error term given by the difference between predicted and real values of US, $(\lambda_{US} - B_{US})$. By Equation [2.9a], CSs and the CX compete to become associated with the US, a device used to explain overshadowing, blocking and conditioned inhibition (Chapter 3).

By Equation [2.9a], $V_{CS,US}$ increases whenever X_{CS} is active and $\lambda_{US} > B_{US}$ (dV > 0 in Figure 2.3) and decreases when $\lambda_{US} < B_{US}$ (dV < 0 in Figure 2.3). That is, according to Equation [2.9a] (also Equation [2.9b]), periods of acquisition and extinction occur within the same "acquisition" trial. Acquisition occurs in those periods when X_{CS} and the US temporally overlap, and extinction occurs in those periods when X_{CS} is active in the absence of the US. Asymptotic learning is reached when the amounts of acquisition and extinction are similar within an acquisition trial.

28 Attentional and associative mechanisms

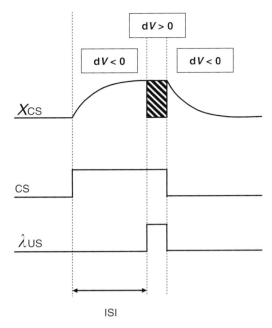

Figure 2.3 $V_{CS,US}$ increases whenever X_{CS} is active in the presence of the US ($\Delta V > 0$) and decreases when the US is absent (dV < 0). ISI: interstimulus interval.

An important consequence Equation [2.9a] is that excitatory $V_{CX,US}$ associations tend to be weaker than $V_{CS,US}$ associations. That is, $V_{CX,US}$ and $V_{CS,US}$ associations increase during the period of time when the τ_{CX} and τ_{CS} are active in the presence of the US (i.e. the term $[\lambda_{US} - B_{US}]$ is positive, $\Delta V > 0$), and decrease when the US is absent (i.e. the term $[0 - B_{US}]$ is negative, $\Delta V < 0$). Then, because the CS and the CX compete to become associated with the US, if the CS is more salient than the CX, the CS–US association will increase more than the CX–US association. In addition, because the τ_{CX} is active for a longer time (the whole duration of the ITI) than τ_{CS}, the excitatory $V_{CX,US}$ association will decrease more than the $V_{CS,US}$ association in the absence of the US. In sum, because $V_{CS,US}$ increases more than $V_{CX,US}$ during the US presentation, and decreases less in its absence, $V_{CX,US}$ will be generally weaker than $V_{CS,US}$. This property is important to explain why responding at the end of conditioning tends to be CX-independent and why the CX becomes inhibitory during extinction (see Chapter 9).

In contrast to the Rescorla–Wagner model, and in order to prevent the extinction of conditioned inhibition (Zimmer-Hart & Rescorla, 1974) or the generation of an excitatory CS by presenting a neutral CS with an inhibitory CS (Baker, 1974), our model assumes that B_{US} (and B_{CS}) takes on only positive values (when $B_{US} < 0$ then $B_{US} = 0$). Because B_{US} does not take negative values, inhibitory associations are not extinguished by CS or CX presentations. Inhibitory CX–US associations are established in extinction, during the period of time when the CS predicts the presence of the US, which is absent. These inhibitory CX–US associations do not change when the CX representation is active for the duration of the intertrial interval (ITI), in the absence of the US. Therefore inhibitory CX–US associations are independent of the ITI, but excitatory CX–US associations are not; a fact that has two important consequences. One is that inhibitory CX–US associations tend to be stronger than excitatory ones, a property important in extinction in which responding becomes CX-dependent (see Chapter 9). Another consequence is that US presentations are more effective in decreasing the magnitude of inhibitory CX–US associations than in increasing the magnitude of excitatory CX–US associations (see Chapter 3); a property that is important when the effects of context reinforcement following extinction, weak conditioning (inflation), and partial reinforcement (see Chapter 9, section on "Reinstatement") are compared. Finally, it is worth noticing that even if inhibitory CS–US and CX–US associations do not decrease with CS or CX presentations, attention to the CS or the CX might decrease with repeated presentations, thereby making the inhibitory association difficult to detect (see Chapter 9).

Supernormal conditioning

Although B_{US} does not take negative values, the model can still describe, like the Rescorla–Wagner model, supernormal conditioning (Rescorla, 1971b). Conditioning in an inhibitory CX (or in the presence of an inhibitory CS) will keep the sum $X_{CS}V_{CS-US} - X_{CX}V_{CX-US} < 0$, and $B_{US} = 0$, until $X_{CS}V_{CS-US} = X_{CX}V_{CX-US}$. Only at that point will the term (US – B_{US}) start decreasing. Because CS–US associations increase while the term (US – B_{US}) > 0, the CS–US association will become stronger in the presence of an inhibitory CS (or in an inhibitory CX) than in the presence of a neutral CS (or in a neutral CX). In short, the model describes supernormal conditioning when the prediction of the US, B_{US}, stays at zero for a longer time in an inhibitory CX than when conditioning takes place in a neutral CX. (See Chapters 3 and 9.)

Limiting term for $V_{CS,US}$ associations

According to Equation [2.9a], changes in associations $V_{CS,US}$ are also proportional to $(1 - |V_{CS,US}|)$. This term (a) limits $V_{CS,US}$, $-1 < V_{CS,US} < +1$, and (b) makes changes to $\Delta V_{CS,US}$ relatively slow when $V_{CS,US}$ gets close to 1 or −1 (and relatively fast when $V_{CS,US}$ is close to 0). When $(\lambda_{US} - B_{US}) > 0$, it is $\Delta V_{CS,US} > 0$, and when $(\lambda_{US} - B_{US}) < 0$ it is $dV_{CS,US}/dt < 0$. At the same time, when $V_{CS,US}$ approaches either +1 or −1, the limiting term $(1 - |V_{CS,US}|)$ makes $dV_{CS,US}/dt = 0$.

One justification of the limits imposed on associations $V_{CS,US}$ is simply the intuition that the neurophysiological representation (for example, number of postsynaptic receptors, available presynaptic neurotransmitter) of both excitatory and inhibitory magnitudes of $V_{CS,US}$ cannot surpass certain values. In addition, because a CS with a weak X_{CS} will result in a weak B_{US} (see Equation [2.10a], below) and a very strong $V_{CS,US}$ according to Equation [2.9a], a non-salient CS would accrue unlikely powerful associations. Limits on associations were proposed by Blough (1975, page 20) in order to avoid unrealistic perfect discrimination in his simulated psychometric function. A similar constraint was introduced in a classical conditioning model by Schmajuk and DiCarlo (1992) and Schmajuk et al. (1996), and later used in the conditioning models presented by Brandon, Vogel and Wagner (2002) and Le Pelley (2004). This term is important in explaining the experimental results addressed in Chapter 3. Furthermore, by limiting $V_{CS,US}$ the model is able to explain maximality effects on blocking (Beckers et al., 2005), that is, the fact that blocking is present only when the prediction of the US by the blocking stimulus has reached a maximum value and is unable to completely predict the actual US (see Chapter 16).

Chaining in mediated acquisition, extinction and cognitive mapping

Because the outputs of Node 1 ($\tau_{CS} + B_{CS}$) and Node 2 (X_{CS}) are active even when the CS is absent, but predicted by other CSs or the CX (see Equation [2.3]), all associations established by the CS can change even in its absence. This mechanism explains mediated acquisition (Holland, 1990) and mediated extinction (Shevill & Hall, 2004, see Chapter 9). Furthermore, this mechanism allows for the generation of inferences, which are used to describe sensory preconditioning (by combining, or chaining, CS_1–CS_2 and CS_2–US associations, see Chapter 16) and second-order conditioning (by combining, or chaining, CS_2–US and CS_1–CS_2 associations). Chaining is an important feature of the model that explains a type of reinstatement after extinction, as it will be seen in the corresponding section (Chapter 9). Furthermore, Schmajuk and Thieme (1992) showed that latent learning and detour tasks in complex mazes can be described by a neural cognitive map based on $Place_1$–$Place_2$ associations that can be chained to infer the path to the goal.

Changes in CS–CS associations

As shown in Figure 2.1, the associative system also stores (a) associations of each X_{CS} with its corresponding CS, e.g. $V_{CS1-CS1}$; (b) associations between X_{CS} with other CSs, e.g. $V_{CS1-CS2}$ (see Rescorla & Durlach, 1981); (c) associations of X_{CS} with the context (CX), V_{CS-CX} (see Rescorla, 1984); (d) associations of X_{CX} with

An attentional–associative model of conditioning   31

the CS, $V_{CX\text{-}CS}$ (see Marlin, 1982); and (e) associations of X_{CX} with the US, $V_{CX\text{-}US}$ (see Baker et al., 1981). Changes in the CS_1–CS_2 associations, $V_{CS1,CS2}$, are given by

$$dV_{CS1,CS2}/dt = K_7 X_{CS1}(\lambda_{CS2} - B_{CS2})(1 - |V_{CS1,CS2}|), \qquad [2.9b]$$

where X_{CS1} is the internal representation of CS_1, λ_{CS2} is the intensity of CS_2, B_{CS2} is the aggregate prediction of CS_2 by all X_{CS} active at a given time. By Equation [2.9b], $V_{CS1,CS2}$ increases whenever X_{CS1} is active and $\lambda_{CS2} > B_{CS2}$ and decreases when $\lambda_{CS2} < B_{CS2}$. $V_{CS1,CS1}$ is the association of CS_1 with itself. Like $V_{CS,US}$, $V_{CS1,CS1}$ and $V_{CS1,CS2}$ also vary between 1 and −1.

Aggregate predictions of the US and CS

The output of Node 3 in Figure 2.2 is the *aggregate prediction* of the US by all CSs with representations active at a given time, B_{US}, given by

$$B_{US} = \Sigma_{CS} B_{CS,US} = \Sigma_{CS} X_{CS} V_{CS,US}, \qquad [2.10a]$$

where $B_{CS,US}$ is the prediction of the US by CS and $V_{CS,US}$ is the association of X_{CS} with the US. B_{US} is used to compute $dV_{CS,US}$ in Equation [2.9a], and determines the magnitude of the CR. As mentioned, although $V_{CS\text{-}US}$ can be either positive (excitatory) or negative (inhibitory), B_{US} is always positive.

Interestingly, Equation [2.10a] allows the model to describe that *extensive conditioning* will increase CS–US associations, but also decrease Novelty′ and attention to the CS, with the consequent decrease in CR responding during the late trials of acquisition; a result reported by Pavlov (1927) and Sherman and Maier (1978).

As shown in Figure 2.2, the *aggregate prediction* of CS by all CSs with representations active at a given time, B_{CS}, given by

$$B_{CS} = \Sigma_i B_{CSi,CS} = \Sigma_i X_{CSi} V_{CSi,CS}, \qquad [2.10b]$$

where $B_{CSi,CS}$ is the prediction of CS by CS_i and $V_{CSi,CS}$ is the association of X_{CSi} with CS. B_{CS} is used to compute $\Delta V_{CS1,CS2}$ in Equation [2.9b], and reaches the feedback system to add to τ_{CS}. Aggregate predictions of the CS (B_{CS}) and the CX (B_{CX}) are computed as B_{US}, and cannot become negative either.

Learning (storage) and performance (retrieval)

Because the rate of changes in associations $V_{CS,US}$ and $V_{CS1,CS2}$ is directly proportional to X_{CS} (or X_{CS1}) (Equations [2.9a] and [2.9b]), X_{CS} controls the storage (formation or read-in) of CS_1–CS_2 and CS–US associations. Because the magnitude of the aggregate predictions B_{US} and B_{CS} is proportional to X_{CS} (Equations [2.10a] and [2.10b]), X_{CS} also controls the retrieval (activation or read-out) of CS_1–CS_2 and CS–US associations. Because attentional memory z_{CS} controls the magnitude of the internal representation X_{CS} (see Equation [2.3]), it indirectly controls storage and retrieval of CS_1–CS_2 and CS–US associations. Simultaneous control of memory storage and retrieval, a property that characterizes memory storage in neural networks, is a most important feature of the SLG model. This property, however, is not an additional assumption, but a simple and direct consequence (an emergent property) of the mechanism by which neural networks store and retrieve information. Furthermore, this property makes the SLG model different from most other models of classical conditioning, in which attention controls only the storage of associations. A similar concept was used in Wagner's (1981) SOP model, and was later incorporated in the McLaren and Mackintosh (2000) and Kruschke (2001) models.

CR strength

The output of Node 3 is B_{US}, which controls the conditioned response (CR), according to a non-linear function

$$CR = B_{US}^2/(K_{11}^2 + B_{US}^2). \quad [2.11]$$

According to this sigmoid function, the CR (a) is small for small values of B_{US}, (b) increases rapidly for intermediate values of B_{US}, and (c) has a maximum value of 1.

Although not included in most simulations presented in this book, the original model assumes that the OR inhibits the CR, that is, that the attention directed to a novel stimulus decreases the strength of the CR elicited by the target CS.

$$CR = (1 - K_{10}OR)B_{US}^2/(K_{11}^2 + B_{US}^2). \quad [2.12],$$

This assumption allows the model to describe (a) external inhibition, i.e. decrements in conditioned responding when an environmental change is introduced (Pavlov, 1927); and (b) decreased CR responding after extensive training in one environment followed by testing in a second environment (e.g. Penick & Solomon, 1991; see also McLaren et al., 1989, page 118).

However, as observed by Schmajuk and Larrauri (2006), Equation [2.12] is not applicable when a suppression paradigm is used, because both appetitive and aversive behaviors are affected, and the effects possibly cancel out. Equation [2.12] does not apply either (a) when the conditioned response is freezing, because both the CR and the OR work in the same direction; or (b) when the licking time in the presence of CS is measured, because the OR equally affects both the appetitive and aversive behavior. Equation [2.12] is valid, however, when direct measures of behavior are reported using eyeblink conditioning, pecking in autoshaping or elevation scores. For these latter cases, computer simulations showed that similar results were obtained when the OR is considered or ignored, except for the cases mentioned above, and the case of the partial reinforcement extinction effect (PREE) that the model approximates resorting to the opposing effect of the OR on the CR (see Chapter 9).

Summary

The SLG model, which portrays behavior on a moment-to-moment basis, is characterized by the following features: (a) attentional control of the formation of CS–US and CS–CS associations; (b) attention is determined by Novelty', which is in turn a function of CS–US, CX–CS and CS–CS associations; (c) competition among CSs to become associated with other CSs and the US; and (d) retrieval of CS–US and CS–CS associations is controlled by the magnitude of the attention to the CSs.

Appendix 2.1 Parameter values

Unless otherwise indicated, simulations in Chapters 3 to 10 and Chapters 15–16 adopted the same parameters values: $K_1 = 0.2$, $K_2 = 2$, $K_3 = 0.4$, $K_4 = 0.1$, $K_5 = 0.02$, $K_6 = 0.005$, $K_7 = 0.005$, $K_8 = 0.005$, $K_9 = 0.75$, $K_{10} = 0$, and $K_{11} = 0.15$. $K_{12} = 1$ in the normal case, 0.65 after haloperidol and 1.5 after nicotine administration (see Chapter 6). This set of parameters is identical to that of the original version of the model (Schmajuk et al., 1996; Schmajuk, Buhusi & Gray; 1998; Buhusi, Gray & Schmajuk, 1998; and Schmajuk, Cox & Gray, 2001), except for $K_{10} = 0$ (originally 0.7), which eliminates the inhibitory effect of the OR on the CR. As mentioned, the original K_{10} value might lead to incorrect results when freezing, licking time or suppression ratios are used (see Schmajuk & Larrauri, 2006, for a complete discussion). When simulations were used to verify the specific contributions of the associative mechanisms (Chapter 9), we assumed a fixed value for variables z_{CS} and z_{CX}, set to 0.5.

In several studies, we verified that the model is robust within a considerable range of parameter values. For instance, in the case of extinction (Chapter 9), we established that the model robustly describes acquisition, extinction, spontaneous recovery and renewal, with ten-fold increments or decrements in K_5, K_6 and K_7; parameters that control learning and attention. The values of K_5 and K_6 were varied simultaneously by the same amount, because together they determine the rate of change in z_{CS}. For renewal simulations, the upper limit for K_7 was a two-fold increase.

A version of the SLG program is available at www.duke.edu/~nestor.

3

Simple and compound conditioning

We have previously shown (Schmajuk, 1997), that the SLG model addresses different basic conditioning paradigms, including (a) simultaneous, delay and trace conditioning, and the effect of varying the interstimulus interval (Smith, 1968); (b) the effect of increasing US duration (Burkhardt & Ayres, 1978); (c) backward conditioning (Siegel & Domjan, 1971; Heith & Rescorla, 1973); (d) second-order excitatory (Kamil, 1969) and inhibitory conditioning (Rescorla, 1976); (e) sensory preconditioning (Brogden, 1939); and (f) partial reinforcement with different percentages of reinforced trials (Gormezano & Moore, 1969). According to the model, interstimulus interval (ISI, see Figure 2.3) effects are explained in terms of the shape of the trace τ_{CS} (see Figure 2.3 and Equation [2.1]), the effect of increasing US duration in terms of the increase in the size of the temporal overlap between X_{CS} and λ_{US} (see Equation [2.9a]), backward conditioning in terms of the combination (chaining) of CS–CX and CX–US associations (see Equation [2.3]), second-order conditioning and sensory preconditioning in terms of the combination (chaining) of CS_1–CS_2 and CS_2–US associations. The model correctly describes the competition between the excitation due to chaining and the conditioned inhibition (see below), produced by the alternated presentations of CS–CX (or CS_1–CS_2) and CX–US (or CS_2–US) trials, present in backward conditioning, second-order conditioning and sensory preconditioning.

Recent simulations show that the SLG model also describes the effect of massed vs. spaced trials (e.g. Spence & Norris, 1950) in terms of the increased CX–US associations (that compete with CS–US associations) when massed trials are used. The model also describes the deleterious effect on conditioning of increasing the CS duration with spaced trials in terms of the decrease in CS–US associations (Holland, 2000).

In the rest of this chapter we apply the SLG model to the description of compound conditioning. First, we succinctly address studies of relative validity, potentiation, extinction of conditioned inhibition and conditioned inhibition as a "slave process." Then, we analyze in detail the associative changes during compound conditioning when the elements of the compound have different initial associative values.

Compound conditioning

We have also shown (Schmajuk, 1997) that the SLG model addresses compound conditioning paradigms, such as overshadowing (Pavlov, 1927) and blocking (Kamin, 1968, 1969a,1969b), using the competitive mechanism of Equation [2.9a]. In Chapter 8, we will analyze how the model describes other compound conditioning paradigms, including recovery from overshadowing, recovery from blocking, backward blocking and recovery from backward blocking. Here, we report simulation results for relative validity, potentiation and some important properties of inhibitory conditioning.

Relative validity

Wagner et al. (1968) reported that responding stimulus X in a "correlated" group (which received AX+, AX+, BX−, and BX− trials) was weaker than in an "uncorrelated" group (which received AX+, AX−, BX+, and BX− trials). Computer simulations show that the SLG model describes "relative validity." Following 120 correlated or uncorrelated trials (CSs salience 1, duration 20 time units; US = 2, duration 5 time units), responding to X in the correlated group was weaker (CR_X = 0.48) than in the uncorrelated group (CR_X = 0.67).

Potentiation

Some experiments (e.g, Galef & Osborne, 1978; Durlach & Rescorla, 1980) found that the presence of a visual cue during a taste aversion experiment resulted in a visual aversion stronger than that obtained with a tasteless solution. That is, the presence of another cue seems to potentiate, rather than overshadow the CS–US association. The SLG model is able to describe potentiation. In agreement with Durlach and Rescorla's (1980) suggestion, according to the model, CS–CS associations are partly responsible for the effect. More detailed simulation studies are needed, however, to further understand how the model explains the phenomenon.

Extinction of inhibitory conditioning

As mentioned, the SLG model describes conditioned inhibition (Pavlov, 1927). According to the model, and in agreement with data in rats (Zimmer-Hart &

Rescorla, 1974), presentations of the inhibitory CS alone do not extinguish inhibitory CS–US associations (because B_{US} cannot become negative in Equation [2.9a]). Inhibitory conditioning can be extinguished, nevertheless, by reinforced presentations of the inhibitory CS. Also, using human participants, Melchers, Wolff & Lachnit (2006) reported the extinction of conditioned inhibition by simple presentation of the inhibitor. Although the model cannot eliminate inhibitory CS–US associations without US presentations, it still can describe extinction of condition inhibition in terms of a decreased attention to the CS when it is repeatedly presented by itself. A similar attentional decrement is used by the model to approximate the counteraction between two inhibitory procedures, as reported by Urcelay and Miller (2008).

Interestingly, Lysle and Fowler (1985) found that (a) extinction of the excitatory CS decreases the retardation of conditioning of the inhibitory CS, and (b) retardation can be increased by presentations reconditioning the extinguished excitatory CS. Lysle and Fowler interpreted these results in terms of a decrement in the inhibitory association of the inhibitory CS, and proposed that inhibitory associations are a *slave* process of excitatory ones. Preliminary simulations show that the SLG model is able to explain these results (Schmajuk & Kutlu, 2009). According to the model, a conditioned inhibitor shows retardation in conditioning, compared to a neutral CS, because of its (a) initial inhibitory association and (b) decreased attention. Attention to both the excitor and the inhibitor decreases when they are repeatedly active together in the absence of the US (as shown later in this chapter). Presentation of the conditioned excitor alone increases Novelty because the inhibitor is absent and, therefore, the expectation of the US is large when the US is absent. Because the excitor is strongly associated with the conditioned inhibitor, the representation of the inhibitor is active even when it is not presented, and attention to the inhibitor increases, thereby decreasing retardation. However, when the US is presented with the extinguished excitor, Novelty is smaller than in the previous case, because the expectation of the US approximates the actual US, and attention to the inhibitor is relatively small, which results again in retardation of conditioning.

Associative changes during compound conditioning

In a series of experiments, Rescorla (2000, 2001, 2002) reported that, against the predictions of the Rescorla and Wagner (1972) and other error-correction models, the associative changes undergone by two CSs of equal salience presented in compound depends on the value of their initial associations with the US.

According to the Rescorla and Wagner model (1972), changes in the associations V_{CS} between conditioned stimuli (CS) and the unconditioned stimulus (US) are proportional to a common error term

$$DV_{CS} = KCS(\lambda_{US} - \Sigma V_{CS}), \qquad [3.1]$$

where DV_{CS} represents the change in V_{CS} in one trial, λ_{US} represents the actual value of the US, and ΣV_{CS} represents the sum of the associations of all present CSs with the US. The larger the error term, the larger the changes in the associative values of the CSs.

Rescorla analyzed the effect of (a) reinforced and nonreinforced presentations of a compound in which CS_A was excitatory and CS_B was inhibitory (Rescorla, 2000), (b) reinforced and nonreinforced presentations of a compound in which CS_A was excitatory and CS_B was neutral (Rescorla, 2001), and (b) weak reinforcement and partial reinforcement of a compound in which CS_A was excitatory and CS_B inhibitory (Rescorla, 2002). Based on those studies, Rescorla suggested that the cumulative increment in V_A during reinforced AB compound presentations, ΔV_A, was smaller than the total increment in V_B during those reinforced presentations, ΔV_B. In contrast, when the AB compound was not reinforced, cumulative decrements in V_A, $-\Delta V_A$, were larger than those in V_B, $-\Delta V_B$. Briefly, $\Delta V_A < \Delta V_B$ during AB+ presentations, but $-\Delta V_A > -\Delta V_B$ during AB− presentations. Notice that the cumulative increment ΔV_{CS} during compound trials is obtained by adding the changes undergone by V_{CS} on each trial, DV_{CS}, over all those compound trials, that is $\Delta V_{CS} = \Sigma DV_{CS}$.

Rescorla (2000) used an experimental procedure that allows the comparison of the associative changes of two stimuli with different initial associative strength, such as an excitatory CS_A and an inhibitory CS_B. The procedure consists of measuring responding to the excitatory CS_A in compound with an inhibitory CS_D, and to the inhibitory CS_B in a compound with an excitatory CS_C. According to Rescorla (2000, page 429), the method avoids making a commitment to any particular function that maps learning values into performance. Using this measuring procedure, conditioned responding (CR) to the AD and BC compounds are given by

$$CR_{AD} = V_A + \Delta V_A - V_D, \text{ and} \qquad [3.2]$$
$$CR_{BC} = -V_B + \Delta V_B + V_C, \qquad [3.3]$$

where V_A, V_B, V_C and V_D are the associations of CS_A, CS_B, CS_C and CS_D with the US formed before compound training. At the beginning of compound training $-V_B = -V_D$, $V_C = V_A$. The cumulative changes in CS_A-US and CS_B-US associations, V_A and V_B, during compound training are indicated by ΔV_A and ΔV_B. In contrast, V_C and V_D do not change during compound training. Therefore, if as predicted by error-correction models $\Delta V_B = \Delta V_A$, it should be $CR_{AD} = CR_{BC}$, which is in contrast with the experimental results showing that $CR_{AD} < CR_{BC}$. According to Equations [3.2] and [3.3], $CR_{AD} < CR_{BC}$ is the consequence of $\Delta V_B > \Delta V_A$ when the AB compound is reinforced, or $-\Delta V_A > -\Delta V_B$ when the AB compound is not reinforced.

In order to solve the failure of error-correction rules, Rescorla (2000, page 436) suggested, among other possibilities, that associative changes ΔV_{CS} might be governed by a constrained error term, given by the common error term $(\lambda_{US} - \Sigma V_{CS})$ multiplied by the discrepancy between each individual CS–US association and the US strength, $(\lambda_{US} - V_{CS})$. A similar concept had been incorporated in the models proposed by Blough (1975), Schmajuk and DiCarlo (1992), Schmajuk, Lam and Gray (SLG, 1996), Brandon, Wagner and Vogel (2002) and Le Pelley (2004). For example, Brandon et al. (2002) incorporated a limiting term into the original standard operating procedures (SOP) model (Wagner, 1981). According to this rule, and assuming the positive limit is 1 and the negative limit is −1, V_{CS} increases according to $\Delta V_{CS} \sim (\lambda_{US} - \Sigma V_{CS})(1 - V_{CS})$, when $\lambda_{US} - \Sigma V_{CS} > 0$, and decreases according to $\Delta V_{CS} \sim (\lambda_{US} - \Sigma V_{CS})(V_{CS} + 1)$, when $\lambda_{US} - \Sigma V_{CS} < 0$. Therefore, when $V_A > V_B$, and AB is reinforced, $\Delta V_B > \Delta V_A$, and when AB is not reinforced, $-\Delta V_A > -\Delta V_B$. In both cases, the model would correctly describe the $CR_{AD} < CR_{BC}$ result.

In this chapter, we show that Schmajuk et al.'s (1996) attentional–associative model of classical conditioning presented in Chapter 2 – which includes a constrained error term – is also able to explain Rescorla's (2000, 2001, 2002) data. Interestingly, the model explains the $CR_{AD} < CR_{BC}$ results in attentional terms, even without predicting $\Delta V_B > \Delta V_A$ when the AB compound is reinforced, or $-\Delta V_B < -\Delta V_A$ when the AB compound is not reinforced.

Attention and error-correcting rules

The SLG model can be applied to Rescorla's (2000) experimental procedure, which, as mentioned, allows the comparison of the associative changes of an excitatory CS_A and an inhibitory CS_B during compound conditioning.

According to Equation [2.9a], changes in all V_{CS} associations are proportional to (a) representation the X_{CS}, (b) the common error term $(\lambda_{US} - B_{US})$, and to (c) the limiting term $(1 - |V_{CS}|)$. Therefore, even if two CSs are of equal salience, their associative changes when trained in compound depend on the value of their internal representations X_{CS}, which might be different. If two CSs, CS_A and CS_B, are reinforced and $(\lambda_{US} - B_{US}) > 0$, it will be $\Delta V_A > 0$ and $\Delta V_B > 0$. If initially $|V_A| > |V_B|$, the limiting term $(1 - |V_{CS}|)$ will make $\Delta V_A < \Delta V_B$. Similarly, if CS_A and CS_B are not reinforced and $(0 - B_{US}) < 0$, it will be $\Delta V_A < 0$ and $\Delta V_B < 0$. If initially

$|V_A| > |V_B|$, the limiting term $(1 - |V_{CS}|)$ will make $-\Delta V_A < -\Delta V_B$. In short, by itself, when $|V_A| > |V_B|$, the limiting term $(1 - |V_{CS}|)$ will make positive and negative changes in V_A always smaller than changes in V_B, that is, $\Delta V_B > \Delta V_A$ and $-\Delta V_B > -\Delta V_A$. In contrast, other constrained rules mentioned above yield $\Delta V_B > \Delta V_A$ when the AB compound is reinforced, and $-\Delta V_A > -\Delta V_B$ when AB is not reinforced. Notice, however, that because ΔV_{CS} is proportional to X_{CS}, the SLG model can still predict $-\Delta V_A > -\Delta V_B$, if $X_A > X_B$.

According to Equation [2.9a], ΔV_{CS} is proportional to (a) the limiting term $(1 - |V_{CS}|)$, and (b) the representation X_{CS}. At the beginning of compound AB conditioning, X_{CS} is defined by the previous, individual reinforcement histories of A and B. According to Equation [2.10a], following $AB+$ presentations, responding during testing is proportional to

$$CR_{AD} = X_A V_A + X_A \Delta V_A - X_D V_D, \text{ and} \qquad [3.4]$$
$$CR_{BC} = -X_B V_B + X_B \Delta V_B + X_C V_C. \qquad [3.5]$$

In these equations, X_A, X_B, X_C and X_D are the representations of CS_A, CS_B, CS_C and CS_D; V_A, V_D, V_B and V_C the associations formed before compound training, and ΔV_A and ΔV_B the cumulative changes in V_A and V_B during compound $AB+$ training. In contrast, V_C and V_D do not change during compound $AB+$ training. Because Novelty′ and attention to A and B decrease with enough $AB+$ training, X_A and X_B also decrease. With enough compound training, both X_A and X_B might become sufficiently small so that responding can be roughly estimated by the limits of CR_{AD} and CR_{BC}, as X_A and X_B tend to 0:

$$\lim CR_{AD} (X_A \to 0) = -X_D V_D \text{ and} \qquad [3.6]$$
$$\lim CR_{BC} (X_B \to 0) = +X_C V_C. \qquad [3.7]$$

Therefore, in contrast to Equations [3.2] and [3.3], our model predicts that with enough compound training the relation $CR_{AD} < CR_{BC}$ might be independent of the relative values of ΔV_A and ΔV_B. A similar analysis is possible for the case of nonreinforced $AB-$ presentations.

In the simulations that follow, we show how (a) the limiting term $(1 - |V_{CS}|)$, which controls ΔV_{CS} (Equation [2.9a]); and (b) the representation X_{CS}, which controls both ΔV_{CS} (Equation [2.9a]) and B_{US} (Equation [2.10a]), determine the value of the CR in a number of experiments that challenge the predictions of simple error-correction rules. As explained above, attentional mechanisms (controlling ΔV_{CS} and the CR) might be more important than the limiting term controlling ΔV_{CS} in describing the results. If that is the case, the relation $CR_{AD} < CR_{BC}$ might not provide unequivocal information about the relation between ΔV_A and ΔV_B.

In every simulation, the CSs are 15 time units (t.u.) in duration and of salience 1, the CX salience is 0.5, the intertrial interval (ITI) is 1000 t.u., the

US duration is 5 t.u. and overlaps with the last 5 t.u. of the CSs, and the US salience is 1 (except when indicated otherwise). Because the parameters of function defining B_{US} (see Equation [2.10a]) might depend on the specific experimental preparation (e.g. magazine entries, key pecks), our simulations simply assume that CR= B_{US}, whenever B_{US} > 0, and CR = 0 otherwise. All computer simulations were carried out with the same set of parameter values (presented in Chapter 2, Appendix 2.1).

Reinforced and nonreinforced presentations of an excitatory–inhibitory compound

Rescorla (2000, Experiments 1, 1a and 5) studied the effect of reinforced and nonreinforced presentations of a two-element compound consisting of a previously established excitatory CS_A and a previously established inhibitory CS_B on ΔV_A and ΔV_B.

Reinforced compound

Experimental data

In Rescorla's (2000) Experiment 1a, rats received magazine-approach conditioning consisting of 100 pseudorandom presentations of A+, C+, X+, XB−, and XD− trial types; 4 AB+ compound presentations; and 2 test trials each of AD and BC, presented in counterbalanced order, preceded by 4 AB+ trials. Figure 3.1 (upper panel) shows that magazine entries during presentation of the AD compound were fewer than during presentation of the BC compound. Similar results were obtained using an autoshaping procedure (Rescorla, 2000, Experiments 3 and 4). According to Rescorla, these results suggest that AB+ presentations yielded a greater increase in the inhibitory V_B than in the excitatory V_A, that is, $\Delta V_B > \Delta V_A$.

Rescorla (2000) reasoned that the formation of A–B associations during AB+ presentations would obscure the conclusions that can be drawn from the tests. This is so because, whereas A–B associations can increase responding to the inhibitory B through B–A–US chaining, that is not the case for the inhibitory D, which had not been associated with the excitatory C. In order to reduce this potential problem, in Experiment 5, rats received 12 sessions each with 8 presentations of each A+, AB−, C+, or CD− trial types, followed by 12 AB+ presentations, and by 2 AD and BC test trials. This procedure makes D–C–US chaining possible, thereby approximating the excitatory contributions of A to CR_{BC}, and of C to CR_{AD}.

Computer simulations

Simulations for Experiment 1a consisted of 70 alternated presentations of A+, C+, X+, XB− and XD− trials; followed by 20 AB+ presentations; and

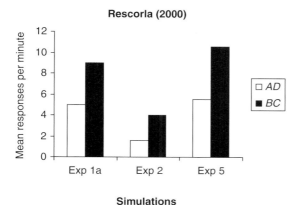

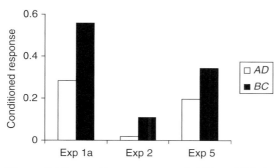

Figure 3.1 Responding to compounds (AD) and (BC). Bars represent the average responding to compounds tested in different orders. Compounds contain one previously excitatory stimulus (A or C), and one previously inhibitory stimulus (B or D), after AB+ or AB− presentations. *Upper panel*: Data from Rescorla's (2000) Experiment 1a, Experiment 2 and Experiment 5. *Lower panel*: Computer simulations with the Schmajuk et al. (1996) attentional–associative model.

by 4 alternated AD and BC test trials, counterbalanced for order. Simulations for Experiment 5 differed from Experiment 1a in that pre-training contained 72 alternated presentations of A+, C+, AB− and CD− trials. Figure 3.1 (lower panel) shows that the simulated results match well the results of Experiment 1a and Experiment 5, that is, responding in Group AD is weaker than in Group BC in both cases. However, the model does not capture the fact the responding in Experiment 5 is stronger than in Experiment 1a.

According to the SLG model, the results of Experiment 1a are explained as follows. Figure 3.2 (upper panel) shows the CRs, associations of stimuli A, C, B and D with the US (V_{CS}), as well as their internal representations (X_{CS}), at the beginning of the AB+ training (Pre), and during AD and BC test trials. At the beginning of compound training, A and C are excitatory, while B and D are inhibitory. Internal representations X_A and X_C are relatively weak because A

42 Attentional and associative mechanisms

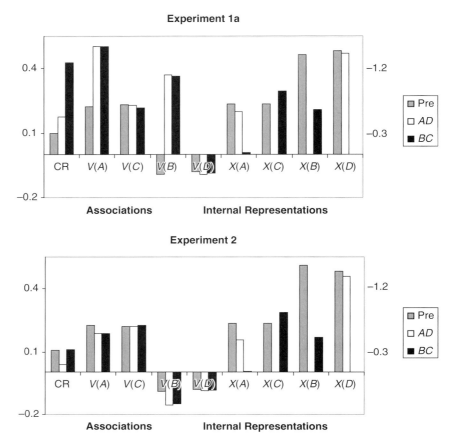

Figure 3.2 Simulated strength of the conditioned response (CR), associations $V_{A\text{-US}}$, $V_{C\text{-US}}$, $V_{B\text{-US}}$, and $V_{D\text{-US}}$, and internal representations X_A, X_C, X_B and X_D, before compound training (Pre) and during AD and BC testing. *Upper panel*: Simulation of Rescorla's (2000) Experiment 1a. *Lower panel*: Simulation of Rescorla's (2000) Experiment 2. V's scales are on the left axis, X's scales on the right axis.

and C undergo simple conditioning. In contrast, X_B and X_D are strong due to the relatively large value of Novelty' when the excitatory stimulus X is alternately presented alone with the US or with B and D in the absence of the US. At the beginning of compound training, it is $CR_{AB} = 0.10$ and at the end it is $CR_{AB} = 0.29$. During AB+ presentations, because $|V_A| > |V_B|$ and therefore $(1 - |V_A|) < (1 - |V_B|)$ in Equation [2.9a], V_A increases less than V_B, which becomes excitatory. The increment in V_A is smaller than in V_B (0.30 and 0.43, respectively), i.e. $\Delta V_A < \Delta V_B$. In addition, during the AB+ period, X_A and X_B decrease (from 0.71 to 0.21 and from 1.39 to 0.21, respectively) because Novelty' and attention to A and B decrease over compound trials. During testing, X_A and X_B increase again, respectively to 0.60 and 0.62. Consequently, responding during BC presentations is strong because both V_B and V_C are excitatory (0.36 and 0.22, respectively) and X_C (0.89) is relatively strong. In contrast, responding to AD is relatively weak because although V_A is excitatory (0.50), its corresponding X_A is relatively small (0.60), and V_D is inhibitory (−0.09) and X_D (1.40) is strong. Notice that repeated testing preserves the difference between AD and BC responding because Novelty' increases, thereby augmenting attention to all stimuli. Experiment 5 – modeled after Experiment 1a – in which both CS_A-CS_B and CS_C-CS_D associations can be formed and are appropriately captured by the model, the result is explained in the same terms applied to Experiment 1a.

In sum, in the case of reinforced presentations of an AB compound in which A is excitatory and B inhibitory, associative and attentional mechanisms acting during performance converge to explain the result $CR_{AD} < CR_{BC}$.

Nonreinforced compound

Experimental data

Rescorla's (2000) Experiment 2 was identical to Experiment 1a, except that rats received nonreinforced AB− presentations instead of reinforced AB+ presentations. Figure 3.1 (upper panel) shows that magazine entries during presentation of the AD compound were fewer than during presentation of the BC compound. Rescorla (2000) obtained similar results using autoshaping procedures (Experiment 6). In Rescorla's view, these results suggest that AB− presentations yielded a greater decrease in the excitatory V_A than in the inhibitory V_B, that is, $-\Delta V_A > -\Delta V_B$.

Computer simulations

Experiment 2 was identical to that of Experiment 1a, except that 20 AB− compound presentations were used. Figure 3.1 (lower panel) shows that the simulated results match well the results of Experiment 2, that is, responding in Group AD was weaker than in Group BC.

According to the model, the results of Experiment 2 are explained as follows. Figure 3.2 (lower panel) shows that during AB− presentations, V_A and V_B decrease. At the beginning of AB− training, $CR_{AB} = 0.10$, and at the end, $CR_{AB} = 0.003$. However, in contrast to Rescorla's (2000) suggestion that $-\Delta V_A > -\Delta V_B$, because $|V_A| > |V_B|$ and therefore $(1 - |V_A|) < (1 - |V_B|)$ in Equation [2.9a], the model shows $-\Delta V_A < -\Delta V_B$ (0.04 and 0.06, respectively). During the AB− period, representations X_A and X_B decrease (from 0.71 to 0.21 and from 1.52 to 0.26, respectively) because Novelty' and attention decrease over trials. As shown in Figure 3.2 (lower panel), because Novelty' increases during AD and BC testing, X_A and X_B increase again, respectively to 0.47 and 0.53. Consequently, responding to AD is weaker than responding to BC because (a) excitatory V_A and V_C are very similar and $X_A < X_C$ (0.88); and (b) even when B is more inhibitory than D (−0.15 vs. −0.08), it is $X_B < X_D$ (0.53 vs. 1.38). Again, repeated testing preserves the $CR_{AD} < CR_{BC}$ result because Novelty′ increases, thereby augmenting attention to all stimuli.

In sum, in the case of nonreinforced presentations of an AB compound in which A is excitatory and B inhibitory, the model shows that $-\Delta V_B > -\Delta V_A$ and, therefore, the result $CR_{AD} < CR_{BC}$ is explained in terms of attentional mechanisms acting during performance.

Reinforced and nonreinforced presentations of an excitatory–neutral compound

Blocking

Rescorla (2001, Experiment 1) studied the effect of presenting an excitatory–neutral compound with the US on ΔV_A and ΔV_B. Experiment 1 consisted of a total of 144 pseudorandom presentations of A+ and C+ trials; 8 B− and C− trials; 8 AB+ compound trials; 2 test trials each of AD and BC, presented in counterbalanced order, preceded by 4 AB+ trials. As shown in Figure 3.3 (upper panel), magazine entries during presentation of the AD compound were fewer than during presentation of the BC compound. According to Rescorla, these results suggest that AB+ presentations yielded a greater increase in the neutral V_B than in the excitatory V_A, that is, $\Delta V_B > \Delta V_A$.

Simulations for Experiment 1 consisted of 72 alternated presentations of A+ and C+ trials; 10 alternated presentations of B− and D− trials; 20 AB+ presentations; and 4 alternated presentations of AD and BC test trials counterbalanced for order. Figure 3.3 (lower panel) shows that the simulated results match the results of Experiment 1 well. According to the model, during AB+ presentations, V_A and V_B increase (from 0.41 to 0.44 and from 0.01 to 0.08, respectively). In agreement with Rescorla's (2000) suggestion that $\Delta V_A < \Delta V_B$, the model shows that $\Delta V_A = 0.04$ and $\Delta V_B = 0.07$ because $(1 - |V_A|) < (1 - |V_B|)$ in Equation [2.9a]. During the AB+ period, representations X_A and X_B decrease (from 0.69 to 0.51 and from 0.83 to 0.58, respectively) because Novelty′ and attention decrease over trials. Consequently, CR_{AD} is weaker than CR_{BC} because, even when V_A is slightly stronger than V_C, X_A of the excitatory A is smaller than X_C of the excitatory C; and whereas D remains neutral, B has become slightly excitatory during AB+ presentations.

Conditioned inhibition

Rescorla (2001, Experiment 2) also studied the effect of presenting an excitatory–neutral compound without the US on ΔV_A and ΔV_B. Experiment 2 was identical to Experiment 1, except that during the compound-conditioning phase, the AB compound was nonreinforced. Figure 3.3 (upper panel) shows

Simple and compound conditioning   45

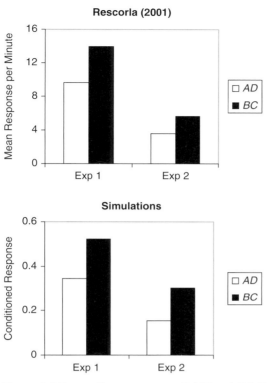

Figure 3.3 Responding to compounds (*AD*) and (*BC*). Bars represent the average responding to compounds tested in different orders. Compounds contain one previously excitatory stimulus (*A* or *C*), and one previously neutral stimulus (*B* or *D*), after *AB+* or *AB–* presentations. *Upper panel*: Data from Rescorla's (2001) Experiment 1 and Experiment 2. *Lower panel*: Computer simulations with the Schmajuk *et al.* (1996) attentional–associative model.

that magazine entries during presentation of the *AD* compound were fewer than during presentation of the *BC* compound. According to Rescorla, these results suggest that *AB–* presentations yielded a greater change in the excitatory V_A than in the neutral V_B, that is, $-\Delta V_A > -\Delta V_B$.

Simulations for Experiment 2 were identical to those of Experiment 1, except that *AB–* trials replaced *AB+* trials. Figure 3.3 (lower panel) shows that the simulated results match the results of Rescorla's (2001) Experiment 2 well. According to the model, during *AB–* presentations, V_A decreases and V_B becomes slightly inhibitory (from 0.40 to 0.27 and from 0.0 to –0.17, respectively). In contrast to Rescorla's (2000) suggestion, it is $-\Delta V_A < -\Delta V_B$ because $V_A > V_B$, and thus $(1 - |V_A|) < (1 - |V_B|)$ in Equation [2.9a]. During the *AB–* period, X_A decreases from 0.81 to 0.21, and X_B from 0.90 to 0.21. Therefore, in the case of nonreinforced presentations of an *AB* compound in which *A* is excitatory and *B* neutral (conditioned inhibition), attentional mechanisms explain the result $CR_{AD} < CR_{BC}$.

Intermediate and partial reinforcement of an excitatory–inhibitory compound

Intermediate reinforcement of the compound

Rescorla (2002, Experiment 1) studied the effect of a procedure in which animals first received *A*-strong US presentations alternated with *AB*– presentation, followed by *AB*-intermediate US presentations, on ΔV_A and ΔV_B. Experiment 1 consisted of a total of 64 pseudorandom presentations of *A*++, *AB*–, *C*++, and *CD*– trial types; a total of 64 *AB*+ trials; 2 presentations of *A*, *B*, *C* and *D* (counterbalanced for order) preceded by 8 *AB*+ trials; and 2 *AD* and *BC* test trials counterbalanced for order, preceded by another 8 *AB*+ presentations. The strong US consisted of 5 food pellets and the weak US consisted of 1 pellet. Figure 3.4 (upper panel) shows that $CR_{AD} < CR_{BC}$, which suggests a greater change in the inhibitory V_B than in the excitatory V_A, $\Delta V_A < \Delta V_B$. Rescorla (2002) reported similar results using magazine-approach conditioning (Experiment 1) and an autoshaping procedure (Experiment 2).

In both Experiments 1 and 2, Rescorla found significantly increased responding to *A* compared to *C*, i.e. $CR_A > CR_C$. But only in Experiment 2 did he also report increased responding to *B* compared to *D*, i.e. $CR_B > CR_D$ (see left upper panel in Figure 3.5). According to Rescorla (2002, page 166), the increase in associative strength of *A*, $\Delta V_A > 0$, can be considered an example of superconditioning, in which the presence of the inhibitory stimulus *B* increased the effectiveness of the weaker US during *AB*+ presentations.

Simulations for Experiment 1 consisted of 48 alternated presentations of *A*++, *AB*–, *C*++, and *CD*– trials; 50 *AB*+ trials; 2 test presentations of *A*, *B*, *C* and *D* trials counterbalanced for order; and 4 *AD* and *BC* test trials counterbalanced for order, preceded by 5 additional presentations of *AB*+ trials. Simulation parameters were identical to the previous simulations, except that the strong US was of strength 2 (the weak US was of strength 1, as before.) Figure 3.4 (lower panel) shows that the simulated results match the results of Experiment 1 well. In addition, Figure 3.5 (left lower panel), the model predicts that $CR_A > CR_C$ (which is in agreement with Rescorla's, 2002, Experiments 1 and 2 results) and $CR_B > CR_D$ (which agrees with Rescorla's, 2002, Experiment 2 results).

According to the model, during *AB*+ presentations, it is $\Delta V_A < \Delta V_B$ (0.23 and 0.35, respectively). Notice that the model captures the experimental result that both $\Delta V_A > 0$ and $\Delta V_B > 0$, that is, superconditioning of *A*. This is due to the fact that V_B became negative during pretraining with *A*++ and *AB*–, thereby reducing the magnitude of B_{US} during *AB*+ presentations, and making $\Delta V_A > 0$ even with a reduced US strength (see Equation [2.9a]). In addition, during the *AB*+ period,

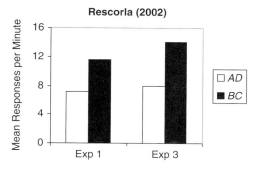

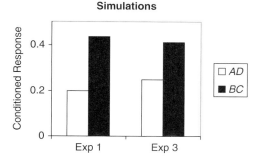

Figure 3.4 Responding to compounds (AD) and (BC). Bars represent the average responding to compounds tested in different orders. Compounds contain one previously excitatory stimulus (A or C), and one previously inhibitory stimulus (B or D), after AB intermediately reinforced or AB partial reinforcement. *Upper panel*: Data from Rescorla's (2002) Experiment 1 and Experiment 3. *Lower panel*: Computer simulations with the Schmajuk *et al.* (1996) attentional–associative model.

X_A and X_B decrease (from 1.44 to 0.43 and from 1.27 to .043, respectively). In sum, in the case of intermediate reinforcement of an AB compound in which A is excitatory and B inhibitory, associative and attentional mechanisms concur to explain the result $CR_{AD} < CR_{BC}$. Notice that, although during the phase of individual testing, attention to CS_A, CS_B, CS_C and CS_D increases, additional compound AB+ trials reduce attention to CS_A and CS_B, thereby enhancing the $CR_{AD} < CR_{BC}$ result.

Partial reinforcement of the compound

Rescorla's (2002) Experiment 3 also studied the effect partial reinforcement of an excitatory–inhibitory compound on ΔV_A and ΔV_B. Experiment 3 consisted of a total of 80 pseudorandom presentations of A+, AB−, C+ and CD− trial types; a total of 224 presentations of random AB+/AB− trials; 2 presentations of A, B, C and D counterbalanced for order and preceded by 8 AB+/AB− trials;

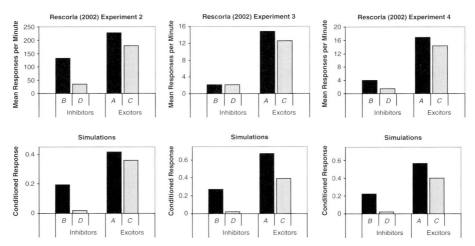

Figure 3.5 Responding to inhibitors B and D, and excitors A and C. Bars represent average responding to the CSs tested in different orders. *Left upper panel*: Mean key pecks per minute, data from Rescorla's (2002) Experiment 2. *Middle upper panel*: Mean magazine entries per minute, data from Rescorla's (2002) Experiment 3. *Right upper panel*: Mean magazine entries per minute, data from Rescorla's (2002) Experiment 4. *Lower panels*: Corresponding computer simulations with the Schmajuk *et al.* (1996) attentional–associative model.

and 2 *AD* and *BC* test trials counterbalanced for order, preceded by another 8 *AB+* or *AB–* trials. As shown in Figure 3.4 (upper panel), magazine entries during presentation of the *AD* compound were fewer than during presentation of the *BC* compound. According to Rescorla, the fact that $CR_{AD} < CR_{BC}$ suggests that *AB+/AB–* presentations yielded a greater change in the inhibitory V_B than in the excitatory V_A, $\Delta V_A < \Delta V_B$. As shown in Figure 3.5 (middle upper panel), magazine entries during separate presentations of the excitors (*A* and *C*) and inhibitors (*B* and *D*) show increased responding to *A* compared to *C*, a result that suggests superconditioning of *A*; and no increased responding to *B* compared to *D*, which suggests that *B* decreased its inhibitory association without becoming excitatory.

Simulation of Experiment 3 consisted of 72 alternate presentations of *A+*, *AB–*, *C+* and *CD–* trials; 140 alternated *AB+/AB–* presentations; 2 test presentations of *A*, *B*, *C* and *D* counterbalanced for order; 10 additional *AB+/AB–* trials; and 4 *AD* and *BC* test trials counterbalanced for order. Figure 3.4 (lower panel) shows that the simulated results match the results of Experiment 3 well. In addition, Figure 3.5 (middle lower panel) shows that, in agreement with Rescorla's (2002, Figure 4) results, the model correctly predicts that $CR_A > CR_C$, but incorrectly predicts $CR_B > CR_D$, because V_B becomes excitatory.

According to the model, during *AB+/AB−* presentations, V_A and V_B increase. Notice that $\Delta V_A < \Delta V_B$ (0.09 and 0.27, respectively) because $|V_A| > |V_B|$ (0.43 vs. 0.17), and therefore $(1 - |V_A|) < (1 - |V_B|)$ in Equation [2.9a]. During the *AB+/AB−* period, both X_A and X_B decrease (from 1.24 to 0.74 and from 1.18 to 0.74, respectively). In sum, in the case of partial reinforcement of an *AB* compound in which *A* is excitatory and *B* inhibitory, associative and attentional mechanisms combine to produce the result $CR_{AD} < CR_{BC}$.

Series of reinforced and nonreinforced presentations of the excitatory-inhibitory compound

Rescorla (2002, Experiment 4) assessed the effect of successive series of reinforced and nonreinforced compound presentations without the close intermixing of partial reinforcement used in Experiment 3. The intention was to find out whether repeated conditioning and extinction of the compound "B would show such substantially greater changes than A that they (A and B) would attain similar associative strengths" (Rescorla, 2002, page 170), as predicted by a modified single error term model in which stimuli with stronger associations with the US increase less and decrease more than those with weaker associations (see section on "Alternative approaches"). Experiment 4 consisted of initial conditioning similar to that of Experiment 4: 14 days with 16 presentations of either *AB+* or *AB−* of trials decided randomly for each day; 2 presentations of *A*, *B*, *C* and *D* test trials counterbalanced for order, preceded by 8 *AB+* trials; 2 *AD* and *BC* test trials counterbalanced for order, preceded by another 8 *AB+* trials; another 2 *AD* and *BC* test trials preceded by 24 *AB−* trials. Figure 3.6 (upper panel) shows that the results confirmed those of Experiment 3, but in this case the compound had been reinforced or not reinforced in daily blocks, instead of being intermixed within a day. Against the predictions of some modified models, even with repeated conditioning and extinction, responding to *AD* and *BC* retained their separation. In addition, as shown in Figure 3.5 (right upper panel), Rescorla reported significant differences between responding to the excitors, $CR_A > CR_C$, and inhibitors, $CR_B > CR_D$.

Simulations for Experiment 4 consisted of 72 alternate presentations of *A+*, *AB−*, *C+* and *CD−* trials; 14 alternated blocks of either 10 *AB+* or 10 *AB−* trials; 2 test presentations of *A*, *B*, *C* and *D* trials counterbalanced for order and preceded by 10 additional *AB+* trials; 4 *AD* and *BC* test trials counterbalanced for order, preceded by 10 *AB+* trials; and another 4 *AD* and *BC* test trials counterbalanced for order, preceded by 20 *AB−* trials. As shown in Figure 3.6 (lower panel), simulations match the experimental results well. According to the model, after the last block of 10 *AB+* presentations, it is $\Delta V_A < \Delta V_B$ (0.20 and 0.33, respectively) and X_A and X_B decrease (from 1.35 to 0.27 and from 1.28 to 0.27, respectively). After the last block of 20 *AB−* presentations, it is $-\Delta V_A < -\Delta V_B$ (−0.38 and −0.43,

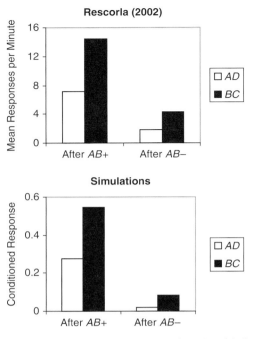

Figure 3.6 Responding to compounds (AD) and (BC). Bars represent the average responding to compounds tested in different orders. Compounds contained one previously excitatory stimulus (A or C), and one previously inhibitory stimulus (B or D), after (a) the last series of AB+ presentations and (b) the last series of AB− presentations. *Upper panel*: Data from Rescorla's (2002) Experiment 4. *Lower panel*: Computer simulations with the Schmajuk et al. (1996) attentional–associative model.

respectively) and X_A and X_B decrease (from 1.83 to 0.26 and from 1.86 to 0.25, respectively). In addition, as shown in Figure 3.5 (right lower panel), the model correctly describes that $CR_A > CR_C$ and $CR_B > CR_D$, which is in agreement with Rescorla's (2002) Experiment 4 results.

Briefly, the results following AB+ and AB− blocks of trials are explained as the joint effect of both associative mechanisms during compound training ($\Delta V_A < \Delta V_B$ after compound reinforcement, but $-\Delta V_A < -\Delta V_B$ after compound nonreinforcement), and attentional ($X_A < X_C$ and $X_B < X_D$) mechanisms during testing.

Discussion

Rescorla (2000, 2001, 2002) studied the effect of reinforced and non-reinforced presentations of an AB compound in which A and B had different initial associations with the US. In all cases, he reported that responding to AD (with D inhibitory) was weaker than responding to BC (with C excitatory).

According to Rescorla, these results show that changes in A–US and B–US associations are different, specifically that AB+ presentations yield $\Delta V_B > \Delta V_A$, whereas AB− presentations yield $-\Delta V_A > -\Delta V_B$, which contradicts the predictions of models of conditioning that rely only on a single, common error-correcting term. These models predict that the associative changes of two conditioned stimuli presented in compound are independent of the value of their initial associations.

Computer simulations show that the SLG model is able to explain those results. Computer simulations with the model show that when the compound is continuously, weakly or partially reinforced (Rescorla, 2000, Experiment 1a; Rescorla, 2002, Experiment 1; Rescorla, 2002, Experiments 3 and 4), changes in the associative values of A are smaller than in those of B, $\Delta V_A < \Delta V_B$ (similar to Rescorla's suggestion). When the compound is not reinforced (Rescorla, 2000, Experiment 2; Rescorla, 2001, Experiment 2), these changes are similar to those of the reinforced cases $-\Delta V_A < -\Delta V_B$ (in contrast to Rescorla's suggestion). In addition, the model correctly describes that compound reinforcement increases responding to both the excitor and inhibitor compared with responding to the nonreinforced CSs (Rescorla, 2002, Experiments 2 and 4).

The model explains the $CR_{AD} < CR_{BC}$ results as the consequence of (a) associative and attentional mechanisms acting during compound training (Equation [2.9a]), and (b) attentional mechanisms controlling performance during testing (Equation [2.10a]). According to the model, the relation between ΔV_A and ΔV_B during compound training is controlled by the values of $|V_A|$, $|V_B|$, X_A and X_B achieved during pretraining. In addition, V_A and V_B are modulated by X_A and X_B during testing to yield the $CR_{AD} < CR_{BC}$ results. This explanation differs from those provided by other "constrained models", which show $\Delta V_A < \Delta V_B$, when the compound is reinforced, but $-\Delta V_A > -\Delta V_B$ in the cases in which the compound is not reinforced. Independently of the specific explanations provided by our model for each experiment, our simulations suggest that the effect of attentional mechanisms, acting before and during compound training, might be overlooked when Rescorla's results are interpreted only in terms of associative changes.

It is important to point out that computer simulations show that the $CR_{AD} < CR_{BC}$ result is reliably produced by the model, even when using different (a) CS durations, (b) US intensities, and (c) numbers of pre-compound and compound trials. In a way, the robustness of the results should not be surprising, because this is expected whenever X_A and X_B decrease compared to X_C and X_D, as demonstrated in Equations [3.6] and [3.7].

Furthermore, the range of number of trials for which the results are obtained is very large. For instance, when the number of pre-compound and compound trials in the simulations are identical to the number used in the experiments, simulations show that the model still yields $CR_{AD} < CR_{BC}$, but in this case the model exaggerates the experimental results and predicts $CR_{AD} \sim 0$ (see Equations [3.6] and [3.7]). This explains why, in order to correctly approximate the empirical data, the number of pretraining trials in our simulations was always smaller than the number of corresponding experimental trials.

Existing support

Interestingly, the model's explanation of the results in terms of attentional mechanisms is supported, in part, by existing literature. For instance, the model suggests that, during the second phase of Rescorla's (2001, Experiment 1) blocking experiment, (a) the associative competitive mechanism leaves less room for the blocked stimulus to increase its association with the US, and (b) the attentional mechanism decreases Novelty′ and attention to the blocked CS. Such decrease in attention to the blocked CS has been suggested by Mackintosh and Turner (1971), who analyzed the effect of blocking on the attention to the blocked CS. In their study, an experimental group received *A*+ trials followed by *AB*+ trials, whereas its control group received only *A*+ trials. When both groups later received *AB*− trials, *B* became a conditioned inhibitor at a slower rate in the experimental than in the control group. Mackintosh and Turner (1971) suggested that attention to *B* had decreased during the second phase of blocking. Similarly, reduced attention to the blocked cue was reported by Kruschke and Blair (2000) during forward and backward blocking. As expected, the SLG model can replicate the Mackintosh and Turner results.

Also, in the case of Rescorla's (2001, Experiment 3) conditioned inhibition experiment, the model suggests that attention to *A* and *B* decreases during *AB*− presentations that follow the initial *A*+ and *B*− trials. A similar decrease in attention to the *AB* compound, during *A*+/*AB*− alternated trials (instead of sequential blocks, as in the Rescorla studies), was suggested by Pearce and Hall's (1979) experiments. Furthermore, the decreased attention to the *AB* compound might extend to its components *A* and *B* (which were tested separately in the Rescorla studies), as suggested by Rudy *et al.*'s (1976, Experiment 1) data showing significant latent inhibition to an element of a preexposed compound.

Even if the above experiments support the notion that attention has a role during the formation of associations (i.e. CS associability), they say nothing about the model's suggestion that attentional variables also control the strength of the response. Such a view finds some support in Pavlov's (1927) and Sherman and Maier's (1978) data, showing that responding during extensive conditioning first increases but later decreases, or Blaisdell, Gunther and Miller's (1999) data, showing that extinction of the blocking CS produced a recovery of the response to the blocked CS; results that can be interpreted (as our model does) in terms of a changed X_{CS}, but a constant V_{CS}, of the tested CS. Below, we delineate two experiments that test the model's prediction that the strength of CR_{AD} and CR_{BC} is under attentional control.

Neurophysiological basis

Interestingly, recent neurophysiological evidence seems to support error-correcting mechanisms. For instance, Schultz and Dickinson (2000) reviewed how neurons within several brain structures appear to code prediction errors in relation to positive and negative reinforcement, CSs, and responses. In some cases, dopamine, norepinephrine and nucleus basalis neurons broadcast these error signals to different brain structures. In other cases, these error signals are coded and broadcast within certain structures (e.g. cerebellum). Furthermore, neurophysiological data also support the existence of attentional mechanisms controlled by novelty. In this regard, Fiorillo *et al.* (2003) reported that in addition to coding the discrepancy between predicted and actual reward, the response of dopamine neurons "covaried with uncertainty and consisted of a gradual increase in activity until the potential time of reward." This coding of uncertainty, closely related to the Novelty' variable in our model, suggests a possible role for dopamine signals in attentional mechanisms. We will address similar issues in Chapter 4.

Alternative approaches

Brandon *et al.* (2002) incorporated a limiting term into the original SOP model (Wagner, 1981). According to this rule, and assuming the positive limit is 1 and the negative limit is −1, V_{CS} increases according to $\Delta V_{CS} \sim (\lambda_{US} - \Sigma V_{CS})(1 - V_{CS})$, when $\lambda_{US} - \Sigma V_{CS} > 0$, and decreases according to $\Delta V_{CS} \sim (\lambda_{US} - \Sigma V_{CS})(V_{CS} + 1)$, when $\lambda_{US} - \Sigma V_{CS} < 0$. Even if this rule similar to our Equation [2.9a] (in which ΔV_{CS} is zero when $V_{CS} = 1$ or $V_{CS} = -1$), for Brandon *et al.* ΔV_{CS} is zero when $V_{CS} = 1$ and $\lambda_{US} - \Sigma V_{CS} > 0$ or when $V = -1$ and $\lambda_{US} - \Sigma V_{CS} < 0$. Therefore, when $V_A > 0$, $V_B < 0$, and *AB* is reinforced (Rescorla, 2000, Experiments 1a and 5), V_B increases more than V_A, because $\Delta V_B \sim (1 - (-V_B))$, $\Delta V_A \sim (1 - V_A)$, and $V_A > V_B$, it is $\Delta V_B > \Delta V_A$, and the model correctly describes the $CR_{AD} < CR_{BC}$ result. For the case of *AB* nonreinforced presentations (Rescorla, 2000, Experiment 2), because $\Delta V_A \sim -(V_A + 1)$, $\Delta V_B \sim -(-V_B + 1)$, and $V_A > V_B$, it is $\Delta V_A > \Delta V_B$, the model correctly describes the $CR_{AD} < CR_{BC}$ result. Brandon *et al.* (2003) also reported that the constrained SOP rule correctly describes Rescorla's (2001) Experiments 1 and 3, but that it encounters some problems addressing Rescorla's (2002) Experiments 3 and 4.

According to Le Pelley (2004), CS–US associations, V_{CS}, increase according to $\Delta V_{CS} \sim (1 - V_{CS} + \overline{V}_{CS}) |\lambda_{US} - (\Sigma V_{CS} - \Sigma \overline{V}_{CS})|$, when $\lambda_{US} - (\Sigma V_{CS} - \Sigma \overline{V}_{CS}) > 0$; whereas CS–US antiassociations, $\overline{V}_{CS}$, increase according to $\Delta \overline{V}_{CS} \sim (1 - \overline{V}_{CS} + V_t) |\lambda_{US} - (\Sigma V_{CS} - \Sigma \overline{V}_t)|$, when $\lambda_{US} - (\Sigma V_{CS} - \Sigma \overline{V}_{CS}) < 0$. When $V_A > 0$, $V_B < 0$, and the animals receive AB reinforced presentations (Rescorla, 2000, Experiments 1a and 5), V_B increases more than V_A, because $\Delta V_B \sim (1 + \overline{V}_B)$ and $\Delta V_A \sim (1 - V_A)$, it is $\Delta V_B > \Delta V_A$. Therefore, the model correctly describes the $CR_{AD} < CR_{BC}$ result. For the case of AB nonreinforced presentations (Rescorla, 2000, Experiment 2), V_A decreases more than V_B, because $\Delta \overline{V}_A \sim (1 + V_A)$ and $\Delta \overline{V}_B \sim (1 - \overline{V}_t)$, it is $\Delta \overline{V}_A > \Delta \overline{V}_B$, so the model correctly describes the $CR_{AD} < CR_{BC}$ result. In addition, the model describes Rescorla's (2001) Experiments 1 and 3, and Rescorla's (2002) Experiments 1, 3, and 4.

According to Harris (2006), the associative strength of a CS is represented by an array of CS–US associations. When the excitatory A and inhibitory B are reinforced together, half of As already acquired associative strength is reduced and, therefore, many US elements that receive few connections from the stronger A elements can support further conditioning. In contrast, because B is not associated with the US, it can acquire the greater portion of the available associative strength. Similarly, when the AB compound is not reinforced, because the A elements are better connected than the B elements to the US elements with the strongest associative input, they undergo the largest decrease in associative strength. Similarly, Harris' (2006) model describes Rescorla's (2000) Experiment 2, Rescorla's (2001) Experiment 3, and Rescorla's (2002) Experiment 4 (but in this last case, only under the assumption of a random difference in salience between A and B).

Two unique predictions of the model

The SLG model and the models proposed by Brandon et al. (2002), Le Pelley (2004) and Harris (2006) explain the effect of reinforced presentations of the compound as the consequence of changes in V_A and V_B similar to those expected by Rescorla ($\Delta V_B > \Delta V_A$). However, in the cases when the compound is not reinforced, the SLG model explains the $CR_{AD} < CR_{BC}$ result in terms of attentional mechanisms at work during compound conditioning and testing. For one of those cases (Rescorla, 2000, Experiment 2), we analyzed the effect of replacing the constraining term in Equation [2.9a], $(1 - |V_{CS}|)$, by the one suggested by Brandon et al. (2003), i.e. $(1 - V_{CS})$ when $\lambda_{US} - \Sigma V_{CS} > 0$ and $(V_{CS} + 1)$ when $\lambda_{US} - \Sigma V_{CS} < 0$. Computer simulations with our model show similar results with either constraining term, which suggests that in our model the specific constraining term does not determine the $CR_{AD} < CR_{BC}$ relationship. The results also suggest that, without controlling for the values of the attentional variables, our model cannot generate predictions to decide the correct form of the constraining term.

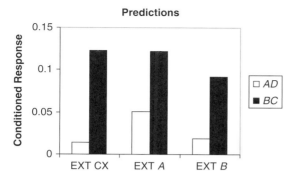

Figure 3.7 Responding to compounds (AD) and (BC). The proposed experiments are similar to those of Rescorla's (2000) Experiment 1a with the addition of 1 nonreinforced presentation of the CX (EXT CX), CSA (EXT A), or CSB (EXT B), before 4 test AD and BC trials. Bars indicate average responding over 4 trials in 2 different orders.

However, based on its attentional control of associative changes and performance, our model generates predictions that distinguish it from alternative models. For instance, in the case of Rescorla's (2000) Experiment 2, the model shows $-\Delta V_B > -\Delta V_A$ (because $|V_B| < |V_A|$) and the result $CR_{AD} < CR_{BC}$ is explained in terms of mechanisms that decrease attention during AB– training. During testing, the decreased X_A of the excitatory CS_A reduces responding to CR_{AD}, and the decreased X_B of the inhibitory CS_B increases responding to CR_{BC}. Therefore, we propose to carry out an experiment that includes a nonreinforced presentation of CS_A following compound training and before the test trials (Group EXT A in Figure 3.7). Under this condition, our model predicts that attention the excitatory CS_A will increase, and therefore X_A and CR_{AD} will increase. We also propose another experiment that includes a nonreinforced presentation of CS_B following compound training and before the test trials (Group EXT B in Figure 3.7). Under this condition, our model predicts that attention to the inhibitory CS_B and X_B will increase, thereby decreasing CR_{BC}.

Using the same design used for Rescorla's (2000) Experiment 2, Figure 3.7 shows that simulations for group EXT CX, in which the animal is placed in the context for 1 trial, reproduce Rescorla's results. Simulations for the group EXT A show increments in CR_{AD}, and simulations for group EXT B show decreased responding to CR_{BC}. The increased responding to the excitatory CS_A is similar to an extinction burst, i.e. the transient increase in responding during the initial sessions of extinction (e.g. Thomas & Papini, 2001). Because extinction bursts are mostly seen in autoshaping, we suggest running the proposed experiments using this preparation, similar to Rescorla's (2000) Experiment 4.

Summary

We have previously shown (Schmajuk, 1997), that the SLG model addresses a large number of the properties of simple conditioning. The model describes several compound conditioning paradigms, including relative validity (Wagner *et al.*, 1968) and potentiation (Galef & Osborne, 1978). Although the model cannot eliminate inhibitory CS–US associations without US presentations, it still can describe extinction of conditioned inhibition in terms of a decreased attention to the CS when it is repeatedly presented by itself, as reported by Melchers *et al.* (2006). Interestingly, the model seems able to describe Lysle and Fowler's (1985) results showing that (a) extinction of the excitatory CS decreases the retardation of conditioning of the inhibitory CS, and (b) retardation can be increased by reconditioning the extinguished CS.

Importantly, the model can describe a series of experimental results regarding the effects of reinforcing, partially reinforcing or not reinforcing an *AB* compound, in which *A* is excitatory and *B* is either inhibitory or neutral (Rescorla, 2000, 2001, 2002).

4

The neurobiology of fear conditioning

Gray (1975; Hebb, 1949; McNaughton, 2004) suggested that in order to relate brain and behavior, one should first develop a "conceptual nervous system" to handle behavioral data, and second find out whether brain structures and neural elements carry out the operations described by the conceptual system. The present chapter shows that variables representing "neural activity" in the SLG model (see Chapter 2) are consistent to brain responses reported by Dunsmoor et al. (2007) during a human fear conditioning task. In the Dunsmoor et al. (2007) neuroimaging study, different patterns of neural activity were revealed to CSs that predicted an aversive US on all trials, half the trials or no trials. Computer simulations with the SLG model demonstrate that (a) activity in the amygdala and anterior cingulate is well characterized by the prediction of the US by the CS and the CX, B_{US}, (b) activity in the dorsolateral prefrontal cortex (dlPFC) and anterior insula is well described by the representation of the CS, X_{CS}, and (c) the skin conductance response (SCR) is a nonlinear function of B_{US}. It is important to notice that variables B_{US} and X_{CS} represent "neural activities" (see Figure 2.2), and not the strength of their related synaptic associations, $V_{CS\text{-}US}$ and z_{CS}, which cannot be appreciated by functional magnetic resonance imaging (fMRI) methods.

Results

Experimental data

Dunsmoor et al. (2007) demonstrated patterns of learning-related activity within several brain regions in a fear-conditioning task that varied the CS–US pairing rate, while in fMRI, subjects were presented with three auditory CSs of 10 s each, which coterminated with a 500 ms 100 dB aversive white noise on

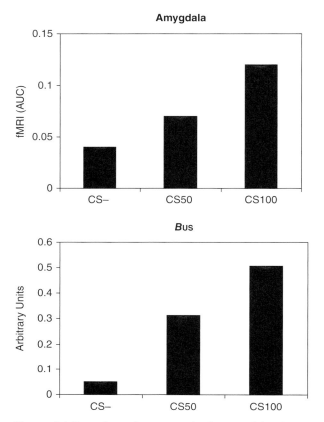

Figure 4.1 Hemodynamic response in the amygdala. *Upper panel*: Area under the hemodynamic response curve (AUC) within the bilateral amygdala. Data from Dunsmoor et al. (2007). *Lower panel*: Computer simulations of the B_{US} variable for CS−, CS50 and CS100, with the SLG model.

100% (CS100), 50% (CS50) or 0% (CS−) of trials. The conditioning session included 40 trials of each CS (120 total). Fear conditioning was evaluated by the skin conductance response (SCR) and ratings of expectancy for receiving the US.

The authors reported two distinct patterns of brain activity to CSs that varied as a function of the CS–US pairing rate. The magnitude of activity in the amygdala and anterior cingulate cortex was greatest to a CS that coterminated with the US on 100% of trials (CS100), while activity to a partially paired CS (CS50) fell at an intermediate level between the CS100 and an unpaired control stimulus (CS−) (see Figure 4.1, upper panel).

Activity observed within these regions was suggested by the authors as reflecting the strength of the CS–US association. A separate pattern of activity was observed within the dorsolateral prefrontal cortex and insula. The

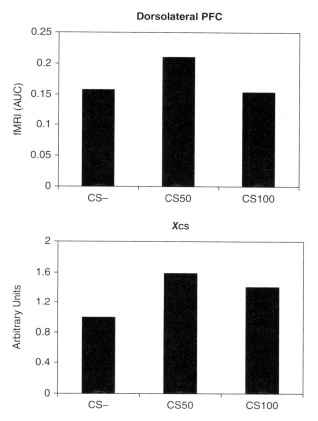

Figure 4.2 Hemodynamic response in the dorsolateral PFC. *Upper panel*: Area under the hemodynamic response curve (AUC) within the dorsolateral prefrontal cortex (dlPFC). Data from Dunsmoor et al. (2007). *Lower panel*: Computer simulations of the X_{CS} variable for CS–, CS50 and CS100, with the SLG model.

magnitude of activity within these regions was greatest to the partially paired CS50 than to either the CS100 or CS–. Heightened activity to the CS50 within the dlPFC and insula was suggested as reflecting the uncertainty for receiving the US (see Figure 4.2, upper panel).

Finally, Dunsmoor et al. (2007) reported that the magnitude of the SCR was proportional to the probability of reinforcement (see Figure 4.3, upper panel).

Simulation

Computer simulations with the SLG model consisted of presentations of three different CSs paired with the US on 0% (CS–), 50% (CS50) or 100% of the trials (CS100). The CS salience was 1 and CSs were 20 time units in duration;

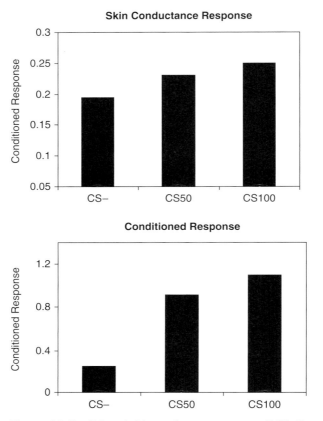

Figure 4.3 Conditioned skin conductance response (SCR). *Upper panel*: Data from Dunsmoor *et al.* (2007). *Lower panel*: Computer simulations of the CR for CS−, CS50 and CS100, with the SLG model.

the US had strength 2 and overlapped with the last 5 time units of the CSs; the context salience was 0.1. Generalization between CSs was achieved by including an additional CS, CSg, with salience 0.1, representing elements common to all CSs. The duration of the intertrial interval (ITI) was set to 3000 time units, in order to ensure that the traces of the different CSs became associated with their corresponding level of Novelty′, independent of the Novelty′ corresponding to other CSs. The model received 64 CS−, 64 CS100 and 64 CS50 alternated trials. As mentioned, we hypothesized that (a) activity in the amygdala and anterior cingulate can be characterized by the prediction of the US by the CS and the CX, B_{US}, (b) activity in the dlPFC and insula can be described by the representation of the CS, X_{CS}, and (c) the SCR is a nonlinear function of B_{US}.

Computer simulations using the SLG model show that the prediction for the US, B_{US}, increases linearly with the CS–US pairing rate; such that the continuously paired CS elicits the strongest association, the partially paired CS an intermediate amount of association and the unpaired control stimulus the lowest amount of association (Figure 4.1, lower panel). These results derive from the process of acquisition to CSs with variable CS–US reinforcement. That is, acquisition to a continuously paired CS increases in associative strength on each paired trial, whereas a partially paired CS alternates between gaining and losing associative strength over the course of paired and unpaired trials. Notably, although the CS– received no reinforcement with the US, B_{US} is still greater than zero for this stimulus, because CSg and the CX provide excitatory associations with the US. Overall, the pattern of responses revealed by the simulation is in line with those obtained from the amygdala and anterior cingulate cortex (Dunsmoor et al., 2007).

In addition, the internal representation of each CS (X_{CS}) is proportional to the level of Novelty′ detected at the time the CS trace is active (Figure 4.2, lower panel). For instance, Novelty′ is minimal when the CS predicts either that the US will be present on 100% of the trials, or present on 0% of the trials, and the US is consistently either present or absent (the subject is never surprised). In contrast, Novelty′ is maximal when the CS predicts that the US will be present on 50% of the trials, and the US is alternatively present or absent (the subject is always surprised). This pattern of simulated results seems to be consistent with the pattern of brain activity obtained from the dlPFC and insula (Dunsmoor et al., 2007).

Finally, Figure 4.3 (lower panel) demonstrates a pattern of simulated CRs in line with the SCRs and US expectancy exhibited by human participants during fear conditioning. Because the CR in the present model is a nonlinear function of B_{US}, CR = $f(B_{US})$ where f represents a sigmoid function (Equation 2.11), the CRs obtained by the model show a pattern similar to that of B_{US} (Figure 4.1).

Discussion

We showed that the SLG model is capable of describing (a) activity in the amygdala and anterior cingulate cortex, (b) activity in the dorsolateral prefrontal cortex (dlPFC) and insula, and (c) the SCR during fear conditioning in humans (Dunsmoor et al., 2007).

The amygdala is an important area for conditioned fear learning and is consistently implicated in (a) storing CS–US associations (LeDoux, 2000), and (b) controlling the CR (Knight et al., 2005). The simulations shown in Figure 4.1 appear to capture both these functions of the amygdala: (a) predicting the US, B_{US}, based on X_{CS}–US associations, V_{CS-US}, and (b) using this prediction to generate the CR. Likewise, the anterior cingulate has been shown to respond to paired, relative to unpaired, CSs in brain imaging studies of classical fear conditioning (Buchel et al., 1998). Reciprocal connections between the amygdala and the anterior cingulate may facilitate heightened responses to stimuli associated with an aversive outcome (Devinsky, Morrell & Vogt, 1995). However, the overall pattern of simulated results is not exclusive to the SLG model, as most associative learning models (e.g. Rescorla & Wagner, 1972; Mackintosh, 1975; Pearce & Hall, 1980) predict that differential reinforcement schedules affect the associative value of a CS.

Dunsmoor et al. (2007) characterized the pattern of responses in the dlPFC and insula as reflecting uncertainty for receiving the US, as activity was greater to the partially paired CS50 than to the CSs with more predictable outcomes (CS− and CS100). The simulations shown in Figure 4.2 support this suggestion. Specifically, the internal representation, X_{CS}, and attention, z_{CS}, were greatest for the CS partially paired with the US. These results are also in line with the Pearce–Hall model (1980) of associative learning, which predicts that attention is greater to CSs with more uncertain outcomes. The results are not in agreement with other models, such as Wagner's (1981) SOP, which expects similar responses to CS−, CS50 and CS100 because they are all based on similar CX–CS associations in all cases.

Uniquely in the SLG model, X_{CS} is proportional to the short-term memory trace of the CS, τ_{CS}, modulated by the magnitude of attention z_{CS}. Therefore, X_{CS} is an attention-modulated, sustained activity that is closely related to a "working memory" process. This aspect of the model converges with previous findings that the dlPFC is involved in holding the representation of a stimulus in working memory (Fuster, 1973; D'Esposito et al., 2000).

The simulation shown in Figure 4.3 appears to capture the physiological SCR data reported by Dunsmoor et al. (2007). Notice that although the internal representation of the CS, X_{CS}, is reduced for the 100% reinforced CS (Figure 4.2, lower panel), both the B_{US} (Figure 4.1, lower panel) and the CR (Figure 4.3, lower panel) are strongest to the most predictive CS. According to the SLG model, as X_{CS} keeps decreasing with an increased number of 100% reinforced trials, the CR will decrease with extended training. Such a decreased responding has been reported during classical conditioning in animals (Pavlov, 1927; Sherman & Maier, 1978).

Brain circuitry

Is there a brain circuitry that can provide a substrate for the functional characterization of the amygdala and anterior cingulated in terms of B_{US}, and of the dlPFC and insula in terms of X_{CS}? Projections from the ventral tegmental area (VTA), through the thalamus, to the amygdala and the PFC would provide Novelty′ information to these areas to control, respectively, the formation of

X_{CS}-US associations and the activation of working memory, X_{CS}. Gray, Buhusi and Schmajuk (1997) suggested that the activity of dopaminergic (DA) cells in the VTA represent Novelty' as defined in the SLG model (see Chapter 6); an assumption in line with reports that DA codes for novel stimuli (Williams, Rolls, Leonard & Stern, 1993; Horvitz, 2000; Legault & Wise, 2001; Fiorillo, Tobler & Schultz, 2003; Schmajuk, Larrauri, De la Casa & Levin, 2009).

Summary

The present chapter shows that variables representing "neural activity" in the attentional-associative model of classical conditioning presented in Chapter 2 correspond to the activity reported by Dunsmoor *et al.* (2007) in different areas of the brain (amygdala, anterior cingulated cortex, dlPFC and insula) during human fear conditioning.

As mentioned, only the Pearce–Hall (1980) model is also able to describe activity in the dlPFC and insula under different rates of reinforcement. However, the SLG model offers advantages over this competitor when addressing the Dunsmoor *et al.* (2007) data: (a) it relates the temporal course of X_{CS} to previous findings showing that the dlPFC holds the representation of a stimulus in working memory, and (b) it demonstrates that the decreased X_{CS} amplitude is compatible with an increased CR. In addition, as will be explained in Chapter 5, the SLG model offers numerous advantages over the Pearce–Hall model when applied to latent inhibition data, such as the disruption of latent inhibition by the presentation of an unexpected CS, or the omission of an expected CS, the restoration of the orienting response by the omission of an expected CS, motivational effects on latent inhibition and recovery from latent inhibition. Furthermore, as described in Chapter 8, only the SLG model is able to describe recovery from blocking, backward blocking, the recovery from backward blocking and the recovery from overshadowing. Finally, as presented in Chapter 9, only the SLG model can describe spontaneous recovery following extinction.

5

Latent inhibition

In this chapter, we explain how the SLG model (Schmajuk et al., 1996; Schmajuk & Larrauri, 2006; Schmajuk, 2002) introduced in Chapter 2 describes the multiple properties of latent inhibition (LI). We also discuss alternative approaches to LI.

Simulation procedures

In all the simulations presented in this chapter, different measures of conditioning were simulated as follows. When a conditioned emotional response paradigm was simulated, CR strengths were transformed into suppression ratios. Suppression ratios were calculated with the equation $A/(A + B)$, where A represents the appetitive responding (e.g. bar pressing for food, water licking) when the CS is present, and B represents the appetitive responding during the preceding non-CS period of equal duration. We assume that responding during the CS period is given by $\beta - CR(CS)$, and responding during the preceding non-CS period is given by $\beta - CR(CX)$, where β is proportional to the intensity of the appetitive behavior. Therefore, the suppression ratio was calculated by $(\beta - CR(CS))/((\beta - CR(CX)) + \beta - CR(CS)) = (\beta - CR(CS))/(2\beta - CR(CS) - CR(CX))$. In order to avoid unrealistic negative suppression ratios, the value of β was arbitrarily set to the maximum value of the CR(CS). Notice that when $V(CX,US)$ is close to or smaller than 0, then the suppression ratio is well approximated by $(\beta - CR(CS))/(2\beta - CR(CS))$. Percent CRs, time to complete a number of licks and freezing were simulated as proportional to the strength of the CR.

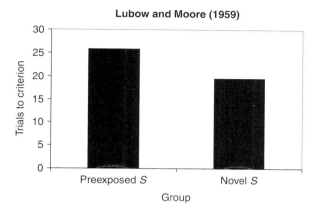

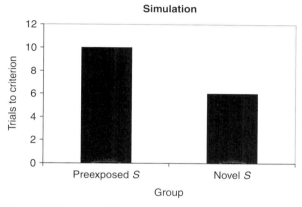

Figure 5.1 Latent inhibition of excitatory conditioning. *Upper panel*: Number of trials to criterion (10 conditioned responses) during conditioning of a preexposed CS (10 preexposures) and a non-preexposed CS. Data from Lubow and Moore (1959). *Lower panel*: Number of trials to criterion (CR=0.5) for a preexposed CS (Preexposed S Group) trained interspersed with a novel CS (Novel S Group), following 40 preexposure trials. The intertrial interval (ITI) was 200 time units (t.u.), the CX salience was 0.1, the CSs lasted 20 t.u. with intensity 1.0, the US was 1.8, overlapping the last 5 t.u. of the CS. Parameter values used in the simulations are listed in Chapter 2.

Application of the SLG model to latent inhibition

The effect of different preexposure procedures on the retardation of conditioning

Experimental results show that preexposure to a given CS, but not to the context or a different CS, retards excitatory conditioning to that specific CS (Lubow & Moore, 1959, Experiment 1; Reiss & Wagner, 1972, Experiment 1), CS preexposure retards inhibitory conditioning to that CS (Rescorla, 1971; Reiss & Wagner, 1972, Experiment 2), and that CS-weak US presentations retard CS-strong

66   Attentional and associative mechanisms

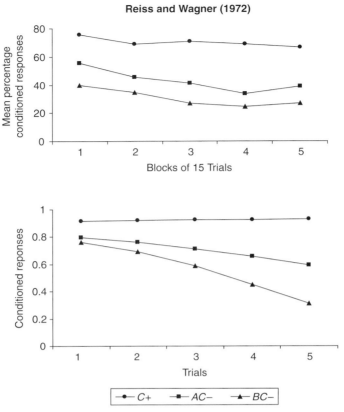

Figure 5.2 Latent inhibition of inhibitory conditioning. *Upper panel*: Mean percentage conditioned eyelid responses over 5 trials to (a) the reinforced C, (b) the A–C nonreinforced compound, and (c) the B–C nonreinforced compound, after 1380 nonreinforced presentations of A, 12 nonreinforced presentations of B, followed by 120 reinforced presentations of C. Data from Reiss and Wagner (1972). *Lower panel*: Conditioned responses during the last 5 (out of 7) trials to (a) C reinforced trials, (b) A–C nonreinforced trials, and (c) B–C nonreinforced trials, after 120 nonreinforced presentations of A, no nonreinforced presentations of B, and 10 reinforced presentations of C. The intertrial interval (ITI) was 200 time units (t.u.), the CX salience was 0.1, the CSs lasted 20 t.u. with intensity 1, the US was 1.8, overlapping the last 5 t.u. of the CS.

US conditioning (Hall & Pearce, 1979). In addition, preexposure to one CS element of a compound prevents that CS from overshadowing the other CS (Carr, 1974).

The model describes LI of excitatory conditioning (Figure 5.1), LI of inhibitory conditioning (Figure 5.2), as well as the Hall and Pearce (1979) effect (Figure 5.3). According to the SLG model, LI is manifested because: (a) CS preexposure reduces Novelty', thereby reducing z_{CS} and X_{CS} (in Figure 5.4 compare X_{CS} for

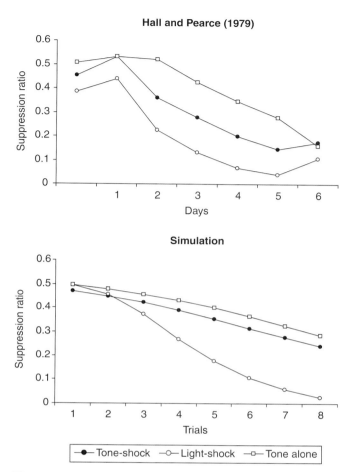

Figure 5.3 Conditioning following training with a weak US. *Upper panel*: Supression ratio during 6 conditioning sessions to a tone with a strong shock, following 66 trials with (a) a tone followed by a weak shock (Tone–shock group), (b) the overhead light followed by a weak shock (Light–shock group), or (c) the Tone alone (latent inhibition). Data from Hall and Pearce (1979). *Lower panel*: Suppression ratio during 8 CS–US conditioning trials with a strong US, after (a) 10 Tone–weak US trials (Tone–shock group), (b) 10 Light–weak US trials (Light–shock Group), and (c) 10 Tone-alone trials. The intertrial interval (ITI) was 200 time units (t.u.), the CX salience was 0.1, the Light and Tone CSs lasted 20 t.u. with intensity 1.0, the weak US was 0.3 and the strong US was 1.8, overlapping the last 5 t.u. of the CSs.

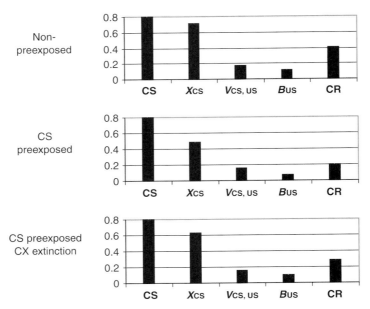

Figure 5.4 SLG model. Values of the variables representing the CS, X_{CS}, V_{CS-US}, and the CR (during a test trial) for the non-preexposed, CS preexposed, and CS preexposed followed by CX extinction cases. Y axes in arbitrary units.

CS preexposure vs. non-preexposure); (b) A small X_{CS} implies a slow rate of acquisition of excitatory or inhibitory $V_{CS,US}$ during conditioning, i.e. the storage (formation or read-in) of $V_{CS,US}$ is slower after CS preexposure (in Figure 5.4 compare $V_{CS,US}$ for CS preexposure vs. non-preexposure); (c) A small X_{CS} implies a small value of $B_{US} = X_{CS} V_{CS,US}$, i.e. retrieval (activation or read-out) of $V_{CS,US}$ decreases after CS preexposure (in Figure 5.4 compare B_{US} for CS preexposure vs. non-preexposure); (d) A small B_{US} generates a small CR (in Figure 5.4 compare CR for CS preexposure vs. non-preexposure). A similar explanation applies to the case of CS-weak US presentations retarding CS-strong US conditioning. The SLG model is able to describe Carr's (1974) results in terms of (b), that is, an acquisition deficit to the preexposed CS.

The effects of different parameters of preexposure on the strength of LI

Experimental data show that CS preexposure first facilitates the CR (freezing, Fanselow, 1990), and that facilitation is followed by LI (Kiernan & Westbrook, 1993; Prados, 2000). Also, LI increases with increasing number of CS preexposures (Lantz, 1973, Experiment 1), with increasing CS duration (Westbrook, Bond & Feyer, 1981, Experiment 3), increasing the total CS-preexposure time (number of CS preexposures multiplied by CS duration; Ayres, Philbin, Cassidy

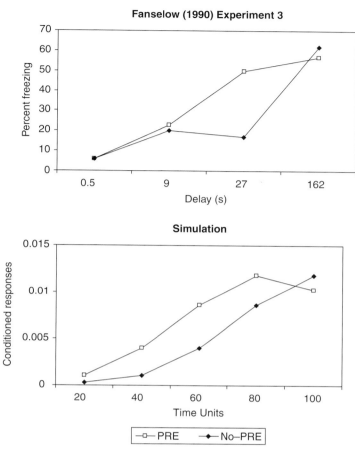

Figure 5.5 Conditioning as a function of the time spent in the context preceding the US presentation. *Upper panel:* Percent freezing following 0.5, 9, 27 and 162 s in the context preceding presentation of the US with (PRE) or without (No PRE) 120-s preexposure to the context. Data from Fanselow (1990, Experiment 3). *Lower panel:* Simulated peak CR amplitude after 1 CX–US conditioning trial following 20, 40, 60 or 80 time units in the context (with general context features $CX_g = 0.5$, and a specific context features $CX_c = 0.5$, see Chapter 8) with (PRE) or without (No PRE) 20 time units of preexposure to the context. The US was 5 t.u. long and of intensity 2.

& Belling, 1992), increasing CS intensity (Schnur & Lubow, 1976, Experiment 2; Crowell & Anderson, 1972, Experiment 1; but see Solomon, Brennan & Moore, 1974, Experiment 1), and increasing intertrial interval durations (Lantz, 1973, Experiment 2; Schnur & Lubow, 1976, Experiment 1; but see DeVietti & Barrett, 1986; Crowell & Anderson, 1972).

The model replicates Fanselow's (1990) data well, showing that a short preexposure facilitates conditioning (Figure 5.5), Lantz's (1973, Experiment 1)

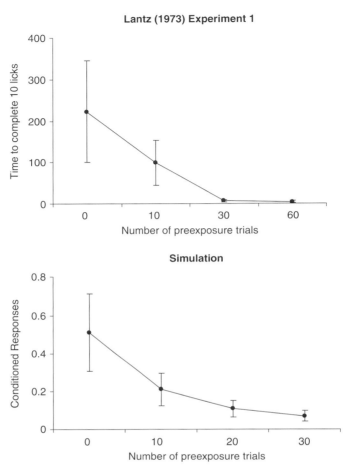

Figure 5.6 Latent inhibition as a function of the number of CS preexposures. *Upper panel*: Time to complete 10 licks after the offset of a tone previously paired with a shock US in groups receiving either 0, 10, 20 or 30 preexposure trials. Data from Lantz (1973, Experiment 1). *Lower panel*: Simulated peak CR amplitude after 5 conditioning trials following 0, 10, 20 or 30 CS preexposure trials. The intertrial interval (ITI) was 200 time units (t.u.), the CX salience was 0.1, the CS lasted 20 t.u. with intensity 1, and the US was 1.8, overlapping the last 5 t.u. of the CS.

data showing the LI increases with increasing number of CS preexposures (Figure 5.6), Westbrook *et al.*'s (1981, Experiment 3) data showing that LI increases with increasing CS duration (Figure 5.7), Ayres *et al.*'s (1992) data demonstrating that LI increases with increasingly total CS-preexposure time (05.8), Crowell and Anderson's (1972, Experiment 1) data showing that LI increases with increasing CS intensity (Figure 5.9), as well as Lantz's (1973, Experiment 2) results showing stronger LI with increasing intertrial interval

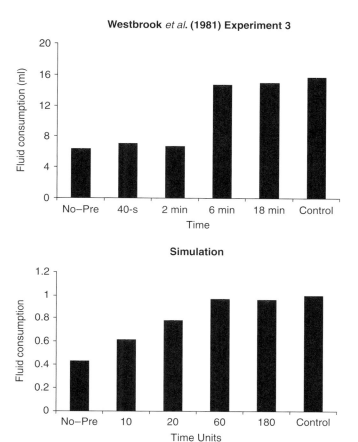

Figure 5.7 Latent inhibition as a function of CS duration. *Upper panel*: Mean consumption of water in the presence of an odor after access to the water in the presence of the odor, followed by an injection of LiCl, except in the Control Group. groups were not preexposed (No Pre), or preexposed to the odor for 40 s, 2, 6 or 18 min during a 18-min period of access to the water. Data from Westbrook *et al.* (1981, Experiment 3). *Lower panel*: Simulated fluid consumption (1-CR) after 20 conditioning trials and 20 CS preexposure trials, with CSs of 0 (No Pre), 10, 20, 60 and 180 t.u. long. Fluid consumption of the Control Group is 1. The intertrial interval (ITI) was 400 time units (t.u.), the CX salience was 0.1, the CS always had intensity 1, and the US was 1.0, overlapping the last 5 t.u. of the CS.

durations (Figure 5.10). According to the SLG model, all these parametric effects can be described by one simple rule: if the manipulation reduces Novelty', it will increase LI. The facilitation of fear acquisition by CS preexposure, an emergent property of the model, is explained because attention to the CS has first to increase before CS–US associations can be formed and freezing observed. Additional CS presentations decrease Novelty' and attention to the CS, thereby

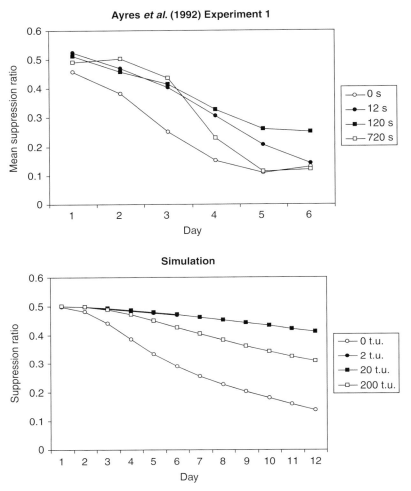

Figure 5.8 Latent inhibition and total CS–preexposure duration. *Upper panel*: Acquisition of conditioned suppression to a 120-s CS following preexposure to CSs of 0, 12, 120 or 720 s with a total CS–preexposure time of 14,400 s. Data from Ayres *et al.* (1992, Experiment 1). *Lower panel*: Suppression ratio during 12 CS–US conditioning trials following preexposure to CSs of 0, 2, 20 or 200 t.u. and a total CS-preexposure of 400 t.u. The intertrial interval (ITI) was 400 time units (t.u.), the CX salience was 0.1, the CS had duration 20 t.u. during conditioning and always had intensity 1, the US was 1, and overlapped the last 5 t.u. of the CSs.

resulting in LI. Interestingly, whereas the increase in attention that follows the initial CS presentation is similar to sensitization, the subsequent decrease in attention to the CS is similar to habituation (see Groves & Thompson, 1970). Despite these similarities, habituation of the OR and latent inhibition seem to be different processes (see section on "Orienting response and LI").

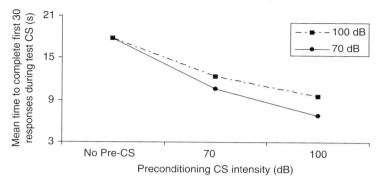

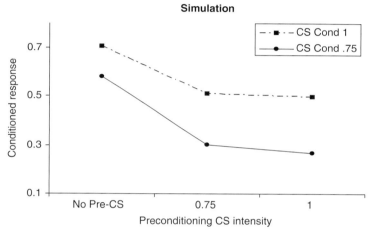

Figure 5.9 Latent inhibition as a function of CS intensity during preexposure and conditioning. *Upper panel*: Mean time to complete 30 lick responses after 10 reinforced trials with CS intensities of 70 or 100 dB, following no preexposure or 3 preexposures with a CS of intensities of either 70 or 100 dB. Data from Crowell and Anderson (1972, Experiment 1). *Lower panel*: Conditioned responses after 20 preeposure trials and 40 conditioning trials, either with no CS, a CS of intensity 0.75 or 1.0 during preexposure, and a CS of intensity 0.75 or 1.0 during conditioning. The intertrial interval (ITI) was 400 time units (t.u.), the CX salience was 0.1, the CSs lasted 25 t.u. with intensity 1, and the US overlapped the last 5 t.u. of the CS.

The consequences of presenting a novel event, or the omission of an expected event

It has been reported that LI is disrupted by the presentation of a surprising event (Lantz, 1973, Experiment 3; Rudy, Rosenberg & Sandell, 1977; Best, Gemberling & Johnson, 1979), as well as by the omission of an expected event (Hall & Pearce, 1982), following CS preexposure.

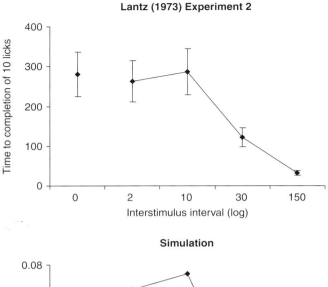

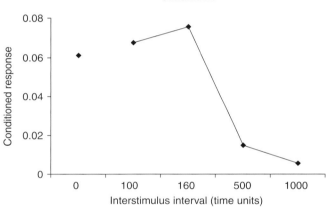

Figure 5.10 Latent inhibition as a function of the interstimulus interval duration. *Upper panel*: Time to complete 10 licks after the offset of a tone previously paired with a shock US in groups receiving CS preexposure with either 0, 2, 10, 30, or 150 (log s) interstimulus (intertrial) interval. Data from Lantz (1973, Experiment 2). *Lower panel*: Simulated peak CR amplitude after 5 conditioning trials following with 100, 160, 500, and 1000 t.u. interstimulus (intertrial) interval (ITI). The CX salience was 0.1, the CSs lasted 20 t.u. with intensity 1, the US had intensity 1.8, and overlapped the last 5 t.u. of the CS.

Figure 5.11 shows that the model replicates well the Lantz (1973, Experiment 3) data. In this case, the SLG model (but not the Pearce–Hall model) is able to describe the result, because Novelty' can be increased by the presentation of a unfamiliar CS. Correspondingly, Figure 5.12 shows that the model describes the Hall and Pearce (1982) data because Novelty' also increases when an expected US is omitted.

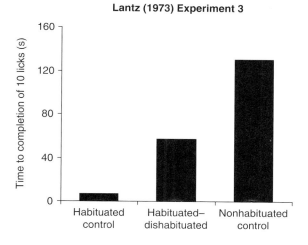

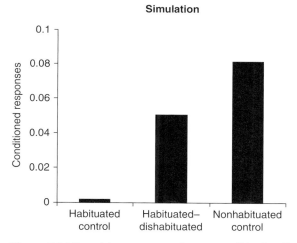

Figure 5.11 Surprising events previous to conditioning. *Upper panel*: Time to completion of 10 licks after 1 tone–US pairing following (a) 60 tone presentations (Habituated Control Group), (b) 60 tone presentations and 1 light presentation (Habituated–Dishabituated Group), or (c) exposure to the context. Data from Lantz (1973, Experiment 3). *Lower panel*: Simulated peak CR amplitude after 2 tone–US conditioning trials following (a) 20 tone presentations (Habituated Control Group), (b) 20 tone presentations and 1 light CS presentation preceding the tone CS by 20 t.u. during conditioning (Habituated–Dishabituated Group), or (c) 20 exposures to the context and 1 light CS presentation preceding the tone CS during conditioning (Nonhabituated–Control Group). The ITI was 500 t.u., the CX salience was 0.1, the CSs were 20 t.u. long and of intensity 1, and the US was 5 t.u. long and of intensity 2.

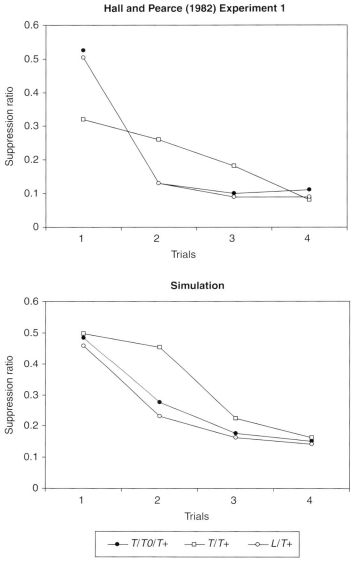

Figure 5.12 Omission of an expected event following CS preexposure. *Upper panel*: Suppression ratios during four tone-reinforced trials with a strong shock US, following (a) two tone-nonreinforced trials with weak shock US and two tone-nonreinforced trials (Group T/T0/T+), (b) two tone-reinforced trials with weak shock US (Group T/T+), and (c) two light-reinforced trials with weak shock US (Group L/T+). Data from Hall & Pearce (1982). *Lower panel*: Suppression ratio during four blocks of three CS-strong US conditioning trials, after (a) 10 tone-weak US and 10 tone-alone trials (T/T0/T+ Group), (b) 10 tone-alone trials (T/T+), and (c) 10 light-weak US trials (L/T+ Group). The intertrial interval (ITI) was 100 time units (t.u.), the CX salience was 0.1, the light and tone CSs lasted 20 t.u. with intensity 1.0, the weak US strength was 0.5 and the strong US was 2, overlapping the last 5 t.u. of the CSs.

The consequences of preexposing to different combinations of CSs

In some respects similar to the results described in the preceding paragraph, experimental results demonstrate that preexposing CS(A) and CS(B) in compound attenuates LI to CS(A) or CS(B) (overshadowing of LI; Mackintosh, 1973; Rudy, Krauter & Gaffuri, 1976; Holland & Forbes, 1980; Honey & Hall, 1989), LI to CS(A) is attenuated with simultaneous but not sequential presentations of CS(A) and CS(B) (Honey & Hall, 1988, Experiment 1), preexposing CS(B) followed by simultaneous preexposure of CS(A) and CS(B) in compound tends to preserve LI to CS(A) (blocking of LI; Reed, 1991), and that preexposure to CS(A)–CS(B) is less effective than alternated preexposure to separate presentations of CS(A) and CS(B) in producing LI to the CS(A)–CS(B) compound (Holland & Forbes, 1980; but see Baker, Haskins & Hall, 1990, Experiment 1). Schmajuk *et al.* (1996) showed that the SLG model is able to describe most of these results.

Figure 5.13 shows that the model replicates well the Honey and Hall (1989, Experiment 1b) data. In this case, the SLG model is able to describe the result because Novelty′ is increased by the absence of an expected CS.

The effects of contextual manipulations on the strength of LI

Experimental results show that context preexposure prior to CS preexposure facilitates LI (Hall & Channel, 1985a, Experiments 1, 2 and 3), a phase of exposure to the context alone interposed between CS preexposure and conditioning in the same context might slightly decrease LI (e.g. Wagner, 1979; Baker & Mercier, 1982), or slightly facilitate LI (Hall & Minor, 1984, Experiment 5), LI is disrupted by a change in the context from the CS preexposure phase to the conditioning phase (Wickens, Tuber & Wickens, 1983, Experiment 3; Hall & Channel, 1985a, Experiment 3), and this attenuation of LI occurs even when conditioning occurs in an already familiar context (Hall & Minor, 1984; Hall & Channel, 1985a; Lovibond, Preston & Mackintosh, 1984). Schmajuk *et al.* (1996) showed that the SLG model is able to describe these results.

Figure 5.14 shows that the model describes the Hall and Channel (1985a, Experiment 3) results, showing that LI is attenuated by a change in CX from the preexposure to the conditioning phase.

The effect of preexposure to a pair of CSs

Although CS preexposure typically yields LI, preexposure to a pair of CSs might facilitate performance on subsequent discrimination tasks. This phenomenon is known as perceptual learning. Experimental results show that CS preexposure in the rats' home cage produced perceptual learning, whereas CS preexposure in the training environment produced LI (Channel & Hall, 1981, Experiment 1). However, preexposure to intramaze cues resulted in perceptual

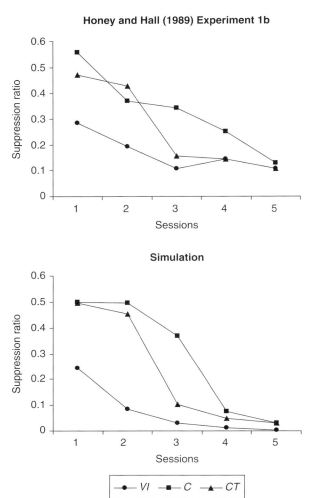

Figure 5.13 Compound preexposure. *Upper panel*: Suppression ratios during 5 click-reinforced trials following (a) context preexposure (Group VI), (b) 48 preexposures to the click (Group C), and (c) 48 preexposures to the click and a tone (Group CT). Data from Honey and Hall (1989, Experiment 1b). *Lower panel*: Suppression ratio during 5 blocks with 3 CS_C–US conditioning trials, after (a) 60 presentations in the CX (VI Group), (b) 60 CS_C nonreinforced trials (C Group), and (c) 60 CS_C–CS_T nonreinforced trials (CT Group). The intertrial interval (ITI) was 200 time units (t.u.), CX salience was 0.1, the Light and Tone CSs lasted 20 t.u. with intensity 1.0, the US was of strength 2, overlapping the last 5 t.u. of the CSs.

learning only when these cues were presented in the same context during preexposure and discrimination training (Trobalon, Chamizo & Mackintosh, 1992). Schmajuk *et al.* (1996) showed that the SLG model is able to describe these results.

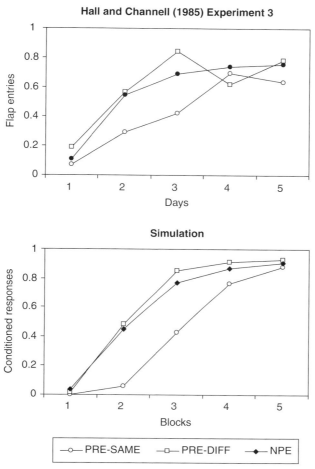

Figure 5.14 Latent inhibition is context specific. *Upper panel*: Flap entries during 40 trials of conditioning (8 each day) following 144 light preexposure trials either in (a) the context of preexposure (PRE-SAME) or (b) in another context (PRE-DIFF), and (c) when the light was not preexposed (NPE). Data from Hall and Channell (1985b, Experiment 3). *Lower panel*: Simulated peak CR amplitude over 5 blocks (of 3 conditioning trials) following (a) 15 CS presentations in the context of conditioning (PRE-SAME), (b) 15 CS presentations in a different context (PRE-DIFF), or (c) 15 exposures to the context (NPE). The ITI was 2000 t.u., the CX salience was 0.5, the CS was 20 t.u. in duration and of intensity 1, and the US was 5 t.u. in duration and of intensity 2.

Orienting response (OR) and LI

Kaye and Pearce (1984) showed that whereas CR acquisition is slowed down by CS preexposure, the OR to the CS during acquisition is stronger in preexposed (Group N in Experiment 2) than in non-preexposed (Group C in Experiment 2) animals. This result is well explained by the Pearce–Hall (1980) model, in terms of the larger difference between the US and the weak preexposed CS–US association, than between the US and the strong non-preexposed CS–US association. Similarly, the SLG model correctly describes the Kaye and Pearce (1984) results in terms of the smaller novelty of the preexposed CS during conditioning. Because lower Novelty' translates into lower attention to the CS, acquisition is slower and the OR stronger in the preexposed group.

Wilson et al. (1992) reported that percent ORs decreased with randomly alternated presentation of light–tone–food trials and light–tone alone trials, but it was restored when the nonreinforced trials contained only the light. Figure 5.15 shows that the SLG model reproduces the Wilson et al. (1992) data demonstrating the restoration of the OR. Although in line with the basic notions of the Pearce and Hall (1980) model, this result cannot be replicated by that model, because it does not include CS–CS associations.

Experimental data also show that LI can be impaired by contextual changes that fail to produce dishabituation of the OR (Hall & Channell, 1985b), and can remain undisturbed after a period of time that produces dishabituation of the OR (Hall & Schachtman, 1987). Schmajuk et al. (1996) showed that the SLG model can approximate both results. According to the model, the OR (proportional to Novelty') has a cumulative effect on attention z_{CS} (see Equation [2.2] in Chapter 2). Therefore, the dissociation between LI and the OR is the consequence of the relative independence of z_{CS} and the OR at a given time. For example, in the Hall and Channell (1985b) experiment, z_{CS} increases during alternated CS preexposure in two different contexts and, even if the OR is relatively small at the beginning of conditioning, the increased z_{CS} results in fast conditioning and absence of LI. Symmetrical to the idea that decrements in OR take time to be reflected in z_{CS}, Hall and Schachtman's (1987) results can be explained in terms of the delay for increments in the OR to change z_{CS}.

Motivational effects on LI

Experimental data show that preexposure to a CS results in LI only when the reinforcer is relevant to the motivational state in which CS preexposure was conducted. Killcross and Balleine (1996) reported that a CS preexposed when a rat is hungry (or thirsty) results in LI when the rat, which is both hungry and thirsty, is conditioned with food US (or saline US). Computer simulations with the SLG model show that the model is able to describe this phenomenon when the motivational states of the animal are represented as "internal" contexts that are associated with their corresponding goals. When food (liquid) US is used during conditioning, Novelty' is smaller for the CS preexposed in the hunger (thirst) state, which is associated with food (liquid), than for the CS

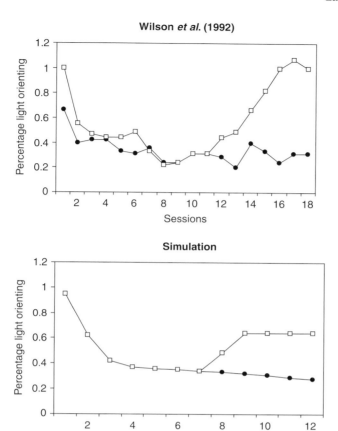

Figure 5.15 Restoration of the light orienting response. *Upper panel*: Light orienting expressed as a percentage of a total number of responses (magazine approaching and light orienting) during 10 light-tone-food sessions alternated with 10 light-tone sessions, and 8 additional same sessions (Control Group) or 8 light sessions (Experimental Group). Data from Wilson *et al.* (1992). *Lower panel*: Orienting responses, expressed as a percentage of the sum of the CR and the OR. Simulations consisted of (a) 12 CS_1-CS_2 presentations alternated with 12 CS_1-CS_2 presentations, followed by (b) 12 CS_1-CS_2-US presentations, alternated with (c) 12 CS_1 presentations. The ITI was 500 t.u., the CX salience was 0.1, the CSs were 10 t.u. in duration and of intensity 1, and the US was 5 t.u. in duration and of intensity 2.

preexposed in the thirst (hunger) state, which is associated with liquid (food). Because a smaller Novelty' is associated with slower conditioning, the model is able to describe the experimental data showing that preexposure to a CS results in LI only when the US is relevant to the motivational state in which CS preexposure was conducted. Figure 5.16 shows that the model reproduces Killcross and Balleine's (1996) data well.

The effects of postconditioning manipulations

Experimental data show that LI is attenuated by extensive exposure to the training context in the absence of the US (Grahame, Barnet, Gunther & Miller, 1994), by US presentations in another context (Kasprow et al., 1984), or by a single presentation of taste B (or taste compound BX) following preexposure and conditioning to taste compound AX (Killcross, 2001). The SLG model is able to describe these results.

Figure 5.17 shows that the model correctly replicates the Grahame et al. (1994) result. According to the model, extinction of the context following conditioning manipulations increases Novelty' as well as the magnitudes of z_{CS}, X_{CS} and the CR without increasing $V_{CS,US}$ (in Figure 5.04 compare V_{CS-US} for CS preexposure vs. CS preexposure–CX extinction). In other words, even if CS preexposure results in a decreased $V_{CS,US}$ value and CR magnitude during conditioning (LI, in Figure 5.4 compare V_{CS-US} for CS preexposure vs. non-preexposure), the magnitude of the CR can be increased by increasing z_{CS} and X_{CS} after conditioning (attenuation of LI, in Figure 5.4 compare X_{CS} for CS preexposure vs. CS preexposure–CX extinction).

The effects of the passage of time

LI can decrease (Kraemer et al., 1991; De La Casa & Lubow, 2000, 2002) or increase (De La Casa & Lubow, 2000, 2002; Wheeler et al., 2004) with the passage of time. This latter effect, referred to as super-LI, is well described by the model, and seen only when (a) the number of preexposure trials is relatively large, (b) the delay period is introduced between conditioning and testing phases instead of between preexposure and conditioning phases, (c) when all experimental stages are conducted in different contexts, and (d) the US is relatively strong (De La Casa & Lubow, 2002).

Figure 5.18 shows that the model correctly replicates the De La Casa and Lubow (2002) results for different US intensities and short and long delays between conditioning and testing. According to the SLG model, super-LI is not due to a further decrease in attention to the flavor in the PRE group, but to an increased attention to the stimulus representing the water and related stimuli, which becomes a conditioned inhibitor. When the US is relatively weak, introducing a delay between conditioning and testing decreases LI, because attention to the water barely changes, as a result of a relatively small

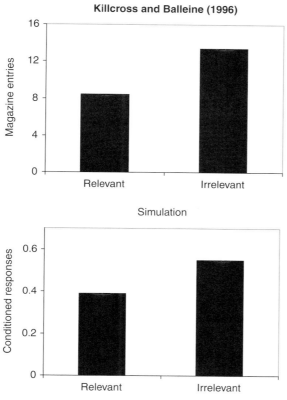

Figure 5.16 Role of motivation on latent inhibition. *Upper panel:* Mean magazine entries during conditioning trials of CS_1 preexposed thirsty and reinforced with either saline (Relevant Group) or pellets (Irrelevant Group) during 15 conditioning trials. Conditioning was preceded by 36 CS_1 preexposure trials while thirsty, and 36 CS_2 preexposure trials while hungry. Data from Killcross and Balleine (1996, Experiment 1). *Lower panel:* Average simulated peak CR amplitude for Relevant and Irrelevant Groups during conditioning. Simulations consisted of (a) 20 presentations of a CX (0.5) representing hunger, together with a 20 t.u. CS representing food (intensity 1), alternated with 20 presentations of another CX (0.5) representing thirst, together with a 20 t.u. CS representing water, (b) 10 presentations of the CX (0.5) representing hunger, with a 20 t.u. CS (intensity 1), alternated with 10 presentations of the CX (0.5) representing thirst with a different 20 t.u. CS, and (c) 10 presentations of both contexts (hunger and thirst) together the 20 t.u. CS representing water, either (a) with the CS preexposed with thirst (Relevant) or (b) with the CS preexposed with hunger (Irrelevant). The ITI was 200 t.u.

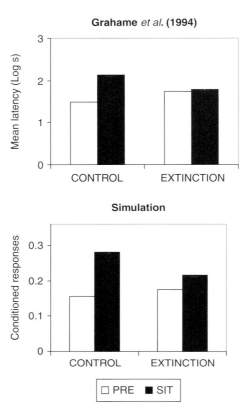

Figure 5.17 Recovery from latent inhibition. *Upper panel*: Mean latencies to complete the first 5 cumulative s of drinking in a test context after 120 preexposures to the CS (PRE Groups) or the CX (SIT Groups), followed by 4 reinforced-CS presentations, and 5 days in the home cage in the Control Groups and 5 days of CX extinction in the Extinction Groups. Data from Grahame *et al*. (1994, Experiment 1, Figure 3). *Lower panel*: Simulated results showing the strength of the conditioned response to the CS during 1 test trial, following 75 CS– trials (PRE) or 75 CX–trials (SIT), 45 CS–US trials, and 120 CX– trials. The ITI was 600 t.u., the CX salience was 0.1, the CSs were 10 t.u. in duration and of intensity 0.6, and the US was 5 t.u. in duration and of intensity 1.

increment in Novelty′. In turn, the small increase in Novelty′ is the consequence of having only the unfulfilled expectation of the training context, the flavor, and of a weak US, when the water is presented in their absence in the home cage. Because attention to the water does not increase during the delay, the water–US association becomes only barely inhibitory, and super-LI is absent. In addition, the increase in Novelty′ causes attention to the flavor to increase and, therefore, LI decreases.

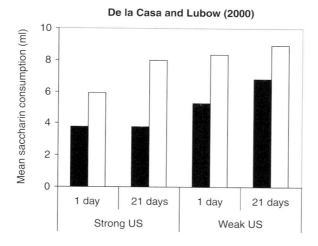

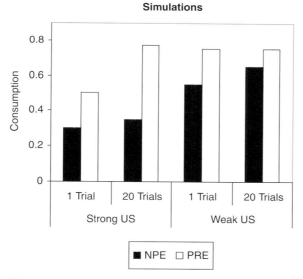

Figure 5.18 Super-latent inhibition after a post-conditioning delay with a strong or a weak US. *Upper panel*: Mean saccharin consumption in groups that received no preexposure (NPE) or 4 preexposure days (PRE) followed by 1 day (Short Delay) or 21 days (Long Delay) in the home cage after conditioning using a Strong or a Weak US. Data from Lubow and De la Casa (2000, Experiment 2). *Lower panel*: Five-trial average of simulated consumption after no preexposure (NPE) or 30 preexposure trials (Long PRE) and 1 (Short Delay) or 33 trials (Long Delay) after conditioning using a Strong or a Weak US. The ITI was 500 t.u., the Home and Training CX salience were 0.1, the CSs were 25 t.u. in duration and of intensity 1, and the US was 5 t.u. in duration and of intensity 1 (strong) or 0.8 (weak).

Alternative approaches to latent inhibition

Theorizing about LI has taken two opposite views, one suggesting that CS preexposure disturbs the storage of CS–US associations, another proposing that it hinders the retrieval of CS–US associations. According to the storage position, CS preexposure disrupts the formation of CS–US associations during conditioning by either (a) decreasing the associability of the CS (e.g. Lubow, Weiner & Schnur, 1981; Mackintosh, 1975; Wagner, 1978; Moore & Stickney, 1980; Pearce & Hall, 1980; Weiner, 1990; Schmajuk & DiCarlo, 1991a; Frey & Sears, 1978), or (b) fostering the formation of CS–no consequence associations that interfere with the subsequent establishment of CS–US associations (Revusky, 1971; Testa & Ternes, 1977). According to the retrieval position, CS preexposure disrupts the subsequent retrieval of CS–US associations (e.g. Kasprow, Catterson, Schachtman & Miller, 1984; Kraemer, Randall & Carbary, 1991). That is, whereas the first approach suggests that LI can be explained in terms of a mechanism operating during memory storage, the second approach proposes that LI is the result of a mechanism operating during memory retrieval (Bouton, 1993; Spear, 1981; Spear, Miller & Jagielo, 1990). As mentioned above, according to the SLG model, LI is the consequence of both a decreased storage *and* a decreased retrieval of CS–US associations.

Modulation of storage of CS–US associations

Most theories of classical conditioning assume that temporal contiguity between CS_i and the US leads to the formation of CS_i–US associations. Different rules have been proposed to describe changes in CS_i–US associations. According to Dickinson and Mackintosh (1978), these rules either assume variations in the associability or effectiveness of CS_i, or variations in the effectiveness of the US.

Variations in the associability of the CS

Attentional theories assume that the associability or effectiveness of CS_i to form CS_i–US associations depends on the magnitude of the "internal representation" of CS_i. In neural network terms, attention may be interpreted as the modulation of the CS representation that activates the presynaptic neuronal population involved in associative learning.

Mackintosh's (1975) attentional theory suggests that the associability of CS_i increases when it is the best predictor of the US, and decreases otherwise. CS_i preexposed animals show LI because both the CS_i and the context are equally poor predictors of the US. Moore and Stickney (1980) and Schmajuk and Moore (1989) generalized Mackintosh's approach to include the predictions of other CSs, and assumed that CS_i associability increases when CS_i is the best predictor of other CSs or the US, but decreases otherwise.

In contrast to Mackintosh's (1975) view, Pearce and Hall (1980) suggested that CS_i associability increases when CS_i is a poor predictor of the US, i.e. when CS_i has been recently followed by the unexpected presentation of the US. The model describes LI by assuming that during preexposure the (initially large) associability of CS_i declines.

According to Grossberg's (1975) neural attentional theory, pairing of CS_i with a US causes both an association of the sensory representation of CS_i with the US (conditioned reinforcement learning) and an association of the drive representation of the US with the sensory representation of CS_i (incentive motivation learning). Sensory representations compete among themselves for a limited-capacity short-term memory activation that is reflected in the CS_i–US associations. CSs with larger incentive motivation associations accrue CS–US associations faster than those with smaller incentive motivation. Schmajuk and DiCarlo (1991a) applied Grossberg's (1975) model to describe LI. They assumed that the initial value of incentive motivation decreases during CS preexposure, thereby retarding the acquisition of CS–US associations.

Lubow, Weiner and Schnur (1981; Lubow, 1989) presented a conditioned attention theory (CAT) of LI. According to CAT, attention to CS_i is a response, R_i, that occurs when CS_i is presented. The same laws of conditioning govern the acquisition of CR and R_i. The theory assumes that (a) R_i declines with repeated nonreinforced CS_i presentations and increases with reinforced CS_i presentations, (b) R_i may become conditioned to other CSs, and (c) R_i is correlated with CS_i associability. Levels of R_i that are higher than the level of R_i on the first presentation of CS_i represent *attention* to CS_i, whereas levels that are lower represent *inattention* to CS_i. Lubow (1989) showed that the model generates numerous predictions regarding (a) the conditioning of inattention to CS_i when CS_i is presented in isolation, and (b) the modulation of attention to CS_i when CS_i is presented with other CSs.

Weiner (1990) proposed that LI is the result of the animals maintaining, during conditioning, the small CS_i associability achieved when experiencing a CS–no US relationship during preexposure. As explained in Chapter 6, Weiner (2003) described how different brain structures and neurotransmitters participate in the control of LI.

In Le Pelley's (2004) hybrid model, latent inhibition is the result of a reduced salience associability, σ, during preexposure, because the US is not presented and not predicted. According to Harris's (2006) model, when a familiar stimulus is presented, the elements activated by its onset activates the later elements, thereby preventing the later elements from entering the attention buffer. The stimulus elements can be associatively activated by the context, which results in context-specific latent inhibition.

To the extent that LI involves CS preexposure, it can be successfully described by models that depict variations in CS effectiveness. However, in order to describe arbitrary input–output functions in classical conditioning (e.g. blocking, overshadowing), rules that assume variations in the effectiveness of the US are necessary.

Variations in the effectiveness of both the CS and the US

Some classical conditioning theories have combined variations in the effectiveness of both the CS and the US. For example, Frey and Sears (1978) proposed a model of classical conditioning that assumed variations in the effectiveness of both the CS and the US: the internal representation of CS_i is modulated by its association with the US, $V_{i,US}$. Nonreinforced presentations of CS_i yield LI because they decrease initial values of $V_{i,US}$. Ayres, Albert and Bombace (1987) proposed a real-time model of conditioning that combines the Frey and Sears (1978) attentional rule with the Rescorla and Wagner (1972) model.

Wagner (1978) suggested that CS_i–US associations are determined by (a) the effectiveness of the US, $(US - \Sigma_j V_{j,US} CS_j)$, as in the Rescorla–Wagner model, and (b) the effectiveness (associability) of the CS_i, $(CS_i - V_{i,CX} CX)$, where CX represents the context and $V_{i,CX}$ the strength of the CX–CS_i association. CS preexposure causes the CS to be predicted by the context ($V_{i,CX}$ increases) and, therefore, CS_i associability decreases. Similar ideas are incorporated in Wagner's (1981) SOP theory. According to SOP, acquisition is explained in terms of the excitatory association formed between a CS and a US when their representations are both in the A_1 state. Therefore, LI is the consequence of the CS being represented in A_2 when predicted by the context, thereby limiting the CS A_1 representation and retarding its association with the US.

Schmajuk and Moore (1988) offered a real-time attentional–associative model of classical conditioning that incorporates CS–CS as well as CS–US associations. In the model, CS salience (assumed to modulate the rate of CS–CS and CS–US associations) is determined by CS novelty, i.e. the absolute value of the difference between its predicted and observed amplitude, $|CS_i - \Sigma_j V_{j,i} CS_j|$. The model describes LI because $|CS_i - \Sigma_j V_{j,i} CS_j|$ and CS salience decrease during CS preexposure. McLaren, Kaye, and Mackintosh (1989; McLaren & Mackintosh, 2000) proposed a similar model that describes perceptual learning and LI, in which the formation of CS_i–CS_j and CS_i–US associations are modulated by the associability of CS_i, $(CS_i - \Sigma_j V_{j,i} CS_j)$.

Modulation of retrieval of the CS–US associations

The preceding section depicts theories that describe the modulation of storage of CS–US associations. These theories, which explain LI as the consequence of impaired acquisition, find difficulty in explaining data showing that responding to the CS can be increased by postconditioning procedures such as extinction of the training context, or testing after long retention intervals. Presumably these results are better explained by theories that assume that CS preexposure produces a deficit in performance despite the fact that CS–US associations had been adequately learned. Notice that both the SOP (Wagner, 1981) and the SLG (Schmajuk et al., 1996) models assume not only modulation of storage, but also modulation of retrieval of the CS–US association. In the SOP model, this retrieval deficit following CS preexposure is due to the fact that the CS is predicted by the CX and, therefore, the CS representation in A_1 decreases, and so does the activation of the (already weak) CS–US association.

LI as a retrieval failure is well addressed by Miller and Schachtman's (1985) comparator hypothesis. According to this hypothesis, conditioned responding is assumed to be the result of a comparison between CS–US associations and CX–US associations. The context is the context in which the CS was trained. During testing, presentation of the CS activates the US directly, and indirectly through the combination of CS–CX and CX–US associations. When the direct activation of the US representation is stronger than the indirect activation of the US representation, excitatory responding is expected. When the indirect activation is stronger than the direct activation, inhibitory responding is expected. Therefore, excitatory responding to a CS decreases with increasing CS–CX and CX–US associations.

In terms of the comparator hypothesis, LI is the result of increased CS–CX associations generated during CS preexposure. Although CS–US associations develop during conditioning, increased CS–CX associations will activate CX–US associations, thereby decreasing conditioned responding. Extinction of the training context will decrease the indirect activation of the US representation, thereby increasing conditioned responding and decreasing LI. The hypothesis predicts recovery from LI after a retention interval (Kraemer et al., 1991) by assuming that CX–US associations (but not CS–US associations) decrease during the interval.

Interference with the formation of CS–US associations

As an alternative to the notion that LI results from a decrease in CS associability following CS preexposure, it has been proposed that CS preexposure fosters the formation of CS–no consequence associations that interfere with the subsequent establishment of CS–US associations.

Revusky (1971) has suggested that the formation of an association between the CS and any other event interferes with the ability of the CS to form other associations. In the case of LI, the CS becomes associated to other events during preexposure, and these associations reduce the ability of the CS to form associations with the US. Gordon and Weaver (1989) suggested that CS–no consequence associations are encoded along with the preexposure context. When animals are conditioned in a novel context, they fail to retrieve the CS–no consequence association and, therefore, LI is attenuated.

Testa and Ternes (1977) contended that conditioning is determined by the conditional ocurrence of the US in the presence of the CS, p(US/CS). CS preexposure decreases p(US/CS), and therefore results in weaker conditioning.

Hall (1991) presented a hybrid theory of LI that extends the original Pearce and Hall (P-H) (1980) model. According to Hall (1991, page 137), CS preexposure not only produces a loss of CS associability, but also allows the formation of potentially interfering associations. Interference and low associability retard the formation of CS–US associations during conditioning. Interfering associations depend on the preexposure context and, therefore, LI is attenuated by changes in the context.

Learned irrelevance

As mentioned in Chapter 1, in addition to LI, other procedures that affect the rate of conditioning include US preexposure and learned irrelevance (the uncorrelated presentation of the CS and the US). Bennett *et al.* (1995) reported that, even if learned irrelevance is obtained by CS and US separate presentations, the phenomenon is not the sum of exposure to the CS and the US. We ran simulations with the SLG model, in which 5 CS–US conditioning trials, following either (a) 50 CX− trials (conditioning), (b) 50 CX+ and CX− alternated trials (preexposure to US = 1, duration 5 time units), (c) 50 CS− and CX− alternated trials (LI, CS = 1, duration 10 time units), or (d) 50 CX+ and CS− alternated trials (learned irrelevance). The model was able to replicate the results: CR = 0.092 for conditioning, CR = 0.062 for US preexposure, CR = 0.073 for LI, and CR = 0.021 for learned irrelevance. That is, the maginitude of learned irrelevance (0.092 − 0.021 = 0.071) is more than the added magnitude of the effects of separate CS and US presentations ([0.092 − 0.062] + [0.092 − 0.073]) = 0.048). More detailed and complete studies are needed, however, to determine the ability of the model to capture other properties of the phenomenon (e.g. Bonardi *et al.*, 2005).

Summary

This chapter shows that the SLG model describes a large number of the behavioral properties of LI. These results and previously reported simulated

Table 5.1. *Properties of latent inhibition described with the SLG model.*

1. LI of excitatory or inhibitory conditioning
CS(A) preexposure and excitatory conditioning
CS(A) preexposure and inhibitory conditioning
CS(A)-weak US pairings

2. Parameters of LI
Number of CS preexposures
CS duration
Total duration
CS intensity
ITI duration

3. Combinations of CSs during preexposure
Simultaneous CS(A)–CS(B) preexposure
Many CS(B) preexposures, CS(A)–CS(B) preexposure
CS(B), CS(A) preexposure
CS(A)–CS(B) preexposure, CS(A)–CS(B) conditioning
CS(A), CS(B) preexposure, CS(A)–CS(B) conditioning
CS(A) preexposure, added event
CS(A) preexposure, omitted event

4. Contextual manipulations
Context, CS(A) preexposure
CS(A) preexposure, context
Change context after CS(A) preexposure
CS(A) preexposure, other context
Context, CS(A) preexposure, new context
Change context and cuing
Changes in CS and context (SoEo)
Changes in CS and context (SnEo)
Changes in CS and context (SoEn)
Changes in CS and context (SnEn)

5. Perceptual learning
Discrimination, same context
Discrimination, different context
Maze discrimination, same context
Maze discrimination, different context

6. Orienting response and latent inhibition
Dishabituation by context change

7. Post-conditioning manipulations
Extinction of the training context
US presentations in a different context
Delayed testing/ super-latent inhibition

results are listed in Table 5.1. In addition, the model seems able to describe some aspects of learned irrelevance.

The SLG model seems to offer some advantages over competing theories. Although the SLG model shares with the Pearce and Hall (1980) theory the idea that attention to the CS is regulated by the mismatch between the predicted and actual values of the US, the fact that in the SLG model CS novelty also controls attention, confers it with special properties. For example, only the SLG model can describe the reported LI disruption by presentation of a surprising CS (Lantz, 1973), the effect of ommiting a CS preceded by another CS on the orienting response (Wilson *et al.*, 1992), or the effect of preexposing to a compound but conditioninig to a CS (Honey & Hall, 1989). In addition, because in the SLG model attention to a CS controls not only learning but also performance, the SLG, but not the Pearce and Hall model, describes the attenuation of LI by postconditioning exposure to the context alone (which increases Novelty´ and attention to the CS). Finally, because the Pearce and Hall model is not a real-time model, it cannot describe the effect of changing CS duration, interstimulus intervals or intertrial intervals.

Although both the SLG model and Wagner's SOP model describe behaviour, in real time the effect of postconditioning manipulation on performance (both describe the decrements in LI following contextual exposure of omission of a familiar CS before testing), only the former is capable of describing the deleterious effect of the presentation of a surprising CS on latent inhibition (Lantz, 1973). Also, the SLG model is the only one, at this point, that has been shown to explain super-LI.

6

The neurobiology of latent inhibition

Following the ideas proposed in Chapter 4, this chapter applies the "conceptual nervous system" provided by the SLG model in order to establish brain–behavior relationships during latent inhibition. Here (a) we describe possible neural substrates for some of the variables in the SLG model; (b) based on those assumptions we use of the model to explain the effects of manipulations of the dopaminergic [DA] system, the hippocampus and the nucleus accumbens on LI; and (c) we apply the model to the description of some of the positive symptoms of schizophrenia.

The neural substrates of latent inhibition

In order to extend the application of the SLG model from the purely behavioral domain to the neurophysiological domain, we defined a mapping function between psychological and neurophysiological spaces that establishes where psychological variables are represented in the brain. Schmajuk, Cox and Gray (2001; Schmajuk et al., 2004) mapped nodes and connections in the SLG network onto the brain circuit, as shown in Figure 6.1. The participation of the brain regions in this circuit in LI, a combination of those proposed by Gray et al. (1997) and Weiner and Feldon (1997), was later confirmed by Puga et al. (2007) using cytochrome oxidase histochemistry in mice. Importantly, in line with our approach, according to Puga et al. (2007) their data suggest that LI is the consequence of a reduced processing of the CS.

Control of the formation of CX–CS and CS–CX associations in the cortex: hippocampus proper (HP)

Jarrard and Davison (1991) suggested that it was important to distinguish the effects of localized excitotoxic hippocampal lesions (hippocampus

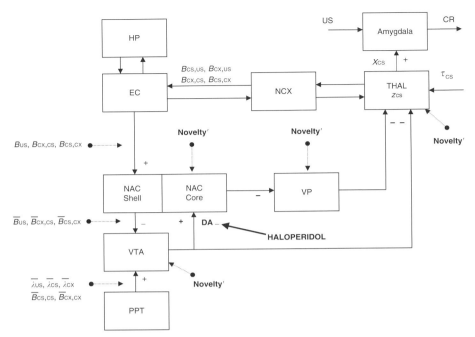

Figure 6.1 Mapping of the variables in the model onto the neural circuit involved in latent inhibition. Brain regions: EC: entorhinal cortex, HP: hippocampus proper, NAC: nucleus accumbens, NCX: neocortex, PPT: pedunculopontine tegmental nucleus, THAL: thalamic areas, VP: ventral pallidum, VTM: ventral tegmental area, DA= dopamine. Model variables: B_{US}: aggregate prediction of the US; $B_{CX,CS}$: prediction of the CS by the context; $B_{CS,CX}$: prediction of the context by the CS; $B_{CS,CS}$: prediction of the CS by itself; $B_{CX,CX}$: prediction of the context by itself; $\overline{B}_{US}$, $\overline{B}_{CX,CS}$, $\overline{B}_{CS,CX}$, $\overline{B}_{CS,CS}$ and $\overline{B}_{CX,CX}$: corresponding average predictions; $\overline{\lambda}_{US}$: average observed value of the US; $\overline{\lambda}_{CS}$: average observed value of the CS; $\overline{\lambda}_{CX}$: average observed value of the CX; X_{CS}: internal representation of the CS; z_{CS}: attentional memory of the CS.

proper lesions, HPLs) and extended aspiration lesions (hippocampal formation lesions, HFLs). Following that suggestion, Schmajuk and Blair (1993) assigned several computational functions to the hippocampus and proposed that different hippocampal regions might process information in different ways. One of those functions is the computation of the aggregate prediction of the US (B_{US}), the prediction of the CSs by the CX (B_{CX-CS}), and the prediction of the CX by the CSs (B_{CS-CX}). As explained below, we assumed that this function is performed in the entorhinal cortex (EC).

Another hippocampal function is the control of the formation of cortical associations between the CX with the CS, $V_{CX,CS}$, the association of the CX with the CS, $V_{CS,CX}$, or the association between two CSs, $V_{CS1,CS2}$. Based on data showing that HPLs spare configural discriminations but impair place learning,

Schmajuk and Blair (1993) assumed that the control of the modification of cortical associations (and configurations, see Chapter 13) is mediated by the HP. In turn, this function would be modulated by the medial septum, whose activity gradually decreases as the animal acquires classical associations. Because the HP is damaged following both HPLs and HFLs (that also affect the dentate gyrus, subiculum, presubiculum and the entorhinal cortex), $V_{CX,CS}$, $V_{CS,CX}$ and $V_{CS1,CS2}$ do not change following either of those lesions.

Computation of the predictions of CSs and US: entorhinal cortex (EC) and subiculum (SUB)

Berger and Thompson (1978a, Berger, Rinaldi, Weisz & Thompson, 1983) reported that pyramidal activity in cornu ammonis (CA) regions CA1 and CA3 is positively correlated with the topography of the CR. Berger, Clark, and Thompson (1980) found that the activity correlated with the CR is present also in the EC and amplified over trials in CA1 and CA3 regions. We assumed that, like the CR, activity in the HP and the EC is proportional to the aggregate prediction of the US, B_{US}. Although B_{US} and CR have similar topographies, the B_{US} representation in the HP and EC does not control the CR, which is regulated by different brain regions depending on the type of conditioning (e.g. fear, nictitating membrane, taste aversion).

Coutureau *et al.*'s (1999) data showing that entorhinal, but not hippocampal or subicular excitotoxic, lesions disrupt LI in rats, provide support to the notion the the EC might be the subregion of the HF where B_{US}, B_{CX-CS}, and B_{CS-CX} are computed in the LI brain circuit. Such computation would be carried out by adding the different components of the prediction, namely the prediction of the US by the CS, $B_{CS,US}$, and the prediction of the US by the context, $B_{CX,US}$. In order to compute the aggregate prediction of the CX and the CS, B_{CX} and B_{CS}, it was assumed that the EC receives information from the neocortex (NCX) about the prediction of the CS by the context, $B_{CX,CS}$, and the prediction of the CX by the CS, $B_{CS,CX}$. This information is sent from the EC to the shell of the nucleus accumbens (NAC).

Oswald *et al.* (2002) reported that excitotoxic lesions of the hippocampus disrupted both the orienting response and LI, lesions of the entorhinal cortex only disrupted LI and lesions of the subiculum (SUB) disrupted just the orienting response. These results suggest that, although the SUB, like the EC, might also compute the prediction of the CS by the context, $B_{CX,CS}$, and the prediction of the CX by the CS, $B_{CS,CX}$, projections from the EC and the SUB reach different regions of the NAC (Totterdell & Meredith, 1997; Groenewegen *et al.*, 1987) which would respectively control attention (and therefore LI) and the orienting response.

96 Attentional and associative mechanisms

Components of Novelty´: nucleus accumbens (NAC)

In the model, z_{CS} is determined by the value of *Novelty´*: $Novelty´ = f(\Sigma |\bar{\lambda}_k - \bar{B}_k|)$, i.e. a function of the sum of the absolute values of the differences between the averages of actual, $\bar{\lambda}_k$, and predicted, $\bar{B}_k$ (see Equations [2.7] and [2.8] in Chapter 2). *Novelty´* is maximal during partial reinforcement, and minimal with continuous or no reinforcement.

Dopaminergic projections from the ventral tegmental area (VTA) reach the NAC (Gray et al., 1997). Carelli and Deadwyler (1994) recorded the firing patterns of NAC neurons in the rat during water-reinforced behavior (see Figure 6.2). They reported that both shell and core NAC neurons exhibited three firing patterns: (a) preresponse (PRE) cells showed an anticipatory increase in firing, (b) reinforcement excitatory (RF_E) cells showed increased activity following the reinforced response, and (c) reinforcement inhibitory (RF_I) cells showed decreased activity following the reinforced response. Figure 6.2 also shows computer simulations of the activity of PRE, RF_E, and RF_I cells during the tenth trial of CS–US conditioning. The activity of PRE cells is described by $PRE = Novelty_I = \Sigma_k[\bar{B}_k - \bar{\lambda}_k]+$, where the + sign indicates that only the positive values of the bracketed term is considered. The activity of RF_E cells is described by $RF_E = Novelty_E = \Sigma_k[\bar{\lambda}_k - \bar{B}_k]+$, and the activity of RF_I cells by $RF_I = 1 - Novelty_E = 1 - \Sigma_k[\bar{\lambda}_k - \bar{B}_k]+$. Figure 6.2 shows a strong similarity between experimental cumulative single unit recordings obtained by Carelli and Deadwyler (1994) and simulated activities. Differences between modeled and recorded neural activity levels are most important preceding and following the presentation of the CS in PRE cells.

Computation of X_{CS}: ventral pallidum (VP), thalamus (THAL), amygdala (Amyg), and prefrontal cortex (PFC)

As mentioned, neural activity coding information about $Novelty_E$ and $Novelty_I$ seems to be present in the NAC. Cells in the core of the NAC inhibit the VP, which we assume to have a baseline activity equal to 1. Activity in the VP is therefore described as $(1 - Novelty´)$ with $Novelty´ = Novelty_E + Novelty_I$. Although it is unclear which pathways convey information from the VP to the thalamic nuclei that relay impulses originating in peripheral sense organs to the appropriate sensory regions of the cerebral cortex and the amygdala, Gray et al. (1997) suggested that the output of the VP indirectly inhibits the thalamus. Therefore, we assumed that the activity of the THAL is given by $(1 - (1 - Novelty´)) = Novelty´$ (see Figure 6.1).

Consequently, signals proportional to *Novelty´* might reach the THAL and become associated to the short-term memory trace of the CS, τ_{CS} (see Figure 6.1, and Equation [2.2] in Chapter 2). The output of the THAL might be proportional

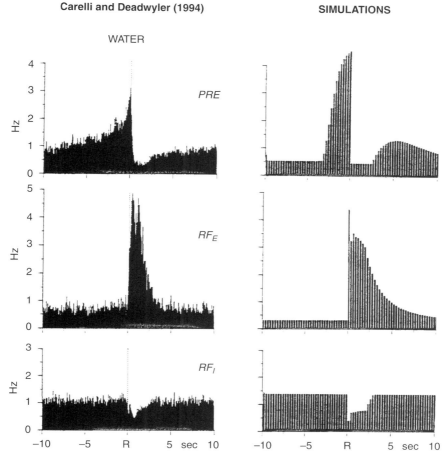

Figure 6.2 Accumulated unit histograms in the nucleus accumbens. *Left panel*: [Data from Carelli & Deadwyler, 1994, Figure 6]. Composite histograms of normalized firing of PR, RF_E, and RF_I cells for water reinforcement. *Right panel*: Simulated cell activity. For preresponse cells PRE = $\Sigma_k[\overline{B}_k - \overline{\lambda}_k]^-$, reinforcement excitatory cells $RF_E = [\overline{\lambda}_k - \overline{B}_k]+$, and reinforcement inhibitory cells $RF_I = 1 - [\overline{\lambda}_k - \overline{B}_k]+$, following 10 training trials.

to the internal representation of the CS, X_{CS}. In turn, X_{CS} would control the modification of CS–CS associations in association cortices, and CS–US associations in the amygdala (Killcross, Robbins & Everitt, 1997).

Even if X_{CS} might be computed in the THAL and then projected to the amygdala, it is also possible that this computation is accomplished in other brain areas. One possibility is that the X_{CS} representation is computed in the VTA and provided to the amygdala through DA projections (Oades & Halliday, 1987). This view receives support from simulations (Schmajuk, Gray & Larrauri, 2005) showing that the firing pattern in VTA during classical conditioning (Fiorillo, Tobler & Schultz, 2003) is well described by X_{CS}. Similarly, DA projections from the VTA to the prefrontal cortex (Schultz, 1998) might provide X_{CS} representations to the dorsolateral prefrontal cortex (Dunsmoor et al., 2007). As shown in Figure 4.2 (Chapter 4), activity in the dorsolateral prefrontal cortex (Dunsmoor et al., 2007) is also well represented by X_{CS}. Furthermore, activity in both areas, as well as X_{CS}, is maximal during partial reinforcement, and minimal with continuous or no reinforcement.

Independently of whether it is assumed that DA represents Novelty' or X_{CS}, dopaminergic drugs will affect X_{CS}, either directly or by way of modifying the magnitude of *Novelty'*. As shown below, computer simulations under either assumption yield similarly correct results.

Dopaminergic involvement in latent inhibition

It has been suggested (e.g. Gray, Feldon, Rawlins, Hemsley & Smith, 1991; Weiner, 1990) that indirect DA agonists increase, and DA receptor antagonists decrease, attention to the preexposed CS, respectively impairing or facilitating LI. This attentional view was challenged by Killcross, Dickinson and Robbins (1994a, 1994b) who reported that the diminishing effects of amphetamine on LI could be reversed by reducing, and the enhancing effects of α-flupenthixol on LI could be averted by increasing, the intensity of the US used in conditioning. According to Killcross et al., these results would suggest that the abolition of LI by amphetamine and the enhancement of LI by haloperidol and α-flupenthixol are not mediated by changes in attentional processes, but rather by changes in the consequences of reinforcers. Because in the SLG model Novelty' (a) controls attention, and (b) is sensitive to the magnitude of the US, the model is consistent with, but still different from, both views.

Schmajuk et al. (1998) analyzed the assumption that, in the framework of the SLG model, indirect DA agonists (e.g. amphetamine and nicotine) increase, and DA receptor antagonists (e.g. haloperidol and α-flupenthixol) decrease, the effect of Novelty' on attention (see Figure 6.1). Under this assumption, the model correctly describes most of the empirical data available for DA agonists: (a) the impairment of LI by amphetamine when a strong US is used (Killcross, Dickinson & Robbins, 1994a), (b) the impairment of LI by amphetamine when a flashing-houselight (nonsalient) CS is used (Ruob, Elsner, Weiner & Feldon, 1997), (c) the impairment of LI by amphetamine administration when a short CS is used (Weiner et al., 1997). It also correctly describes most of the data for the effects of DA antagonists: (d) the facilitation of LI by α-flupenthixol when a weak US is used (Killcross, Dickinson & Robbins, 1994b), (e) the facilitation of LI by haloperidol when a flashing house-light (nonsalient) CS is used (Ruob et al., 1997), (f) the facilitation of LI by haloperidol with a strong US (Ruob, Weiner & Feldon, 1998, Experiment 1), and (g) the facilitation of LI by haloperidol with extended conditioning (Ruob et al., 1998, Experiment 2).

Furthermore, under the closely related assumption that DA codes for X_{CS} (see Figure 6.1), Schmajuk et al. (2004) showed that the SLG model explains why LI is impaired by indirect DA agonists such as amphetamine (e.g. Solomon et al., 1981; Warburton, Mitchell & Joseph, 1996; Weiner, Lubow & Feldon, 1984) or nicotine (Gray et al., 1994; Joseph et al., 1993) and enhanced by haloperidol (e.g. Weiner & Feldon, 1997). In addition to these "typical" data, the model describes how nicotine facilitates (Rochford et al., 1996) and haloperidol can impair (e.g. Buhusi et al., 1999) LI.

Nicotine administration before preexposure

Nicotine directly activates nicotinic cholinergic receptors ($\alpha 7$ and $\alpha 4\beta 2$ subtypes, as well as other subtypes with uncharacterized function) and indirectly causes the stimulation of many other receptors, including DA and serotonin, through the release of acetylcholine. Based on this indirect stimulation of DA, Schmajuk et al. (2004) applied the model to describe the effects on LI of (a) nicotine administration during preexposure when only the CS is present, and (b) changing the duration of the CS under nicotine administration.

Experimental data

Weiner et al. (1984) showed that administration of amphetamine (an indirect DA agonist) impairs LI when administered before preexposure and conditioning, but not when given before preexposure alone. Similarly, Joseph, Peters and Gray (1993) reported that subcutaneous administration of the cholinergic agonist nicotine (another indirect DA agonist) disrupts LI when given before both preexposure and conditioning, or before conditioning alone, but not before preexposure alone (Figure 6.3, left upper panel). In contrast, Rochford, Sen and Quirion (1996) reported that intraperitoneal nicotine facilitates LI when dispensed before both preexposure and conditioning, conditioning or preexposure alone (Figure 6.3, right upper panel). Similarly, Gould, Collins and Wehner (2001) found that intraperitoneal administration of nicotine prior to CS preexposure increased LI.

Although Joseph et al. (1993) and Rochford et al. (1996) use otherwise similar procedures, their parameters were considerably different. Whereas Joseph et al. used 40 preexposures of a tone CS (5 s, 2.5 Hz, 15 dB above a 80 db background noise, at 60 s intervals), Rochford et al. used 60 presentations of a tone CS (60 s, 2.8 kHz, 80 db, at 30 s intervals). It should noted that Rochford et al. were able to

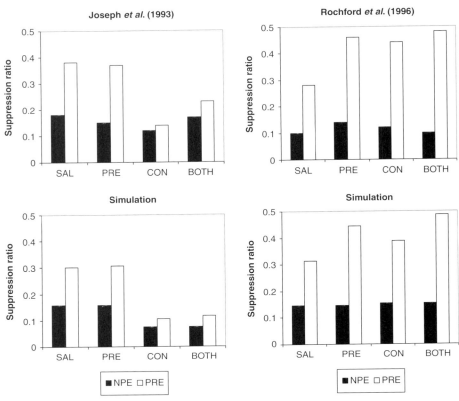

Figure 6.3 Effect of nicotine administration on latent inhibition with short- and long-duration CS. *Left upper panel*: Data from Joseph et al., 1993 (short CS). *Left lower panel*: Simulated results show the suppression ratio ($\beta = 0.53$) after (1) 20 presentations of the CS (PRE) or the context (NPE) and (2) 10 reinforced presentations of the CSs (CX = 0.1, CS duration 25 time units, US duration 5 time units, CS amplitude 1, US amplitude 1.8, intertrial interval = 4000 time units). Correlation coefficient ($r = 0.92$, df = 6, $p < 0.05$). PRE: preexposed group. NPE: non-preexposed group. *Right upper panel*: Data from Rochford et al., 1996 (long CS). *Right lower panel*: Simulated results show the suppression ratio ($\beta = 0.53$) after (1) 20 presentations of the CS (PRE) or the context (NPE) and (2) 13 reinforced presentations of the CSs (CX = 0.1, CS duration 40 time units, US duration 5 time units, CS amplitude 1, US amplitude 1.8, intertrial interval = 4000 time units). Correlation coefficient ($r = 0.99$, df = 6, $p < 0.05$). PRE: preexposed group. NPE: non-preexposed group.

replicate Joseph et al.'s results for the effect of nicotine administration before preexposure and conditioning using, as Joseph et al. did, 40 preexposures and a 5-s CS. Using mice instead of rats, Gould et al. (2001) obtained facilitation using only 10 presentations of a 50-s broadband clicker; a disparity that they attributed to species differences.

Simulated results

In our simulations, the value of X_{CS} is increased (K_{12}=1.5 in Equation [2.2] in Chapter 2) after nicotine administration. In agreement with Joseph *et al.*'s (1993) data, computer simulations (Figure 6.3, left lower panel) demonstrate that administration of nicotine before conditioning, or both preexposure and conditioning, but not before preexposure, using a relatively short CS, impairs LI (r = 0.92, df = 6, p < 0.05). The simulations for nicotine administration during both preexposure and conditioning also agree with Rochford *et al.*'s (1996) data using 40 preexposures to a 5-s CS. Also in agreement with Rochford *et al.*'s (1996) data using 60 preexposures to a 60-s CS, computer simulations shown in Figure 6.3 (right lower panel) demonstrate that administration of nicotine before conditioning, preexposure and conditioning or preexposure, using a relatively long CS, facilitates LI (r = 0.99, df = 6, p < 0.05).

Discussion

Based on the Joseph *et al.* results, one could conclude that nicotine acts only during the US presentation. This result can be explained by theories (e.g. Killcross, Dickinson & Robbins, 1994a,1994b) that suggest that DA codes for the intensity of the US, which is increased by nicotine administration. Alternatively, the result can be explained in terms of nicotine increasing DA release, thereby augmenting the value of X_{CS} during conditioning and, thereby, impairing LI.

One the other hand, the Rochford *et al.* and Gould *et al.* data support the view that DA cells code for the attentionally modulated CS, but not the view that DA codes for the intensity of the US. According to our model, nicotine administration combined with long CS duration results in facilitation of LI because it enhances X_{CS} during preexposure, which results in the fast building of CS–CS and CS–context associations. These associations produce a decrement in Novelty´, which in turn decreases z_{CS} and X_{CS} (even if X_{CS} first tends to be increased by the drug), thereby facilitating LI. This facilitatory effect is absent in the case of a short CS, because CS–CS and CS–CX associations (that facilitate LI) decrease, but CS–US associations (that impair LI) increase, with decreasing CS durations.

According to the SLG model, it is possible for nicotine to facilitate LI by increasing X_{CS}, and DA release in the NAC. In contrast to this explanation, Gray, Mitchell, Joseph, Grigoryan, Date and Hedges (1994) suggested that, although the impairment of LI by nicotine is probably due to DA release in the NAC, LI facilitation might be due to other actions of nicotine elsewhere in the brain, and might be acting via different neurotransmitter pathways. One such neurotransmitter might be serotonin. For example, Killcross *et al.* (1997) reported that

5HT1a antagonists enhance LI following an amount of CS preexposure known to be insufficient to produce an LI in control animals. Killcross *et al.* (1997, page 57) suggested that this facilitation is the consequence of an antagonism of the inhibitory serotoninergic autoreceptors of the mesolimbic 5HT system originating in the medial raphe nucleus, which activates the HP. This hypothesis is in agreement with the reported LI attenuation following specific indoleamine-depleting 5,7-DHT lesions of the fornix-fimbria (Cassaday *et al.*, 1993). In terms of the SLG model, nicotine would facilitate LI through a serotoninergic mechanism that activates the HP, thereby increasing $V_{CX,CS}$, $V_{CS,CX}$, B_{CS}, B_{CX}, and decreasing Novelty', z_{CS} and X_{CS}.

In sum, the model accounts for experimental data showing that, when a relatively long CS is used, nicotine administration before preexposure, conditioning or both, facilitates LI. Also, the model explains why, with a relative short CS, nicotine does not affect LI when administered before preexposure, but it impairs LI when given before conditioning, or before both preexposure and conditioning.

The results seem to support an attentional role for DA. The fact that nicotine has an effect during preexposure, when only the CS is active, supports the view that it is modifying the internal representation of the CS. Had nicotine represented the US by releasing DA, as proposed by alternative theories, administration during preexposure would have resulted in impaired LI. Also, the fact that nicotine effect changes from neutral to enhancing depending on the CS duration, provides additional support for the attentional view.

Although the model can describe the deleterious effects of the indirect DA agonists amphetamine and nicotine on LI, it does not contain the detailed pharmacological mechanisms needed to explain the preservation of LI following administration of the direct DA agonist apomorphine (Feldon, Shofel & Weiner, 1991).

Hippocampal involvement in latent inhibition

Experimental data on hippocampal involvement in LI show contradictions that, for a long time, were difficult to explain. For example, ibotenate lesions of the hippocampus proper (HPL; regions CA1 and CA3) might impair (Han, Gallagher & Holland, 1995), preserve (Honey & Good, 1993), or even facilitate (Reilly, Harley & Revusky, 1993) LI. Similarly, aspiration or electrolytic lesions of the hippocampal formation (HFL; hippocampus proper, dentate gyrus, entorhinal cortex, subiculum and presubiculum) might not affect (Gallo & Candido, 1995) or facilitate LI (Purves, Bonardi & Hall, 1995). Buhusi, Schmajuk and Gray (1998) showed that the SLG model accounts for these results by taking into account not only the effect of the lesions on the variables of the model presumably computed by the hippocampus (see Figure 6.1), but also the behavioral procedure used to obtain LI.

Lesions of the hippocampus proper (HPLs)

In the context of the SLG model, the effect of selective HPLs is described by assuming that cortical CS_1-CS_2 associations, $V_{CS1,CS2}$, CS_2-CS_1 associations, $V_{CS2-CS1}$; CX-CS associations, $V_{CX,CS}$; or CS-CX associations, $V_{CS,CX}$ cannot be modified after the lesions. Formally, changes in $V_{CS2,CS1}$ and $V_{CS1,CS2}$ are given by

$$dV_{CS2,CS1}/dt = 0, dV_{CS1,CS2}/dt = 0 \quad [6.1a]$$

Changes in $V_{CX,CS1}$ and $V_{CS,CX}$ are given by

$$dV_{CX,CS}/dt = 0, dV_{CS,CX}/dt = 0 \quad [6.1b]$$

It is assumed that CS_1-CS_1 associations, which produce habituation to CS_1 by predicting itself, remain unaffected (see Equation [2.9b]). In the absence of CS_2-CS_1 associations, $V_{CS2,CS1}$, changes in the CS_1-CS_1 association, $V_{CS1,CS1}$, are given by

$$d(V_{CS1,CS1})/dt = K_7 X_{CS1}(\lambda_{CS1} - X_{CS1}V_{CS1,CS1})(1 - |V_{CS1,CS1}|). \quad [6.2]$$

Lesions of the hippocampal formation (HFLs)

Consistent with the assumption that the hippocampal formation also computes aggregate predictions, HFLs can be described by assuming that aggregate prediction $B_{CS,US}$ is equal to zero,

$$B_{CS,US} = 0. \quad [6.3]$$

As mentioned, in the case of the LI circuit, $B_{CS,US}$ could be computed in EC. In the absence of the aggregate prediction $B_{CS,US}$, changes in the CS-US association, $V_{CS,US}$, are given by

$$d(V_{CS,US})/dt = K_7 X_{CS}(\lambda_{US} - X_{CS}V_{CS,US})(1 - |V_{CS,US}|). \quad [6.4]$$

In addition, because HFLs also damage the hippocampus proper, it is assumed that in HFL animals changes in cortical CS_2-CS_1 and CX-CS_1 associations, $V_{CS2,CS1}$ and $V_{CX,CS1}$, are given by Equations [6.1a] and [6.1b].

Effects of selective lesions of the hippocampus

Impairment of LI

Experimental data

Han, Gallagher and Holland (1995) reported that ibotenate lesions of the hippocampus impair LI in rats. The within-subject (WS) procedure consisted

104 Attentional and associative mechanisms

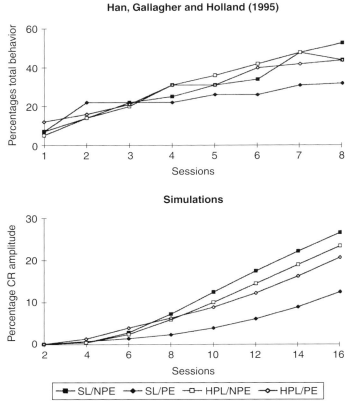

Figure 6.4 Impairment of LI after hippocampus proper lesions (HPLs). *Upper panel*: Data from Han, Gallagher and Holland (1995). *Lower panel*: Simulated percentage CR amplitude for sham- (SL) or hippocampus proper- (HPL) lesioned animals during CS_A presentation (PE) or CS_B presentation (NPE) in a within-subject (WS) procedure with 20 preexposure sessions followed by 16 conditioning sessions. Correlations beween simulations and data are statistically significant ($r = 0.88$, $df = 22$, $p < 0.05$.)

of 5 days of eight 10-s presentations of one of two visual CSs, followed by conditioning with food for both CSs. Han *et al.*'s (1995) results are displayed in the upper panel of Figure 6.4, which shows percent of total behavior elicited in sham lesion (SL) and HPL groups by the presentation of the *A* preexposed CS and the *B* nonpreexposed CS.

Simulated results

The lower panel in Figure 6.4 shows computer simulations of the percentage of responding to the preexposed CS and nonpreexposed CS after 20 CS preexposure trials followed by 40 alternated conditioning trials to the preexposed and nonpreexposed CS. In agreement with Han et al.'s (1995) data, HPL impairs LI.

According to the model, HPLs impair LI using a WS procedure because Novelty' decreases at a slower rate in HPL than in SL animals. Notice that CX–CS associations allow SL animals to detect the omission of both A and B when presented with the US on alternated trials, thereby increasing Novelty' and attenuating LI.

Preservation of LI

Experimental data

Honey and Good (1993, Experiment 2) reported that ibotenate lesions of the hippocampus do not impair LI, but abolished the contextual specificity of LI, using an autoshaping procedure with rats. The between-subject (BS) procedure consisted of 72 preexposure trials to a tone CS (PE group) or to the context (NPE group), followed by food conditioning for both groups. The upper panel in Figure 6.5 shows response rate for preexposed and nonpreexposed groups after SLs or HPLs. Similarly, Shohamy et al. (2000) found that LI was preserved after neurotoxic HPLs.

Simulated results

The lower panel in Figure 6.5 shows computer simulations of the percent conditioned response after 40 CS preexposure trials followed by 5 conditioning trials for SL and HPL animals. In agreement with experimental data, HPL preserves LI.

According to the model, HPLs preserve LI using a BS procedure because $CS_1 - CS_1$ associations can change and therefore Novelty' decreases, slowly but sufficiently, to yield LI.

Facilitation of LI

Experimental data

Reilly et al. (1993) showed that ibotenate lesions of the hippocampus (HPLs) enhance LI in a conditioned taste aversion preparation in rats. Reilly et al. (1993) used a "between-subject with interspersed water presentations" (BW) procedure in which, after a water deprivation schedule, animals were either preexposed (PE) to flavored water A or continued to receive water (nonpreexposed – NPE) for 4 days, for 30 min each day. Afterwards, animals received 30-min conditioning trials (in which flavor A was followed by a lithium chloride injection) on days 5 and 7. Days 6 and 8 were recovery days, and animals were allowed 30-min access to water. Day 9 was a test day and animals received flavored water but no injection. Surprisingly, HPL animals, but not SL animals,

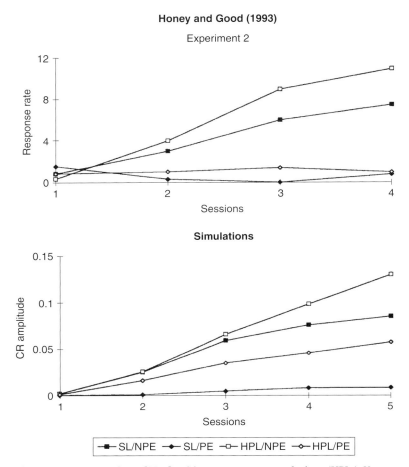

Figure 6.5 Preservation of LI after hippocampus proper lesions (HPLs). *Upper panel*: Data from Honey and Good (1993, Experiment 2). *Lower panel*: Simulated peak CR amplitude for sham- (SL) or hippocampus proper- (HPL) lesioned animals for the NPE and PE groups in a between-subject procedure (BS) with 40 preexposure sessions followed by 5 conditioning sessions. Correlations beween simulations and data are statistically significant (r = 0.92, df = 18, p < 0.05.)

show LI. The upper panel in Figure 6.6 shows percent of amount of flavored water consumed by preexposed and nonpreexposed groups after SLs or HPLs, as reported by Reilly *et al.* (1993).

Simulated results

The lower panel in Figure 6.6 shows computer simulations of percent of amount of flavored water consumed by PE and NPE groups after SLs or HPLs. The BW procedure consisted of 100 water W deprivation sessions, followed by 50 CS

The neurobiology of latent inhibition 107

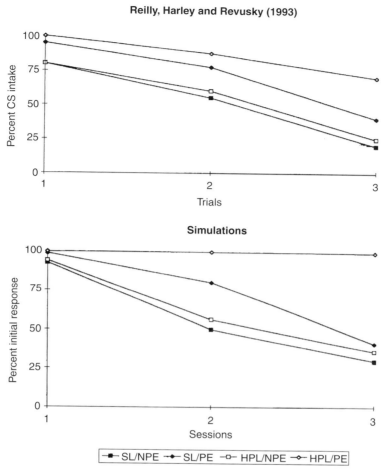

Figure 6.6 Facilitation of LI after hippocampus proper lesions (HPLs). *Upper panel*: Data from Reilly, Harley and Revusky (1993). *Lower panel*: Simulated percentage of initial intake for sham- (SL) or hippocampus proper- (HPL) lesioned animals for the NPE and PE groups in a between-subject with interspersed water presentations (BW) procedure with 50 preexposure sessions followed by conditioning sessions. Correlations beween simulations and data are statistically significant (r = 0.94, df = 10, p < 0.05.)

A preexposure sessions, followed by 2 conditioning sessions and one test session. In agreement with Reilly *et al.*'s (1993) experimental data, HPL facilitates LI.

According to the model, HPL facilitates LI using a BW procedure, because the absence of $V_{CX,A}$ and $V_{CX,W}$ associations implies that LI is not disrupted in HPL animals by the omission of the expected flavored (*A*) or unflavored (*W*) water presentations, as occurs in SL animals (see Hall & Pearce, 1982).

Discussion

The present section shows that, under the mapping described in Figure 6.1, the SLG model is able to describe the apparently conflicting database related to hippocampal lesions and haloperidol administration on LI. These results suggest that the notion that LI (a label that encompasses many alternative procedures) is, or is not affected by hippocampal lesions, should be replaced by a detailed analysis of how Novelty' (or X_{CS}), the key variable that controls LI in the SLG model and can be experimentally evaluated by measuring the magnitude of the OR, varies as a function of the experimental procedure, type of lesion and drug administration. Furthermore, the model is able to explain how animals with hippocampal lesions do not disply the contextual specificity of LI shown by normal animals, that is, the lesioned animals still show LI after context changes. It is interesting to notice that Weiner's (2003) "two-headed" latent inhibition model of schizophrenia describes (a) the disruption of LI under conditions that lead to LI in normal rats, and (b) the abnormal preservation of LI under conditions that disrupt LI in normal rats, but does not seem able to explain the reported LI facilitation.

Effects of reversible inactivation of the entorhinal cortex

As mentioned, Coutureau *et al.* (1999) showed that entorhinal, but not hippocampal or subicular excitotoxic lesions, disrupt LI in rats.

Experimental data

More recently, Seillier *et al.* (2007) reported that LI was unaffected in rats with reversible inactivation of the entorhinal cortex, by intracerebral microinfusion of tetrodotoxin (TTX), during conditioning. In contrast, LI was disrupted when the reversible inactivation took place during preexposure only (see Figure 6.7, upper panel). Preexposure consisted of 20, 20-s tone presentations, and conditioning of 4 tone–shock US presentations.

Simulated results

Computer simulations consisted of 20 preexposure and 5 conditioninig trials with CS salience 1 and duration 20 time units, US intensity 1 and duration 5 time units, and CX salience 0.1. As shown in Figure 6.7 (lower panel), the model correctly describes those effects. According to the model, when the entorhinal cortex is inactive during preexposure, CX–CS associations are not formed, attention to the CS is strong during conditioning and LI is absent. Instead, when the entorhinal cortex is inactive during conditioning, attention z_{CS} to the CS decreases during preexposure, and even when Novelty' is stronger in this case than in the normal case, the effect of preexposure lasts for some

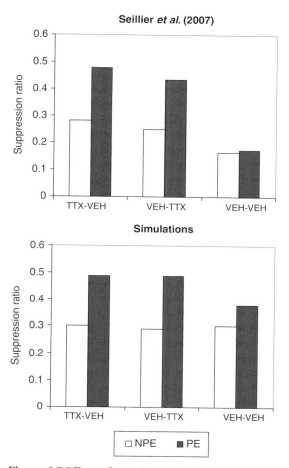

Figure 6.7 Effects of entorhinal cortex inactivation on LI. *Upper panel*: Data from Seillier *et al.* (2007). *Lower panel*: Simulated suppression ratio for the NPE and PE groups with entorhinal cortex inactivation with tetrodotoxin (TTX) during preexposure (TTX–VEH), conditioning (VEH–TTX) or with no inactivation (VEH–VEH). Correlations beween simulations and data are statistically significant (r = 0.81, df = 4, p < 0.05.)

time during conditioning and LI is present. The model predicts that differences between groups will become apparent with an increasing number of conditioning trials.

Nucleus accumbens involvement in latent inhibition

As mentioned above, we assumed that the variable Novelty′ is represented in the core of the NAC by DA (see Figure 6.1). By applying this assumption, Schmajuk *et al.* (2001) showed that the SLG model correctly describes: (1)

the impairment of LI by lesions of the shell of the nucleus NAC (Weiner, Gal, Rawlins & Feldon, 1996, Experiment 2), (2) the restoration of LI by haloperidol following lesions of the shell (Weiner *et al.*, 1996, Experiment 2), (3) the preservation of LI by lesions of the core of the NAC (Gal & Weiner, 1998), and (4) the facilitation of latent inhibition by combined shell-core lesions and by core lesions with extended conditioning (Gal & Weiner, 1998).

Effect of shell lesions and haloperidol administration on latent inhibition

Here we illustrate one of the results mentioned in the previous paragraph.

Experimental data

Weiner *et al.* (1996, Experiment 2) reported that lesions of the NAC shell impair LI, and that this impairment is eliminated by administering 0.2 mg/Kg of haloperidol at the start of the preexposure and conditioning phases. LI was evaluated by the amount of suppression of drinking in the presence of tone CS (conditioned emotional response). In contrast, Pothuizen *et al.* (2006) found that LI was preserved after shell lesions when a conditioned taste aversion paradigm was used. The upper panel in Figure 6.08 shows mean number of licks (average of 3 blocks of 60 s each, as communicated in the report of Weiner *et al.*) during CS presentation for PE and NPE groups after shell lesions either under vehicle (VEH), or haloperidol (HAL) administration before preexposure and conditioning, following 30 CS preexposure trials and 2 reinforced trials, as reported by Weiner *et al.* (1996, Experiment 2).

Simulated results

In our simulations, Novelty' (given by Equation [2.8] in Chapter 2, Novelty = $|\bar{\lambda}_{US} - \bar{B}_{US}| + |\bar{\lambda}_{CS} - \bar{B}_{CS}| + |\bar{\lambda}_{CX} - \bar{B}_{CX}|$) becomes Novelty = $|\bar{\lambda}_{US} - 0| + |\bar{\lambda}_{CS} - \bar{B}_{CS,CS}| + |\bar{\lambda}_{CX} - \bar{B}_{CX,CX}|$ following shell lesions (equivalent to HFLs). Also, in Equation [2.2] (see Chapter 2) $K_{12} = 0.65$ following haloperidol administration. The lower panel in Figure 6.8 shows simulated number of licks (average of 3 trials). Because shell lesions decrease the value of aggregate predictions B_{CS}, B_{CX} and B_{US}, they increase Novelty' relative to normal animals. Increased values of Novelty' translate into increases in z_{CS} during preexposure, thereby facilitating learning in the PE group and impairing LI in shell-lesioned animals. Haloperidol decreases the effect of Novelty' on z_{CS}, thereby reestablishing LI.

Both experimental and simulated results show that shell lesions impair LI, but the impairment disappears with haloperidol administration, assumed to decrease Novelty'. Notice that the simulations show a reversal in the responding of the PE and NPE groups displayed in the experimental results.

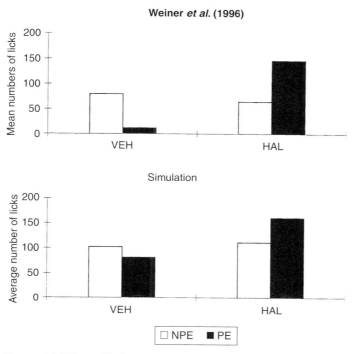

Figure 6.8 Effects of lesions of the shell of the nucleus accumbens and haloperidol administration. *Upper panel*: [Data from Weiner *et al.*, 1996, Experiment 2.] Mean number of licks during CS presentation for PE and NPE groups after shell lesions either under vehicle (VEH) or 0.2 mg haloperidol (HAL) administration before preexposure and conditioning, following 30 CS preexposure trials and 2 reinforced trials. *Lower panel*: Average simulated number of licks during 3 CS reinforced presentations for PE and NPE groups in a between-subject procedure after 110 CS preexposure trials. For VEH is $K_{12} = 1$, for HAL is $K_{12} = 0.65$. Correlations beween simulations and data are statistically significant ($r = 0.96$, $df = 2$, $p < 0.05$.)

Discussion

In the sections on Hippocampal involvement in latent inhibition, we showed that the model describes the apparently contradictory effects on LI that follow hippocampal lesions. For instance, the SLG model explains why whereas a relatively short (225 s) total preexposure time (e.g. Schmajuk *et al.*, 1994) impairs LI, LI is preserved, and is context independent, with a relatively long (720 s) total preexposure time (e.g. Honey & Good, 1993). In terms of the model, a sufficiently long total time of preexposure would increase $V_{CS,CS}$ association, and the $B_{CS,CS}$ prediction, thereby decreasing Novelty' and producing LI. This LI is independent of $V_{CX,CS}$ associations and the $B_{CX,CS}$ prediction (context-independent LI).

In the model (see Figure 6.1), lesions of the hippocampus and the NAC shell are both assumed to eliminate the $B_{CX,CS}$ prediction. Therefore, explanations similar to those provided for the effect of hippocampal lesions can be applied to the effect of NAC shell lesions. For example, the model is able to describe the impairment of LI reported by Weiner *et al.* (1996, Experiment 2). The model also correctly describes the facilitation of latent inhibition by combined shell-core lesions with extended conditioning (Gal & Weiner, 1998).

Furthermore, based on its predictions for the effect of hippocampal lesions on LI, the model *predicted* the preservation of LI in conditioned taste aversion. Pothuisen *et al.* (2006) recently reported results that support these predictions.

In addition, the model expects that in a similar manner to animals with hippocampal lesions, animals with shell lesions would show (a) normal LI with increasing the number of preexposure trials, and (b) LI impervious to context changes. Whereas the first prediction still waits to be tested, the latter prediction finds some support in Westbrook *et al.*'s (1997) results showing context-independent LI following morphine administration into the NAC.

Latent inhibition and schizophrenia

Comprehensive reviews of the biological basis of schizophrenia point to abnormalities in the limbic system of schizophrenic patients and suggest that these abnormalities are concentrated in the hippocampal formation. The notion of an anomalous hippocampus is further supported by data showing that many hippocampal-dependent behaviors are impaired in schizophrenic patients and that some of these impaired behaviors can be normalized by neuroleptic administration (Schmajuk, 2001).

Hippocampal dysfunction and schizophrenia

As described before, a decreased or defective hippocampal function might be present in schizophrenia. As mentioned in the sections on Hippocampal involvement in latent inhibition, HFLs can be described by making $B_{US} = 0$, $B_{CX,CS} = 0$ (because $V_{CX,CS} = 0$) and $B_{CS,CX} = 0$ (because $V_{CS,CX} = 0$). In addition, subjects with either HFLs or HPLs cannot modify their CS_1–CS_2 associations.

Gray, Feldon, Rawlins, Hemsley and Smith (1991) suggested that the DA overactivity in the NAC reported in schizophrenia (Snyder, 1980; Csernansky & Bardgett, 1998; O'Donnell & Grace, 1998; Lipska, Jaskiw, Chrapusta, Karoum & Weinberger, 1992; Saunders, Kolachana, Bachevalier & Weinberger, 1998) is a consequence of impaired hippocampal and temporal lobe function. Supporting this notion, Lipska, Jaskiw *et al.* (1992) communicated that bilateral ibotenic acid of the ventral hippocampus (but not the dorsal hippocampus, see Lipska, Jaskiw et al., 1991) in male rats resulted in increased DA levels in the NAC on the 28th (but decreased DA levels on the 14th day) after the lesion. Consistent with the notion that increased levels of DA result in decreased DA receptor density, Perry, Luchins and Schmajuk (1993) reported that aspiration lesions of the dorsal and ventral hippocampus reduced DA receptor density in the NAC on the 21st day after the lesion. In the same line, Saul'skaya and Gorbachevskaya (1998) found that bilateral ibotenic acid lesions of the hippocampus in rats resulted in a higher-level and longer-lasting release of DA in the NAC during CR generation. Agreeing with these results, Figure 6.1 indicates that a weakened hippocampal input to the NAC shell decreases inhibition on the VTA, thereby increasing DA release in the NAC core. This increased DA in the NAC would be counteracted by DA antagonists.

If schizophrenia is related to a decreased or defective HF input to the NAC, with the resulting increase in X_{CS} and Novelty' coded by DA, blocking the activity of DA will tend to reinstate normal levels of attention. When Novelty' is increased by hippocampal lesions or schizophrenia, administration of DA blockers results in a normalized value of attention.

Absence of LI in schizophrenia

Human experimental data

Seidman (1983) suggested that positive symptoms (delusions, hallucinations, stereotypy) originate from pathophysiological dysfunction in limbic, midbrain and upper brainstem regions. Among the positive symptoms, absence of LI is considered as a failure to filter out irrelevant information. Baruch et al. (1988; see also Gray, Hemsley & Gray, 1992; Lubow, Weiner, Schlossberg & Baruch, 1987) reported that LI is absent in acute schizophrenics tested within the first week of the beginning of a schizophrenic episode, but it is present in chronic, medicated schizophrenics. During the preexposure phase, acute schizophrenics, chronic schizophrenics and normal subjects were either exposed to a white noise CS while monitoring a list of nonsense syllables, or were preexposed to the nonsense syllables alone. During the conditioning phase, the white noise signaled increments in a visually displayed number, which is equivalent to a US. Whereas preexposed normals and chronic schizophrenics learned this CS–US association more slowly than nonpreexposed subjects, thereby displaying LI, acute schizophrenic patients failed to show the effect.

Figure 6.9 (upper panel) shows the experimental results reported by Baruch et al. (1988). The panel presents the percentage of subjects who reached the conditioning criterion (by predicting for five consecutive times the counter-increment US each time the white noise CS was presented) during the first 20 presentations

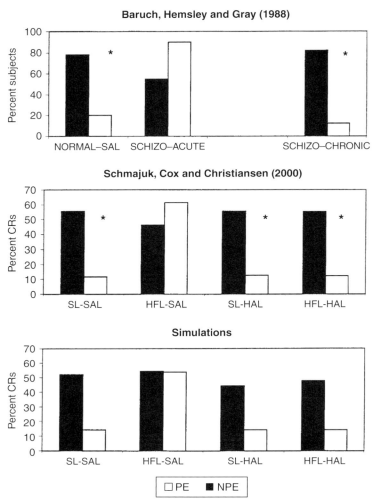

Figure 6.9 Latent inhibition in schizophrenia, hippocampal lesioned animals and the SLG neural network model. *Upper panel*: Experimental results from Baruch et al. (1988). Cumulative percentage of subjects responding to the CS on the first 20 trials of the experiment for NORMAL, SCHIZO-ACUTE, and SCHIZO-CHRONIC groups. *Middle panel*: Experimental results from Schmajuk et al. (2000). Percentage CRs on the 4th day of conditioning for SL-SAL, HFL-SAL, SL-HAL, and HFL-HAL groups under PE and NPE conditions. *Lower panel*: Simulated results from Schmajuk et al. (2000). Simulated percentage CRs on the 4th day of conditioning for SL-SAL, HFL-SAL, SL-HAL, and HFL-HAL groups under PE and NPE conditions. SCHIZO: schizophrenic patients. NORMAL: normal subjects. SL: sham lesion, HFL: hippocampal formation lesion, SAL: saline, HAL: haloperidol, PE: CS preexposed groups, NPE: non-preexposed groups. Correlations between simulations and (a) human data ($r = 0.91$, $df = 4$, $p < 0.05$) and (b) animal data ($r = 0.96$, $df = 6$, $p < 0.05$) are statistically significant.

of the white noise. The panel shows that, whereas normal (NORMAL) and chronic schizophrenics (SCHIZO-CHRONIC) showed LI, i.e. the percentage of subjects responding is greater in the non pre-exposed (NPE) than in the pre-exposed (PE) condition, this phenomenon was absent in acute schizophrenics (SCHIZO-ACUTE).

Animal experimental data

Schmajuk, Christiansen and Cox (2000) extended the results first reported by Schmajuk, Lam and Christiansen (1993) by examining, in the rat eyeblink response preparation, the effect of haloperidol administration on the impairment of LI produced by aspiration lesions of the hippocampus (HFLs). During the preexposure phase, rats with hippocampal or control lesions were either exposed to a tone or allowed to sit in the training apparatus. During the conditioning phase, the tone was paired with an air puff to the eye after the animals were injected with either saline or haloperidol. Although rats with HFLs and injected with saline did not show LI, the phenomenon was reinstated in HFL animals receiving haloperidol injections.

Figure 6.9 (middle panel) shows the experimental results from Schmajuk *et al.* (2000). The panel shows the percentage of CRs on the fourth day of conditioning for the SL and HFL groups, for the PE and NPE conditions, under saline (SAL) and haloperidol (HAL) drug treatments. According to the figure, whereas the SL group showed LI under SAL and HAL conditions, HFL animals demonstrated LI only under HAL administration.

Simulated results

Figure 6.9 (lower panel) shows simulated results. The panel displays percentage of CRs on the fourth day of conditioning for the SL and HFL groups under SAL or HAL administration for PE and NPE groups. In agreement with the experimental results shown in the middle panel, the lower panel shows that SL yield LI both under SAL and HAL conditions, but HFLs yield LI only under HAL administration. In our simulations, in Equation [2.2] (see Chapter 2), $K_{12} = 0.65$ after haloperidol administration.

The simulated results are explained as follows. After preexposure, because $\bar{B}_{CS}$ and $\bar{B}_{CX}$ increase faster in the SL group than in the HFL goup, Novelty′ is smaller in the SL–SAL–PE case than in the HFL–SAL–PE case. Consequently, LI is impaired in the HFL group treated with saline. Administration of haloperidol decreases the effect of Novelty′ on attention during part of the preexposure trials and all conditioning trials. Consequently, HFL subjects treated with haloperidol display LI.

The same brain abnormalities that explain the absence of LI can explain the presence of hallucinations. In the case of an increased DA input to the attentional system, attention to both the traces of perceptual inputs (τ_{CS}) and the predictions of these inputs (B_{CS}) increases. The increased prediction of an absent stimulus through its association to another stimulus (see Figure 2.2) would be experienced as a hallucination, which is the perception of events that do not have an external source. In support of this idea, Silbersweig, Stern *et al.* (1995) reported activation of the thalamus, the putamen, the NAC, the hippocampus and cingulate gyri during hallucinations.

Discussion

The data analyzed in this section seem to support the view that positive symptoms in schizophrenia are linked to hippocampal dysfunction. This abnormality is not necessarily primary to the disease, but might constitute the aftermath of a degenerative process that follows the onset of the pathology.

The notion of a hippocampal dysfunction is supported by the morphological changes in the hippocampal formation reported in schizophrenic patients. In addition to the morphological data, the case receives support from studies showing that similar behavioral deficits are exhibited by schizophrenics and by animals with hippocampal lesions. Actually, several of these deficits in schizophrenics were predicted by observing the effects of hippocampal lesions in animals, as is the case of the blocking effect (see Schmajuk & Tyberg, 1991, page 75) later confirmed by Jones *et al.* (1992). Furthermore, the hippocampal dysfunction view is also supported by the fact that some of these behavioral deficits, such as the absence of LI, are reversed by neuroleptics in both schizophrenics and HFL animals.

Summary

The results presented in this chapter seem to confirm the competence of the SLG model as a tool to link the behavioral aspects of LI (as shown in Chapter 5) and its neurophysiological basis. The model also provides a mechanistic explanation for the biological basis of positive symptoms in schizophrenia. Below we summarize the material presented in the different sections of this chapter, together with related simulation results reported by (a) Buhusi *et al.* (1998), regarding the effects of hippocampal lesions on LI; (b) Schmajuk *et al.* (1998), regarding the effect of DA agonists and antagonists on LI; (c) Schmajuk *et al.* (2001), regarding the effect of lesions of the shell and core of the nucleus accumbens combined with haloperidol administration on LI; and (d) Schmajuk *et al.* (2005), regarding the effect of dopaminergic drugs administered during preexposure.

The section on "The neural substrates of latent inhibition" presents experimental data that suggest that LI is controlled by a circuit that involves the hippocampus, the EC, the shell and core of the NAC, and the mesolimbic DA projection from the VTA to the NAC. Different nodes and connections in the SLG model are mapped onto this brain circuit. Critically, variables X_{CS} (its attentionally modulated component) and Novelty' are mapped onto the DA projection to the NAC, and neural activity in the NAC is assumed to be proportional to (the attentional component of) X_{CS} and (the excitatory and inhibitoty components of) Novelty'. The output of the core of the NAC acts on different brain circuits, including the VP, THAL and Amygdala, that control conditioning.

The section on "Dopaminergic involvement in latent inhibition" analyzes the assumption that indirect DA agonists (e.g. amphetamine and nicotine) increase the effect of the DA projection from the VTA on the NAC, thereby increasing the value of Novelty' and X_{CS}. Correspondently, DA receptor antagonists (e.g. haloperidol and α-flupenthixol) decrease this effect, thereby decreasing the effect of Novelty' and X_{CS}. Consequently, when Novelty' is modified, the SLG model describes (a) the impairment of LI by amphetamine when a strong US is used, (b) the impairment of LI by amphetamine when a nonsalient CS is used, (c) the impairment of LI by amphetamine administration when a short CS is used, (d) the facilitation of LI by α-flupenthixol when a weak US is used, (e) the facilitation of LI by haloperidol when a nonsalient CS is used, (f) the facilitation of LI by haloperidol with a strong US, and (g) the facilitation of LI by haloperidol with extended conditioning. Most interestingly, the model describes the "atypical" results of nicotine-induced facilitation and haloperidol-induced impairment of LI.

The section on "Hippocampal involvement in latent inhibition" analyzes the assumption that hippocampal lesions alter the computation of the predictions of the CS and US, thereby changing the value of Novelty's. Under this assumption, the model describes: (a) the impairment of LI after HFLs using a between-subject procedure, (b) the impairment of LI and nonhabituation of the OR after HFLs using a between-subject procedure, (c) the facilitation of LI after HFLs in a within-subject procedure with interspersed water presentations, (d) the impairment of LI after HPLs in a within-subject procedure, (e) the preservation of LI after HPLs in a between-subject procedure, (f) the absence of contextual effects on LI after HPLs in a within-subject with context change procedure, and (7) the facilitation of LI after HPLs in a between-subject with additional cue procedure. These results suggest that the notion that LI (a label that encompasses many alternative procedures) is, or is not affected by hippocampal lesions, should be replaced by a detailed analysis of how Novelty', the key variable that controls LI in the SLG model and can be experimentally evaluated by measuring the magnitude of the OR, varies as a function of the experimental procedure, type of lesion and drug administration.

The section on "Effects of reversible inactivation of the entorhinal cortex" illustrates how the model is able to replicate Seillier et al.'s (2007) results showing that LI was unaffected in rats with reversible inactivation of the entorhinal cortex during conditioning, but disrupted when the reversible inactivation took place during preexposure only.

The section on "Nucleus accumbens involvement in latent inhibition" examines the assumption that the predictions of CSs and the US are conveyed to the VTA area through the shell of the NAC, and that the DA output of the VTA excites the core of the NAC. It shows that the model correctly describes: (a) the impairment of LI by lesions of the shell of the NAC, (b) the restoration of LI by haloperidol following lesions of the shell, (c) the preservation of LI by lesions of the core of the NAC, (d) the facilitation of LI by core lesions with extended conditioning, and (e) the facilitation of LI by combined shell-core lesions with extended conditioning. Most interestingly, based on its predictions for the effect of hippocampal lesions on LI, the model correctly *predicted* the results recently reported by Pothuisen et al. (2006) regarding the preservation of LI in a conditioned taste-aversion paradigm, following shell lesions.

The section on "Latent inhibition and schizophrenia" shows how the SLG model explains the absence of LI, regarded as a positive symptom of schizophrenia, as caused by a hippocampal dysfunction and ameliorated by DA blockers acting on the NAC.

It is important to notice two instances in which the SLG model is able to explain the apparently conflicting data: (a) how nicotine administration can impair or facilitate LI depending on the duration of the CS; and (b) how hippocampal lesions can impair, preserve or facilitate LI depending on different aspects of the experimental design.

7
Creativity

In this chapter, we apply the SLG model presented in Chapter 2 to the theoretical analysis of creativity. Creativity can be defined as a psychological process that produces original and appropriate ideas. A rather large number of theories have been proposed to account for this process, including Guilford's (1950) psychometric theory, Wertheimer's (1959) Gestalt theory, Mednick's (1962) and Eysenck's (1995) associative theories, Campbell's (1960) Darwinian theory, Amabile's (1983) social–psychological theory, Sternberg and Lubart's (1995) investment theory, and Martindale's (1995) cognitive theory. Other approaches, such as artificial intelligence models (Boden, 1999; Partridge & Rowe, 2002) also contribute to our understanding of creativity.

Mednick (1962) defined creative thinking as the combination of different associations. This combination might result from (a) contiguity, the accidental or planned temporal proximity between the elements of the association; (b) generalization, the sharing of common factors by the elements of the association; or (c) mediation, the simultaneous activation of both elements of the association. Mednick suggested that differences in creativity depend on the strength of the associations that enter in the combinations. In a similar vein, Eysenck's (1995, page 81) theory stipulates that (a) cognition requires associations, (b) differences in intelligence depend on the speed to build these associations, (c) differences in creativity depend on the range of associations considered in problem solving, and (d) a comparator is needed to eliminate wrong solutions.

Eysenck (1995, page 279) suggested that creativity increases when normal mechanisms that limit the formation of associations, such as those at work during latent inhibition, are disrupted. In support of Eysenck's (1995) ideas, Carson, Peterson and Higgins (2003) reported that more creative persons show lower latent inhibition scores than normal ones.

120   Attentional and associative mechanisms

The SLG model incorporates most of the notions about formation and combination of associations of Mednick's (1962) and Eysenck's (1995) theories. The objective of this chapter is to explore the possibility that, because it precisely describes the rules by which stimuli become associated and combined, the SLG model can also describe some aspects of the creative process. Furthermore, as mentioned in Chapter 6, the model describes the impairment of latent inhibition by hippocampal lesions (Buhusi *et al.*, 1998) and administration of indirect agonists (Schmajuk *et al.*, 1998). In both cases, Schmajuk *et al.* (1998) hypothesized that the effects are the consequence of an increase in dopamine (DA) release in the nucleus accumbens. Therefore, the SLG model quantitatively describes two procedures (hippocampal lesions and amphetamine administration) that increase DA levels, impair latent inhibition and, as suggested by Eysenck (1995), might increase the formation of associations, thereby improving creativity.

In this chapter, we explain how the SLG model can be applied to divergent thinking and remote associate tests of creativity. Then, computer simulations are compared with experimental results obtained in creative and normal individuals. Data include experiments on latent inhibition, overinclusion, divergent thinking and remote associates. Finally, following the approach used in Chapters 4 and 6, variables in the model are mapped on different brain regions and neurotransmitters involved in creativity.

Application of the model to creativity

As mentioned, Carson *et al.* (2003) reported that, as predicted by Eysenck (1995), creative individuals show impaired latent inhibition. Therefore, because the SLG model correctly describes the impairment of latent inhibition under the premise of an increased effect of Novelty′ on attention (see Chapter 6), it was hypothesized that similar increased novelty and impaired latent inhibition are present in creative people. As mentioned in Chapter 2, attentional memory, z_{CS}, is the association between the output of the feedback system ($\tau_{CS} + B_{CS}$) and Novelty′. Therefore, an increased attention amplifies both the trace of the CS (τ_{CS}) and the strength of the predictions of the CS (B_{CS}). Also recall that attentional memory (z_{CS}) controls the magnitude of the representation X_{CS}, thereby controlling both the modification (acquisition or extinction) and retrieval of CS–Other CS and CS–US associations. Therefore, creative people would learn faster, extinguish faster and retrieve memories more readily than normal people. In our simulations, it is specifically assumed that the effect of Novelty′ on z_{CS} was twice as strong in creative than in normal people.

Application to the SLG model to creative tasks

Schmajuk, Aziz and Bates (2009) suggested how the mechanisms involved in forming and combining associations in the SLG model apply to creative tasks.

Formation of associations

Mednick (1962; Higgins, Mednick & Thompson, 1966) characterized paired associates in terms of their frequency in the Kent–Rosanoff Word Association Test (Russell & Jenkins, 1954). For instance, presentation of the word table elicits the words chair, food, desk, top and legs in (respectively) 83, 4, 2, 1.4 and 1.1 percent of the participants in the test. Low and high frequency associates are respectively considered remote and close-paired associates. Mednick (1962) suggested that, whereas in a normal person the associative hierarchy is steep (associative strength decreases fast and the remote associations are weak), in a creative individual the hierarchy is flat (associative strength decreases slowly from chair to top).

Although accounting for the frequency of paired associates is well beyond the scope of the present study, the model is used to approximate how the eliciting word (table) activates the response words (chair–food–desk–top–legs). In order to approximate the frequency of pair-associates in the Word Association Test, a simulation was carried out assuming that table and chairs are stimuli experienced together many times, and then one additional stimulus (food–desk–top–legs) at a time was added until all the stimuli were experienced. According to the model, as more stimuli are added, they are able to predict a new added stimulus faster and, therefore, the novelty of the added stimulus decreases faster. Because the novelty of an added stimulus is smaller than the novelty of first-learned stimuli, recently experienced stimuli acquire weaker associations than the previous ones. Figure 7.1 shows the associative hierarchy that results from this assumption. It presents the value of the predictions (interpreted as response strengths) of the different stimuli when the stimulus "table" is presented, following 20 table–chair training trials, 1 table–chair–food training trial, 1 table–chair–food–desk training trial, 1 table–chair–food–desk–top training trial, 1 table–chair–food–desk–top–legs training trial, and 1 table test trial. All stimuli were 10 time units (t.u.) long and their intensity 1, and the ITI was 500 t.u.

The approximate method utilized to account for the frequency of paired associates is not critical for our argument. Similar results were obtained when alternative assumptions about how frequencies of paired associates are generated, e.g. forming paired associates and then combining these associations when the

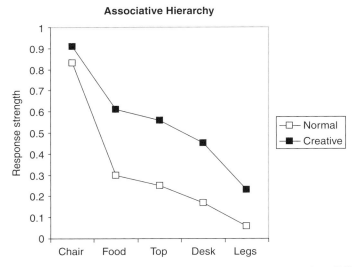

Figure 7.1 Associative hierarchy. Simulated response strength to different word stimuli.

eliciting word is presented. It is important, however, that in agreement with Mednick's suggestion, the associative hierarchy decays more slowly, and remote associates are stronger, in creative than in normal individuals. Nevertheless, in contrast to Mednick's view, due to the increased attention to the presented stimulus, close associates are also slightly stronger in creative individuals.

As mentioned above, in addition to learning by contiguity, Mednick (1962) suggested that presentation of a word might elicit another word because the concepts share common elements (generalization), or the words become associated through the mediation of common elements that activate both words (or concept) simultaneously. The model describes generalization between words W_1 and W_2 if they are both associated to a common element Wg, because presentation of W_1 activates the representation of Wg, which in turn activates the presentation of W_2. In addition, the model describes mediation because presentation of Wg activates the representation of both W_1 and W_2.

Combining associations

In the SLG model, combination of associations is accomplished by the sequential activation of the representations of stimuli presented separately. For example, the food on the tabletop (as in the preceding example) might activate the representation of a desktop, and therefore, of a desk, and its legs. Combination of associations is a procedure used in divergent thinking and remote associates tests, as well as in problem solving.

Simulation parameters

Simulations for normal people with the SLG model were run with parameter values identical to those used in previous chapters: $K_1 = 0.2$, $K_2 = 2$, $K_3 = 0.4$, $K_4 = 0.1$, $K_5 = 0.02$, $K_6 = 0.005$, $K_7 = 0.005$, $K_8 = 0.005$, $K_9 = 0.75$, $K_{10} = 0$, and $K_{11} = 0.15$. Simulations for creative people assumed that the effect of Novelty' on attentional memory, z_{CS}, is twice as strong in creative individuals than in normal persons, that is $K_{12} = 2$.

Results

The results of computer simulations with the SLG model were compared with experimental data illustrating that creative people show deficits in latent inhibition and overinclusion tests, but superior performance in divergent thinking and remote associates tests.

Latent inhibition

Experimental data

In Carson *et al*'s (2003) latent inhibition task, the experimental group was preexposed (PRE) to an auditory target CS, randomly superimposed on a background of nonsense syllables. For the non-preexposed (NPE) control group, the target CS was absent in this first phase. For both groups, the conditioning phase consisted of the presentation of the target CS preceding yellow disks (US) displayed on a video screen. Participants were asked what auditory stimulus preceded the appearance of the yellow disks. Figure 7.2 (upper panel) shows Carson *et al*.'s latent inhibition scores for low (Normal) and high creative (Creative) achievement questionnaire (CAQ) participants. Participants in the Normal PRE group took more trials to learn the association than those in the Normal NPE group, thereby showing latent inhibition. That was not the case in the Creative group.

Simulated results

Figure 7.2 (lower panel) shows simulations obtained with the model for creative and normal individuals, with or without preexposure to the target CS. Simulations consisted of 50 preexposure and 10 conditioning trials, ITI duration 500 time units (t.u.), overlapping CSs duration 20 t.u., CS intensities were 1, and the intensity of the CX was set to 0.1. An arbitrary response level (representing the auditory CS–visual CS association) was chosen, and the number of trials to reach that level for the different groups was determined.

124 Attentional and associative mechanisms

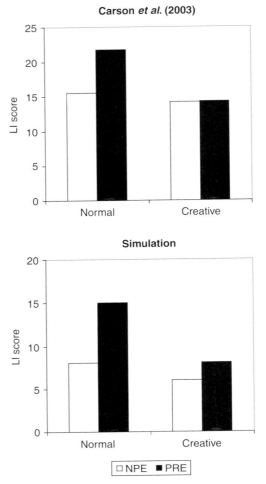

Figure 7.2 Latent inhibition in creative and normal people. Trials to criterion (latent inhibition, LI, scores) for non-preexposed (NPE) and preexposed (PRE) groups. *Upper panel*: Data from Carson *et al.* (2003). *Lower panel*: Simulated latent inhibition results.

Figure 7.2 shows that, as in the experimental data, the simulated PRE normal group showed a larger number of trials to criterion than the NPE normal group (15 vs. 8). In contrast, simulated PRE and NPE creative groups showed a similar number of trials to criterion (8 vs. 6). Notice that in the experiments, creativity is determined by CAQ scores, but our simulations only provide an approximation to divergent thinking scores. However, as reported by Carson *et al.* (2003, Table 1), CAQ scores and scores in several subtests of the divergent thinking test are significantly correlated.

Overinclusion

Experimental data

Overinclusive thinking refers to the thought disorder characterized by an inability to preserve conceptual boundaries (Andreasen & Powers, 1974). Andreasen and Powers (1975) found no difference between creative writers and manics, who in both cases tended to show relatively strong overinclusion. Because Andreasen and Powers (1975) did not report data for normal subjects, Figure 7.3 (upper panel) presents data from Andreasen and Powers' (1974) study for manics and normal participants. Manics (and creative) individuals are more overinclusive than normal people. In associative terms, overinclusive thinking can be described as increased generalization (Eysenck, 1995, page 247). The idea of increased generalization is supported by Ward's (1969) report that creative people pay attention to more incidental cues as material for novel ideas.

Simulated results

Figure 7.3 (lower panel) shows computer simulations of generalization that consisted of 10 trials of CS_1–US training with a CS representing an element common to CS_1 and CS_2, 1 CS_1 test trial and 1 CS_2 test trial. ITI duration 500 time units (t.u.), CSs duration 10 t.u., CS intensities were 1, CS common element intensity was 0.5, US duration 5 t.u. overlapping with the last 5 t.u. of the CSs, US intensity was 1 and the intensity of the CX was 0.1. According to the model, creative individuals show more generalization than normal individuals; a result in agreement with Andreasen and Powers' (1974) report. Interestingly, whereas Eysenck (1995, pages 253 and 279) suggested that the absence of cognitive (latent) inhibition is the cause of the flat generalization gradient and overinclusion shown by creative people, in our view, increased effect of Novelty' on attention is the common cause for both impairments.

Divergent thinking

Experimental data

Carson et al. (2003) examined different measures of creativity and their relation to latent inhibition. These measures include a creative achievement questionnaire (CAQ, Carson et al., 2003), a creative personality scale (CPS, Gough, 1979), and divergent thinking tasks (Torrance, 1968). Because our model cannot be readily applied to a CAQ or CPS, this analysis will concentrate on the Torrance (1968) divergent thinking task. In this task, an eliciting word is presented and divergent thinking scores result from the combination of (a) fluency

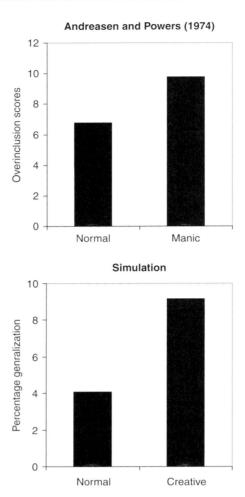

Figure 7.3 Overinclusion in creative and normal people. *Upper panel*: Data from Andreasen and Powers (1974). *Lower panel*: Simulated generalization results.

(number of responses), (b) originality (unusual responses within the experimental sample), (c) flexibility (number of different categories of responses and category changes), and (d) elaboration (details in the responses). Figure 7.4 (left panel) shows the values of the low and high creativity (CAQ) scores (Carson et al., 2003, page 503). These scores were significantly correlated with their corresponding divergent thinking scores.

Simulated results

Figure 7.4 (right panel) shows simulated creativity scores, given by combining equal proportions of fluency and originality in a divergent thinking test.

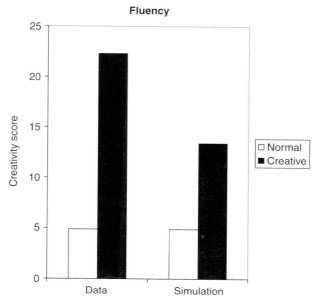

Figure 7.4 Creativity scores in creative and normal people. *Left*: Data from Carson *et al.* (2003) show scores from the creative achievement questionnaire (CAQ). *Right*: Simulated creativity score computed by combining fluency (summed response strength of chair, food, top and desk) and originality (activation of legs) in equal proportions. Response strengths are shown in Figure 7.1.

Fluency was computed as the sum of the strengths of the responses "chair", "food", "top", and "desk" shown in Figure 7.1. Originality was computed as the strength of the (relatively unusual) response "legs." Although it is possible to arbitrarily attribute a category to each of the responses and then determine the number of categories and category changes, the resulting simulated flexibility would be simply proportional to a sum of the simulated fluency and originality and, therefore, was not included in our results. Also, although it could be assumed that detail in a response is proportional to its strength, elaboration was not included in our results. Figure 7.4 shows that the simulation approximates the experimental data.

Remote associates test

Experimental data

Mednick (1962) proposed to measure creativity using a remote associates test (RAT). In this test, participants are presented with three words (e.g. gown, club and mare) and have to generate the word that is common to all of them

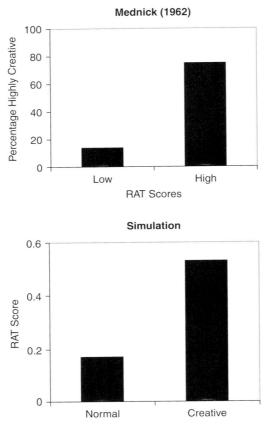

Figure 7.5 Remote associates test (RAT) in creative and normal people. *Upper panel*: Percentage of people rated high on research creativity as a function of RAT scores (data from Mednick, 1962). *Lower panel*: Simulated RAT score.

(night). Mednick (1962, page 228) reported that RAT scores of a group of first year psychology graduate students were highly correlated with their research creativity judged by faculty supervisors. As shown in Figure 7.5 (upper panel), 75 percent of the high RAT scorers and 14 percent of the low RAT scorers were rated highly creative.

Simulated results

Figure 7.5 (lower panel) shows RAT score given by the strength of the response to the word *night* following 1 night–club training trial, 1 night–gown training trial, and 1 night–mare training trial. The ITI duration was 500 time units (t.u.), CSs duration 10 t.u., and all CS intensities were 1. During testing, the activation of the word night by joint presentation of the three words was evaluated. Figure 7.5 (lower panel) shows that creativity and RAT scores are positively correlated; a result in agreement with (but not identical to) Mednick's (1962) data.

Discussion

Under the premise that creative individuals display increased attention to stimuli present when novelty is large, the SLG model (see Chapter 2) provides a mechanistic explanation for why creative people show deficits in latent inhibition and overinclusion tests, but superior performance in divergent thinking and remote associates tests.

Our premise implies that creative people learn faster and retrieve memories faster than normal people. The notion seems related to some ideas offered by Wallach and Wing (1969; Wallach, 1970), who suggested that creative individuals show higher "cognitive energy."

Impaired latent inhibition and overinclusion

Eysenck (1995) suggested that the absence of latent inhibition in creative individuals is the cause of the flat generalization gradient and overinclusion. Deficits in latent inhibition are the consequence of creative people's psychoticism, a temperamental dispositional trait that renders them vulnerable to functional psychosis and cognitive deficits similar to those of psychotic patients (Eysenck, 1995, page 8). In our simulations, creative individuals (psychoticists) show both impaired latent inhibition and overinclusion as the consequence of the increased effect of novelty on attention. De la Casa and Lubow (1994) and Gibbons and Rammsayer (1999) reported that psychoticism is associated with deficits in latent inhibition.

In addition to simulations for latent inhibition in creative individuals, simulations for the blocking effect (Kamin, 1969a,1969b), by which a stimulus associated with a second stimulus blocks the formation of associations of other stimuli with that second stimulus, were also carried out. In contrast to the results for latent inhibition, and in agreement with experimental data showing no systematic relation between psychoticism and blocking (Jones, Gray & Hemsley, 1990), simulations showed no impairment of blocking in creative individuals (psychoticists).

Improved performance in divergent thinking and remote associate tests

Creative people show superior performance in divergent thinking and remote associates tests. In the model, (a) the strength of the predictions of stimuli was interpreted as proportional to the number of responses in the same category (fluency) and different categories (flexibility), and (b) the strength of the prediction of the most remote stimulus was interpreted as proportional to the number of unusual responses (originality). As mentioned, this approach is somewhat similar to Mednick's (1962) associative theory.

Problem solving and response selection

Creative individuals also show superior performance in problem-solving tasks (Getzels & Csikzentmihalyi, 1976). Problem solving has been defined as the process of finding a path (in the problem space) from an initial to a desired state through a directed graph (Winston, 1977, page 90). Tolman (1932, page 177) suggested that a maze is a special case of problem solving. Schmajuk and Thieme (1992) showed that a neural network designed to describe maze navigation could be successfully applied to problem solving. They represented the Tower of Hanoi task in terms of the possible transitions between different ring arrangements in the three poles (see Hampson, 1990), and showed how the network learned to solve the problem in the minimal number of movements.

The Schmajuk and Thieme (1992) system includes (a) an action system, consisting of a goal-seeking mechanism; (b) a goal defined by a motivation block; and (c) a cognitive system, involving a neural cognitive map. The inclusion of a goal-seeking mechanism is important to solve the "problem" of producing responses that meet a criterion. This mechanism is similar to that proposed by Merten (1992) in his model of word associations. Merten (1992) suggested that, before being produced, the chosen response is compared with an established criterion. If the response is rejected, a search for an alternative is started; if approved, the response is emitted. More generally, most creativity tests (divergent thinking, remote associates) establish a criterion that the responses should satisfy: for example, to name, write or draw – without repeating – objects that are circles, or an object that is associated with a set of objects. Therefore, the goal-seeking mechanism in the Schmajuk and Thieme (1992) model could be used to select responses adequate to solve the problem at hand.

The SLG network described in Chapter 2 can be used to implement the cognitive map in Schmajuk and Thieme's problem-solving model. In the network, presentation of a stimulus CS_1 can be regarded as a next possible situation in problem space, and a given stimulus CS_4 can be regarded as the solution to the problem (goal situation). If CS_1 is able to activate the representation of CS_4, then the problem can be solved. Because, as shown in Figure 7.1, creative individuals are able to activate remote representations more strongly than normal people, they will also be better at solving problems. If the activation of CS_4 is present only in creative persons, then it is said that the response is "original."

Furthermore, activation of this remote item can lead to an unexplored, shorter or stronger link that leads to the goal, and therefore to the discovery of new ways of solving problems.

In addition to the activation of remote (original) links, higher levels of attention in creative individuals result in the faster extinction of links that no longer exist. That is, creative people would persevere less than normal people when a known path to the goal is obstructed and a new path should be found. Old solutions would be rapidly abandoned and exploration rapidly resumed.

Comparison with formal models of creativity

Boden (1999) reviewed a number of models of creativity classified as either (a) combinational, or (b) exploratory–transformational. Whereas combinational models create novel products by combining familiar ideas in unusual ways, exploratory–transformational models generate novel products by extending the familiar boundaries of a conceptual space. Although the SLG model is, in itself, a combinational model, when joined to the action system of the Schmajuk and Thieme model, it becomes an exploratory model that explores, and maps, novel regions of the problem space.

Martindale (1995) described a combinational neural network in which each node exchanges information with other nodes while receiving a nonspecific input from the arousal system. Following Hull's (1943) ideas, the net activation of a given node is calculated by combining excitatory and inhibitory inputs (habit strength) and multiplying the sum by the magnitude of the arousal signal (drive). Martindale assumed that with high arousal a few nodes would be active (attention is focused) and, as arousal decreases, more nodes would become active. Martindale and Hasenfus (1978) had reported that creative individuals show low arousal during conceiving a fantasy story (inspiration) and high arousal during writing it out (elaboration). Martindale (1995, page 261) suggested to model this process with a network that goes to a low-arousal state with the purpose of finding a solution and returns to a high-arousal state to verify the quality of the solution. Martindale compared this process to the "simulated annealing" method used in neural network research in order to avoid local minima. Simulated annealing starts with a high activation (temperature) of the network nodes and, then, this activation gradually decreases to find a global solution to the problem. Because, as reported by Martindale and Hasenfus (1978), creative individuals start with low cortical arousal and end with high arousal, Martindale proposed that high temperature corresponds to low cortical arousal.

Partridge and Rowe (2002) run computer programs to compare Martindale's (1995) cortical-arousal theory with Weisberg's (1986) view that the creative process is identical to other normal thinking processes. The programs store and use memories of successful solutions in order to create and discover card-sequence rules. Whereas Martindale's approach generates multiple possible solutions, Weisberg's method stays with one solution until it is sufficiently weakened. Whereas in the first case the activation (temperature) of the nodes is varied as a function of the success in solving the problem at hand, in the second case, connections in the network are weakened if the active node does not lead to the solution of the problem. Interestingly, although both approaches exhibited "significant creative behavior", only the Weisberg approach showed a behavior reminiscent of the illumination stage of creativity (Wallas, 1926). Our model partially agrees with Martindale's view that high node activation is needed to activate its associated nodes. When the activated node matches the required solution, activation decreases. As different nodes are activated, cortical activity increases and decreases.

Gabora (2002) proposed that a variable focus of attention is key to the creative process. Initially, attention is defocused and, therefore, memory activation is flat (many memory locations are active). Gradually, attention becomes focused on the memory location that corresponds to the solution to the problem at hand. In contrast to Gabora's view, we assumed that memory activation and response selection are separate mechanisms. Increased attention to a stimulus results in increased activation of memory locations, and the output of one of these memory locations is selected if it matches the goal of the system.

The neurobiology of creativity

Figure 7.6 (similar to Figure 6.1) shows how attentional and associative functions in the SLG model can be mapped onto different brain areas involved in creative processes (Schmajuk *et al.*, 2001; Nicholson & Freeman, 2002; Weiner & Feldon, 1997; Swerdlow & Koob, 1987). Parts of this brain circuit participate both in latent inhibition (see Chapter 6) and creative processes. The attentional system (nucleus accumbens, ventral pallidum and thalamic areas) modulates the strength of environmental inputs. These attentionally modulated signals enter the associative block (neocortical areas) to become associated with other environmental signals. Attention is regulated by the novelty system (ventral tegmental area), which compares environmental inputs with their predictions provided by the associative system. Formation of associations in cortical areas is controlled by the septo-hippocampal system, which compares environmental inputs and their predictions. Important for this discussion, the Novelty' signal is assumed to be coded by DA release at terminals of the ventral tegmental area, mainly in the nucleus accumbens.

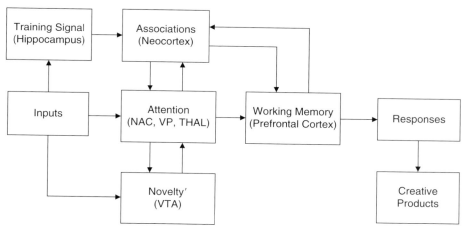

Figure 7.6 Mapping of the variables in the SLG model onto different brain areas involved in creativity. Environmental inputs are compared with their predictions, provided by the Association system located in the neocortex, to compute (a) Novelty´ in the ventral tegmental area (VTA), and (b) a Training Signal in the hippocampus. Novelty´, coded by DA, reaches the attentional system (nucleus accumbens, ventral pallidum, thalamus). The Training Signal reaches the neocortex through the entorhinal cortex. The Association system controls the response system that will determine the creative product. VTA: ventral tegmental area, NAC: nucleus accumbens, VP: ventral pallidum, THAL: thalamus.

Dietrich (2004) suggested that the prefrontal cortex holds the content of consciousness in working memory and instigates the deliberate control of creativity. Working memory is involved in abstract thinking, strategic planning and access to long-term memory. Neocortical (temporal, occipital and parietal) areas are involved in perception and long-term memory. According to Dietrich, the prefrontal cortex (a) evaluates the appropriateness of a conscious idea, (b) maintains attention and retrieves relevant information from long-term memory, and (c) controls action guided by internal goals. Dietrich proposed four types of creativity depending on the processing mode (deliberate or spontaneous) and the knowledge domain (cognitive or emotional). The prefrontal cortex and neocortical areas are involved in deliberate-cognitive processes, the ventromedial prefrontal cortex and the amygdala participate in deliberate-emotional processes, basal ganglia activation of the dorsolateral prefrontal cortex and neocortical areas are involved in spontaneous-cognitive processes, and the dorsolateral prefrontal cortex and the amygdala are involved in spontaneous-emotional processes. Figure 7.6 shows a connection between the ventral tegmental area and the prefrontal cortex (through the nucleus accumbens, ventral pallidus and medio-dorsal thalamus), which might be responsible for the novelty-related activity reported in the prefrontal cortex (Fletcher *et al.*, 2001).

Dopamine and creativity

As mentioned, Schmajuk *et al.* (1998) showed that the impairment of latent inhibition following administration of indirect DA agonists is well described by the SLG model by assuming, as in the present chapter, an increased effect of Novelty′ on attention. Schmajuk *et al.* (1998) hypothesized that this effect is mediated by an increase in DA release in the nucleus accumbens. This hypothesis suggests that (a) increased DA release in the nucleus accumbens is larger in creative than in normal individuals when novelty is detected in the environment; and (b) creativity could be increased by indirect DA agonists, such as (the direct cholinergic agonist) nicotine. Some support for this view is provided by Schrag and Trimble's (2001) report that creativity might be unmasked during treatment of cognitive impairment in patients with Parkinson's disease (using the cholinesterase inhibitor rivastigmine, which increases the neurotransmitter available at the cholinergic synapse).

A possible link between DA and creativity had been proposed before. For example, Ashby, Isen and Turken (1999) proposed a neuropsychological theory that assumes that positive affect, which improves performance on many cognitive tasks including creative problem solving, is associated with increased DA levels in the anterior cingulate cortex. In a similar vein, Flaherty (2005) presented a model of idea generation and creative drive, in which the mesolimbic DA controls novelty-seeking and creative drive by acting on the temporal and frontal lobes. Flaherty (2005, page 149) suggested that increased DA might explain the higher baseline level of arousal shown by creative individuals (Martindale, 1999), as well as the absence of latent inhibition reported by Carson *et al.* (2003). In support of the DA view, Flaherty (2005, page 150) indicates that while DA agonists can induce hypomania and hallucinations (Peet & Peters, 1995), DA antagonists might suppress free associations underlying creativity.

Hippocampus, creativity and schizophrenia

As mentioned in Chapter 6, in addition to administration of indirect DA agonists, excitotoxic lesions extending from the entorhinal cortex to the ventral subiculum (Yee, Feldon & Rawlins, 1995) and hippocampal aspiration lesions (Schmajuk, Lam & Christiansen, 1993) also impair latent inhibition. Moreover, patients with schizophrenia, a disorder in which hippocampal dysfunction seems to be present (Schmajuk, 2001), also show absence of latent inhibition. Buhusi *et al.* (1998) and Schmajuk *et al.* (2000) showed that, under the premise of deficits in the storage or modification of CS–Other CS associations in cortical areas after (excitotoxic) hippocampal lesions, the SLG model correctly predicts impairments in latent inhibition following lesions and in schizophrenia. Notice that the deficiency in forming CS–Other CS associations captures memory deficits following hippocampal lesions well (Shimamura & Squire, 1984). Poor CS–Other CS associations are also consistent with memory deficits (e.g. Lencz et al., 2006), intelligence deficits (e.g. Aylward, Walker & Bettes, 1984), and hypoactivity in brain areas such as the associative frontal, parietal or temporal cortices (Kishimoto et al., 1998) reported in schizophrenic patients.

According to the model, impairments in the formation of CS–Other CS associations in schizophrenia result in a decrease prediction of the CS and, thereby, in an increased Novelty'. Again, this increased Novelty' results in an increased DA release, presumably in the nucleus accumbens, and the concomitant attentional deficits (Dykes & McGhie, 1976). The effect of this increased DA can be blocked by administration of haloperidol, which reinstates the latent inhibition lost after hippocampal lesions (Schmajuk et al., 2000) or in schizophrenic patients (Lubow et al., 1987; Baruch et al., 1988; Gray et al., 1991).

Also, according to the model, impaired formation of CS–Other CS associations in schizophrenic patients should also result in poor performance in divergent thinking tests, i.e. poor creativity. This observation is consistent with Burch et al.'s (2006) study demonstrating that people with schizotypal personality show impaired latent inhibition but not increased creativity, a result that could be also explained in terms of the relatively low IQ scores (and related weak CS–Other CS associations) of their sample. Interestingly, similar to what has been shown in schizophrenic patients, Dickey et al. (2007) reported that female subjects with schizotypal personality disorder presented bilaterally relative small hippocampal volumes.

Furthermore, the need for preserved CS–Other CS associations in creative processes is consistent with data showing that bipolar disorder is better correlated with creativity than schizophrenia (Flaherty, 2005, page 148; Andreasen & Powers, 1974). Even if patients with bipolar disorder show hippocampal abnormalities (Strasser et al., 2005), disturbances in dopaminergic transmission (Goodwin et al., 1970), and some cognitive impairment (e.g. Atre-Vaidya, et al., 1998), their cognitive performance is better (and presumably the strength of their CS–Other CS associations stronger) than that of schizophrenic patients (Krabbendam et al., 2005; Santosa et al., 2006; but see Sass, 2001–2002). In sum, even if schizophrenics, people with schizotypal personality, and creative individuals might be similar in their attentional deficits, the model suggests that when cognitive deficits are present they might preclude patients from behaving creatively.

Summary

In this chapter, we applied the SLG model to the description of the mechanisms participating in creative processes. Under the premise that attention to novel stimuli is increased in creative people, the model explains why creative people show improved (a) divergent thinking (fluency and originality), (b) performance in remote associations tests, and (c) problem solving; but impaired (d) latent inhibition, and (e) generalization (overinclusion). The increased attentional processing might be linked to an increased dopamine release in the nucleus accumbens. The approach suggests a possible integration of experimental and theoretical studies on classical conditioning and creativity.

8

Overshadowing and blocking

In this chapter, we show how the SLG model applies to recovery from overshadowing, the interaction between overshadowing and latent inhibition, recovery from blocking, the inability of a blocked CS to become a blocker of another CS, backward blocking and recovery from backward blocking.

In the last decades, new phenomena have been presented that challenge traditional theories of classical conditioning. Among other observations, Matzel, Schachtman and Miller (1985; Kaufman & Bolles, 1981) established that extinction of the overshadowing CS results in the recovery of the response to the overshadowed CS; Blaisdell, Gunther and Miller (1999) showed that extinction of the blocking CS results in the recovery of the response to the blocked CS; Pineño, Urushihara and Miller (2005) reported that a time delay interposed between the last phase of backward blocking and testing results in the recovery of the response to the blocked CS; Grahame, Barnet, Gunther and Miller (1994) reported that extinction of the context following latent inhibition (LI) results in the recovery of the response of the target CS; De la Casa and Lubow (2000, 2002) demonstrated that a delay interposed between conditioning and testing results in an increased LI effect (super-LI); and Blaisdell *et al.* (1998) showed that LI and overshadowing counteract each other.

Recovery from overshadowing

In Chapter 3, we mentioned that the competitive rule in the SLG model describes overshadowing (Pavlov, 1927) and relative validity (Wagner *et al.*, 1968). Here we describe recovery from overshadowing, a phenomenon that the model explains in attentional terms.

138   Attentional and associative mechanisms

Experimental data

Matzel *et al.* (1985) reported that extinction of the overshadowing CS results in the recovery of the response to the overshadowed CS. In Experiment 2, rats in the OVER group received four presentations of a tone and a flashing light together with a footshock. Rats in the acquisition (OVER-CONTROL) group received four presentations of a flashing light together with the footshock, interspersed with four presentations of a flashing light alone. Finally, rats in the EXTINCTION group received the same treatment of the OVER group. After this initial training, group EXTINCTION received 18 nonreinforced presentations of the tone, whereas groups OVER and OVER-CONTROL stayed in the conditioning chambers for an equivalent length of time.

The upper panel of Figure 8.1 shows the experimental results expressed as the mean latency to complete 25 licks in the presence of the CS. Because the US is an aversive shock, a stronger CR is reflected in a longer latency. The OVER group shows overshadowing because its response is smaller than the response of the OVER-CONTROL group, but this effect decreases in the EXTINCTION group.

Simulated results

In the OVER case, CS_1 and CS_2 were presented for 20 trials, followed by 50 nonreinforced presentations in the CX. In the OVER-CONTROL case, CS_1 was presented with the US for 20 trials, followed by 50 nonreinforced presentations of the CX. In the EXTINCTION case, CS_1 and CS_2 were presented for 20 trials, followed by 50 presentations of CS_2.

The lower panel of Figure 8.1 shows the simulated results. As in the experimental case, the extinction of the conditioning context results in the elimination of overshadowing. A correlation test ($r = 0.99$, $df = 1$, $p < 0.05$) shows that the simulated results approximate the experimental data well.

Figure 8.2 shows the trial-by-trial values of the relevant variables in the model for the OVER group (upper panel) and the EXTINCTION group (lower panel). During the first phase of training (Trials 1–20), Novelty, and attention to CS_1 and CS_2, first increase and then decrease. During these trials, CS_1–US and CS_2–US associations, which compete to predict the US, increase. During the second phase of the experiment (Trials 21–70), Novelty decreases rapidly in the overshadowing group (upper panel). In the extinction group, because CS_2 and the US are predicted by CS_1 and the context but now are absent, Novelty, and attention to the present CS, z_{CS_1}, first increase and decrease. Attention to the absent CS, z_{CS_2}, increases until the CS_1–CS_2 associations are extinguished and then remains constant. The CS_1–US association and CR strength strongly decrease. The CS_2–US association extinguishes slightly.

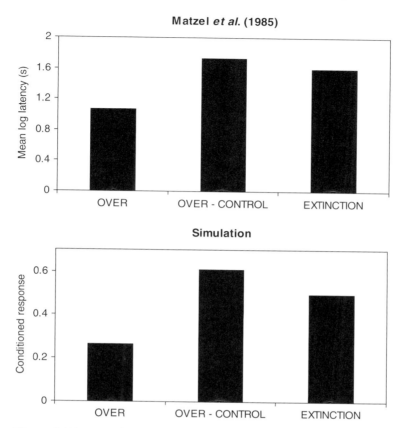

Figure 8.1 Recovery from overshadowing. *Upper panel*: Data from Matzel et al., 1985, Figure 2. Mean latencies to complete the first 5 cumulative seconds of drinking in the test context. *Lower panel*: Simulation. Strength of the conditioned response (r = 0.99, df = 1, p < 0.05).

Because attention to CS_2, z_{CS2}, is much larger in the EXTINCTION case than in the OVER case, the CR in the EXTINCTION group is larger than in the OVER group and similar to that of the OVER-CONTROL group. Therefore, overshadowing is not manifested.

Comparison with other models

According to the comparator model (Miller & Schachtman, 1985; Miller & Matzel, 1988; Denniston, Savastano & Miller, 2001), overshadowing is the consequence of strong CS_1-CS_2 and CS_2-US associations, which increase the indirect representation of the US, thereby attenuating responding to CS_1. Extinction of CS_1 following training results in a decreased CS_1-CS_2 association and the attenuation of overshadowing because it reduces the indirect representation of the US.

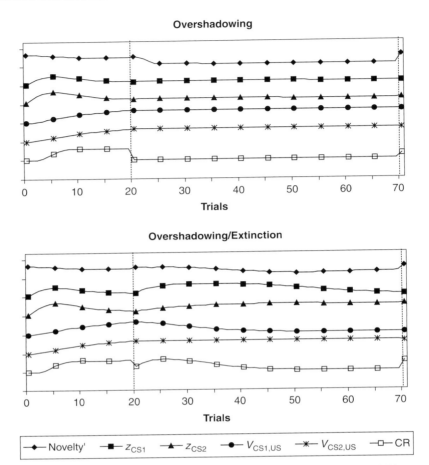

Figure 8.2 Recovery from overshadowing. Values of Novelty'; z of CS_1 and CS_2, z_{CS1}, z_{CS2}; CS_1-US association, $V_{CS1\text{-}US}$; CS_2-US association, $V_{CS2\text{-}US}$; and conditioned response, CR; at the time unit preceding the presentation of the US, as a function of Trials. CS_1 and CS_2 conditioning Trials 1–20, CS_1 extinction Trials 21–70, and test Trial 71. Notice that the CR is larger in the Overshadowing/Extinction than in the Overshadowing/Control condition.

According to the Van Hamme and Wasserman (1994) version of the Rescorla and Wagner (1972) model, extinction of CS_1 in the case of overshadowing produces an increase in the CS_2-US association, $V_{CS2,US}$, $\Delta V_{CS2,US} = K\alpha(\lambda - \Sigma V) > 0$, because both $\alpha < 0$ and $(\lambda - \Sigma V) = (0 - \Sigma V) < 0$, and therefore $\Delta V_{CS2,US} > 0$. It should be noted in a control group, one in which the animal is presented in the context alone, it is also $\Delta V_{CS2,US} > 0$. However, because ΣV is larger when both CS_1 and the context are active than when only the context is present, $\Delta V_{CS2,US}$ is larger when CS_1 is extinguished than when the context is extinguished.

According to the Dickinson and Burke (1996) version of Wagner's (1981) SOP model, overshadowing is explained in the following terms. The A_2 representation of the US is increased by both CSs. This increment in A_2 decreases the A_1 representation of the US, thereby limiting the growth of the association of the US with CS_1 and CS_2. Extinction of CS_1 activates the A_2 representations of both CS_2 and the US. Because Dickinson and Burke assume that associations between two stimuli in the A_2 state increase, the CS_2–US association will increase and CS_2 will recover from overshadowing.

Latent inhibition and overshadowing counteract each other

Blaisdell et al. (1998) showed that combined LI and overshadowing treatments attenuate the decrease in responding produced by each procedure alone.

Experimental results

The LI followed by overshadowing (PRE + OVER) group received 120 presentations of stimulus X (either a tone or a white noise), followed by 10 reinforced presentations of stimuli X and Y (either the white noise or a tone). The OVER group received 10 reinforced presentations of stimuli X and Y. The PRE group received 120 presentations of X, followed by 10 reinforced presentations of X. The acquisition (ACQ) group received 120 presentations of Y, followed by 10 reinforced presentations of X.

The upper panel in Figure 8.3 shows the mean time to complete 5 cumulative seconds of licking in the presence of the target conditioned stimulus; a measure of its level of conditioning. The results show that the preexposure and overshadow procedures produce more conditioning when given together than when separated.

Simulated results

The PRE + OVER case received 20 presentations of CS_1 followed by 8 reinforced presentations of CS_1 and CS_2. The OVER case received 8 reinforced presentations of CS_1 and CS_2. The PRE case received 20 presentations of CS_1, followed by 8 reinforced presentations of CS_1. The ACQ case received 20 presentations of CS_1, followed by 8 reinforced presentations of CS_2.

The lower panel of Figure 8.3 shows the simulated results. As in the experimental case, preexposure and overshadow procedures produce more conditioning when given together than when given apart. A correlation test ($r = 0.98$, df = 2, $p < 0.05$) shows that the simulated results approximate the experimental data well.

142 Attentional and associative mechanisms

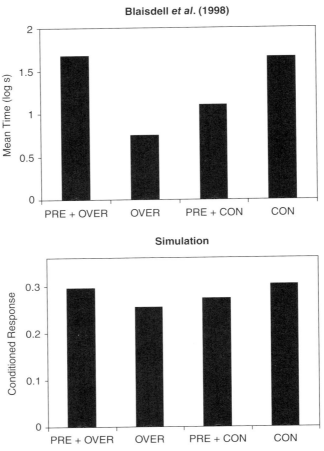

Figure 8.3 Latent inhibition and overshadowing counteract each other. *Upper panel*: Data from Blaisdell *et al.* (1998, Figure 5). Mean latencies to complete the first 5 cumulative seconds of drinking in the test context. *Lower panel*: (Simulation). Strength of the conditioned response (r = 0.98, df = 2, p < 0.05).

Figure 8.4 shows the trial-by-trial values of the relevant variables in the model for the Preexposure and Overshadowing (Panel A), Overshadowing (Panel B), Preexposure and Conditioning (PRE, Panel C), and Conditioning (Panel D) groups. The relationships between the CSs of the different groups are explained as follows.

Responding in the conditioning (CON) group is stronger than in the preconditioning and conditioning (PRE + CON) group

In the case of the PRE + CON (i.e. LI) group, CS_1–CS_1 and CS_1–CX associations are formed during preexposure. These associations make the value

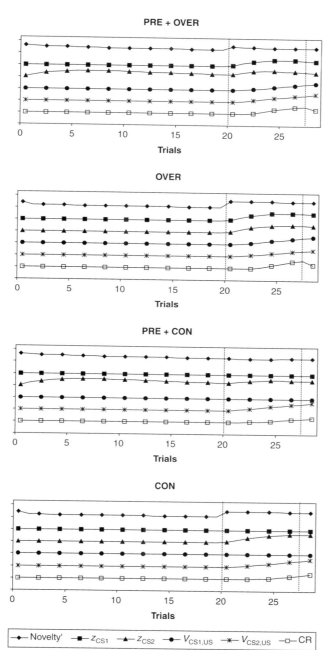

Figure 8.4 Latent inhibition and overshadowing counteract each other. Values of Novelty'; z of CS_1 and CS_2, z_{CS1}, z_{CS2}; CS_1-US and CS_2-US associations, V_{CS1-US} and V_{CS2-US}; and conditioned response, CR; at the time unit preceding the presentation of the US, as a function of Trials. CS_1 preexposure Trials 1-20, CS_1 and CS_2 conditioning Trials 21-28, and test Trial 29.

144   Attentional and associative mechanisms

of Novelty′ in the preexposed group smaller than the value of Novelty′ in the CON group during conditioning (see Figure 8.4), which results in a smaller z_{CS1}, the retardation of the formation of the CS_1-US association, and a smaller CR than that of the CON group.

Responding in the conditioning (CON) group is stronger than in the overshadowing (OVER) group

In the model, overshadowing is the result of the competition between CS_1 and CS_2 to gain association with the US, which reduces the CS–US association and the magnitude of the CR, relative to those of the CON case (see Figure 8.4).

Responding in the preexposure and overshadowing (PRE + OVER) group is stronger than in the preexposure and conditioning (PRE + CON) group

Responding in the PRE + OVER case increases because the added introduction of CS_2 during conditioning increases Novelty′ and attention to CS, z_{CS1}, consequently increasing the CS_1-US association and the magnitude of the CR relative to that of the PRE + CON group. Furthermore, CS_1 testing in the absence of CS_2 and the US, increases Novelty′ more in the PRE + OVER group than in the PRE + CON group, further increasing z_{CS1} and the relative advantage of the CR in the PRE + OVER group (see Figure 8.4).

Responding in the preexposure and overshadowing (PRE + OVER) group is stronger than in the overshadowing (OVER) group

In the case of the PRE + OVER group, CS_1–CS_1 and CS_1–CX associations are formed during preexposure. As mentioned above, these associations make the value of Novelty′ in the preexposed group smaller than its value in the nonpreeexposed group during conditioning, thereby reducing z_{CS1}, retarding the formation of the CS_1-US association, and decreasing the magnitude of the CR. Even if Novelty′ first increases and then decreases during preexposure (see Figure 8.4), at the beginning of conditioning, attention to CS_1, z_{CS1} is larger in the PRE + OVER group than in the OVER group. Therefore, CS_1 has an higher initial z_{CS1} in the PRE + OVER than in the OVER group, which results in a larger CR in the PRE + OVER group.

According to the model, the results described in this section can be obtained in a limited region of parameters. For example, too many preexposure trials will reduce z_{CS1} and the PRE + OVER group will not generate a CR stronger than that of the OVER group. Also, too few conditioning trials will not yield enough overshadowing or latent inhibition, and too many conditioning trials will not yield latent inhibition.

Comparison with other models

These results cannot be explained by Van Hamme and Wasserman's (1994) theory because the model cannot address LI. Nor can they be explained by the Dickinson and Burke (1996) model because (a) the representation of CS_1 is decreased by its activation in the A_2 state by being predicted by the context, and (b) the representation of the US in the A_1 state is decreased by its activation in the A_2 state by both CS_1 and CS_2. Therefore, the CR generated by group PRE + OVER will be smaller than those in the PRE + CON and the OVER groups.

In contrast, Blaisdell *et al.* (1998) argued that increased responding to CS_1 is due to the decreased effective activation of the CX–US link, caused by the interference from the strong CS_2–US association formed during overshadowing. Consequently, during testing the context, instead of CS_2, it is activated by CS_1 and serves as the comparator stimulus. Therefore, preexposure should result in less overshadowing. Savastano, Arcediano, Stout and Miller (2003) showed how a mathematical version of the extended hypothesis addresses the case in which preexposure and overshadowing cancel each other.

Attenuation of blocking

As mentioned, blocking refers to the decreased CR to a CS, when that CS is conditioned to a US in the presence of another CS that was previously conditioned to the same US (Kamin, 1968, 1969a,1969b; Moore & Schmajuk, 2008). In addition to blocking of excitatory conditioning, represented by A+/AB+, Suiter and LoLordo (1971) reported blocking of inhibitory conditioning, described by C+/CA–/CAB–. Some aspects of adequate control groups for blocking were recently addressed Taylor *et al.*'s (2008; but see Guez & Miller, 2008) studies in rats. The competitive rule in the SLG model, embedded in Equation [2.9a], describes blocking of both excitatory and inhibitory conditioning.

Manipulations that reduce blocking have also been studied. For instance, Dickinson, Hall and Mackintosh (1976) demonstrated that blocking was attenuated both by (a) the addition of an unexpected second shock, and (b) the omission of an expected shock. Schmajuk (1997, page 111) showed that the SLG model can explain both results. Whereas the addition of a second shock is handled well by the Rescorla-Wagner (1972) model because increasing the US increases the error term ($\lambda - \Sigma V$), the SLG and the Pearce-Hall (1980) models can describe both results in terms of the increment in attention to both CSs when the expected US is not presented.

More recently, Witnauer, Urcelay and Miller (2008) showed that blocking was attenuated when two (CS_A and CS_B) instead of one (CS_A), blocker CSs were presented with CS_X during training. Computer simulations show that the SLG model can approximate these results in terms of the increased Novelty' (attention to CS_X and responding) as a consequence of the absence of CS_A, CS_B, and the US during testing in the experimental group, but only of CS_A and the US during testing in the control group.

Recovery from blocking by extinction of the blocker

Blaisdell, Gunther and Miller (1999) found that extinction of the blocking CS results in the recovery of the response to the blocked CS. However, Miller, Schachtman and Matzel (1988) and Holland (1999) reported a failure to obtain recovery.

In this section we offer simulations for the Blaisdell *et al.* (1999) and Holland (1999) experiments, and try to solve the apparent contradiction between their respective data. Furthermore, Williams (1996) reported that, contrary to the prediction of the comparator hypothesis, a blocked CS was incapable of blocking another CS, which suggests that the blocked CS–US association was actually decreased during the first blocking procedure.

Experimental data 1

As mentioned, Blaisdell *et al.* (1999) reported that extinction of the blocking CS results in the recovery of the response to the blocked CS. In Experiment 3, rats in the BLOCKING groups received twelve reinforced presentations of a tone (or a white noise) followed by four reinforced presentations of the same CS (the tone or the white noise) and a click train. Importantly, the tone was less salient than the click (6 dB vs. 8 dB above background noise). Rats in the OVERSHADOWING groups received twelve reinforced presentations of a tone or a white noise, followed by four reinforced presentations of a different CS (the white noise when the first reinforced CS was a tone, and a tone when the first CS was the white noise) and a click train. The BLOCKING–EXTINCTION and OVERSHADOWING–EXTINCTION groups received the treatments indicated above followed by 800 nonreinforced presentations of the white noise or the tone in a different context. The BLOCKING–CONTROL and OVERSHADOWING–CONTROL groups received the treatments indicated above, followed by equivalent exposure to a different context. Testing was carried out in the extinction context.

The upper panel of Figure 8.5 shows the experimental results expressed as the mean latency to complete 5 cumulative seconds of licking in the presence of the click train. As before, because the US is an aversive shock, a stronger CR is reflected in a longer latency. The BLOCKING–CONTROL group shows blocking because its CR is smaller than that of the OVERSHADOWING–CONTROL group. Instead, blocking is absent in the EXTINCTION groups.

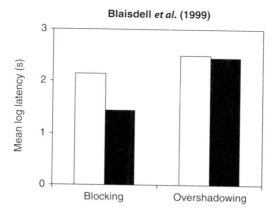
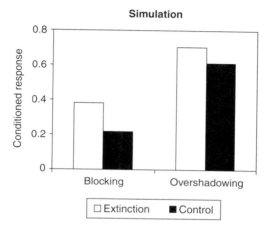

Figure 8.5 Recovery from blocking. *Upper panel*: Data from Blaisdell et al. (1999, Figure 4). Mean latencies to complete the first 5 cumulative seconds of drinking in the presence of the blocked stimulus. *Lower panel*: Simulation. Strength of the conditioned response to the blocked stimulus (r = 0.94 , df = 2, p = 0.057).

Simulated results 1

In the BLOCKING case, CS_1 was presented with the US for 100 trials, and CS_1 and CS_2 were presented with the US for another 20 trials. In the OVERSHADOWING case, the US was presented (with and without CS_3) for 100 trials, and CS_1 and CS_2 were presented with the US for another 20 trials. In addition, both EXTINCTION groups received 70 nonreinforced trials with CS_1 in a different context. The CONTROL groups received 70 nonreinforced trials in a different context. The intensity of CS_1 was 0.5 and the intensity of CS_2 was 1.

The lower panel of Figure 8.5 shows the simulated results. As in the experimental case, the extinction of the blocking CS results in a decreased blocking.

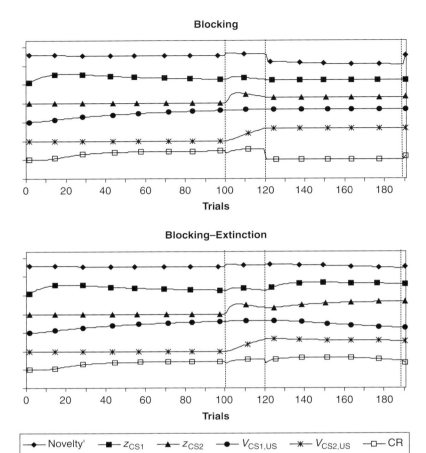

Figure 8.6 Recovery from blocking. Values of Novelty'; z of CS_1 and CS_2, z_{CS1}, z_{CS2}; CS_1-US association, V_{CS1-US}; CS_2-US association, V_{CS2-US}; and conditioned response, CR; at the time unit preceding the presentation of the US, as a function of Trials. CS_1 conditioning Trials 1–100, CS_1 and CS_2 conditioning Trials 101–120, CS_1 extinction Trials 121–190, and test Trial 191. Notice that the CR is larger in the Blocking/Extinction than in the Blocking/Control condition.

A correlation test (r = 0.94, df = 2, p = 0.057) shows that the simulated results approximate the experimental data well. Similar results were obtained when the same context was used in all phases of the simulation, and when CS_3 was not used in the OVERSHADOWING case.

Figure 8.6 shows the trial-by-trial values of the relevant variables in the model for the BLOCKING group (upper panel) and the BLOCKING–EXTINCTION group (lower panel). During the first phase of training in both groups (Trials 1–100), Novelty and attention to CS_1 first increase and then decrease. During these trials, the CS_1–US association and the CR increase. During the second phase of the experiment (Trials 101–120), Novelty briefly increases and then decreases, while the same happens with attention to CS_1 and CS_2.

During the third part of the experiment, in the BLOCKING–CONTROL group (upper panel), Novelty rapidly decreases in the new context, and attention to CS_1 and CS_2 do not change. In the BLOCKING–EXTINCTION group (lower panel), Novelty increases because the CS_2 and the training context are predicted by CS_1, but are absent. Attention to CS_1 and CS_2 also increase. Finally, during the test trial, because of its increased z_{CS2}, CS_2 yields a large CR.

Because attention to CS_2, z_{CS2}, is much larger in the BLOCKING–EXTINCTION group than in the BLOCKING–CONTROL group, the CR in the first group is larger than in the latter group, and similar to that of the OVERSHADOWING group. Therefore, blocking is not manifested.

Extending the Blaisdell et al. (1999) results showing that extinction of the blocking CS results in the recovery of the response to the blocked CS, Piñeño et al. (2006) showed that blocking was also attenuated by extensive training of the blocker–US association following the compound–US conditioning phase. The SLG model describes this result in attentional terms similar to those applied for the Blaisdell et al. experiment.

Experimental data 2

In contrast to Blaisdell et al.'s (1999) and Matzel et al.'s (1985) results, Holland (1999) reported that extinction of the blocking or the overshadowing CS does not result in the recovery of the response to the blocked or overshadowed CS. Data also showing that extinction does not always cause unblocking, which undermines the notion that blocking is a retrieval deficit, was also offered by Rauhut, McPhee, DiPietro and Ayres (2000) and McPhee, Rauhut and Ayres (2001).

In Holland's (1999) Experiment 6, rats in the BLOCKING groups received 64 reinforced presentations of A (a noise or a light) followed by 32 reinforced presentations of A and X (a light or a noise). Rats in the OVERSHADOWING groups received 64 reinforced presentations in the context, followed by 32 reinforced presentations of A and X. Rats in the ACQUISITION groups received 64 reinforced presentations in the context, followed by 32 reinforced presentations of X. Following each treatment, the EXTINCTION groups received 512 nonreinforced presentations of A, whereas the CONTROL groups received equivalent exposure to the same context. Testing consisted of 8 averaged presentations of X.

The upper panel of Figure 8.7 shows the experimental results expressed as food cup behavior during the house light X CS presentations. Because the US is an appetitive US, a stronger CR is reflected in stronger responding. There are

150 Attentional and associative mechanisms

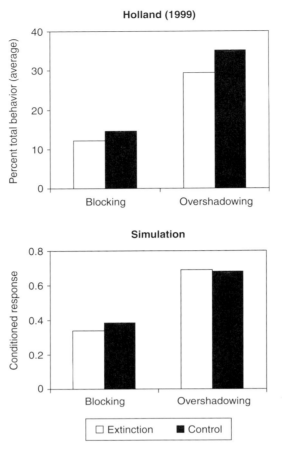

Figure 8.7 Absence of recovery from blocking. *Upper panel*: Data from Holland, (1999, Figure 6). Average food-cup behavior during the presentation of the house light or noise. *Lower panel*: Simulation. Strength of the conditioned response to the blocked stimulus includes the OR term (r = 0.97 df = 2, p < 0.05).

no differences between CONTROL and EXTINCTION groups for the BLOCKING and OVERSHADOWING cases.

Simulated results 2

Blaisdell et al.'s (1999) Experiment 3 and Holland's (1999) Experiment 6 differ in many regards, including the preparation (conditioned emotional response and appetitive, respectively), the type of responses (water licking vs. headjerking to the tone CS and rearing to the light CS), the type of US (footshock vs. food), the type and intensity of the CS (82 dB tone and 84 dB clicker vs. 78 dB white noise and 6 W light), the intensity of the background noise (76 dB vs. 70 dB), the duration of the intertrial interval (12 min vs. 8 min), and the number of trials in each phase of the experiment (12, 4 and 800 vs. 64, 32 and 512). An additional difference is that, whereas Blaisdell *et al.* used a different context for the extinction and testing phases, Holland employed the same context in all phases. In an attempt to identify the parameters responsible for the differences in the reported results, we simulated Holland's (1999) Experiment 6 using parameters identical to those used to simulate Blaisdell *et al.*'s Experiment 3, with the exception of (a) all the phases were run in the same context (instead of the different context used for extinction and testing), (b) a shorter intertrial interval was used (800 vs. 2000 time units), and (c) 8 extinction trials were used.

The lower panel of Figure 8.7 shows the simulated results. Extinction of CS_1 has no effect on the difference between the blocked and overshadowed groups. A correlation test ($r = 0.97$ df = 2, $p < 0.05$) suggests the simulated results approximate the experimental data well.

Figure 8.8 shows the trial-by-trial values of the relevant variables in the model for the BLOCKING–CONTROL and BLOCKING–EXTINCTION groups. Figure 8.8 shows that attention to CS_1 and CS_2 increases both in the CONTROL and EXTINCTION groups, thereby decreasing the difference between their CRs, as shown in Figure 8.7. Because attention to CS_1 and CS_2 increase in the CONTROL groups in the Holland simulation, recovery from blocking (and overshadowing) is absent or weak in the Holland case. This increase in attention in the CONTROL groups is due to two factors. One, the training context alone activates the representations of CS_1 and CS_2, even in the absence of CS_1, and cause them to become associated with the increased Novelty detected in the context in the absence of those two CSs. Two, CX–CS_1 and CX–CS_2 associations are relatively strong because the ITI is short and, consequently, there is less time to extinguish them when the CX is presented by itself, in the absence of CS_1 and CS_2.

In conclusion, the model suggests that two key differences between the Holland (1999) and the Blaisdell *et al.* (1999) experiments, (a) using the same or different contexts for acquisition and extinction, and (b) using different intertrial intervals (8 min vs. 12 min), are responsible for the different reported results.

Experimental data 3

Although the attenuation of blocking by extinction of the blocker seems to support the "limited" comparator hypothesis (Miller & Schachtman, 1985) view that blocking is a performance deficit, we showed that those results are well explained by the attentional mechanism acting during performance in the SLG model. The question of whether a storage deficit is also at work in blocking, as assumed by the SLG model and several others, was addressed by Williams (1996). He used a discriminated operant task in which food reinforcement

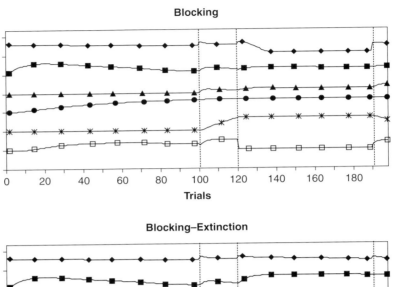

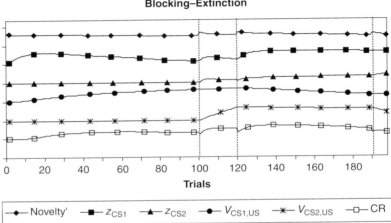

Figure 8.8 Absence of recovery from blocking. Values of Novelty'; z of CS_1 and CS_2, z_{CS1}, z_{CS2}; CS_1-US association, V_{CS1-US}; CS_2-US association, V_{CS2-US}; and conditioned response, CR; at the time unit preceding the presentation of the US, as a function of Trials. CS_1 conditioning Trials 1–100, CS_1 and CS_2 conditioning Trials 101–120, CS_1 extinction Trials 121–190, and test Trials 191–198. Notice that the CRs are more similar in the Blocking/Extinction than in the Blocking/Control condition.

was presented only in the presence of certain stimuli. In Phase 1, rats in the BLOCKING group received 180 reinforced presentations with the noise as discriminant stimulus. In Phase 2, rats received 150 reinforced presentations of noise plus the houselight. Following testing to extinction of each stimuli, animals were retrained with 30 presentations of the stimuli used in Phases 1 and 2. Then, rats received 150 presentations of the houselight and a tone. Then the houselight and the tone were tested without reinforcement, until extinction was achieved. Rats in the CONTROL group received training similar to that of

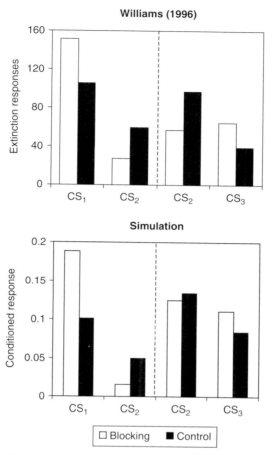

Figure 8.9 Blocking of blocking. *Upper panel*: Data from Williams (1996, Figure 1). Mean numbers of responses to each CS during the test sessions. The left side of the panel shows the results at the end of Phase 2, in which the CSs were a noise (CS_1) and the house-light (CS_2). The right side of the panel shows the results at the end of Phase 3, in which the CSs were the house-light (CS_2) and a tone (CS_3). *Lower panel*: Simulation. Strengths of the conditioned response to each CS, including the OR. The left side of the panel shows the results at the end of Phase 2, for CS_1 and CS_2. The right side of the panel shows the results at the end of Phase 3, for CS_2 and CS_3. Correlation coefficient ($r = 0.82$, $df = 6$, $p < 0.05$).

the BLOCKING group, except that they received 180 reinforced presentations with a clicker as discriminant stimulus, in Phase 1.

The upper panel in Figure 8.9 shows the number of responses to extinction; a measure of the level of conditioning, for the noise, the houselight and the tone. During Phase 2, the noise blocks the houselight in the BLOCKING group, and overshadows it in the CONTROL group. During Phase 3, the houselight blocks the tone in the CONTROL group, but not in the BLOCKING group. The results suggest that the houselight has acquired a weaker association with the US after being blocked by the noise, than after being overshadowed by the tone.

Simulated results 3

In the BLOCKING case, CS_1 was presented with the US for 10 trials, then CS_1 and CS_2 were presented with the US for 10 trials, followed by CS_2 and CS_3 presented with the US for 10 trials. In the CONTROL case, the US was presented by itself for 10 trials, then CS_1 and CS_2 were presented with the US for 10 trials, followed by CS_2 and CS_3 presented with the US for 10 trials.

The lower panel of Figure 8.9 shows the simulated results. As in the experimental case, during Phase 2, CS_1 blocks in the BLOCKING group, and overshadows it in the CONTROL group. During Phase 3, CS_2 blocks CS_3 in the CONTROL group, but not in the BLOCKING group. Our numeric results show that CS_2 acquired a smaller association with the US after being blocked by CS_1 than after being overshadowed by CS_3. A correlation test (r = 0.82, df = 6, p < 0.05) shows that the simulated results approximate the experimental data well.

The explanation for Williams' (1996) results is based on the competition between CSs to become associated with the US, a property that the SLG network shares with the Rescorla–Wagner (1972) model. Simply put, the blocked CS_2, because of its relative weak association with the US, cannot block CS_3.

Comparison with other models

Blaisdell et al. (1999) suggested that their results show that, as with LI, blocking might not be a failure to acquire a CS–US association. According to the comparator hypothesis, blocking is the consequence of a strong CS_1–US association, acquired during both phases of the experiment, which makes the indirect representation of the US (CS_2–CS_1 and CS_1–US) stronger than the direct one (CS_2–US), therefore attenuating responding to CS_2. Extinction of CS_1 results in a decreased CS_1–US association and the recovery from blocking, because it reduces the indirect representation of the US.

According to the SLG model, recovery from blocking in the Blaisdell et al. experiment can be explained in terms of the increased attention to the blocked CS when the blocker CS is extinguished. This increased attention yields a larger CR and the recovery from blocking. Also, the lack of recovery from blocking and overshadowing in the Holland (1999) experiment is the consequence of using the same acquisition and extinction contexts, as well as a shorter ITI. These parameters result in an increased attention to the blocked/overshadowed CS even when the blocker/overshadowing CS is not extinguished. This change minimizes the difference between control and extinction groups, and the recovery effects disappear.

According to Williams (1996), his results cannot be explained by the comparator hypothesis (Miller & Schachtman, 1985), because the houselight–US associations formed in Phase 2 should be similar in both groups and, therefore, responding to the houselight in Phase 3 should also be similar. Williams (1996) indicated that his results can be explained by a model (e.g. Rescorla & Wagner, 1972) that assumes that the houselight–US association is smaller in the BLOCKING than in the CONTROL group. In the SLG model, houselight–US associations are also controlled by a real-time competition rule, which, like the Rescorla–Wagner model, also expects that houselight–US associations should be smaller in the BLOCKING than in the CONTROL group. However, Holland (1999) proposed that it is possible for a modified version of the comparator hypothesis to address the Williams's (1996) result, by assuming that the comparator process "governs not only response performance but also other expressions of a cue's associative strength, including the ability to block." According to Denniston et al. (2001), Williams's (1996) results can be explained by the extended comparator hypothesis. In very simple terms, responding to Y (the tone) is strong in the BLOCKING group because X (the houselight) fails to modulate conditioned responding to Y (the tone) as a result of A (the noise) having being established as a better predictor of the US than X (the houselight).

According to the Van Hamme and Wasserman (1994) version of the Rescorla and Wagner (1972) model, extinction of the blocker CS in the case of blocking produces an increase in the blocked CS–US association, $V_{CS,US}$, $\Delta V_{CS,US} = -KCS(\lambda - \Sigma V) > 0$, because both $-K < 0$ and $(\lambda - \Sigma V) = (0 - \Sigma V) < 0$, and therefore $\Delta V_{CS,US} > 0$. As in the case of overshadowing, the extinction of the blocker CS produces more recovery than the extinction of the context.

According to the Dickinson and Burke (1996) version of Wagner's (1981) SOP model, blocking is explained in the following terms. The A_2 representation of the US is increased by the blocker CS. This increment in A_2 decreases the A_1 representation of the US, thereby limiting the growth of the association of the US with the blocked CSs. Extinction of the blocker CS activates the A_2 representations of both CS_2 and the US, thereby increasing the CS_2–US association, and CS_2 will recover from blocking.

Backward blocking and recovery from backward blocking

Whereas in blocking, CS_1–US training precedes CS_1-CS_2–US training, in backward blocking CS_1-CS_2–US trials precede CS_1–US trials. First reported in causal judgments in humans (Shanks, 1985), Miller and Matute (1996) demonstrated that, under special conditions, backward blocking could be also obtained in animals. In addition, Pineño et al. (2005) demonstrated spontaneous recovery from backward blocking following a retention interval.

Experimental data

Miller and Matute (1996) hypothesized that failures to obtain backward blocking in animals were due to the fact that "cues that are of high biological significance are resistant to reductions in their biological significance and consequently relatively immune to cue competition." Therefore, their Experiment 2 includes sensory preconditioning in the two first phases of the experiment. The backward blocking (BB) group received 4 presentations of *AX* followed by *B*, 4 presentations of *A* followed by *B*, and 4 presentations of *B* followed by a footshock US (B becomes a surrogate US). One control group received 4 presentations of *A* followed by *C* during Phase 2. Another control group just remained in the training cage during Phase 2. In the experiment, *A* was a buzzer, *X* a click train, and *B* and *C* counterbalanced tone and white noise.

Using a similar design, but a different number of trials in Phase 2, Pineño et al. (2005) demonstrated spontaneous recovery from BB following a retention interval. In this case, the BB group received 4 presentations of *AX* followed by *O*, 20 presentations of *A* followed by *O*, and 4 presentations of *O* followed by a footshock US (*B* becomes a surrogate US). The CONTROL group received 20 presentations of *B* followed by *O* either before or after receiving 4 presentations of *AX* followed by *O*, and 4 presentations of *O* followed by a footshock US afterwards. Testing to *A* and *X* was preceded by no delay or a 15-day delay. Subjects in the DELAY condition were handled 3 times per week for 30 seconds and were maintained on the water deprivation schedule. The results show response recovery to the blocked stimulus after the delay period.

The upper panel in Figure 8.10 shows that the presentation of *A* followed by *O* after the presentation *AX* followed by *O* (BB group), decreased responding to *X* compared to the case when *B* was followed by *O* (CONTROL group) in the non-delay case. In the DELAY case, however, responding to *X* was similar in the BB and CONTROL groups.

Simulated results

Computer simulations show that the SLG model can reproduce the Miller and Matute (1996) Experiment 2 results. According to the model, responding in the BB group was weaker (CRX = 0.0002) than in both CONTROL groups (CRX = 0.0008)

The lower panel in Figure 8.10 shows simulated results for the Pineño et al. (2005) experiment. Columns represent responding to *X* on a test trial following

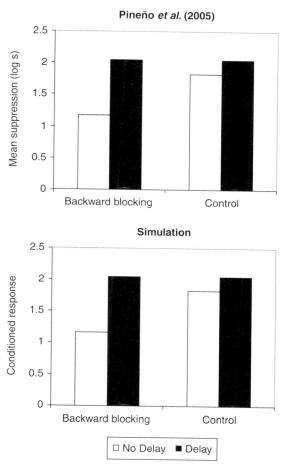

Figure 8.10 Backward blocking. *Upper panel*: Data from Pineño et al. (2005, Figure 3). Mean latencies to complete the first 5 cumulative seconds of drinking in the presence of the backward blocked stimulus. *Lower panel*: Simulated results showing the strength of the conditioned response to X during 1 test trial, following 100 X–A–O trials, 20 A–O trials, 150 O–US trials, and 20 delay trials. Simulation parameters were X: 30 to 40 time units, salience 1; A, B: salience 0.6, O: salience 0.6, US: 35 to 40 time units, strength 1; Context saliences 0.1/0.05/0.05; Intertrial interval 600 time units.

100 A–X–O trials in Context 1, 20 A–O trials or B–O trials in Context 1, 150 O–US trials in Context 1, and 20 delay trials in Context 2, which generalizes to Context 1 (see section on simulation procedures in Chapter 9). The salience of stimulus A (representing the 80 dB tone or the 80 dB white noise) was 0.6, the salience of stimulus X (representing the 80 dB clicker) was 1, and the salience of O (a 150 W flashing light) was 0.6. When presented together, A, X, B, O and the US temporally overlap. A correlation test ($r = 0.98$ df = 2, $p < 0.05$) suggests the simulated results approximate the experimental data well.

Pineño et al. (2005) also established that the A–O association did not decrease during delay; a result also shown by our model. Similar results are obtained when the number of trials is kept constant for three phases (Phase 1, 100 trials; Phase 2, 20 trials; Phase 3, 150 trials; Phase 4, 20 trials) and one of the other phases is varied within the following ranges: Phase 1 from 65 to 117 trials, Phase 2 from 18 to 32 trials, Phase 3 from 116 to over 600 trials, and Phase 4 from 2 to over 600 trials.

Acquisition of backward blocking

BB is explained in the following terms. During the A–X–O presentations on Phase 1 (Trials 1–100), both the CONTROL and the BB groups build $V_{A,A}$, $V_{A,O}$, $V_{A,X}$, $V_{X,X}$, $V_{X,O}$, $V_{X,A}$, $V_{O,O}$, $V_{O,A}$ and $V_{O,X}$ associations. On Phase 2 (Trials 101–120), the BB group experiences A–O trials, and Novelty' increases because both A and O predict X (through $V_{A,X}$ and $V_{O,X}$ associations) and X is absent. In contrast, on Phase 2, the CONTROL group experiences B–O trials, and Novelty' increases more than in the BB group because B is not predicted and O predicts A and X (through $V_{O,A}$ and $V_{O,X}$ associations) and A and X are absent. In both groups, X is predicted by O through $V_{O,X}$ associations and, because Novelty' is greater in the control than in the BB group, attention to X, z_X, increases more in the control group than in the BB group. On Phase 3 (Trials 121–270), both groups receive O–US trials, and build the O–US association, $V_{O,US}$. According to the model, the CR to X is proportional to (a) the attention to X, z_X, (b) the X–O association, $V_{X,O}$, and (c) the $V_{O,US}$ association (CR ~ $z_X V_{X,O} z_O V_{O,US}$). Because z_X is greater in the CONTROL group than in the BB group, z_X(CONTROL) > z_X(BB), backward blocking is present (CR(CONTROL) > CR(BB)).

Just as with backward blocking, the model is also able to show forward blocking using preconditioning of A and X with conditioned stimulus O. As mentioned in the previous section, according to the SLG model, in forward blocking two mechanisms are at play: the BLOCKED CS (a) gains a smaller association with the outcome O than the CONTROL (overshadowed) CS, and (b) accrues less attention than the CONTROL CS. In contrast, in backward blocking, (a) although BLOCKED and CONTROL (overshadowed) CSs gain the same association with the outcome O, (b) the BLOCKED CS accrues less attention than the CONTROL (overshadowed) CS. In both cases, however, responding to the BLOCKED CS is weaker than responding to the CONTROL (overshadowed) CS.

Recovery from backward blocking

Recovery from BB is explained as follows. On Phase 3 (Trials 121–270), both groups receive O–US trials. We will analyze first what happens at the beginning of this phase. In the BB group, Novelty′ increases because O predicts A and X (through $V_{O,A}$, and $V_{O,X}$ associations), and they are absent. In contrast, in the control group, Novelty′ increases more than in the BB group because O predicts A, X and B (through $V_{O,A}$, $V_{O,X}$ and $V_{O,B}$ associations), but they are absent. So, at the beginning of this phase there is more Novelty′ in the CONTROL than in the BB group. Because Novelty′ is initially larger, O–US learning in the CONTROL group is faster than in the BB group, and Novelty′ decreases faster too. So, by the end of Phase 3, attention to the Context 1, z_{CX1}, is smaller in the control than in the BB group.

During the delay period, in the home cage context, CX_2, we assumed some generalization to CX_1. In this situation, CX_1 predicts all the stimuli proportionally to z_{CX1}, which, as explained above, is smaller in the CONTROL than in the BB group. Then, because the predictions of all stimuli in the CONTROL group are smaller than those predictions in the BB group, in the absence of all those stimuli, Novelty′ is smaller in the CONTROL than in the BB group. Therefore, since CX_1 predicts O through $V_{CX1,O}$ associations, the prediction of O, B_O, becomes associated with Novelty′, and attention z_O will also be smaller in the CONTROL than in the BB group ($z_O(BB) > z_O(CONTROL)$). Remember that, as explained above, CR~ $z_X V_{X,O} z_O V_{O,US}$. Because z_X is greater in the CONTROL group than in the BB group, $z_X(CONTROL) > z_X(BB)$, backward blocking is present before the delay (CR(CONTROL) > CR(BB)). After the delay, however, the larger z_O in the BB group than in the CONTROL group, $z_O(BB) > z_O(CONTROL)$, results in the elimination of backward blocking (CR(CONTROL) ~ CR(BB)).

A proposed modification of the SLG model

As mentioned above, in our simulations we assumed that A, X and O temporally overlapped, whereas in the Pineño et al. experiment, stimulus O followed A and X. Although our model suggests that O–X associations are most important to obtain BB, the present version of the model does not have a mechanism to explain the acquisition of such an association when X precedes O. However, evidence published by Ward-Robinson and Hall (1996), demonstrated backward sensory preconditioning, a paradigm in which presentation of A preceding X, followed by A–US presentations, results in X being able to produce a CR. Therefore, we propose to modify the model to describe backward sensory preconditioning in terms of a backward O–X association. In Equation 2.9a, we would replace the actual value of the predicted X_{τ_X}, originally used as the "teaching signal" to build O–X associations, by its trace, τ_X. This change would allow the model to build O–X associations even when X precedes O, because their traces overlap. Computer simulations run under this assumption successfully yield BB.

Novel prediction

In addition to describing the existing data, the SLG model predicts that extinction of A in a different context will result in recovery from BB. Computer simulations showed increased responding to X, following 100 A–X–O trials in Context 1, 20 A–O trials or B–O trials in Context 1, 150 O–US trials in Context 2, and 20 A trials in Context 2. Although no generalization between contexts was assumed in this case, similar results are obtained when this assumption is made.

Comparison with other models

According to the comparator hypothesis (Blaisdell et al., 1998), BB, like forward blocking, is the consequence of a strong A–O association, acquired during both phases of the experiment, which makes the indirect representation of the O (through X–A and A–O links) stronger than the direct one (X–O link), therefore attenuating responding to X. Whereas most traditional models (e.g. Rescorla & Wagner, 1972; Mackintosh, 1975; Pearce & Hall, 1980; Wagner, 1981) can explain forward blocking, they do not address BB. According to the Dickinson and Burke (1996) version of Wagner's (1981) SOP model, BB is explained in the following terms. During the second phase of the experiment, the presence of A which has a within-compound association with X, will lead to a decrement in the X–O association in the experimental group. Because A is absent during this second phase for the control group, the X–O association will not change. Even though it can explain the occurrence of BB, the Dickinson and Burke (1996) model cannot explain spontaneous recovery from BB. Pineño et al. (2005) suggested that spontaneous recovery from BB can be explained in terms of the comparator hypothesis, if it is assumed that the A–X association weakens over the delay period, and provided experimental evidence for this assumption. A decreased A–X association results in a decreased indirect representation of O (through X–A and A–O associations), therefore recovering responding to X. Significantly, the SLG model describes the attenuation of the A–X association during both the A–O trials and the delay period.

Interestingly, a model presented by Kruschke (2001; Kruschke & Blair, 2000) can also explain both BB and its spontaneous recovery in ways similar to those offered by the SLG model. The model assumes error-driven attentional shifting that accelerates learning of new associations, but also protects previously learned associations from retroactive interference. According to this model, whereas forward blocking is the consequence of a reduced X–O association, BB is the consequence of inattention to X. If attention increases during the delay period, responding to X will recover.

As mentioned, Miller and Matute (1996) hypothesized that failure to obtain backward blocking in animals, without using sensory preconditioning, was due to the high biological significance of the US. Our model agrees with the idea that there is something different about the US in a BB arrangement; the US cannot predict X directly. The model suggests that stimulus O should predict X. In order to do this, O should either overlap with X (as in the simulations shown in Figure 8.10) or a backward association should be present (as in the proposed modification of the model). However, the SLG model could explain BB with a shock US under the assumptions that (a) the "teaching signal" used to build CS–US associations is a trace of the US and not the US itself, and (b) the US can become a predictor of CSs such as X.

Discussion

In this chapter, we showed how the SLG model described in Chapter 2 describes (a) overshadowing and recovery from overshadowing (Matzel et al., 1985), (b) the counteraction between latent inhibition and overshadowing (Blaisdell et al., 1998), (c) the conditions under which recovery from blocking by extinction of the blocker is obtained (Blaisdell et al., 1999; Holland, 1999), (d) the inability of a blocked CS to become a blocker of another CS (Williams, 1996), and (e) backward blocking and recovery from backward blocking following a retention interval (Pineño et al., 2005).

Apparently the SLG model is the only one able, at this point, to give an account for the opposite results reported by Blaisdell et al. (1999) and Holland (1999). Although the SLG model seems to correctly address the listed experiments, some results were more robust than others. For instance, even if latent inhibition, blocking overshadowing and their recovery can be obtained under a large number of simulated experimental conditions, the LI and overshadowing counteraction can be obtained under the conditions specified above. Interestingly, this might reflect a real property of the paradigm, as experimental results indicate, and the model successfully mimics, that this counteraction is not obtained with the longer CS durations used in taste-aversion preparations (e.g. Ishii et al., 1999; Nakajima & Nagaishi, 2005; Nagaishi & Nakajima, 2008). Furthermore, when attenuation of latent inhibition by overshadowing was obtained, Ishii et al. (1999, Experiment 2) showed that the effect could be accounted for by the restoration of the salience of the target CS by the surprise introduction of the overshadowing stimulus, as the SLG model would predict.

Significantly, while the SLG is an attentional–associative model, the other theories are associative-only models. Whereas the SLG model explains most of the present results as retrieval effects, the competing models explain the results as the consequence of an associative change. That is, whereas the SLG model explains the results in terms of a change in attention to already established associations that remain mostly unchanged, the alternative models modify existing associations, either the ones directly controlling the CR, or those used in the comparisons. Attentional–associative models are favored over purely associative ones by neurophysiological data showing that attentional and associative mechanisms can be found in different brain regions (see Chapter 4) and are differentially affected by drugs (see Chapter 6, for the case of LI).

In spite of the above difference, all the models agree on the importance of the CS–CS associations. In the SLG model, CS–CS and CS–CX associations are used to (a) compute the value of Novelty'; and (b) activate the representation of an absent CS, which then becomes associated with the value of Novelty', thereby changing attention to the missing CS. In the comparator hypothesis, CS–CS and CS–CX associations are needed to compute the comparison value that determines the CR strength. According to Dickinson and Burke (1996), in the Van Hamme and Wasserman (1994) version of the Rescorla–Wagner (1972) rule, the expectation of an absent CS via its within-compound association with a present CS specifies when an absent CS is allowed to decrease its association with the US. Finally, in the Dickinson and Burke (1996) version of Wagner's (1981) SOP model, presentation of a CS may activate, through CS–CS associations, the representations of other CSs in the A_2 state.

Summary

In this chapter, we showed how the SLG model described in Chapter 2 describes (a) recovery from overshadowing, (b) the counteraction between latent inhibition and overshadowing, (c) the conditions under which recovery from blocking by extinction of the blocker is obtained, (d) the inability of a blocked CS to become a blocker of another CS, (e) backward blocking and recovery from backward blocking following a retention interval. In addition, the SLG model predicts that extinction of the blocker will result in recovery from backward blocking. Apparently, the SLG model is the only one able, at this point, to give an account for the opposite results reported by Blaisdell *et al.* (1999) and Holland (1999) for the recovery of blocking by extinction of the blocker.

In Chapter 16, we will address Beckers *et al.*'s (2005) blocking experiments in humans. These authors showed that the strength of blocking and backward blocking increases when the outcome is submaximal or additive, and information regarding maximality (i.e. maximal possible value of the outcome) and additivity (i.e. the effect of cues is stronger when presented together) is provided.

Appendix 8.1 Simulation results

Different measures of conditioning were simulated as described below.

Drinking rate

In some cases, the experimental data were expressed as the amount of liquid consumed in a given time in the presence of an aversively conditioned Flavor CS. In order to compare simulated and experimental CR values, CR values were converted into amount of liquid consumed, by assuming that amount of liquid is proportional to the net drive to lick. This net drive increases with the level of Thirst, assumed for simplicity to be 1, and decreases with Fear, that is, with the prediction of the aversive US, represented by the CR. Therefore,

$$\text{Drinking rate (mL/s)} = \text{Thirst} - \text{Fear} = 1 - \text{CR} \qquad [8.1A]$$

That is, the amount of liquid consumed by the animal in a given time decreases as the magnitude of the CR increases.

Time to complete a number of licks

In many cases, the experimental data are expressed as the logarithm of the time to complete a certain number of licks in the presence of an aversive CS. In order to compare simulated and experimental CR values, CR values can be converted into log (time) by assuming that the time to complete a given number of licks (or to consume a given amount of liquid) is inversely proportional to the net drive to lick. Therefore,

$$\text{Time licking (s)} = 1[\text{ml}]/(1 - \text{CR})[\text{s/ml}]. \qquad [8.2A]$$

Because the correlation between log (time) and the value of the CR approaches 1 (0.996), when CR varies between 0 and 1, we compare the simulated CR with data values expressed in the logarithmic form.

Orienting response and behavioral inhibition

Although in purely appetitive or aversive behaviors the model assumes that the OR decreases the strength of the CR, because the subject directs its attention to the novel stimulus, it is not clear that this is accurate when appetitive and aversive behaviors are combined. In this case, the strength of the appetitive responding would be proportional to the response to the appetitive stimulus (e.g. licking), minus the conditioned response to the aversive stimulus,

$$R_{APPETITIVE} = CR_{US\text{-}APPETITIVE} - CR_{AVERSIVE} = CR_{US\text{-}APPETITIVE} - B_{US\text{-}AVERSIVE}(1 - K_{10}OR). \qquad [8.3A]$$

According to this equation, $R_{APPETITIVE}$ should *increase* when the OR increases; a result that contradicts the fact that, if distracted by the novel stimulus, the subject should *decrease* its responding to the appetitive stimulus too. Because this analysis exceeds the scope of the present chapter, simulations do not include this inhibition in cases when appetitive and aversive behaviors are combined (i.e. Grahame et al.,

1994; De la Casa and Lubow, 2000; 2002; Matzel *et al.*, 1985; Blaisdell *et al.*, 1998). We still use the inhibitory OR in the cases of appetitive-only protocols (Williams, 1996; Holland, 1999).

Table 8.1. *Summary of paradigms and models' performance.*

Model Paradigm	Comparator Hypothesis (Miller & Schachtman, 1985), Denniston et al. 2001)	Modified Rescorla–Wagner (Van Hamme & Wasserman, 1994)	Modified SOP (Dickinson & Burke, 1996)	Attentional-Associative Model (Schmajuk et al., 1996)
Recovery from Overshadowing (Matzel *et al.* 1985)	YES	YES	YES	YES
Recovery from Blocking (Blaisdell *et al.*, 1999)	YES	YES	YES	YES
Absence of Recovery from Blocking and Overshadowing (Holland, 1999)	NO	NO	NO	YES
Blocking of Blocking (Williams, 1996)	YES	YES	YES	YES
Latent Inhibition and Overshadowing Counteraction (Blaisdell *et al.*, 1998)	YES	NO	NO	YES

9
Extinction

In this chapter, we apply the SLG model to try to determine the mechanisms at work during extinction. Extinction refers to the phenomenon by which nonreinforced presentations of the conditioned stimulus (CS) after conditioning reduce the strength and frequency of the conditioned response (CR) to an arbitrarily small value. A large number of theories have been proposed to account for the extensive information available on extinction of classical conditioning. Associative models (e.g. Mackintosh, 1975; Rescorla & Wagner, 1972) assume that the phenomenon involves the weakening of the association between a CS and the unconditioned stimulus (US). In contrast, other approaches propose that extinction leaves the initial CS–US association intact. For instance, Pavlov (1927, Lecture XXII; see Robbins, 1990, page 236) provided a "new interpretation" of extinction in terms of a decrease in the activation of the cells triggered by the CS (CS representations), without changes in the connecting path between the CS cells and those cells excited by the US. For Rescorla (1974), extinction is the consequence of a decrease in the representation of the US, which controls both the CR *and* changes in the CS–US association. Hull (1943) suggested that extinction is the result of a "reactive inhibition"; a tendency not to repeat the CR when it is produced in the absence of the US.

Some other theories (e.g. Konorski, 1948; Moore and Stickney, 1980; Pearce & Hall, 1980; Schmajuk and Moore, 1988) propose that CS–US inhibitory associations are formed during extinction, which counteract the CS–US excitatory associations formed during acquisition. Competing memory theories (e.g. Gleitman, 1971; Spear, 1971) suggest that acquisition and extinction produce memories of the US and no-US that oppose each other. For Devenport (1998), recent CS–US inhibitory memories are weighted more heavily than old CS–US excitatory memories. According to Bouton (1993, 1994), the context might act as a negative spatial and temporal occasion setter (i.e. indicating that the CS will not be followed by the US, without directly predicting the absence of the US) that needs to be present to activate those inhibitory CS–US associations. For generalization decrement models (Capaldi, 1967), extinction is the result of the change from the conditioning situation produced by the omission of the US. According to Gallistel and Gibbon (2000), responding occurs when the ratio of the rate of reinforcement in the presence of the CS and the background rate exceeds a threshold. In contrast, extinction occurs when the ratio of the estimated interval elapsed since the time of the last reinforcement credited to the CS, and the expected interval between CS reinforcements, is greater than a threshold. The ratio grows in proportion to the duration of the extinction period.

According to Bouton (2002), four experimental results seem to support the notion that extinction does not completely eliminate CS–US associations. These results show that responding to the extinguished CS reappears when the CS is tested (a) some time after extinction (spontaneous recovery), (b) in a context different from that of extinction (renewal), or (c) in the context of extinction following US presentations in that context (reinstatement), and that (d) reacquisition might be fast when the CS is retrained in the context of extinction.

In this chapter, we try to understand the mechanisms at work during these four paradigms. Computer simulations with the SLG model suggest that its combination of attentional and associative mechanisms is required to describe most of the properties of extinction and post-extinction manipulations. However, configural mechanisms are needed to describe the properties of "extinction" cues that precede the target CS presented during extinction (see Chapter 16), and might improve the description of some experimental results regarding the associative properties of the extinction context.

Simulation procedures

Specific and generalized characteristics of contexts

In our simulations, we represented a context as a combination of its generalized aspects and its specific characteristics. This idea is in line with Gonzalez, Quinn and Fanselow's (2003) data, showing that removal of a cue from the context results in a significant generalization decrement in freezing; a result they indicate is consistent with elemental models such as that of Rescorla and Wagner (1972), and also with the associative properties of our model.

In most simulations, different tonic contextual stimuli were used to represent the contexts in which the animals are placed. CXh represents the home cage where animals are housed during the interval between different daily sessions. CXc represents the features of a given experimental chamber. CXg is used to represent the common features of chambers and the home cage, that is, the generalization between all contexts. Notice that associations can be established between the contextual representations CXh and CXg, or CXc and CXg. Using these associations, the model detects when the subject is "moved" from an experimental chamber to the home cage (or vice versa): CXg predicts the context in which the animal was previously located (CXc or CXh) when they are placed in a new location (CXh or CXc). Also, existing CXh–CS, CXg–CS, and CXc–CS associations serve to detect the absence of the CS in a given context.

Simulation values

Whereas model parameters remain fixed, simulation values (e.g. CS intensity and duration, US intensity and duration, ITI durations) vary to reflect the experiments as closely as possible. Here we explain how simulation values of stimuli salience (defined as their capacity to become associated with other stimuli) and durations were selected. Details of the simulation procedures are described in each section.

Our simulations tried to reflect the relative salience of the contexts. Therefore, we used CXh = 0.5 and CXg = 0.4 in all simulations, but CXc differed according to the training apparatus. Because the addition of odors to the cage has been shown to increase contextual conditioning (e.g. McKinzie and Spear, 1995), CXc was relatively large (0.5 or 0.7) when odors were present. When no odor or food was present, and freely moving rats were subjects, CXc was 0.1. When rabbits were immobilized and tightly restrained in a Plexiglas box with Velcro straps secured around the head and muzzle, with a plastic head stage with an attached potentiometer and air nozzle to deliver the US, and a nylon loop stitched to one of the nictitating membranes to record the CR, CXc was 0.7. To our knowledge, there are no direct measures of context salience for the rabbit preparation.

Whereas the US strength was always 2, CS salience was 1 in most simulations. In most cases, our simulations tried to approximate the average ratio between CS and ITI durations used in the experiments presented in a given section of the chapter, thereby permitting to compare simulations dealing with different aspects of the same phenomenon. However, because the model predicts a faster decrease in contextual associations as a function of the ITI than that reported in the literature (e.g. Barela, 1999), in some cases this average was larger than the one used in the experiments. Therefore, in the simulations for extinction, we used a 1/7 ratio, larger than the 1/30 average ratio used in the experiments. In the simulations for spontaneous recovery, we used a 1/12 ratio, identical to the average ratio used in the experiments. In the simulations for renewal, we used a 1/6 average ratio, larger than the 1/10 average ratio used in the experiments. In the simulations for reinstatement, we used a 1/7 average ratio, identical to the average ratio used in the experiments. In the simulations for reacquisition, we used a 1/7 ratio, larger than the 1/20 average ratio used in experiments. Finally, in the simulations for extinction cues, we used a 1/5 ratio, larger than the 1/10 average ratio used in the experiments.

More importantly, simulation results are robust within a broad range of CS and CX intensities and durations that includes the values mentioned in the preceding paragraphs. For instance, similar results are obtained in extinction, spontaneous recovery, renewal and reinstatement when (a) the strength of the US and salience of the CS are either lowered (from 2 to 1, or from 1 to 0.6, respectively), or increased (from 2 to 2.5, or from 1 to 2); (b) the duration of the US and CS is reduced (from 5 to 2 time units (t.u.), or from 20 to 8 t.u., respectively) or increased (from 5 to 10 t.u., or from 20 to 30 t.u., respectively); and (c) CXg salience is varied between 0.1 and 0.7.

Simulation of measures of conditioning

Suppression ratios, percent CRs, time to complete a number of licks and freezing were simulated as described in Chapter 5. In addition, liquid consumption after conditioned taste aversion was simulated as $1 - CR$, where 1 represents the appetitive value of liquids (see Chapter 8). Elevation scores were expressed as the difference between responding in the presence of the CS and in its absence, $CR(CS) - CR(CX)$. Preference for a context was computed as Context Preference $= 0.1 - [V_{CXc,US} + V_{CXg,US}]$, where 0.1 represents the appetitive value of the Skinner box and $V_{CXc,US}$ and $V_{CXg,US}$ are contextual associations with an aversive US.

Simulated results

In this section we describe computer simulations for experiments investigating the properties of (a) extinction, (b) spontaneous recovery and external disinhibition, (c) renewal, (d) reinstatement, and (e) reacquisition.

Extinction

Because many experimental procedures show recovery of the CR after extinction, it is generally accepted that the CS–US association formed during acquisition survives extinction, and that contextual stimuli play an important role in this survival. In addition to the presentation of the CS alone, extinction can be obtained by unpaired CS–US presentations (e.g. Napier, Macrae & Kehoe, 1992) or by presentation of a CS associated with the target CS (mediated extinction; Shevill & Hall, 2004). Also, although many of the experimental results are ambiguous, some data indicate that extinction is faster with short than with long ITIs (Cain, Blouin & Barad, 2003; Moody, Sunsay & Bouton, 2006). In the framework of the model, extinction occurs because the excitatory CS–US associations decrease but, if contextual cues are sufficiently strong, the CS–US association decay is hindered by inhibitory CX–US associations. In addition, attention to the CS and the CX decreases over the course of nonreinforced trials, thereby making the inhibitory CX–US associations difficult to detect.

Extinction bursts

Furthermore, computer simulations show that these attentional and associative mechanisms allow the model to describe extinction bursts, that is, the transient increase in responding during the initial sessions of extinction (Thomas & Papini, 2001). Extinction bursts are shown after 10 acquisition trials (CSs salience 1, duration 20 time units; US strength 1, duration 5 time units, CX salience 0.1).

Contextual associations during extinction

The upper panel of Figure 9.1 shows the model's associative values and CR, and the lower panel shows average Novelty' and the representations of the CX and the CS during conditioning and extinction, just before the time of presentation of the US on reinforced trials. Because in this simulation CXg (0.4) is more salient than CXc (0.1), Figure 9.1 shows only V_{CXg-US} and V_{CS-US}, their representations, X_{CXg} and X_{CS} and Novelty'. As explained in Chapter 2 (section on "Attention"), internal representations X_{CS} and X_{CXg} are proportional to attentions z_{CS} and z_{CXg} which, in turn, are modulated by Novelty'. During acquisition (trials 1–20), the CS–US association (V_{CS-US}) increases and the CX–US associations (V_{CXg-US}) remain close to zero. During extinction (trials 21–55) the association of the CS with the US decreases, but does not vanish. The CS–US association V_{CS-US} is protected by an inhibitory CXg–US association (V_{CXg-US}) that eliminates the CR. In addition, during acquisition (trials 1–20), Novelty' decreases, representation X_{CS} increases, and representation X_{CXg} first increases and then decreases. During extinction (trials 21–55), Novelty' decreases first, followed by decrements in X_{CXg} and, later, by decrements in X_{CS}. So, by the end of extinction, V_{CS-US} is excitatory, V_{CXg-US} is inhibitory and representations X_{CXg} and X_{CS} are small. During the period in the home cage (trials 56–85) and during testing (trials 86–105), increments in both CXg and CS representations lag behind Novelty'.

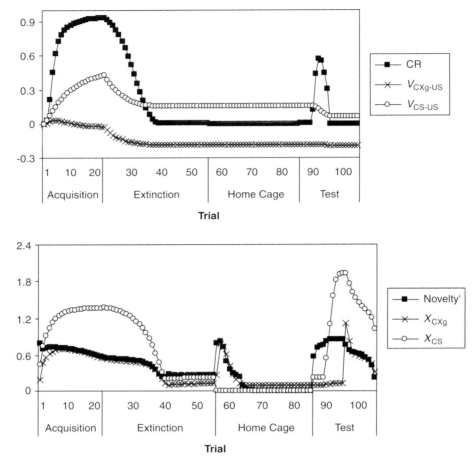

Figure 9.1 Variables of the model during acquisition, extinction and spontaneous recovery. *Upper panel*: Values of the conditioned response, CR; the association between the generalized context and the unconditioned stimulus, V_{CXg-US}; and the CS–US association, V_{CS-US}; as a function of trials. *Lower panel*: Values of average Novelty'; the internal representation of CXg, X_{CXg}; and the CS internal representation X_{CS}; as a function of trials. Conditioning Trials 1–20, Extinction Trials 21–55, Home-cage Time Periods 56–85, and Test Trials 86–105.

This dissociation between the orienting response (proportional to Novelty') and attention has been reported by Hall and Schachtman (1987).

Notice that, because the CS is more salient than the CX, and the trace of the CS is present for a shorter period of time than the trace of the CX (active for the duration of the ITI), contextual associations are almost zero at the end of conditioning (see Chapter 2, section on "Changes in CS–US associations"). Therefore, the CR is context independent following acquisition. In contrast, during extinction, contextual associations are strongly inhibitory and, if attention to the context happens to increase, the CR will be context dependent following extinction (see Chapter 2). Also notice that, because CX–US associations are stronger with a short than with a long ITI, a short ITI will result in a higher Novelty′ when the US is not presented during extinction. Therefore, the increased attention to the CS will facilitate extinction, a result in agreement with Cain *et al.*'s (2003) and Moody *et al.*'s (2006) data.

Interestingly, whereas Novelty′ stays relatively high during conditioning, it decreases to a very small value during extinction. The difference is due to the fact that, during extinction, CSs and CXs can predict themselves and each other and, therefore, perfectly match predicted and actual values. In contrast, with the simulation values used in this case, the US is not perfectly predicted and Novelty′ stays at a relatively high level during acquisition. Simulations with other CS and US durations and intensities show that Novelty′, attention to the CS and CR strength can decrease.

According to our model, a CX–US inhibitory association formed during extinction would protect the complete elimination of the CS–US association (Chorazyna, 1962; Rescorla, 2003; Soltysik, 1985). However, Bouton and King (1983), Bouton and Swartzentruber's (1989) and Richards and Sargent (1983) presented data that suggest that the context does not become inhibitory at the end of extinction. Below we show simulated results that demonstrate, that even when inhibitory CX–US associations are formed, they can be difficult to detect because attention to the inhibitory CXc and CXg is small.

Summation tests following extinction

Bouton and Swartzentruber (1989, Experiment 3; see also Bouton & King, 1983, Experiment 4) evaluated whether the context of extinction becomes inhibitory or not. Rats received training sessions to a CS in CXA. Half of the animals received CS extinction trials in CXB, while the other half received them in CXC. Rats received additional training to another CS in CXA. Both the group extinguished in CXB (Diff) and the group extinguished in CXC (Same) received a summation test with the second CS in CXC. A 60-s light-off CS, 60-s, 3,000-Hz tone CS, 0.6-mA, 0.5-s US, a 20-min average ITI and odors to increase the differences between contexts, were used. As shown in Figure 9.2 (left), a summation test for contextual inhibition revealed no evidence of inhibitory associations, i.e. responding in the context of extinction (Same) was not significantly different from responding in another context (Diff).

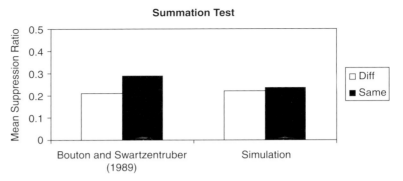

Figure 9.2 Summation test in the context of extinction. Suppression ratios to an excitatory CS when extinction to another CS had taken place in (a) another context (Diff), or (b) that context (Same). *Left*: Data from Bouton and Swartzentruber (1989). *Right*: Computer simulations, $\beta = 1.1$.

Simulations

Conditioning sessions consisted of 20 CS_1–US trials in CXA, followed by 80 CS_1 trials in either CXB (Diff) or CXC (Same). After extinction, 20 CS_2–US trials were simulated in CXA, followed by a summation test, which consisted of one CS_2 trial in CXC. Simulations of 20 home-cage (CXh = 0.5, CXg = 0.4) time periods of duration equal to the ITI were included between the different sessions. The salience of the contexts (CXA, CXB and CXC) was set to CXc = 0.7 and CXg = 0.4 in extinction and test sessions. The CSs had an 11 t.u. duration and salience 0.7, the US was 5 t.u. long with salience 2, and the ITI was 70 t.u. The model correctly describes the statistically non-significant difference in the experimental data (Figure 9.2, right).

Other experiments

Although Bouton and Swartzentruber (1989; Bouton & King, 1983) were not able to detect CX–US inhibitory associations, in some cases inhibitory associations were uncovered. For instance, Cunningham (1979) evaluated the effects of alcohol as a contextual cue on extinction, and found that it became a conditioned inhibitor.

Bouton and Swartzentruber (1986) used a summation test, a supernormal conditioning test and a retardation test to investigate the nature of the contextual associations formed during a contextual discrimination in which a CS was reinforced in CXA but not in CXB. The tests did not reveal contextual inhibitory associations in the nonreinforced context. (These contextual discrimination results are well described by the SD/SLH model, as indicated in Chapter 12.) Furthermore, Maes and Vossen (1996) examined whether external contextual stimuli can acquire inhibitory properties through a simple differential context-reinforcement procedure. Although summation and retardation tests failed to reveal contextual inhibitory associations of the nonreinforced context, they revealed that an acoustic stimulus that was unique to the nonreinforced context did acquire inhibitory associations. Maes and Vossen (1996) proposed that this inhibition "was masked by the concurrent presence of excitation from elements common to both the nonreinforced context and the reinforced context." These results suggest that only some particular contextual stimuli that are attended when the target CS is extinguished, but not the whole context, become inhibitory. When a different excitatory CS is presented in the context, attention might be shifted away from those "unique" inhibitory stimuli and no inhibition is detected. A target CS might attract attention to those unique contextual stimuli simply by proximity (e.g. the shape of the cage near the speaker). According to this view, in summation tests the excitatory CS should be placed close to where the target CS was located during extinction, in order to activate those presumably inhibitory contextual features.

Comparison with other theories

In order to address experimental data showing no detectable CX–US inhibitory associations, Bouton and Swartzentruber (1986) proposed that the context might act as a spatial and temporal negative occasion setter (Holland, 1983). Occasion setters differ from simple CSs in several ways: (a) whereas simple CSs control responding through direct associations with the US, occasion setters enable (set the occasion for) responding based on an association between another CS and the US; (b) occasion setters do not have the summation properties of simple excitatory or inhibitory CSs; and (c) the ability of a CS_1 to modulate another CS_2–US association (occasion setting) is independent of the CS_1–US association. While certainly attractive, Bouton and colleagues indicated that the notion of a context acting only as an occasion setter seems problematic. Although an occasion setter is not supposed to have the summation properties of a simple CS, pairing a CX with a US following extinction affects the ability of the context to modulate responding (Bouton & Bolles, 1979a), a result that suggests that the context acts also as a simple CS. As explained in the corresponding section, this property is needed to explain reinstatement.

Notice that the Rescorla and Wagner (1972) rule describes extinction in terms of CX–US inhibitory associations only under certain conditions (e.g. when CX–US associations are already decreased at the beginning of extinction, or when extinction occurs in a context different from that of acquisition). In those cases, and similar to our model, the CX–US association becomes inhibitory with nonreinforced trials in the presence of the excitatory CS, and therefore protects the CS–US association. However, in contrast to our model, in the absence of an attentional mechanism, the Rescorla–Wagner (1972) rule cannot describe summation results that suggest that the context of extinction is not inhibitory (Bouton & Swartzentruber, 1989).

Facilitation of extinction

Using a rabbit eyeblink preparation, Poulos *et al.* (2006) studied the effect that exposure to the training context has on extinction. Following conditioning, rabbits received one of three procedures. The experimental group (CXT) was returned to the experimental chambers without either CS or US presentations. A control group (HMC) remained in their home cages. Another control group was handled in a different context. All groups received CS-alone extinction sessions. A 600-ms, 1-kHz tone CS, coterminating with a 100-ms, 3-psi, air-puff US, and 30-s average ITI, were used.

Poulos *et al.* (2006) found that exposure to the training context reduced the CR compared to the control groups that either remained in their home cages, or received exposure to handling and a novel context (Figure 9.3, upper panel). Although Poulos *et al.* (2006) found that nonreinforced exposure to the context facilitates extinction, Richards and Sargent (1983) reported no effect of nonreinforced exposure to the context and another CS on the subsequent extinction of a second CS. It has also been reported that reinforced exposure to the context following acquisition facilitates extinction (Kehoe, Weidemann & Dartnall, 2004).

Simulations

In the simulations for Poulos *et al.* (2006), conditioning sessions consisted of 10 CS–US trials in CX_1, followed by 30 trials in CX_1 or in the home cage, and 10 extinction trials in CX_1. Simulations of 20 home-cage (CXh = 0.5, CXg = 0.4) time periods of duration equal to the ITI were included between the different sessions. The salience of the training context was set to CXc = 0.7 and CXg = 0.4. The CS had a 11 t.u. duration and salience 0.7, the US was 5 t.u. long with salience 2, and the ITI was 70 t.u. Figure 9.3 (lower panel) shows that the model correctly replicates the Poulos *et al.* (2006) results (for the model, the control group that remained in the home cage [HMC] and the group that was handled in a different context are equivalent).

According to the model, because the CX is salient and the ITI relatively short, the CX–US association will be significant at the end of conditioning. Therefore, exposure to the context of extinction results mainly in the weakening of the CX–US excitatory association, which leads to a fast acquisition of the CX–US inhibitory association underlying extinction, whereas other mechanisms play

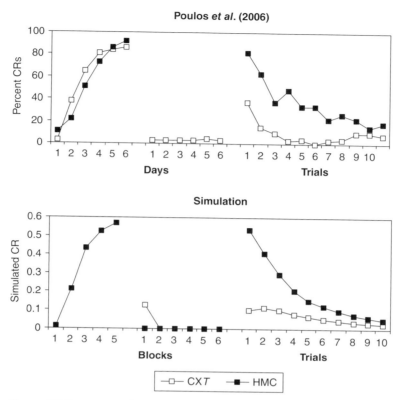

Figure 9.3 Exposure to the context following acquisition facilitates extinction. Percent CRs during acquisition, exposure to the context (CXT) or the home cage (HMC), and extinction. *Upper panel*: Data from Poulos *et al.* (2006). *Lower panel*: Simulations.

a less important role. When an equivalent period is spent in the home cage, the CX–US association is large at the beginning of extinction and takes longer to become inhibitory and decrease responding.

Notice that although the model suggests that CRs are generated during the exposure to the context between acquisition and extinction, such responding is almost absent in the experimental data. This disagreement is explained because CX–US associations do not seem to activate the nictitating membrane CRs, but produce, instead, conditioned emotional responses that modulate the CR strength (see Wagner & Brandon, 1989).

Computer simulations suggest that the strength of the contextual cues (assumed to be 0.7 when the rabbits are immobilized in cages as in the Poulos *et al.* [2006] experiment, and 0.1 when rats are free to move in cages with no scents as in the Richards and Sargent [1983] experiment) might be responsible for the different results. When the strength of CXc is decreased to 0.1 in the simulations for Poulos *et al.* (2006), the model shows no effect of the nonreinforced exposure to the context on the first extinction trials, and some facilitation afterwards. Reciprocally, when the strength of CXc is increased to 0.5 in the simulations for Richards and Sargent (1983), the model shows that extinction of a CS facilitates the subsequent extinction of a second CS. This last prediction can be readily tested by replicating Richards and Sargent's (1983) experiment, with the addition of cage odors to increase the salience of CXc. Such an experiment would also serve to test our assumption that contextual cues are strong in the case of the Poulos *et al.* rabbit preparation.

Although the model can explain the facilitatory effects of nonreinforced exposure to the CX, it cannot explain results showing that reinforced exposure to the context also facilitates extinction (Kehoe et al., 2004, Experiment 2A).

Comparison with other theories

As explained in detail by Poulos *et al.* (2006), some theories only predict the effect of reinforced CX exposures preceding extinction. Comparator models, which assume that responding reflects the difference between the excitatory CS–US associations compared with CX–US associations (Gibbon & Balsam, 1981; Miller & Matzel, 1988), correctly predict the effect of reinforced, but not of nonreinforced, CX exposures. Dickinson and Burke (1996) and Van Hamme and Wasserman (1994) theories, which assume that if the CS and CX have been presented together then increasing (or decreasing) CX–US associations will decrease (or increase) CS–US associations, correctly predict the effect of reinforced, but not of nonreinforced, CX exposures (but only if the CS is tested in a different CX where CX–US association do not increase the CR). Theories that assume that CX–US associations increase the motivation to produce a CR (Wagner & Brandon, 1989) can account for the effect of nonreinforced, but not of reinforced, CX exposures. In contrast, Kehoe *et al.* (2004) suggested that some theories (Konorski, 1967; Cunningham, 1981; Hall, 1996; Holland, 1990) are able to explain the decremental effect of both nonreinforced and reinforced CX exposures. These theories assume that CX–CS associations will activate a representation of the CS during CX exposure and decrease the CS–US association, a phenomenon known as "mediated extinction." In the case of reinforced presentations, Kehoe *et al.* (2004, page 267) proposed that the CS representation would be continuously activated by the CX, thereby lacking a clear onset preceding the US presentation, which has been shown to determine the formation of CS–US associations. Even when CX–CS associations present in our model are able to describe mediated extinction of the CS–US association (Shevill & Hall, 2004), our simulations show that this effect might be small compared with the weakening of the CX–CS association, which leads to a weaker activation of the CS representation, and of the CS–US association.

Discussion

In the SLG model, when the CX is salient, inhibitory CX–US associations are formed during extinction. Protected by those inhibitory CX–US associations, the CS–US association decreases, but is not eliminated. Both the decreased CS–US and the inhibitory CX–US associations then combine to reduce the strength of the CR. The inhibitory CX–US association, however, is hard to detect in a summation test due to a decreased attention to the context. Computer simulations suggest that, if attention to the CX is increased by presentation of a novel CS, the CX–US inhibitory association could be revealed in a summation test. In the "General discussion" (see subsection on "Predictions"), we propose to replicate Bouton and Swartzentruber's (1989) summation experiment introducing a modification that would allow unattended, inhibitory CX–US associations to be expressed.

Interestingly, according to the SLG model, even if inhibitory CX–US (or CS–US) associations cannot be extinguished (a property based on Zimmerhart and Rescorla's [1974] data), the model predicts that sufficiently long exposure to the inhibitory CX (or to an inhibitory CS) should result in decreased attention to the CX (or the CS), thereby decreasing the inhibitory power of the CX. Some data reported by Melchers, Wolff and Lachnit (2006) showing extinction of conditioned inhibition through nonreinforced presentations of an inhibitor CS, seems to support this view. Furthermore, this decrease in attention is used by the model to describe Rescorla's (2000, 2001, 2002) results reported in Chapter 3. Also, as mentioned in Chapter 3, the SLG model can describe Lysle and Fowler's (1985) results showing that (a) extinction of the excitatory CS decreases the retardation of conditioning of the inhibitory CS, and (b) retardation can be increased by receiving presentations of the US under various conditions.

Regarding the facilitation of extinction, although associative mechanisms are sufficient to explain why nonreinforced CX exposures facilitate extinction when contextual cues are salient, attentional mechanisms also contribute by increasing responding in the control groups, which experience a transition between the home-cage and testing contexts. The model cannot robustly explain, however, why many reinforced presentations in the context following conditioning facilitate extinction.

Spontaneous recovery and external disinhibition

This section addresses spontaneous recovery, the phenomenon by which an extinguished CR reappears when the target CS is tested some time

178　Attentional and associative mechanisms

after extinction (Pavlov, 1927; Rescorla, 2004a). It also shows simulations for external disinhibition (Pavlov, 1927), that is, the recovery from extinction obtained by presenting a novel stimulus immediately before a previously extinguished CS. In the framework of the model, spontaneous recovery is explained because Novelty′ increases when the CS is presented in the context of extinction after a period of absence. This increased Novelty′ increases attention to the excitatory CS first, and then to the inhibitory CX, which results in a temporary recovery of responding.

Effects of varying the extinction–test intervals

Robbins (1990, Experiment 1) studied the effect of varying the interval between extinction and testing on the magnitude of spontaneous recovery. He trained pigeons in an autoshaping procedure, followed by CS extinction trials. After extinction, all the birds were tested for spontaneous recovery either 15 min, 1 day, 2 days or 7 days after being returned to the home cage. The CS was a 5-s light, the US was a 5-s access to food and the mean ITI 1 min in duration.

As shown in the upper panel of Figure 9.4, he found that the recovery ratio (responding on the first 3 trials of recovery divided by the total level of responding on the first 3 trials of both extinction and recovery) increases as the interval between extinction and testing increases

Simulations

Twenty acquisition trials were simulated in the experimental chambers (CXc = 0.1 and CXg = 0.4), in which a 20-t.u. CS (salience 1) coterminated with a 5-t.u. US (salience 2), and were followed by 35 extinction trials. The ITI was 240 t.u. To simulate the period spent by the pigeons in the housing chambers before testing, 2, 5, 10, 20, 30 and 50 time periods of duration equal to the ITI used during acquisition or extinction in that context (CXh = 0.5 and CXg = 0.4) were used. Finally, 10 test trials in the experimental context were carried out to assess the spontaneous recovery of the pecking response. As shown in the lower panel of Figure 9.4, the model correctly describes the effect of changing the extinction–test interval on spontaneous recovery.

Figure 9.1 also shows, in addition to acquisition and extinction, spontaneous recovery when the home-cage context is interposed between extinction and testing trials. During the time periods in the home cage (periods 56–85), attention to CXg initially increases due to the novelty produced by the change of the environment (CXg predicts CXc which is not present). Novelty′ later decreases as CXg and CXh (not shown) better predict each other. During test trials (trials 86–105), CXc and the CS are present, and Novelty′ increases because CXg–CS and CXg–CXc associations had decreased in the home cage. This increase in Novelty′

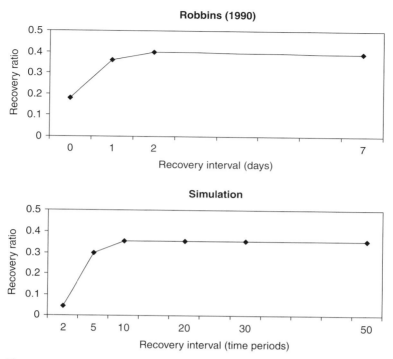

Figure 9.4 Spontaneous recovery. Recovery ratio (responding on the first 3 trials of recovery divided by the total level of responding on the first 3 trials of both extinction and recovery) as a function of the interval between extinction and testing. *Upper panel*: Data from Robbins (1990). *Lower panel*: Simulations. Duration of each time period is equal to the intertrial interval (ITI) used during acquisition or extinction.

triggers an increase in attention to the CS and CXg, which increases their representations, X_{CS} and X_{CXg}. However, X_{CS} increases faster than X_{CXg}, since attention to CXg had decreased more than attention to the CS during the home-cage trials. This a basic property of the model (see Chapter 2, section on "Attention") by which attention to stimuli of long duration (the CX) will decrease more than attention to stimuli of short duration (the CS) when Novelty' is relatively small. Therefore, spontaneous recovery occurs due to a faster increase in the representation of the CS than that of the CXg during testing, allowing the expression of the remaining V_{CS-US} association (which decreases to some extent in the absence of the US). Because attention to CXg, and X_{CXg}, decreases as the extinction–test interval increases, CXg takes longer to inhibit the CS–US association, and spontaneous recovery becomes stronger as this interval increases (see Figure 9.4). In addition, because CXg–CS and CXg–CXc associations decrease to a lower level as the time spent in the home cage increases; Novelty', attention to the CS and activation of the CS–US association become stronger as this interval increases, also contributing to the increased spontaneous recovery.

Using the simulation values of Figure 9.4, simulations shown in Figure 9.5 demonstrate that the model is able to describe spontaneous recovery even without context changes between acquisition and extinction trials, or between extinction and test trials. Very similar results are obtained when simulations assume that either the acquisition–extinction period or the extinction–testing period takes place in a different context (e.g. the home cage.) According to the model, spontaneous recovery without context changes is described as follows. After extinction, attention to the context decreases due to repeated uneventful presentations of the CX, as does the CX–CS association. However, the inhibitory CX–US association remains unchanged, since there are no US presentations in this phase (as explained in Chapter 2, section on "Changes in CS–US associations", the CX is a conditioned inhibitor that does not extinguish when presented in the absence of the US). Subsequent test trials in the same context increase Novelty', since the CX fails to predict the CS occurrences. In turn, this increase in Novelty' activates the representation of the CS, which produces spontaneous recovery, until the representation of the CX increases, and eliminates the CR. As in the case of spontaneous recovery with context changes, because X_{CX} decreases as the extinction–test interval increases, the CX takes longer to inhibit the CS–US association, and spontaneous recovery becomes stronger as this interval increases.

Effects of varying the acquisition–extinction interval

Rescorla (2004b) reported that spontaneous recovery decreases when the training–extinction interval increases from 1 to 8 days. In his experiment, two CSs were successively conditioned to a food US and then simultaneously extinguished. Spontaneous recovery was smaller for the CS conditioned earlier, and had an 8-day delay between conditioning and extinction, than for the one that received extinction trials immediately following acquisition.

In line with Rescorla's (2004b) results, Maren and Chang (2006) reported that a 24 h, but not a 15 min acquisition–extinction interval, eliminated spontaneous recovery of freezing behavior in rats. In contrast with those studies, Myers, Ressler and Davis (2006) reported that spontaneous recovery (in addition to renewal and reinstatement) of fear-potentiated startle was absent with a 10 min acquisition–extinction interval, and became stronger when the interval increased from 1 h to 3 days. Myers *et al.* suggested that the absence of all those effects was due to the elimination of the CS–US associations when extinction took place shortly following conditioning.

Figure 9.5 shows simulations for the combined effect of changing both the acquisition–extinction and the extinction–test intervals, with those intervals

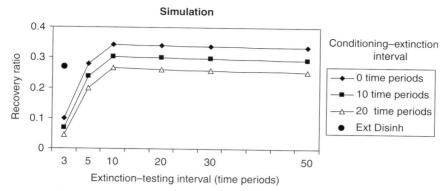

Figure 9.5 Spontaneous recovery and external disinhibition. Simulations of the effects of (a) varying the interval between extinction and testing (3, 5, 10, 20, 30 and 50 time periods), (b) varying the interval between acquisition and extinction (0, 10 and 20 time periods), and (c) applying a novel stimulus preceding the CS presentation (External Disinhibition, Ext. Disinh), on the recovery ratio. Duration of each time period is equal to the intertrial interval (ITI) used during acquisition or extinction.

spent in the training cage. The simulations, which do not intend to replicate any specific experimental data, are in general agreement with Rescorla's (2004b) and Maren and Chang's (2006), but not with Myers et al.'s (2006), data. According to the model, recovery decreases when the acquisition–extinction interval increases from 0 to 20 time periods of duration equal to the intertrial interval (ITI) used during acquisition or extinction, because increasing the interval between conditioning and extinction decreases attention to CXg. Therefore, during extinction the CS–US association will be protected only by the inhibitory CXc–US association (which is equivalent to having a weaker context), resulting in a more pronounced decrement in the CS–US association, and therefore, in a decreased spontaneous recovery.

External disinhibition

Besides occurring spontaneously, recovery from extinction can be obtained by presenting a novel stimulus immediately before a previously extinguished CS; an effect labeled external disinhibition (Pavlov, 1927). Bottjer (1982) studied this phenomenon in pigeons' approach–withdrawal behavior. Figure 9.5 (solid circle) shows that the simulated recovery ratio following presentation of a novel CS preceding the presentation of the target CS is much higher than that obtained simply by spontaneous recovery after the same number (3) of trials following extinction. According to the model, external disinhibition is caused by an increase in Novelty′ in the test phase, that arises when the novel stimulus is presented immediately before the previously extinguished CS. The effect of this increased Novelty′ is similar to that illustrated in Figure 9.1 during the test trials. Novelty′ first increases the representation of the target CS, X_{CS}, and then the representation of the CX, X_{CXg}. The net result is a greater CR than the one that occurs when a novel stimulus does not precede the CS. In a similar way, the model is also able to describe Vervliet *et al.*'s (2005) demonstration that a change in the perceptual characteristics of a stimulus after extinction causes a return in fear in humans.

As mentioned, in the "General discussion" (subsection on "Predictions") we suggest replicating Bouton and Swartzentruber's (1989) experiment in which attention to the CX would be increased by presentation of a novel CS, thereby revealing the CX–US inhibitory association in a summation test. In that case, the novel CS would be presented in the CX instead of preceding the previously extinguished CS, thereby increasing attention to the inhibitory context, which would decrease responding to the excitatory CS. In contrast, in external disinhibition, the novel CS is presented preceding the presentation of the extinguished CS, thereby increasing attention to that CS and activating the remaining excitatory CS–US association. As in the case of spontaneous recovery, an attentional mechanism is needed to describe external disinhibition.

Effects of presentations of another CS during delayed testing

Robbins (1990) also studied the effect of (a) past reinforcement, (b) the time of testing during the testing session, (c) presentation of a reinforced CS during extinction, and (d) reextinction of another CS on spontaneous recovery. In Robbins' (1990) Experiment 2, pigeons received reinforced presentations of CS_1, CS_2 and CS_3, and nonreinforced presentations of CS_4 (all CSs were keylights). Extinction included nonreinforced presentations of the previously reinforced CS_3 and of the nonreinforced CS_4, as well as reinforced occurrences of the previously reinforced CS_1 and CS_2. In the first half of the test session, one group received nonreinforced presentations of the extinguished CS_3 (Group CS+/G1), whereas the other received presentations of the never-reinforced CS_4 (Group CS−/G2). In the second half of the session, pigeons received nonreinforced presentations of CS_4 and CS_3 (Groups CS−/G1 and CS+/G2, respectively). Test trials were intermixed with reinforced presentations of CS_1 and CS_2, which had always been reinforced. Keylight CSs were 5 s in duration, the US a 5-s access to food, and the mean ITI was 1 min in duration.

Figure 9.6 (upper panel) shows the rate of responding during the First and Last nonreinforced presentations of (a) a previously conditioned CS (CS_3, Groups CS+) and (b) a CS that had never been reinforced before (CS_4, Groups CS−). Figure 9.06 also shows the rate of responding during the two parts of the recovery session. During the first part, only Group CS+/G1 showed spontaneous

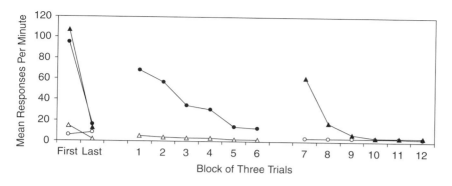
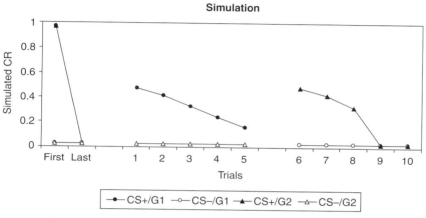

Figure 9.6 Spontaneous recovery. Spontaneous recovery (a) does not require test trials to be presented at the beginning of the session, and (b) occurs in the presence of reinforced presentations of another CS during extinction. *Upper panel*: Data from Robbins (1990). *Lower panel*: Simulations. CS+/G1: conditioned and extinguished CS tested first, CS−/G1: CS that had never undergone acquisition tested second, CS+/G2: conditioned and extinguished CS tested second, CS−/G2: CS that had never undergone acquisition tested first.

recovery. In the second part of the session, spontaneous recovery only occurred in Group CS+/G2. In sum, as shown in Figure 9.6 (upper panel), Robbins (1990) found that spontaneous recovery (a) is obtained only with CSs that have been reinforced in the past, (b) does not require test trials to be presented at the beginning of the session, and (c) occurs in the presence of reinforced presentations of another CS during extinction.

Simulations

The first phase of the simulation consisted of 60 alternated CS_3-US, CS_4-US and CS_1-US trials. For simplicity, stimulus CS_2 (equivalent to the always reinforced CS_1) was not included. The following phase consisted of 105 alternated CS_3-, CS_4-, and CS_1-US trials. After extinction, 30 time periods in the home cage were simulated, followed by two 7-trial sessions, each composed of 5 CS– trials, intermixed with 2 CS_1-US trials. Two simulations were performed for each test session. In the first simulation, the reinforced CS_3 was tested first (Group CS+/ G1), whereas the nonreinforced CS_4 was tested second (Group CS–/ G1). In the second simulation, the order of the CSs was inverted (Groups CS–/G2 and CS+/G2). All CSs were 20 t.u. in duration and salience 1, the US was 5 t.u. long and salience 2, and the ITI was 240 t.u. Figure 9.6 (lower panel) shows that simulated and experimental results are similar.

According to the model, spontaneous recovery is obtained only with a CS that has been previously reinforced and extinguished, because recovery requires the existence of a CS–US association. In the case of a previously reinforced CS, this association is not completely eliminated by nonreinforced presentations of the CS. When the CS was not previously reinforced, both Novelty' and the representation of that CS increase, but no response is generated due to the lack of CS–US association. Also, the model suggests that spontaneous recovery occurs even when the target CS is presented in the second part of the test session, because Novelty' increases, as a consequence of the context not predicting this reinforced CS, but instead predicting the nonreinforced CS presented earlier in the test session. This increase in Novelty' increases the representation of the reinforced CS (X_{CS}) that, as explained above, results in the recovery of the CR. However, the representation of the generalized context (X_{CXg}) starts to increase at the beginning of the test session, due to the initial presentations of the nonreinforced CS. Therefore, in agreement with the data, increased attention to the inhibitory CXg decreases spontaneous recovery faster when the extinguished CS is presented later in the test session.

Furthermore, additional simulations (not shown here) demonstrate that spontaneous recovery can occur to one CS after the recovery and reextinction of another CS, since the occurrence of spontaneous recovery of a CS (CS_1) and its subsequent re-extinction only changes that CS_1-US association, but does not affect the ability of another extinguished CS (CS_2) to undergo spontaneous recovery through the CS_2-US association (Robbins, 1990, Experiment 2). Finally, the presence of a reinforced CS (CS_1) does not interfere with the ability of an extinguished CS (CS_2) to undergo spontaneous recovery, since CS_1 does not affect the CS_2-US association.

Weak conditioning and spontaneous recovery

Using a taste aversion preparation, Rosas and Bouton (1996, Experiment 3) compared the effect of a retention interval after (a) extinction, and (b) conditioning with a weaker US to an intermediate level roughly similar to that observed at the end of extinction. They found that spontaneous recovery was present after acquisition and extinction, but not after weak conditioning for a long retention interval.

In Rosas and Bouton's (1996) Experiment 3, rats received free access to saccharin solution for 30 min. Illness in the experimental groups (EXT-L and EXT-S) was produced by a 2% body-weight injection of 0.15M LiCl. Illness in the control groups (0.5%-L and 0.5%-S) was produced by a 0.5% body-weight injection of 0.15M LiCl. The amount of liquid consumed was measured 1 day after extinction in Group EXT-S, 5 days after conditioning in Group 0.5%-S, 18 days after extinction in Group EXT-L, and 21days after conditioning in Group 0.5%-L.

As shown in Figure 9.7 (upper panel), whereas conditioned and extinguished rats showed spontaneous recovery after a long (18 days) but not a short (1 day) retention interval, no difference was present in rats receiving weak conditioning. These results suggest that the effect of a retention interval on CS responding depends on the history of reinforcement of the CS: spontaneous recovery was present only when the path to a certain level of responding included extinction.

Simulations

The parameters were the same ones used in the simulation of Robbins (1990). Simulations consisted of 20 conditioning and 10 extinction trials in groups EXT-L and EXT-S. Groups 0.5%-S and 0.5%-L received only 5 conditioning trials to the same level of conditioning of the EXT groups at the end of extinction. Eight test trials were given after 10 time periods of duration equal to the ITI in the home cage for the short interval (S) groups, and after 30 time periods of duration equal to the ITI in the long interval (L) groups. Figure 9.7 (lower panel) shows that the model correctly describes the experimental data.

According to the model, in the acquisition–extinction case, the CS is excitatory and CXg is inhibitory at the end of extinction. Attention to the CS and CXg somewhat decreases during extinction, and attention to CXg decreases further during home-cage time periods in the retention interval following extinction. Longer retention intervals result in weaker attention to the inhibitory CXg. Therefore, when the excitatory CS is reintroduced, attention to the CS is high and a CR is produced. The longer the attention to the inhibitory CXg takes to increase, the stronger the spontaneous recovery will be. When the retention interval is short, attention to the inhibitory CXg is relatively large and spontaneous recovery does not occur. Therefore, in the acquisition–extinction case, spontaneous recovery is stronger with a long retention interval. According to the model, in the weak conditioning case, a very weak excitatory CXg–US association is formed. In the absence of an inhibitory CXg–US

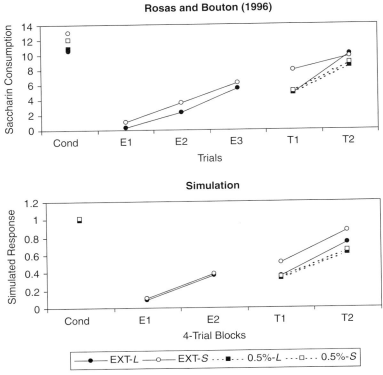

Figure 9.7 Spontaneous recovery is present after acquisition and extinction, but not after weak conditioning. *Upper panel*: Data from Rosas and Bouton (1996). *Lower panel*: Simulations. EXT-L: group that received extinction trials tested after a long extinction–test interval. EXT-S: group that received extinction trials tested after a short extinction–test interval. 0.5% L: group weakly conditioned tested after a long conditioning–test interval. 0.5% L: group weakly conditioned tested after a short conditioning–test interval. Cond: saccharin consumption on the last conditioning trial. E: extinction blocks. T: test blocks.

association, only attention to the CS determines the CR. Attention to the CS increases during acquisition and, in the absence of extinction, stays relatively high independently of the retention interval duration. Therefore, there is no difference in responding between short and long retention intervals in the weak conditioning case.

Effects of extinction cues on spontaneous recovery

Brooks and Bowker (2001) reported that presentation during testing of an extinction cue (EC), i.e. a cue that had been presented preceding the CS during extinction, attenuates spontaneous recovery. In their study, rats received CS–US presentations followed by extinction, in which three out of four

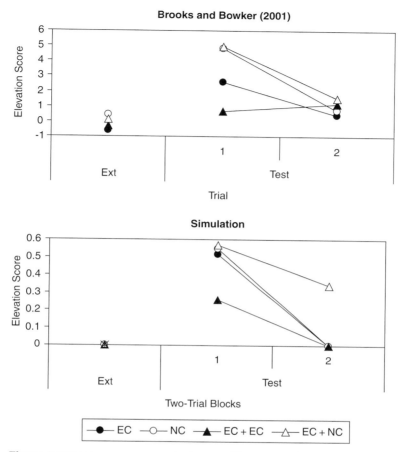

Figure 9.8 Spontaneous recovery is attenuated by presentation of a nonreinforced or reinforced extinction cue. *Upper panel*: Data from Brooks and Bowker (2001). *Lower panel*: Simulations. EC: testing with a nonreinforced extinction cue, NC: testing with no extinction cue, EC + EC: testing with a reinforced extinction cue, EC + NC: reinforced extinction cue not present during testing. Compare with Figure 15.2.

nonreinforced CS presentations were preceded by the EC, which terminated 40 s before the onset of the CS. After extinction, animals in groups EC + NC and EC + EC received EC–US presentations with the US onset coinciding with the EC termination, whereas animals in groups EC and NC spent an equivalent amount of time without any stimuli (retention interval). Finally, rats were tested for spontaneous recovery in the presence (EC) and the absence (NC) of the EC. A 30-s CS, a 30-s EC, food pellets as US and a 270-s mean ITI, were used.

As shown in Figure 9.8 (upper panel), Brooks and Bowker (2001) reported that a group presented with an EC preceding the CS during testing (Group EC) showed less spontaneous recovery than one presented without the cue

(Group NC). In addition, they showed that spontaneous recovery is attenuated even more if a US is presented immediately following the EC before the testing session (Group EC + EC vs. Group EC + NC).

Simulations

Simulations consisted of 10 CS–US trials, 10 cued extinction trials, 20 retention-interval trials, and 2 test trials. Simulations of 5 home-cage (CXh = 0.5, CXg = 0.4) time periods were included between the different sessions. The salience of the training context was set to CXc = 0.7 and CXg = 0.4. Both the CS and the EC had a 20 t.u. duration and salience 1. In cued-CS trials (extinction and testing), the EC coterminated with the onset of the CS. The ITI was 100 t.u.

Figure 9.8 (lower panel) shows that, in contrast to the experimental data, the model predicts responding in Group EC to be almost identical to responding in Group NC. However, the model correctly describes that a reinforced EC (Group EC + EC) attenuates responding when compared with responding in Group EC + NC. According to the model, responding in Group EC is similar to that in Group NC because, during extinction and testing, the trace and internal representation of the EC are relatively weak at the time of the CS presentation. Therefore, during extinction, a relatively modest EC–US inhibitory association is formed, and during testing this association is weakly activated when the CS is present. Also, responding in Group EC + EC is weaker than responding in Group EC + NC because the presence of an excitatory EC preceding the CS presentation during testing results in the formation of inhibitory CX–US associations that inhibit responding to the CS. In Chapter 15, we show that a configural version of the SLG correctly describes these results.

Discussion

Whereas the model describes extinction in terms of an associative competitive mechanism, with an important attentional process also at play, it explains spontaneous recovery in terms of an attentional process.

Computer simulations demonstrate that the model correctly describes that spontaneous recovery is (a) an increasing function of extinction–test time intervals, either when these intervals are spent in the home cage (Robbins, 1990) or in the training context (Pavlov, 1927), and (b) a decreasing function of the acquisition–extinction interval (Rescorla, 2004b). Both phenomena are explained in terms of decrements in attention to the CX. When attention to the CX decreases during the acquisition–extinction interval, the reduced X_{CX} is less able to form inhibitory CX–US associations during extinction, therefore failing to preserve the CS–US association needed for spontaneous recovery. When attention to the CX decreases during the extinction–test interval, the reduced X_{CX} results in less

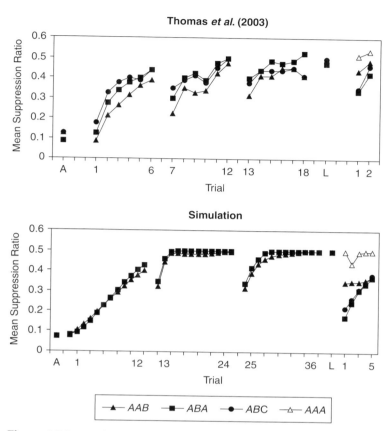

Figure 9.9 Renewal. Renewal is significant in groups *AAB*, *ABC* and *ABA* compared to Group *AAA*. *Upper panel*: Data from Thomas *et al.* (2003). *Lower panel*: Simulations. *A*: last acquisition trial, *L*: last trial of extinction.

activation of the existing inhibitory CX–US associations during testing, which frees the CS–US associations, and results in a strong spontaneous recovery.

Also in agreement with experimental results, the model can reproduce Robbins' (1990) data showing that spontaneous recovery (a) is obtained only with a CS that has been previously reinforced, (b) is present even when the test trials are presented in the middle of the session, (c) occurs despite the presence of another reinforced CS, and (d) occurs to a CS after spontaneous recovery and reextinction of another CS.

In addition, the model correctly describes that spontaneous recovery is present only when the path to a certain level of responding included extinction (Rosas & Bouton, 1996, Experiment 3). Furthermore, the model describes external disinhibition and, as shown in Figure 9.9, the model is capable of describing that spontaneous recovery decreases with repeated extinction sessions (Thomas *et al.*,

2003). However, the model is unable to capture the effect of a nonreinforced extinction cue on spontaneous recovery. As mentioned, Chapter 15 shows how a configural version of the SLG model addresses these data.

Comparison with other models

Robbins (1990) compared predictions of different theories of extinction in light of his results. In his view, the suggestion that recovery reflects the incomplete elimination of early-session cues like handling (e.g. Burstein, 1967; Skinner, 1950) has difficulties explaining his data showing spontaneous recovery in the middle of a test session. Also in Robbins' view, Capaldi's (1967) generalization decrement model and memory retrieval accounts of recovery (e.g. Gleitman, 1971; Konorski, 1967; Spear, 1971) have problems to explain the occurrence of recovery when the conditions of recovery testing are more similar to those of extinction than those of acquisition (Experiment 2). In addition, Robbins suggested that the facts that (a) recovery occurs after extinction in the presence of a US representation (Experiment 2), and (b) recovery to one CS is possible immediately following the extinction of another CS (Experiment 2) do not support Rescorla's (1979) US processing view or Hull's (1943) reactive inhibition theory. In contrast, Robbins proposed that Pavlov's (1927) idea that extinction decreases CS processing, which can recover during a time delay, is adequate to explain his data. Robbins pointed out that slow reacquisition (Bouton, 1986; Bouton & Swartzentruber, 1989; Ricker & Bouton, 1996) also supports Pavlov's view of reduced CS processing.

Common to Pavlov's theory and our model is the notion that attention (processing) to the CS decreases during extinction. In our model, this is because Novelty' decreases as the prediction of the US decreases. Also common to both approaches (see also Sokolov, 1963; Wagner, 1976) is the idea that extinction shares some properties with both habituation and latent inhibition. In our model, although not identical (see Chapter 5, sections on "The effects of different parameters of preexposure on the strength of LI" and on "Orienting response and LI"), both habituation and latent inhibition occur because Novelty' and attention to a stimulus decrease with its repeated presentations. Furthermore, because in our model an attentional mechanism is at work during both extinction and latent inhibition, the model is capable of describing data on spontaneous recovery from extinction (data and simulations shown in this section) and recovery from latent inhibition (simulations shown in Schmajuk & Larrauri, 2006; data from Grahame *et al.*, 1994).

The protection from extinction phenomenon, described by our model, can be correctly described by the Rescorla and Wagner (1972) model and has been previously reported in the literature (Chorazyna, 1962; Rescorla, 2003; Soltysik, 1985).

However, in contrast to our model, the Rescorla and Wagner (1972) model also predicts that a CX–US inhibitory association should decrease during the periods in which the CX is presented alone. Such decrements, although at odds with data on the extinction of conditioned inhibitors (Zimmer-Hart & Rescorla, 1974; Rescorla, 1982; Witcher & Ayres, 1984), would allow the Rescorla–Wagner model to describe spontaneous recovery following periods in which the CS is not presented.

Finally, experimental data are in agreement with Devenport's (1998) suggestion that, because recent CS–US inhibitory memories are weighted more heavily than old CS–US excitatory memories, extending the extinction–test interval should reduce more the recent CS–US inhibitory memory than the old CS–US excitatory memory, thereby increasing recovery. Also, Devenport's (1998) view is supported by experimental results showing that extending the acquisition–extinction interval decreases spontaneous recovery because extending this interval reduces mostly the remote acquisition memories.

Renewal

Renewal refers to the phenomenon by which the extinguished CR reappears when the CS is tested in a context different from that of extinction. The renewal effect, first reported by Bouton and Bolles (1979b), has been demonstrated using three different procedures described below. In the framework of the model, renewal is explained because when the animals are removed from the inhibitory CX of extinction, the remaining CS–US associations are free to generate a CR.

Effects of different procedures and the presence of odor cues

Thomas et al. (2003, Experiment 4) compared the magnitude of the renewal effect using *AAA*, *ABA*, *AAB*, and *ABC* procedures, in a conditioned emotional response paradigm in which rats learned to bar press for a sucrose solution. All groups were conditioned in CX*A*; extinguished in CX*A* for Groups *AAA* and *AAB*, and in CX*B* for Groups *ABA* and *ABC*; and tested in CX*A* for groups *AAA* and *ABA*, in CX*B* for Group *AAB*, and in CX*C* for Group *ABC*. A 2-min light-off CS, a 1-s, 0.5-mA shock US, and a 17.4-min mean ITI, and odors and room locations to differentiate between contexts, were used.

Figure 9.9 (upper panel) shows spontaneous recovery between different extinction sessions, and significant renewal in all groups compared to Group *AAA* in the test session. Transfer of fear from the acquisition context to the extinction context was similar in all groups. Renewal was weaker in Group *AAB* than in the *ABA* and *ABC* groups, which did not differ. Notice that even if suppression ratios for all groups are shown during testing, only groups *AAB*, *ABA* and *ABC* are shown during extinction. The group that received extinction trials in CX*A* was split into groups *AAA* and *AAB* during testing.

In addition, Thomas *et al.* (2003, Experiment 1B) found that *ABA* renewal disappears after eliminating the odor cue that distinguished (in addition to their location in the room) the different contexts.

Simulations

Acquisition consisted of twenty CS–US trials in CX*A*, in which an 11-t.u. CS (salience 1) coterminated with a 5-t.u. US (salience 2). Three extinction sessions were simulated for each group (in CX*A* for Groups *AAA* and *AAB*, and in CX*B* for Groups *ABA* and *ABC*), consisting of 12 CS– presentations, and separated by 20 home-cage trials. Before the start of the extinction phase, one trial of exposure to the context where extinction was going to take place was simulated, as in Thomas *et al.*'s (2003) experimental design. Following extinction, 5 test trials were performed after one trial of preexposure to the test context. The salience of the contexts (CXc) was set to 0.7, and the ITI was 70 t.u.

Figure 9.9 (lower panel) shows that the model correctly describes spontaneous recovery between different extinction sessions, and that renewal was present in all groups compared to Group *AAA*. Like in the experimental data, transfer of fear from the acquisition context to the extinction context is similar in all groups. Also, simulated results show that renewal is stronger in groups *ABA* and *ABC* than in Group *AAB*.

Figure 9.10 shows the values of different variables in the model, just before the time of presentation of the US on reinforced trials. Because in this simulation CXc (0.7) is more salient than CXg (0.4), the upper panel of Figure 9.10 shows the simulated CR and associative values V_{CXc-US} and V_{CS-US} and the lower panel shows Novelty' and representations X_{CXc} and X_{CS} during the last phase of extinction and *ABA* renewal. According to the model, repeated extinction sessions decrease spontaneous recovery because at the beginning of each extinction session, Novelty' increases, activating first the representation of the CS (X_{CS}) and later the one of the CX (X_{CXc}). Increases in X_{CS} and not in X_{CX} not only produce spontaneous recovery, but also a decrease in V_{CS-US} because the US is absent and V_{CS-US} is not protected by V_{CS-US}. Therefore, in trials in which spontaneous recovery is obtained (that is, until X_{CXc} becomes active) there is also further extinction of the CS–US association, which is observed as a decrease in spontaneous recovery in repeated extinction sessions.

According to the model, transfer of fear from the acquisition context to the extinction context was similar in all groups because the CS–US association is much stronger than the CX–US association; a result of the competition between the more salient CS and the CX to gain association with the US (see Chapter 2,

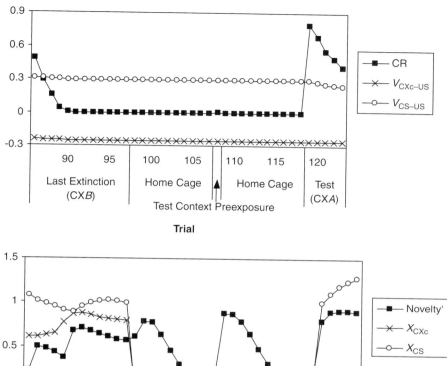
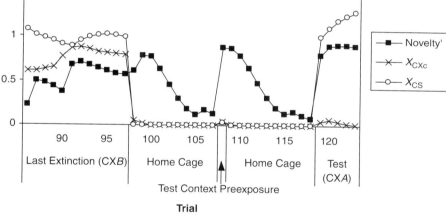

Figure 9.10 Variables in the model during *ABA* renewal. *Upper panel:* Values of the conditioned response, CR; the association between the specific context B and the unconditioned stimulus, $V_{CXc,US}$, and the CS–US association, $V_{CS,US}$; as a function of trials. *Lower panel:* Values of average Novelty′; the internal representation of CXc, X_{CXc}; and CS internal representation X_{CS}; as a function of trials. Trials in the Last Extinction Phase 86–97, Home-cage Time Periods 98–117, and Test Trials 118–137.

section on "Changes in CS–US associations"). While fear transfers well from the acquisition to the extinction context, extinction of fear does not transfer between contexts because it depends on the inhibitory CXc–US association that is specific to the context in which extinction took place, i.e. renewal is present.

The model explains renewal as following. In the *AAA* case, responding is small at testing because the test context becomes inhibitory during extinction trials and only a weak spontaneous recovery is present at the beginning of the testing session. In the *AAB* case, CX*A*c becomes inhibitory during extinction, but since this context is not activated during testing (CX*B*c is), it does not reduce the CR. Similarly, in the *ABA* (or *ABC*) case, CX*B*c becomes inhibitory during extinction trials, but since CX*A*c (or CX*C*c) is activated during test (and not CX*B*c), the CR is not inhibited, and therefore, renewal is obtained.

Importantly, the model correctly describes the different types of renewal (*AAB*, *ABA*, *ABC*) as follows. Responding in the *ABA* and *ABC* cases is stronger than in the AAB case because during extinction Novelty′ is higher when the context is different from that of acquisition (*ABA* and *ABC* cases), which increases the internal representation of the CS. This increased CS representation results in a stronger responding during testing. In short, an attentional mechanism explains the stronger responding in groups *ABA* and *ABC* than in Group *AAB*. Notice that this increased attention also explains the relatively faster extinction in groups *ABA* and *ABC* than in Group AAB. Interestingly, the model predicts that different types of renewal will become increasingly differentiated as the salience of the contexts increases.

However, the model expects *ABA* renewal to be slightly stronger than *ABC* renewal, a difference that is absent in the Thomas *et al.* (2003) data, but had been previously reported by Harris *et al.* (2000, Experiment 1). According to the model, stronger responding in Group *ABA* than in *ABC* is explained in terms of the small excitatory CX*A*c–US associations formed during conditioning that contribute to responding when test trials are conducted in CX*A* (*ABA* case).

Finally, in agreement with Thomas *et al.*'s (2003, Experiment 1B) results, the model expects renewal to disappear when the contextual cues are decreased. According to the model, low salience contexts eliminate *ABA* renewal because the unique features of the chambers (CXc) have a relatively small salience and, therefore, it is CXg that becomes most inhibitory during extinction. Consequently, during test trials the CR is still inhibited by CXg, regardless of the context in which extinction took place. However, spontaneous recovery still happens because of the delayed activation of CXg during testing.

Elimination of renewal by pretreatment of the testing context

Lovibond *et al.* (1984, Experiments 1c and 2) reported that renewal (demonstrated in their Experiments 1a and 1b) was eliminated when the subjects had similar experience of the contexts during all phases of the experiment. They suggested that renewal might be the consequence of the differences in the history of reinforcement and nonreinforcement associated with the contexts. In their experiments, conditioning was followed by extinction either in the same context of conditioning (Group Same, or in a different one (Group Diff). In Experiment 1b, a tone CS (*T*) was conditioned in CX_1, followed by presentations in CX_2. For Group Same, extinction consisted of T– presentations in CX_1, alternated with sessions in CX_2 alone. For Group Diff, extinction consisted of T– presentations in CX_2, alternated with sessions in CX_1 alone. In both groups, T was tested in CX_1. In Experiment 1c, a tone CS (T) was conditioned in CX_1, whereas a light (L) was conditioned in CX_2. For Group Same, extinction consisted of T– presentations in CX1, alternated with L– presentations in CX_2. For Group Diff, extinction consisted of T– presentations in CX_2, alternated with L– presentations in CX_1. Testing in both groups consisted of presentations of T in CX_1 and of L in CX_2. A 90-s, 80-dB tone CS, a 0.5-Hz flashing overhead light CS, followed by a 0.5-s, 0.6-mA shock US, a 15-min mean ITI and odors to distinguish between contexts, were used.

Figure 9.11 (upper panel) shows suppression ratios to a tone CS (T) in two groups, Group Same (Same) receiving conditioning, extinction and testing in the same context, and Group Different (Diff) that received extinction in a different context (Experiment 1b). It also shows suppression ratios of a group (Group Diff Ext) that received extinction in a context different from that of conditioning and was tested in its own conditioning context, in which another CS was previously extinguished (Experiment 1c). The results demonstrate renewal (small suppression ratio) in Group Diff, but the phenomenon disappears in Group Diff Ext in which another CS had undergone extinction. Although Grahame et al. (1990, Experiment 2) reported results similar to those of Lovibond et al. (1984), Harris et al. (2000, Experiment 1) reported that renewal was still present when a CS was tested in a context in which extinction of another CS had taken place. Harris et al.'s (2000) data seem to favor the view that extinction of a CS is context specific, and that a CX–CS configuration might be present. However, according to our model, the increased novelty and attention to the CS when tested in a context different from that of its own extinction might also contribute to the result.

Simulations

Simulations of acquisition in Experiment 1b consisted of 10 CS–US trials in CX_1, followed by 10 CX_2-alone presentations. For Group Same, 10 nonreinforced presentations of the CS in CX_1 and 10 CX_2-alone presentations were simulated. For Group Diff, 10 CX_1-alone presentations were followed by 10 nonreinforced presentations of the CS in CX_2. The test phase consisted of 5 CS trials in CX_1. Acquisition in Experiment 1c consisted of 10 T–US presentations in CX_1 followed by 10 L–US presentations in CX_2. For Group Same (not shown), 10 nonreinforced presentations of T in CX_1 and 10 nonreinforced presentations of L in CX_2 were simulated. For Group Diff Ext, 10 nonreinforced presentations of L in CX_1 and 10 nonreinforced presentations of T in CX_2 were simulated. The

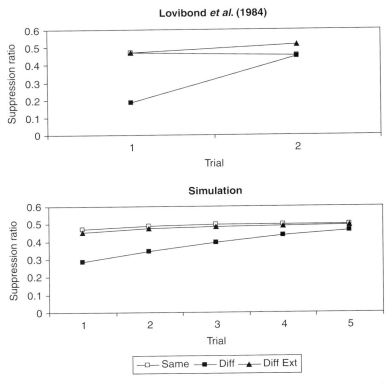

Figure 9.11 Elimination of renewal. Renewal disappears when the subjects have similar experience of the contexts during all phases of the experiment. *Upper panel*: Combined data from Lovibond *et al.*'s (1984) Experiments 1c and 2. *Lower panel*: Simulations. Same: test in the context of extinction, Diff: test in a context different from that of extinction, Diff Ext: test in a context in which another CS had received extinction.

test phase consisted of 5 presentations of T in CX_1. Between the different phases of the experiment, 30 home-cage (CXh = 0.5, CXg = 0.4) time periods were simulated. The salience of the training context was set to CXc = 0.5 and CXg = 0.4. The CS had a 20 t.u. duration and salience 1, the US was 5 t.u. long with salience 2 and the ITI was 100 t.u.

Figure 9.11 (lower panel) shows that the experimental data are well described by the model. According to the model, previous extinction in a given context makes it inhibitory to any CS that has a remaining excitatory association with the US, thereby eliminating responding.

It is interesting to notice that, according to the model, different results are obtained when (a) a previously conditioned CS (e.g. Bouton & Swartzentruber's summation experiment) and (b) a previously trained and extinguished CS (e.g. Lovibond *et al.*'s renewal experiment) are tested in a context in which extinction has taken place. As shown in Figures 9.3 and 9.12, the model correctly describes a small difference in summation, and a significant difference in renewal, between experimental and control groups. According to the model, the inhibitory CX–US association has a stronger effect on the CR when the CS–US association has decreased during extinction and is weak (as in Lovibond *et al.*'s renewal experiment), than when the CS–US association is intact and strong after conditioning (as in Bouton & Swartzentruber's summation experiment).

Elimination of renewal by massive extinction

Denniston *et al.* (2003, Experiment 1) studied the effect of massive extinction on ABC renewal. Acquisition consisted of CS–US pairings in CX*A* for all groups. Extinction for Group Moderate (Ext-Mod) consisted of 160 presentations of CS alone in CX*B*. Group Massive (Ext-Many), received 800 presentations of CS alone in CX*B*. All subjects received exposure to CX*C* where testing was conducted. As a control, Denniston *et al.* (2003) also determined the effect of extinction in the ABB case (Ext-Mod-B). A 10-s white noise CS, a 0.5-s, 1.0-mA shock US, a 180-s (groups Mod) or a 36-s (Group Ext-Many) between-CS mean ITI and odors to increase the differences between contexts, were used.

Figure 9.12 (upper panel) shows that renewal is present after moderate extinction in CX*B* (responding in Group Ext-Mod tested in CX*C* is stronger than in Group Ext-Mod-B tested in CX*B*). Renewal, however, is eliminated by massive extinction in CX*B* (responding in Group Ext-Many is weaker than in Group Ext-Mod). Like Denniston *et al.* (2003), Bouton and Swartzentruber (1989), Rauhut *et al.* (2001) and Tamai and Nakajima (2000) also failed to observe an attenuation of renewal with a moderate number of extinction trials.

Simulations

Simulations of Denniston *et al.*'s (2003) Experiment 1 were carried out using the same parameters than those described for the Thomas *et al.* (2003) case, except that CXc = 0.5. Groups Ext-10 and Ext-10-B received 3 sessions of 10 CS– trials in CX*B*, whereas Group NoExt received an equivalent number of trials of CX*B* alone exposure. Group Ext-80 received 80 CS– trials in each extinction session in CX*B*. Testing was conducted in CX*C* in Groups NoExt, Ext-10 and Ext-80, and in CX*B* in Group Ext-10-B.

As in the experimental results, Figure 9.12 (lower panel) shows simulations with the model demonstrating that renewal is present after 3 sessions of 10 extinction trials in CX*B* (responding in Group Ext-10 is much stronger than in Group Ext-10-B), but is clearly reduced by 3 sessions of 80 extinction trials in CX*B* (responding in Group Ext-80 is weaker than in Group Ext-10). According to the model, during extinction trials, the CS–US association decreases until

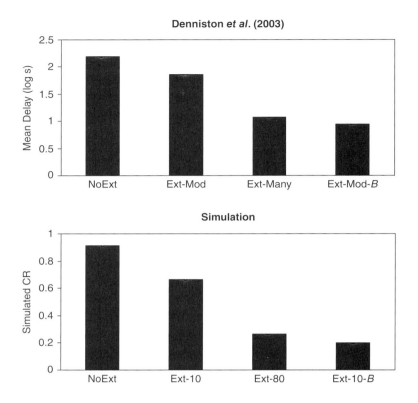

Figure 9.12 Elimination of renewal. Renewal is present after moderate extinction in Context B (responding in Group Ext-Mod is stronger than in Group Ext-Mod-B), but it is eliminated by massive extinction in Context B (Groups Ext-Many and Ext-Mod-B show similar responding). *Upper panel*: Data from Denniston *et al.* (2003). *Lower panel*: Simulations. NoExt: no extinction, Ext-10: 10 trials of extinction in CXB and testing in CXC, Ext-80: 80 trials of extinction in CXB and testing in CXC, Ext-10-B: 10 trials of extinction in CXB and testing in CXB.

the CXBc–US inhibitory association counterbalances the CS–US excitatory association. Therefore, the CS–US association becomes protected from extinction (because the US is not present and not predicted). However, increasing the number of extinction trials decreases attention to the CS, as Novelty′ decreases in those trials. Therefore, representation X_{CS} is very small (z_{CS} becomes negative) and will take time to increase even when Novelty′ becomes relatively large in a different context, leading to a decreased renewal effect.

Effects of an extinction cue on renewal

Brooks and Bouton (1994) reported that presentation of an extinction cue preceding the CS during testing attenuates renewal. In their Experiment 2,

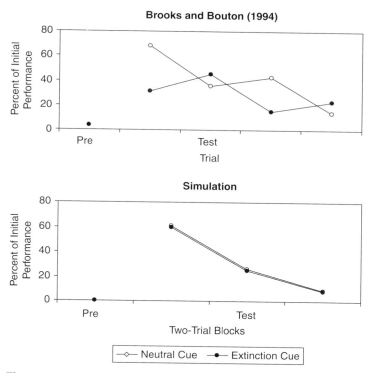

Figure 9.13 Renewal is attenuated by presentation of an extinction cue. *Upper panel*: Data from Brooks and Bouton (1994, Figure 3). *Lower panel*: Simulations. Compare with Figure 15.5.

rats received conditioning sessions of a tone-CS in Context A. Extinction sessions took place in Context B. An extinction cue (EC) preceded the nonreinforced presentation of the CS on three of every four extinction trials. The offset of the house light and a 30-s, 65-dB white noise were used either as the extinction cue or the neutral cue. Test sessions were conducted in Context A, where the CS was preceded either by the extinction cue (Group Extinction Cue) or the neutral cue (Group Neutral Cue). A 30-s extinction cue preceded by a 15-s gap, an 80-dB tone on cued trials, food pellets as US and a 310-s mean ITI, were used. Figure 9.13 (upper panel) shows that renewal is stronger with a neutral CS than with an extinction cue.

Simulations

The parameters used were identical to those described for the Brooks and Bowker (2001) study on spontaneous recovery (Figure 9.8), without the retention interval phase. Computer simulations with the model cannot replicate the experimental results (Figure 9.13, lower panel). According to the model, the absence of a difference in responding when the target CS is preceded by the extinction cue, or by a neutral cue is due to the weak trace and internal representation of the extinction cue, during extinction and testing, at the time of the CS presentation. Even if a closer temporal proximity between the extinction cue and the CS would have resulted in the extinction cue becoming a conditioned inhibitor, which could have contributed to the decreased renewal reported by Brooks and Bouton (1993), such a solution contradicts their results showing that extinction cues show no evidence of having become inhibitory in summation and retardation tests. Chapter 15 addresses this issue with a configural version of the SLG model.

Discussion

The model correctly describes experimental results showing that renewal is absent when the target CS is tested in the context where another CS underwent extinction (Lovibond *et al.*, 1984; Grahame *et al.*, 1990; but see Harris *et al.*, 2000), that *ABA* and *ABC* procedures result in stronger renewal than the *AAB* procedure (Thomas *et al.*, 2003), that decreasing the salience of the context impairs renewal (Thomas *et al.*, 2003) and that renewal is eliminated with extended extinction (Denniston *et al.*, 2003). However, it cannot replicate the results showing that renewal is attenuated by an extinction cue (Brooks & Bouton, 1994). Notice that the model explains renewal in terms of inhibitory associations, and the difference between *ABA* and *ABC* renewal in terms of excitatory associations. In contrast, the difference in renewal obtained with different procedures (*ABA* and *ABC* vs. *AAB*) and the reduction of renewal with extended extinction are both explained in attentional terms.

Comparison with other models

As observed by Tamai and Nakajima (2000) and Thomas *et al.* (2003), the differences in responding in the *AAB* and *ABC* cases contradict the prediction of Bouton's (1993, 1994) model. According to Bouton's (1993, 1994) view, removal from the context of extinction should eliminate the inhibition of the CS–US association and result in identical responding in both cases. As indicated by Thomas *et al.* (2003), the fact that renewal is incomplete (renewed responding does not reach the level of responding by the end of acquisition) also contradicts Bouton's (1993, 1994) model, which implies that it should be complete unless some generalization is assumed between the extinction and test contexts. Such a generalization is included in our model (CXg). Denniston *et al.*'s (2003) results are also opposite to Bouton's (1993, 1994) view, which anticipates that massive extinction would enhance the potential of the context to serve as a negative occasion setter and, therefore, enhance the context specificity of extinction and increase renewal. Denniston *et al.* suggested that an inhibitory

CS–US association is established during massive, but not moderate, extinction, that can be expressed outside of the extinction context.

Thomas and Ayres (2004) proposed a theory of renewal that suggests that (a) nonreinforced presentations of the CS can reduce CS–US associations, (b) these presentations also produce some form of CS–US inhibitory associations that are more specific to the context of extinction than the excitatory ones, (c) excitatory and inhibitory CS–US associations coexist, and (d) the CR is proportional to all co-present CSs–US associations. Points (a), (b), and (d) are shared by our model.

Finally, the Rescorla and Wagner (1972) model also describes renewal in terms of CX–US inhibitory associations that are absent in the testing context. This model can describe the advantage of *ABA* over *ABC* renewal (Thomas et al., 2003) in terms of the remaining excitatory CXAc–US associations in the *ABA* case. As mentioned, the Rescorla and Wagner (1972) rule can only explain *AAB* renewal if the context of acquisition becomes inhibitory during extinction (e.g. when CX–US associations are already decreased at the beginning of extinction.) In this case, *AAB* renewal will be larger with weaker CXAc–US associations at the beginning of extinction. Only if the CXAc–US association is zero, will *AAB* renewal become as strong as *ABC* renewal. Therefore, in general, the Rescorla and Wagner (1972) rule would be able to describe *ABC* renewal being stronger than *AAB* renewal. Notice that our model does not require an explicit extinction of CXAc–US associations for them to become inhibitory, thereby protecting the CS–US association, and producing *AAB* renewal. Thus, our model explains in attentional terms why *AAB* renewal is weaker than *ABC* renewal even when CS–US associations are similar in both cases. Attentional mechanisms are also required to describe the attenuation of renewal by massive extinction (Denniston et al., 2003).

Reinstatement

Reinstatement refers to the phenomenon, first described by Pavlov (1927) and later by Rescorla and Heth (1975), by which the extinguished CR reappears when the CS is tested in the context of extinction following unsignaled US presentations in that context. In the framework of the model, this form of reinstatement is explained because US presentations in the context decrease inhibitory CX–US associations, allowing the remaining CS–US associations to generate a CR.

Effects of reinforced exposures in the context of extinction and testing

In Bouton and Bolles' (1979a) Experiment 2, rats in the Classical Conditioning (CC) and Extinction (E0, E1, and E2) groups received (a) operant training in

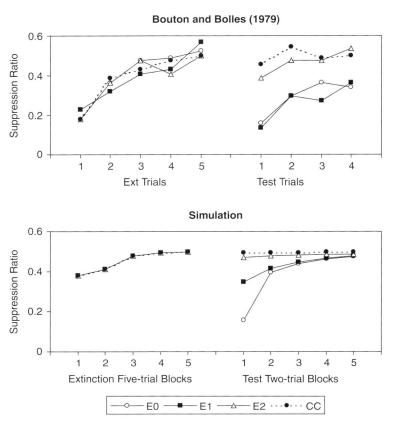

Figure 9.14 Reinstatement following nonreinforced exposures in the context of extinction and testing. Reinstatement is obtained only when the US was presented in the context in which testing occurs (Group E0 vs. Group CC) and prolonged extinction of the context of US presentation and testing degrades reinstatement (Group E0 vs. Group E2). *Upper panel*: Data from Bouton and Bolles (1979a). *Lower panel*: Simulations.

CX*A*, (b) conditioning in CX*B*, (c) extinction in CX*A*, (d) reinstatement consisting of US presentations in CX*A* (Groups E0, E1, and E2) or CX*B* (Group CC), and (e) testing in CX*A*. Rats in the Extinction groups also received either (a) no extinction trials (Group E0), (b) two extinction sessions in CX*A* (Group E1), or (c) six extinction sessions in CX*A* (Group E2), between reinstatement and CS testing in CX*A*. Rats in Group CC were kept in holding cages for an equivalent amount of time. A 60-s tone CS, a 0.5-s, 1-mA footshock US and a 3-min mean ITI, were used.

Figure 9.14 (upper panel) shows that (a) reinstatement is obtained only when the US was presented in the context in which testing occurs (Group E0 vs. Group CC), (b) a prolonged extinction of the context of US presentation and testing degrades reinstatement (Group E0 vs. Group E2), and (c) a brief extinction of the context has little effect on reinstatement (Group E0 vs. Group E1). Importantly, Bouton and Bolles (1979a, page 376) reported that, following presentations of the US in the context, CX–US associations were not detected by bar-pressing rates in the absence of the CS. However, such excitatory contextual associations were later detected using a context-preference test (Bouton & King, 1983).

Simulations

Conditioning sessions consisted of 4 CS–US trials in CXB (conditioning box). Extinction consisted of 25 CS trials in CXA (Skinner box). After extinction, 5 US trials were simulated in CXB (Group CC) or in CXA (Groups E0, E1 and E2). In addition, following the reinstatement shocks, Groups E1 and E2 received, respectively, 1 and 60 trials of exposure to CXA, whereas Groups CC and E0 spent an equivalent amount of time periods in the home cage. The salience of CXA (Skinner box) was set to CXc = 0.5 and CXg = 0.4. The salience of CXB (conditioning box) was set to CXc = 0.1 and CXg = 0.4. The CS had a 10 t.u. duration and salience 0.8, the US was 5 t.u. long with salience 2 and the ITI was 100 t.u. Simulations of 20 home-cage (CXh = 0.5, CXg = 0.4) time periods were included between the different sessions.

Figure 9.14 (lower panel) shows that the model approximates well the Bouton and Bolles (1979a) results: Group CC shows little, and Group E0 shows strong responding to the CS. Whereas a partially reduced reinstatement is present in E1, Group E2 shows no reinstatement.

In addition, like in the Bouton and Bolles (1979a) study, when expressed as suppression ratios, our simulations show no appreciable contribution of the CX–US associations following US presentations in the context. However, as reported by Bouton and King (1983), due to the excitatory associations formed during CX–US presentations, the context is able to clearly activate the US representation.

Because in this simulation CXc (0.5) is more salient than CXg (0.4), the upper panel of Figure 9.15 shows CR and associative values V_{CXc-US} and V_{CS-US}, and the lower panel shows average Novelty' and representations X_{CXc} and X_{CS} during extinction, US presentations in the context and testing in the same context. Two periods in the home cage are also shown. According to the model, reinstatement (Group E0 in Figure 9.14) is explained in the following terms. During extinction, the excitatory CS–US association decreases and the inhibitory CXc–US association increases until they counterbalance each other. When the shock US is administered in the context, the CXc–US association becomes excitatory, which results in a large CR during the test trials. Increased responding during testing is also helped by the increased attention to the CS and the CX as Novelty' increases when (a) the context predicts the US that is not presented,

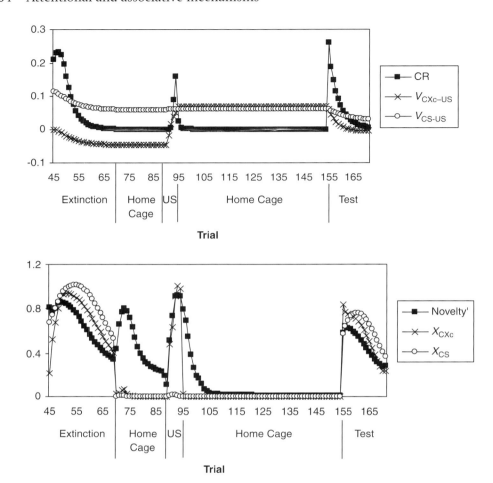

Figure 9.15 Variables in the model during reinstatement. *Upper panel*: Values of the conditioned response, CR; the association between the specific context and the unconditioned stimulus, V_{CXc-US}; and the CS–US association, V_{CS-US}; as a function of trials. *Lower panel*: Values of average Novelty'; the internal representation of CXc, X_{CXc}; and CS internal representation X_{CS}; as a function of trials. Trials in the Extinction Phase 45–69, Home-cage Time Periods 70–89, US Presentation Trials 90–94, Home-cage Time Periods 95–154, and Test Trials 155–174.

and (b) the CS is presented in the context that no longer predicts it. Conversely, when shocks are administered in a context different than that of extinction, the inhibitory CXc–US association developed during extinction does not change, and the CR does not reappear during test trials.

It is worth noticing that the model provides an intriguing explanation for why attention contributes to the reduction of reinstatement with many nonreinforced presentations of the previously reinforced context (Group E2 in Figure 9.14). According to the model, exposure to the CX following presentations of the US partially reduces reinstatement because excitatory CX–US associations will tend to decrease to zero. However, nonreinforced CX presentations will not produce the inhibitory CX–US associations needed to counterbalance the unchanged CS–US excitatory association, and eliminate the CR. This is so because in order to obtain an inhibitory CX–US association, the excitatory CS–US association should be active too, as it was during extinction. In short, the model shows that extinction of the context cannot bring back the situation present at the end of extinction when the inhibitory CX–US association counteracted the excitatory CS–US association. Therefore, Group E2 shows a reduced reinstatement due, in part, to a decreased excitatory CX–US association, and in a smaller part to a relatively small attention to the CS (compared with the large attention to the CS in Group E0, Figure 9.15). Attention to the CS is reduced because Novelty′ does not increase since the CX does not predict the US, which is absent during testing.

The model also addresses Delamater's (1997) results showing that reinstatement (a) is reduced by extended extinction, and (b) is not present if the reinstating US is different from that used during conditioning. In the first case, reinstatement is reduced because attention to the CS decreases during extended extinction. In the second case, reinstatement is US-specific in the case of classical conditioning because the inhibitory $CXc-US_1$ association does not decrease when the representation of CXc is active in the absence of the original US_1, and does not increase when a different US_2 is administered. In contrast, reinstatement was found to be nonspecific in the case of instrumental conditioning (Delamater, 1997) or after retroactive interference in human causal learning (Garcia-Gutierrez & Rosas, 2003). These nonspecific increases in responding could be the result of spontaneous recovery, as suggested by Delamater (1997, page 11), or a degree of generalization between USs.

Effects of reinforced exposures in the context of extinction or testing

As shown in Figure 9.16 (upper panel), Westbrook *et al.* (2002, Experiment 2) reported reinstatement when US delivery and CS testing occur in the same context: when using an *AAAA* procedure (CS conditioning in CX*A*, CS extinction in CX*A*, US presentations in CX*A*, and CS testing in CX*A*) or an *AABB* procedure (CS conditioning in CX*A*, CS extinction in CX*A*, US presentations in CX*B*, and CS testing in CX*B*). In addition, they also reported reinstatement when the US is presented in the context of extinction and tested in a different context (*AAAB* procedure). In contrast, they found no reinstatement when US presentations occur in a context different from that of extinction, and testing occurs in the extinction context (e.g. *AABA*). A 30-s, 78-dB, 10-Hz clicker CS, a 0.5-s, 0.8-mA

206   Attentional and associative mechanisms

footshock US, a 2.5-min mean ITI and odors to increase the differences between contexts, were used.

Furthermore, as shown in Figure 9.17 (upper panel), Westbrook *et al.* (2002, Experiment 3) reported that Group *AAAB* shows stronger freezing than Group *AA–B*, which did not receive US presentations, a result that demonstrates that responding in the *AAAB* group is reinstatement (the effect of the US delivered in CXA) and not renewal (the effect of changing from the context of extinction CXA to CXB).

Simulations

Simulations shown in Figures 9.16 and 9.17 included 10 CS–US acquisition trials and 25 CS extinction trials, in CXA. Presentation of the US and testing occurred in the contexts corresponding to those used in the experiment. The salience of all CXs was set to CXc = 0.7 and CXg = 0.4. The CS had a 20 t.u. duration and salience 1, the US was 5 t.u. long with salience 2 and the ITI was 100 t.u. Simulations of 10 home-cage (CXh = 0.5, CXg = 0.4) time periods were included between the different sessions.

As shown in Figure 9.16 (lower panel), in agreement with Westbrook *et al.* (2002, Experiment 2a), simulations show strong freezing in the *AAAA*, *AAAB*, and *AABB* groups and weak freezing in the *AABA* group. As explained in the previous section, when US presentations and testing occur in the same context (*AAAA* and *AABB* cases), reinstatement is the consequence of (a) the elimination of CX–US inhibitory associations by the US presentations; and (b) the increased attention to the CS and the CX, which activates the remaining CS–US association and the newly formed CX–US association during testing. As mentioned, attention to the CS increases during testing as a consequence of the increased Novelty´, which results from the absence of the US after having been presented in that context, and the unpredicted presence of the CS after being absent in the previous phase in that context.

In the *AAAB* case, in which US presentations occur in the context of extinction (CXA) but the CS is tested in a different context (CXB), reinstatement results from (a) the combination (chaining) of the CS–CXAc association with the now excitatory CXAc–US association, and (b) increased attention to the CS when the animal is moved from CXA to CXB. In contrast, because inhibitory CXAc–US associations are present in CXA, Group *AABA* does not show reinstatement of the freezing response.

Figure 9.17 (lower panel) shows that the model can also describe Westbrook *et al.*'s (2002, Experiment 3) results. These simulated results demonstrate that even when CXAc–US inhibitory associations are absent when the animals respond in CXB (Group *AA–B*), part of the freezing response is due to the

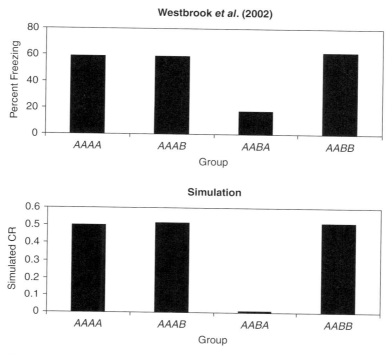

Figure 9.16 Reinstatement following reinforced exposures in the context of extinction or testing. Reinstatement is obtained when the US was presented in the context in which testing occurs (Groups *AAAA* and *AABB*), or (b) in the context of extinction (*AAAB*). *Upper panel*: Data from Westbrook *et al.* (2002). *Lower panel*: Simulations.

combination of CS–CXAc and CXAc–US associations (reinstatement) and not to the decreased inhibition when testing occurs in CXB in the absence of US presentations in CXA (renewal).

Elimination of reinstatement

In addition to Bouton and Bolles' (1979a, Experiment 2) demonstration that context extinction sessions between reinstatement and testing stages eliminate reinstatement, Rescorla and Cunningham (1977) showed that reinstatement could be eliminated by two presentations of a previously conditioned CS in one 60-min period. In Rescorla and Cunningham's (1977) Experiment 1, rats received (a) training in Skinner boxes; (b) conditioning of both a 2-min tone and a 2-min flashlight CS terminating in a 0.5-s, 0.5-mA footshock US; and (c) alternated extinction presentations of each of two CSs. Reinstatement was obtained with US presentations in the context alone. The effect of the US presentations was tested by presenting the tone CS after the

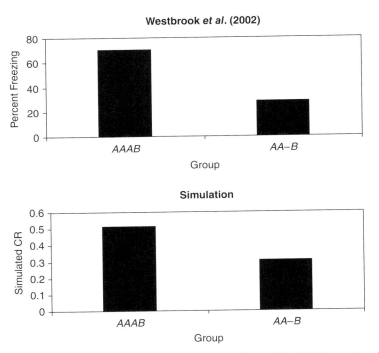

Figure 9.17 Strong responding in Group *AAAB* is mostly reinstatement and not renewal. *Upper panel*: Data from Westbrook *et al.* (2002, Experiment 3). *Lower panel*: Simulations.

rats either (a) stayed in the context, or (b) received nonreinforced presentations of the reconditioned light CS. A 30-min ITI was used.

As shown in Figure 9.18 (upper panel), Rescorla and Cunningham (1977) reported that reinstatement was obtained when the animals had stayed in the context in the absence of a conditioned light CS (Group No CS) but not in its presence (Group Conditioned CS).

Simulations

Computer simulations for the Rescorla and Cunningham (1977) consisted of 20 CS_1–US trials in CX_1, followed by 20 CS_2–US trials in the same CX_1. Extinction sessions were 25 CS_2– trials in CX_1. After extinction, 6 US trials were simulated in the CX_1, followed by 20 extinction trials in CX_1 either in the presence (Group Conditioned CS), or the absence of CS_1 (Group No CS). The salience of CX_1 was set to CXc = 0.1 and CXg = 0.4, to reflect the absence of odors in the cage.

In agreement with the experimental data, Figure 9.18 (lower panel) shows that reinstatement is present in Group No CS but not in Group Conditioned CS. According to the model, exposure to the CX in the presence of a previously conditioned CS reduces reinstatement because it (a) makes the CX–US association

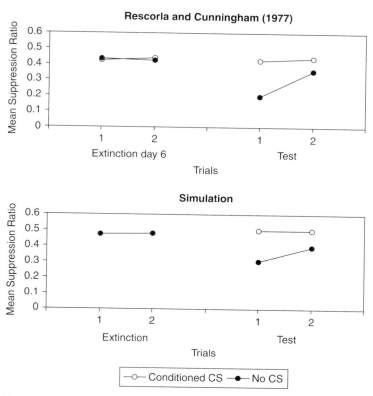

Figure 9.18 Elimination of reinstatement by presentation of a previously conditioned CS between extinction and testing. *Upper panel*: Data from Rescorla and Cunningham (1977). *Lower panel*: Simulations.

inhibitory, and (b) increases Novelty′ and attention to the CX. Notice that in contrast to Bouton and Bolles' (1979a, Experiment 2) context–extinction procedure, the presence of the excitatory CS is capable of producing an inhibitory CX–US association that counterbalances the unchanged the CS–US excitatory association.

Inflation and reinstatement

In order to determine whether having undergone extinction makes a CS especially sensitive to US presentations in the context, Bouton (1984) compared reinstatement with inflation. That is, he compared the effect of presenting a strong US in a context alone after (a) many acquisition and extinction trials of a CS (reinstatement), and (b) a few trials of CS–weak US pairings (inflation effect). In both cases, the same level of conditioning to the CS was attained before the strong US was presented.

Bouton (1984, Experiment 5) found that reinstatement was stronger than inflation, a result that suggests that US presentations have a stronger effect on a conditioned and extinguished CS than on a CS that had been paired with a weak US. In his Experiment 5, rats in the reinstatement groups (Groups *R* and *R*-Control) received CS–US presentations, followed by nonreinforced CS presentations. Rats in the inflation groups (Groups *I* and *I*-Control) received only CS-weak US presentations. In the following session, rats in Groups *R* and *I* received unsignaled presentations of the US in the context of conditioning, while rats in Groups *R*-Control and *I*-Control received those shocks in a different context. Testing for all groups occurred in the context of conditioning.. The CS was a 60-s, 80-dB pure tone, the US was a 3-mA, 0.5-s footshock, the weak US was a 0.3-mA shock and the ITI was 20 min. Odors were used to increase the differences between contexts.

Figure 9.19 (upper panel) shows strong reinstatement when the US is presented in the extinction and test context (Group *R*), but not when the US is presented in a different context (Group *R*-Control). In contrast, US presentations produce similar inflation when they take place in the training context (Group *I*) or when they occur in a different context (Group *I*-Control).

Simulations

Reinstatement was simulated with 5 acquisition trials, 10 extinction trials and 20 presentations of a strong US (salience 2) in the context. Inflation simulations included 5 CS–US (weak US, salience 1) trials. Also, an additional control group was simulated (*I*-No Shock) where only the training context was presented between acquisition and testing (without the US). Ten home-cage time periods were included between the different phases of the experiment. As for Bouton and Bolles' (1979a, 1979b) simulations, the salience of the contexts was set to CXc = 0.5 and CXg = 0.4. The CS had a 10 t.u. duration and salience 0.8, the US was 5 t.u. long with salience 2 and the ITI was 100 t.u.

Figure 9.19 (lower panel) shows that the model approximates the experimental results well, indicating that reinstatement is stronger than inflation. According to the model, the CS–US association is much stronger in the case of acquisition and extinction (reinstatement) than in the case of conditioning with a weak US (inflation), even if, in both cases, similar final levels of responding are attained. Whereas following extinction, responding is proportional to the difference between strong excitatory CS–US associations and strong inhibitory CX–US associations, responding following weak conditioning is simply proportional to the weak CS–US association.

As explained in Chapter 2, section on "Changes in US-US associations", an important consequence of the fact that inhibitory CX–US associations are not

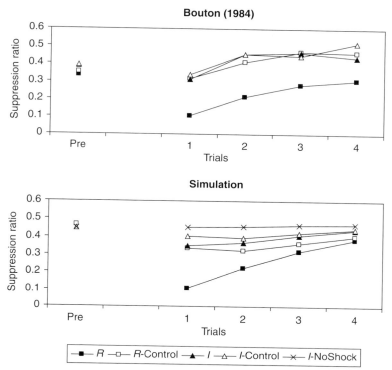

Figure 9.19 Reinstatement is stronger than inflation. *Upper panel*: Data from Bouton (1984). *Lower panel*: Simulations. *R*: reinstatement, *R*-Control: reinstatement control, *I*: inflation with a shock delivered in the context of testing, *I*-Control: inflation with a shock delivered in a different context, and *I*-NoShock: weak conditioning and no shocks delivered.

extinguished by CX-alone presentations (unlike excitatory CX–US associations) is that US presentations are more effective in decreasing the magnitude of inhibitory CX–US associations than in increasing the magnitude of excitatory CX–US associations (see Chapter 3). Therefore, a number of CX–US pairings that result in the elimination of the strong inhibitory CX–US associations in the case of reinstatement only result in the formation of relatively weak excitatory CX–US associations. As a consequence, responding in the case of reinstatement is proportional to the strong CS–US associations, and in the case of inflation, proportional to the sum of the weak CS–US association and the weak excitatory CX–US associations.

As mentioned, because the experiment is designed to attain similar levels of responding after weak conditioning and after acquisition followed by extinction, the weak CS–US associations are equivalent to the strong excitatory CS–US associations minus the strong inhibitory CX–US associations. Therefore, responding in inflation is proportional to the strong excitatory

CS–US associations minus the strong inhibitory CX–US associations (equivalent to the weak CS–US associations) plus the weak excitatory CX–US associations. Notice that adding a weak CX–US association to the strong inhibitory CX–US association always results in an inhibitory CX–US association, which partially counteracts the strong CS–US association. Therefore, responding in inflation (proportional to the strong CS–US association minus an inhibitory CX–US association) will be always weaker than responding in reinstatement (proportional to the strong CS–US associations).

Figure 9.19 also shows that the model approximates the results indicating that reinstatement, but not inflation, is context dependent. Reinstatement occurs when the US is presented in the context of extinction because it eliminates the strong inhibitory CX_c–US associations and releases the also strong CS–US associations. If the US is presented in another context, although CX_g–US associations increase, CX_c–US associations remain strongly inhibitory and there is no reinstatement. In contrast, the model shows that inflation is relatively context independent. Inflation occurs because US presentations result in the formation of weak CX_c–US and CX_g–US associations. Whereas in Group I responding is proportional to weak CS–US, CX_c–US and CX_g–US associations; responding in Group I-Control is proportional to weak CS–US and CX_g–US associations. Because CX_c–US associations are weak, they marginally contribute to the CR, resulting in only a slightly stronger responding in Group I than in Group I-Control.

Interestingly, like reinstatement and unlike inflation, renewal is context dependent: increased responding in a context different from the one of extinction is the result of strong inhibitory CX_c–US associations disappearing when test trials take place in another context.

Partial reinforcement and reinstatement

As described in the previous section, Bouton (1984) suggested that having undergone extinction makes a CS especially sensitive to US presentations in the context. Similarly, Bouton and King (1986) compared the effect of US presentations following partial reinforcement, or following acquisition and extinction (reinstatement). In their Experiment 1, rats in the extinction groups (EXT), received reinforced CS presentations followed by nonreinforced CS presentations. Animals in the partial reinforcement groups (PRF) received cycles of reinforced and nonreinforced CS presentations. After conditioning took place, half of the animals in each group received unsignaled US presentations in the conditioning context (Groups EXT-*S* and PRF-*S*), and the other half received the same US presentations in a different context (Groups EXT-*D* and PRF-*D*). Testing consisted of nonreinforced presentations of the CS in the

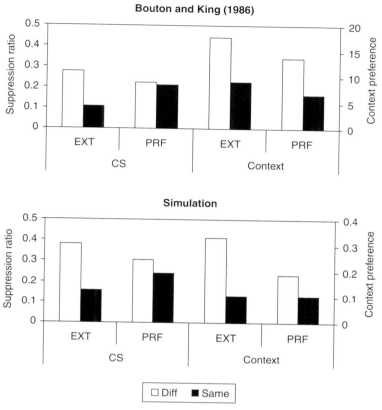

Figure 9.20 Partial reinforcement and reinstatement. Suppression ratios and context preferences following US presentations in the context after acquisition and extinction or partial reinforcement. *Upper panel*: Data from Bouton and King (1986, Experiment 1) showing that US presentations in the context have a greater effect in the group undergoing acquisition and extinction (EXT) than in the group experiencing partial reinforcement (PRF). *Lower panel*: Simulations.

conditioning context. Context preference was evaluated by sampling the rat's location at regular intervals when the animal was able to freely move between the shocked (Same), or not shocked (Diff) Skinner box and a side box. The CS was a 60-s, 80-dB pure tone, presented with a mean ITI of 20 minutes, and the US consisted of a 0.6-mA, 0.5-s electric shock, that coterminated with the CS in reinforced presentations. Odors were used to increase the differences between contexts.

As shown in Figure 9.20 (upper panel), Bouton and King (1986, Experiment 1) reported that (a) reinstatement was present following CS acquisition and extinction (Group EXT), but not following partial reinforcement (Group PRF); and (b) that in both cases animals preferred a context in which they had not received

214 Attentional and associative mechanisms

the shock US. They suggested that the effect of context reinforcement on CS responding depended on the history of reinforcement of the CS.

Simulations

For the extinction groups, simulations included 10 acquisition trials followed by 7 extinction trials. For the partial reinforcement groups, 17 CS-trials were simulated, where only one out of every eight consecutive CS presentations was reinforced. Simulated responses at the end of conditioning for both groups was similar. Trials of exposure to the US were simulated with 10 US presentations. The CS had a 20 t.u. duration and salience 1, and the US was 5 t.u. long with salience 2. The ITI was 100 t.u., and the salience of the contexts was set to $CXc = 0.5$ and $CXg = 0.4$.

As shown in Figure 9.20 (lower panel), computer simulations show that the model is able to replicate the Bouton and King (1986) results. According to the model, the explanation is similar to that of inflation. In the case of partial reinforcement, CS–US and CX–US associations increase during reinforced trials and decrease during nonreinforced ones. In contrast, in the acquisition and extinction case, CS–US associations end strongly excitatory, and CX–US associations end strongly inhibitory. Once again, US presentations in the context result in large decrements in the inhibitory associations of the context of extinction, and small increments in the slightly excitatory associations of the context of partial reinforcement. Therefore, as explained in detail for the case of inflation, responding is stronger after reinstatement than after US presentations following partial reinforcement. In addition, the model is also able to replicate the preference for the nonreinforced context shown in the experiment.

Discussion

According to the model, most of the properties of reinstatement can be explained in terms of associative mechanisms only. However, attentional mechanisms are required to explain that reinstatement is attenuated with increasing number of extinction trials (Delamater, 1997) and contribute to explain why reinstatement is eliminated by exposure to the context (Bouton and Bolles, 1979a).

Several forms of reinstatement

As demonstrated by Westbrook *et al.* (2002), our simulations show that reinstatement can be explained in several ways. When US presentations and testing occur in the same context (e.g. Group *AABB*), reinstatement seems to be the consequence of the elimination of CXc–US inhibitory associations and the increased attention to the CS and to the CXc, which activate the remaining CS–US association and, if formed, the excitatory CX–US association during testing. Regarding the CX–US association, two possibilities exist. In our simulations for Bouton and Bolles (1979a), the initially inhibitory V_{CXc-US} association becomes excitatory and contributes, with the disinhibited residual excitatory CS–US association, to produce the CR. Instead, in our simulations for Rescorla and Cunningham (1977), the initially inhibitory V_{CXc-US} simply decreases or loses its inhibitory value, thereby releasing the CR. Whereas this latter account is similar to that offered by Rescorla and Cunningham (1978), the former is supported by Bouton and King's (1983) data showing that the context might become excitatory.

Instead, when US presentations occur in the context of extinction but the CS is tested in a different context (e.g. Group *AAAB*), simulations suggest that reinstatement results from an increased attention to the CS and the combination (chaining) during testing of CS–CX associations (formed during acquisition and extinction) and CX–US associations (formed during CX–US pairings). In contrast, Westbrook et al. (2002) explained *AAAB* reinstatement as the consequence of mediated acquisition (see Holland, 1990), i.e. during US presentations the CX activates a representation of the CS and this representation becomes associated with the US. In our model, the mechanism that implements chaining also implements mediated acquisition because the CX–CS association activates the CS representation at the time of the US presentation and CS–US associations are formed (see Chapter 2, section on "Attention"). However, computer simulations show that chaining (given by the product $V_{CS-CX}V_{CX-US}$), and not mediated acquisition (the CS representation is activated by a decreasing V_{CX-CS} association which results in a small increment in V_{CS-US}), is the strongest determinant of *AAAB* reinstatement. This result does not preclude the possibility that mediated acquisition and chaining might be equally important.

Several forms of extinction of reinstatement

Extinction of reinstatement can also be explained in alternative ways according to the method used to "erase" it. Simulations for Bouton and Bolles (1979a) suggest that extinction of reinstatement by exposure to the context results in a decreased CX–US association and a decreased attention to the CS, which results in a smaller CR. Instead, simulations for Rescorla and Cunningham (1977) show that extinction of reinstatement when an excitatory CS is present during exposure to the context, is due to an inhibitory CX–US association that counteracts the remaining CS–US association and decreases the CR.

Reacquisition following extinction

Although earlier reports suggested that reacquisition could be fast due to a savings effect (Bouton, 1986), later studies showed that retraining a CS in

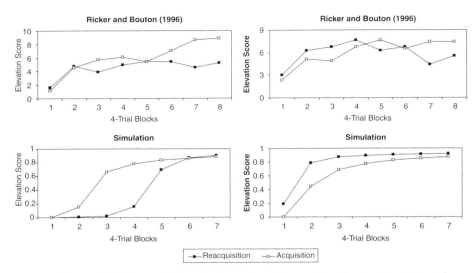

Figure 9.21 Reacquisition rate depends on the number of extinction trials. *Left panels:* Many extinction trials. *Right panels*: Few extinction trials. *Upper panels*: Data from Ricker and Bouton (1996). *Lower panels*: Simulations.

the context of extinction might be slower (e.g. Bouton & Swartzentruber, 1989) or faster (e.g. Napier *et al.*, 1992) than when it is conditioned for the first time. Ricker and Bouton (1996) further examined the issue, and reported that reacquisition is slow or fast depending on the number of acquisition and extinction trials. In the framework of the model, whereas slow reacquisition is the consequence of a small attention to the CS, fast reacquisition is the result of the elimination of the inhibitory CX–US association and the subsequent release of the surviving CS–US association.

Slow or fast reacquisition depends on the number of extinction trials

Ricker and Bouton (1996) studied the effect of the number of extinction trials on reacquisition. In their Experiment 1, a tone was conditioned in Group Reacquisition, and a light off was conditioned in Group Acquisition. In both groups, extinction consisted of 80 (Experiment 1a) or 160 (Experiment 1b) nonreinforced presentations of the corresponding CS. In the following phase, reinforced tone-CS presentations were delivered. The CS was 30 s and the US consisted of food pellets presented at the time of the CS offset. The mean ITI was 10 min. Ricker and Bouton (1996, Experiments 1a and 1b) reported that reacquisition is slow with a relatively large number (160) of extinction trials (Figure 9.21, left upper panel), but fast with a relatively small number (80) of extinction trials (Figure 9.21, right upper panel).

Simulations

Simulations were performed as described for Bouton and Swartzentruber's (1989) experiment, except for the number of extinction trials, which was 30 for Experiment 1a and 70 for Experiment 1b. The model (Figure 9.21, left lower panel) shows that simulated reacquisition to a CS that received 70 extinction trials (Group Reacquisition) is slower than acquisition to a CS in a control group that received conditioning and extinction trials to another CS (Group Acquisition). The model explains slow reacquisition because attention to both the CS and the CX decreases during a prolonged extinction and, therefore, it takes more trials for the US to increase attention, increase CS–US associations and eliminate the CX–US inhibitory association. In contrast, in Group Acquisition, attention to the target CS does not take long to increase and, therefore, CS–US associations form rapidly. Also, Figure 9.21 (right lower panel) shows that with 30 extinction trials, Group Reacquisition conditions faster than Group Acquisition. The model predicts fast reacquisition because the CS–US association survives extinction (savings) and is exposed when US presentations eliminate the CX–US inhibitory associations. In contrast, in Group Acquisition, the associations of the target CS with the US start at zero and take time to increase.

Slow reacquisition with a small number of conditioning trials

In addition to varying the number of extinction trials, Ricker and Bouton (1996) studied the effect of reducing the number of conditioning trials on reacquisition. In their Experiment 4, fewer conditioning trials (8) than in Experiments 1a and 1b (48 and 16 trials, respectively) were used, and extinction consisted of 88 unpaired CS presentations. They reported that reacquisition to a CS that received few acquisition trials is slower than acquisition to a CS in a control group that received simple conditioning (Figure 9.22, upper panel).

Simulations

Simulations for Experiment 4 were identical to those for Ricker and Bouton (1996, Experiments 1a and 1b) except that only 5 acquisition trials and 50 extinction trials were used. The model explains slow reacquisition with few acquisition trials (Figure 9.22, lower panel) because attention to the CS increases with the number of acquisition trials (see Figure 9.1). Therefore, fewer acquisition trials result in the attentional variable z_{CS} starting at a relative lower value and becoming more negative during extinction (see Chapter 2, section on "Attention"). As a consequence, attention to the CS will take longer to increase during conditioning, which results in a slow reacquisition.

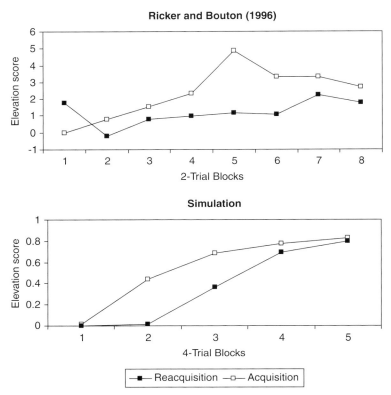

Figure 9.22 Slow reacquisition with a relative small number of acquisition trials. *Upper panel*: Data from Ricker and Bouton (1996). *Lower panel*: Simulations.

Discussion

According to the model, fast rates of reacquisition with few extinction trials are explained in associative terms (surviving CS–US associations are exposed during reconditioning), and slow rates of reacquisition with few acquisition, or many extinction trials are explained in attentional terms (attention to the CS decreases during extinction).

In addition to the Ricker and Bouton (1996) experiments, other studies also reported faster and slower reacquisition. For example, Bouton and Swartzentruber (1989) reported that conditioning to a tone CS is fastest in a group previously conditioned to a light CS, followed by a reacquisition group reconditioned to a previously conditioned and extinguished CS, and slowest in a latent inhibition group conditioned to a previously preexposed CS. Napier *et al*. (1992) reported that reacquisition was faster in a PUP (paired–unpaired–paired) group than in a RRP (rest–rest–paired) group, and faster in the RRP group than in a UUP (unpaired–unpaired–paired) group. Both Bouton and Swartzentruber's (1989) and Napier *et al*.'s (1992) results are correctly described by the model.

Furthermore, Bouton, Woods and Pineño (2004) reported that relatively few reinforced trials during extinction can slow the rate of a rapid reacquisition that follows short extinction. They suggested that rapid reacquisition is obtained when reinforced trials provide a contextual cue that signals other acquisition trials. The model is unable to robustly reproduce these results, because many US presentations result in increments of Novelty′, attention to the CS and fast reacquisition. As observed by Bouton *et al.* (2004), the Pearce and Hall (1980) model offers a similarly incorrect prediction.

Comparison with other models

Ricker and Bouton (1996, page 435) suggested that fast and slow reacquisition could be explained in terms of Capaldi's (1967) sequential learning theory. Whereas reinforced trials become associated with other reinforced trials during conditioning, nonreinforced trials become associated with other nonreinforced trials during extinction. Therefore, in agreement with Ricker and Bouton's experimental data, prolonged conditioning or short extinction favors fast reacquisition, whereas short conditioning and long extinction favor slow reacquisition. In contrast, for our model, short extinction results in surviving CS–US associations that are exposed during reconditioning, and prolonged extinction or short conditioning results in reduced attention to the CS and slow reacquisition.

Notice that faster reacquisition is well described by models (e.g. Kehoe, 1988; Schmajuk & DiCarlo, 1992) that assume a mechanism by which animals "learn to learn." Also notice that models that assume that extinction is the consequence of a buildup of inhibitory CS–US connections that counterbalance non-extinguishable excitatory connections built during acquisition (e.g. Pearce & Hall, 1980) have difficulty in describing reacquisition. In order to describe reacquisition they need to assume that either (a) acquisition keeps increasing, or (b) inhibition decreases. In case (a), unlikely limitless excitatory CS–US associations are required for multiple extinction–reacquisition cycles. In case (b), inhibitory CS–US associations become similar to excitatory CS–US associations in other models: they increase and decrease to describe multiple acquisition–extinction series.

General discussion

In this chapter, computer simulations were used to evaluate how an attentional–associative model describes experimental data on extinction. Analysis of the simulated values of attentional and associative variables in the model allowed us to determine the participation of corresponding mechanisms in the different paradigms. Furthermore, in order to better understand the mechanisms involved in each paradigm, we ran additional simulations with a version of the model in which the attentional variables z_{CS} were fixed and equal to 0.5. Table 9.1 indicates in which cases (a) associative mechanisms are sufficient to explain the results, and (b) a combination of associative and attentional mechanisms best approximates the data.

Table 9.1 shows that some results can be explained by associative mechanisms alone:

Table 9.1. *Associative and attentional mechanisms sufficient to explain different experimental results.*

Experiment	Reference	Only Associative Mechanisms	Attentional and Associative Mechanisms
Summation	Bouton and King (1983) Bouton and Swartzentruber (1989)	NO	YES
Reextinction	Richards and Sargent (1983)	YES	
Facilitation of extinction	Poulos et al. (2006)	YES	
Spontaneous Recovery (Extinction–test interval)	Robbins (1990)	NO	YES
Spontaneous recovery (Acquisition–extinction interval)	Rescorla (2004b)	NO	YES
Spontaneous recovery (Presence of other CSs)	Robbins (1990)	NO	YES
Spontaneous recovery (Extinction cue effects)	Brooks and Bowker (2001)	NO	NO
Spontaneous recovery (Repeated extinction)	Thomas et al. (2003)	NO	YES
External disinhibition	Bottjer (1982)	NO	YES
Renewal (Procedure effects)	Thomas et al. (2003)	NO	YES
Renewal (Context salience effects)	Thomas et al. (2003)	YES	
Attenuation of renewal (Previous extinction)	Lovibond et al. (1984)	YES	
Attenuation of renewal (Massive extinction)	Denniston et al. (2003)	NO	YES
Attenuation of renewal (Extinction cue effects)	Brooks and Bouton (1994)	NO	NO

Table 9.1. (cont.)

Experiment	Reference	Only Associative Mechanisms	Attentional and Associative Mechanisms
Reinstatement	Bouton and Bolles (1979a)	YES	Improved results
Extinction of reinstatement	Bouton and Bolles (1979a)	YES	Improved results
Attenuation of reinstatement (Erasure)	Rescorla and Cunningham (1977)	YES	
Attenuation of reinstatement (Massive extinction)	Delamater (1997)	NO	YES
Reinstatement	Westbrook et al. (2002)	YES	
Reinstatement vs. partial reinforcement	Bouton and King (1986)	YES	
Reinstatement vs. inflation	Bouton (1984)	YES	
Faster reacquisition with few extinction trials	Ricker and Bouton (1996)	YES	
Slower reacquisition with many extinction trials	Ricker and Bouton (1996)	NO	YES
Slower reacquisition with few acquisition trials	Ricker and Bouton (1996)	NO	YES

1. Extinction (a) is the consequence of the decrease in CS–US associations when the CX is non-salient; (b) is the consequence of a decreased CS–US association and an increased CX–US inhibitory association when the CX is salient; (c) is not affected by the order of extinction (Richards & Sargent, 1983) because of the weak inhibitory CX–US associations gained by the non-salient CX; (d) is facilitated by nonreinforced presentations of the CX preceding extinction (Poulos et al., 2006) because the CX–US association is extinguished; and (e) can be mediated by presentation of CS_1 associated with the CS_2 to be extinguished (Shevill & Hall, 2004) because CS_1–CS_2 associations activate CS_2–US associations in the absence of the US.

2. Renewal is (a) obtained with *ABA*, *ABC* and *AAB* procedures (Thomas et al., 2003) because of the elimination of the specific inhibitory characteristics of the extinction context, CXBc or CXAc, (b) stronger with an *ABA* than *ABC* procedures (Harris et al., 2000) because of the residual CXAc–US associations in the testing context, (c) stronger with an *ABC*

than *AAB* procedures (Thomas *et al.*, 2003) because attention to the CS increases when it is extinguished in a context different from that of conditioning, (d) obtained only with salient contexts (Thomas *et al.*, 2003) because only then inhibitory CXc–US associations are sufficiently strong to protect the CS–US association, and (e) disappears when the target CS is tested in a context in which another CS had undergone extinction (Lovibond *et al.*, 1984; Grahame *et al.*, 1990; but see Harris *et al.*, 2000) because that context becomes inhibitory.

3. Reinstatement (a) follows US presentations in the testing context (Bouton & Bolles, 1979a; Bouton & King, 1983), (b) follows US presentations in the context of extinction and testing in a different context (Westbrook *et al.*, 2002), (c) is eliminated by nonreinforced presentations of the reinforced context (Bouton & Bolles, 1979a), (d) is eliminated by presentations of a previously conditioned CS in the context of reinstatement (Rescorla & Cunningham, 1977), (e) is stronger than inflation (Bouton, 1984), and (f) is stronger than the effect of US presentations in the context following partial reinforcement (Bouton & King, 1986).

During reinstatement, administration of the US in the context of testing either decreases the CX–US inhibitory association or makes it slightly excitatory, thereby increasing the magnitude of the CR during testing. Following those US presentations, exposure to an excitatory context reduces reinstatement because it decreases the excitatory CX–US association (attention to the CS also plays a role), which results in a weak CR during testing. Finally, reinstatement is stronger than US presentations following weak conditioning (inflation) or partial reinforcement because US presentations have a stronger effect on eliminating inhibitory CX–US associations and releasing strong CS–US associations (reinstatement) than in increasing the associations formed during weak conditioning or partial reinforcement.

4. Reacquisition can be rapid with relatively few extinction trials (Ricker & Bouton, 1996) because of the remaining CS–US associations uncovered during reconditioning.

Table 9.1 also shows that many results can be explained only when attentional mechanisms are also incorporated:

1. Extinction (a) can result in inhibitory CX–US associations that are not detectable through summation tests (Bouton & Swartzentruber, 1989; also Bouton & King, 1986) because attention to the CS and the CX decrease; and (b) is fast with relatively short ITIs (Cain *et al.*, 2003; Moody *et al.*, 2006) because CX–US associations are stronger with a

short ITI, which results in a higher Novelty′ and increased attention to the CS during extinction.
2. Spontaneous recovery (a) is an increasing function of extinction–test time intervals, either when these intervals are spent in the home cage (Robbins, 1990) or in the training cage (Pavlov, 1927); (b) is a decreasing function of the acquisition–extinction interval for relatively long intervals (Rescorla, 2004b); is obtained (c) only with a previously reinforced CS, even when the test trials are presented in the middle of the session, despite the presence of another reinforced CS, with a CS after spontaneous recovery and reextinction of another CS (Robbins, 1990), and (d) decreases with repeated extinction sessions (Thomas et al., 2003).

During spontaneous recovery, increases in Novelty′ (due either to the change from the home cage to the testing cage, or to the new presentation of the CS during testing) increase attention to the CS faster than attention to the CX. Therefore, the remaining CS–US association is activated, and the CR generated. The effect of varying the extinction-test interval is explained in terms of the decrements in the attention to CXg, which takes longer to become active and inhibit the CS–US association as this interval increases, thereby resulting in a stronger spontaneous recovery. The effect of varying the acquisition–extinction interval is also explained in terms of the decreased attention to CXg, and the consequent lack of protection by the CX of the CS–US association, which decreases during extinction.

3. External disinhibition (Pavlov, 1927; Bottjer, 1982) is the result of the presentation of a novel CS before the presentation of the target CS that increases Novelty′ and attention to the remaining CS–US association, which results in the generation of CR.

4. Renewal (a) is stronger following *ABA* and *ABC* procedures than in an *AAB* procedure, and (b) is eliminated by massive extinction in the extinction context (Denniston et al., 2003). In renewal, inhibitory CX–US associations are formed during extinction but responding recovers when the context is changed during testing. Renewal is relatively strong with *ABA* and *ABC* procedures because the change from the context of acquisition (CXA) to the context of extinction (CXB), increases Novelty′, attention to the CS and responding during testing. Renewal decreases with an increasing number of extinction trials because attention to the CS decreases during the nonreinforced presentation of the CS.

5. Reinstatement (a) is attenuated with increasing number of extinction trials (Delamater, 1997) because prolonged extinction decreases attention to the CS and responding during testing. Notice that attention

to the CS contributes to both reinstatement and the extinction of reinstatement by exposure to the context, which have been explained in terms of changes in CX–US associations.

6. Reacquisition is slow (a) with relatively many extinction trials (Ricker & Bouton, 1996, and (b) with relatively few conditioning trials (Ricker & Bouton, 1996). According to the model, reacquisition is slow when attention to the CS decreases during a prolonged extinction, or when attention to the CS is relatively small at the end of conditioning and further decreases during extinction.

Notice that, even if some of the results can be explained only by associative mechanisms (e.g. renewal, reinstatement), a detailed description of their properties (e.g. renewal with different procedures) or its magnitude (reinstatement, extinction of reinstatement) requires also the presence of attentional mechanisms.

History of reinforcement

Some experiments show that conditioning followed by extinction to a certain level of responding is different from (a) weak conditioning to a similar level, or (b) partial reinforcement. That is, the history of reinforcement of a CS affects the results of (a) the presentation of a US in the context of extinction (reinstatement vs. inflation, Bouton, 1984; reinstatement vs. US presentations after partial reinforcement, Bouton & King, 1986), and (b) the presence of a retention interval between extinction and testing (spontaneous recovery, Rosas & Bouton, 1996).

Bouton and King (1986) suggested that, if different memories are formed during acquisition and extinction, responding might be determined by which memory is active. Therefore, presentation of the US in the context of extinction might retrieve the memory of acquisition, and result in strong responding only after strong conditioning followed by extinction.

For our model, the difference resides in that the CX–US associations become strongly inhibitory during extinction (in contrast to not inhibitory after weak conditioning or partial reinforcement), thereby preserving the strong CS–US association. According to the model, presentations of a US in the context are more effective in decreasing the inhibitory CX–US associations created during extinction (thereby releasing the CS–US association), than in increasing the weak excitatory CX–US associations created during weak conditioning or partial reinforcement. The model also suggests that an extinction–test interval results in spontaneous recovery after acquisition and extinction, but not after weak conditioning. According to the model, attention to the inhibitory CX decreases during the extinction–test interval and allows the CS–US association to be expressed when the CS is reintroduced. In the absence of an inhibitory CX, there is no substantial increase in responding after the acquisition–test interval.

Predictions

In addition to explaining a large number of experimental results, the model is able to generate a number of testable predictions.

1. Summation tests in the context of extinction: effect of the presentation of a novel cue in the context preceding testing

According to the model, the context does not appear inhibitory because of a lack attention to it at the end of extinction. We propose to modify Bouton and Swartzentruber's (1989, Experiment 3; Bouton & King, 1983, Experiment 4) summation experiment to increase the chances that unattended CX–US inhibitory associations will be expressed. In the proposed experiment, rats would receive CS_1–US pairings in CXA, followed by nonreinforced CS_1 presentations in CXB for half of the animals, and in CXC for the other half. Then, rats would receive CS_2–US pairings in CXA. Both the group extinguished in CXB (Different) and the group extinguished in CXC (Same) would be given a summation test in CXC. In the experimental groups, testing would consist of CS_2 presentations preceded by trials in which a novel CS_3 stimulus is presented in CXC alone. This manipulation would increase the representation of CXC through the Novelty' of the unexpected novel CS_3, and possibly allow for the expression of the inhibitory CXC–US association formed during extinction of CS_1. A control group would receive the presentations of the novel CS_3 in a new context (CXD), which would yield results similar to those obtained by Bouton and Swartzentruber (1989). As in the original experiment, odors would be used to increase the differences between contexts.

In order to maximize the chances of detecting the allegedly inhibitory CX–US associations we propose (a) to counterbalance the stimuli used as CS_1, CS_2 and CS_3 (60-s tone, house lights off and a clicker). This is justified because if the extinguished CS_1 happens to be weak relative to the test CS_2, then the CX will acquire only weak inhibitory associations, which might not inhibit the relatively strong CS_2. In addition, we propose (b) to reduce the number of conditioning trials of the excitatory CS used in the summation test that determines the CX–US inhibitory associations, thereby making the CS–US association weaker and more susceptible to be inhibited by the CX. As mentioned in the section on "Elimination of renewal by pretreatment of the testing context", Lovibond *et al.*'s (1984; see also Grahame *et al.*, 1990) data, showing that a presumably inhibitory context has a stronger effect on an extinguished than on a non-extinguished CS, seem to support this view (but see Chapter 12, section on "Transfer effects"). We also propose (c) to reduce the number of extinction trials, thereby avoiding a large decrement in attention to the CX; (d) to verify whether a conditioned inhibitor is capable of inhibiting the excitatory CS used to measure the inhibitory CX–US association; and (e) to place the excitatory CS close to where the target CS was located during extinction in order to activate presumably inhibitory unique contextual features. During testing in CXC, the model expects a large decrease in responding in Group Experimental Same (extinguished in CXC) compared to Group Experimental Different (extinguished in CXB). Because in the groups exposed to CXD attention to CXC is low, small differences in responding are expected between Group Control Same and Group Control Different.

Still, the results of the proposed experiment might not agree with the predictions of the model, but instead resemble those reported by Bouton and Swartzentruber (1989). This latter outcome would favor a version of the model presented in Chapter 15, in which the CX representation is replaced by a CX–CS configural stimulus, similar to that proposed by Bouton (1993, 1994). Computer simulations carried out with that version of the model are comparable to those shown for all the paradigms presented in this chapter.

2. Summation tests in the context of extinction: effects of the number of extinction trials

According to the model, the effect of a novel CS on summation depends on the number of extinction trials. With relatively many (80 in our simulations) extinction trials, attention to the context will decrease, and when a previously reinforced CS is tested in that context, no significant decrease in responding will be present. This expectation matches current available data (Bouton & Swartzentruber, 1989, Experiment 3; Bouton & King, 1983, Experiment 4). Therefore, with relatively many extinction trials, presentations of a novel cue in the context preceding the summation test should result in a significant decrease in responding.

Because attention takes time to decrease, relatively few extinction trials (20 in our simulations) will result in more attention to the context, and the context will appear inhibitory in a summation test. Therefore, with relatively few extinction trials, presentations of a novel cue in the context, preceding the summation test, should result in a relatively small change in responding.

In sum, the model predicts a decreased in responding in a summation test (a) with a novel cue in the context of extinction and a relatively large number of extinction trials, or (b) without a novel cue and a relatively small number of extinction trials.

3. Retardation tests in the context of extinction: effects of the number of extinction trials and presentation of a novel cue in context preceding testing

As mentioned, Bouton and Swartzentruber (1986) used a retardation test to investigate the nature of the contextual associations formed during a contextual discrimination. Because extinction and contextual discriminations might involve different mechanisms, we suggest running a retardation test following extinction with the addition of a novel CS to increase attention to the CX preceding conditioning. According to the model, with relatively many (80 in our simulations) extinction trials, attention to the inhibitory context will decrease, and its conditioning will be retarded. In this case, presentations of a novel cue in the context preceding conditioning should increase attention to the context and result in faster formation of CX–US associations and a significant decrement in retardation.

As in the previous prediction, because attention takes time to decrease, relatively few (20 in our simulations) extinction trials will result in a relatively large attention to the inhibitory context, fast acquisition of CX–US associations, and hence a small retardation. In this case, presentations of a novel cue in the context, preceding the retardation test, should result in a small change in the rate of conditioning.

Notice that, although the model predicts retardation of conditioning of the context after extinction, Bouton and Swartzentruber (1986) reported no such effect for the nonreinforced context in a contextual discrimination. According to the model, whereas in the case of extinction Novelty′ decreases with repetitive presentations of the CS in the CX, Novelty′ stays relatively high with the alternated reinforced and nonreinforced trials used in the discrimination procedure.

In sum, the model predicts relatively rapid conditioning of the context of extinction (a) without a novel cue and a small number of extinction trials, or (b) with a novel cue in that context and a relatively large number of extinction trials.

4. Spontaneous recovery: effect of the presentation of a novel cue in the context preceding testing

According to the model, spontaneous recovery is the result of a rapid activation of the representation of the excitatory CS and the slower activation of the representation of the inhibitory CXg. Spontaneous recovery increases as the extinction–test interval increases (Robbins, 1990) mainly because attention to CXg decreases during this period.

According to the model, with a relatively large number (30 trials in our simulations) of presentations of a novel cue in the context of extinction and testing, attention to the context will increase, and spontaneous recovery will decrease. Interestingly, this prediction of the model seems to be in line with Robbins' (1990) results, showing that spontaneous recovery decreases faster when the target CS is presented late and preceded by another CS, rather than early in the test session with no CS preceding it.

5. Renewal: effect of a time interval between acquisition and extinction

As mentioned in the section on "Spontaneous recovery", when a period of time is interposed between acquisition and extinction, attention to the context decreases, thereby precluding the formation of the inhibitory CX–US associations, which results in the attenuation of the CS–US association and the concomitant decreased renewal. Interestingly, this prediction has been recently confirmed in humans by Huff *et al.* (2007).

6. Renewal: effect of the presentation of a novel CS preceding the CS on the attenuation of renewal by massive extinction

Attenuation of renewal by massive extinction (Denniston *et al.*, 2003) can be partially ameliorated by the presentation of a novel CS preceding CS presentations during testing. According to the model, massive extinction does not decrease the excitatory CS–US association once this association is balanced by the inhibitory CX–US association. However, additional extinction trials will decrease attention to the CS as Novelty' decreases. Therefore, presentation of a novel stimulus preceding each CS presentation should increase Novelty', attention to the CS and partially restore renewal.

7. Reinstatement: effect of presentation of a novel CS in the context of reinstatement before testing on the reduction of reinstatement by nonreinforced presentations of the context

The reduction of reinstatement with nonreinforced presentations in the previously reinforced context (Bouton & Bolles, 1979a) can be attenuated if a novel CS is added during those nonreinforced presentations. According to the model, the effect is due to the increased Novelty' produced by the absence of the novel CS during testing, which increases attention to the CS and the CX.

8. Rate of reacquisition: effect of the presentation of a novel CS preceding the CS on the decreased rate of reacquisition produced by extended extinction

Rate of reacquisition, decreased by extended extinction (Ricker & Bouton, 1996), should increase by presentation of a novel CS preceding each CS presentation. According to the model, the novel CS will increase Novelty', attention to the target CS and speed up reacquisition.

Notice that in the experiments in which responding is expected to increase after the presentation of the novel cue, but not in the experiments in which responding is expected to decrease, a general increase in arousal is a possible confounding factor. However, under the assumption that arousal influences overall responding, scores that reflect the difference in responding in the presence and the absence of the CS (e.g. suppression ratio or elevation score) will tend to ameliorate this effect. However, the problem cannot be completely eliminated, because arousal might have different effects depending on the baseline level of responding.

Some limitations of the model

Even if successfully accounting for the above-mentioned experimental results, the SLG model is unable to describe the effects of an extinction cue on spontaneous recovery (Brooks & Bowker, 2001) or renewal (Brooks & Bouton, 1994). In addition, the model (a) can only approximate the partial reinforcement extinction effect. Furthermore, at least with the present model parameters and simulation values, it has difficulties in explaining (b) the facilitatory effect on extinction of reinforced presentations of the context following acquisition (Kehoe et al., 2004, Experiment 2a); and (c) the slowing effect of occasional reinforced trials during extinction on the rate of rapid reacquisition (Bouton et al., 2004), within the broad range of simulation values used in all other cases.

Partial reinforcement extinction effect (PREE) refers to a phenomenon in which extinction is slower following partial than continuous reinforcement (Boughner & Papini, 2006, for a recent study). Several theories have been successful at accounting for the PREE. Amsel (1958) suggested that it is the result of the association formed between the frustration experienced during the nonreinforced acquisition trials and the US presented on the reinforced trials. For some time during extinction, animals expect that frustration will be followed by the US, and continue to respond. Daly and Daly (1982) show how frustration principles can be combined with the Rescorla and Wagner (1972) model to explain PREE. Similar to frustration theory, Capaldi's (1967) generalization decrement model expects responding to continue after partial reinforcement because the situation during extinction resembles the situation during acquisition. According to Gallistel and Gibbon (2000), conditioned responding depends on ratio comparisons between expected and observed rates of reinforcement. Therefore, PREE is the consequence of this ratio being larger in the continuously, than in the partially reinforced group. Our model can approximate the PREE only under the assumption that the strength of the CR decreases with increasing OR values (see Chapter 2, "CR strength"). In that case, because the Novelty' and OR experienced during extinction by the group receiving continuous reinforcement is larger than the Novelty' and OR experienced by the group receiving partial reinforcement, responding would be weaker after continuous than after partial reinforcement.

The role of configuration

A configural stimulus can be defined as one that is active in the presence of a given combination of active, or inactive input stimuli (e.g. CS_1 and CS_2 are present, and CS_3 is absent). Some models (e.g. Wagner, 1992; Wilson & Pearce, 1990; Schmajuk, Lamoureux & Holland, 1998; Sidman, 1986) suggest that configuration is needed to explain the properties of occasion setters.

Contexts

As mentioned, Bouton and Swartzentruber (1986) proposed that the context might act as a spatial and temporal negative occasion setter. Because occasion setters are not supposed to have the summation properties shown in reinstatement experiments (Bouton & Bolles, 1979a), the context might also act as a simple CS that acquires direct associations with the US. In line with this idea, the SLG model suggests that during extinction the CX acts mostly as a simple stimulus that acquires an inhibitory association with the US. As indicated in Chapter 12, the occasion setting properties of the context might be needed, however, to explain contextual discrimination (Bouton & Swartzentruber, 1986), for which summation, supernormal conditioning and retardation tests suggest that the context is not inhibitory, and that attention to the context is not decreased.

Extinction cues

As mentioned, the SLG model is unable to describe the effects of a nonreinforced extinction cue on spontaneous recovery (Brooks & Bowker, 2001) or renewal (Brooks & Bouton, 1994). We will address these issues in Chapter 15.

Purely attentional solutions

As mentioned before, Pavlov (1927) described extinction in purely attentional terms, a notion shared by Robbins (1990), who indicated that this view can explain spontaneous recovery and slow reacquisition. In a related vein, Kamprath and Wotjak (2004) proposed that conditioning is the result of a nonassociative sensitization process combined with a CS–US associative component. In their view, extinction is the result of a habituation process that abolishes sensitization without affecting the CS–US component.

We carried out computer simulations with our model, under the assumptions that (a) CX–US associations do not become inhibitory, and (b) the rate of decrease in attention is relatively fast ($K_6 = 0.01$ in Equation 2.2). With the exception of using a shorter (5 t.u.) CS, in order to slow down the rate of extinction and allow attention z_{CS} to decrease faster than CS–US associations, the designs of the simulated experiments were identical to those described in this chapter. Under these conditions, the model was able to describe (a) spontaneous recovery; (b) *ABA* renewal; (c) reinstatement and its extinction; and (d) reacquisition being rapid with few, and slow with many extinction trials. Surprisingly, however, just as in the version of the model tested in this study, even in the absence of absence of inhibitory CX–US associations, the model predicted a slightly larger responding to a previously conditioned CS in the nonextinguished context. Although able to describe most of the extinction data (including summation), the model was not able to describe contextual discrimination, for which some type of inhibitory or conditional responding mechanism is needed. It is possible to envision a modified attentional version of our model in which a sustained high level of Novelty' enables the formation of CX–US inhibitory associations.

Competing theories of spontaneous recovery, renewal, reinstatement and reacquisition

Several theories provide approximate accounts for the experimental results addressed in this chapter. For Pavlov's (1927) theory, spontaneous recovery is the consequence of an increase in the magnitude of the CS representation due to the passage of time. Because the CS representation decreases during extinction, it expects a slow reacquisition. For Rescorla's (1974) theory of decreased US processing, spontaneous recovery and reinstatement are the result of the recovery of the US processing. This theory expects reacquisition to occur at the same rate as another, never-extinguished CS (see Rescorla & Cunningham, 1978, page 383). For Hull (1943), spontaneous recovery follows the dissolution of transient reactive inhibition during the time when the animal is not responding. Even when renewal and reinstatement do not appear to be predicted by Hull's view, it seems to anticipate fast reacquisition. For inhibitory theories of extinction (Konorski, 1948), spontaneous recovery, renewal, reinstatement and fast reacquisition might be the result of inhibitory connections being more recent and, therefore, more changeable than the excitatory ones. For the competing memories theories (Gleitman, 1971; Spear, 1971), spontaneous recovery, renewal, reinstatement and fast reacquisition are the result of the older memories of acquisition interfering with the more recent ones of extinction. For generalization decrement models (Capaldi, 1971), renewal, reinstatement and fast reacquisition, but not spontaneous recovery, are the consequence of making the situation similar to that experienced during acquisition. Finally, although the extended comparator hypothesis (Denniston, Savastano & Miller, 2001; Stout & Miller, 2007) is able to describe extinction, the hypothesis cannot be applied to some of the results analyzed in the present study (spontaneous recovery, external disinhibition and renewal).

Bouton (1993, 1994) offered a hypothetical memory structure for extinction. He suggested that (a) excitatory CS–US associations increase during acquisition,

(b) CS–US associations do not decrease during extinction, (c) inhibitory CS–US associations are formed during extinction, (d) the CS and the spatial or temporal CX are combined into a CX–CS gate, (e) the context acquires a negative occasion setting function through the CX–CS gate during extinction, and (f) CX–CS gate-US associations do not change unless both the CS and the CX are simultaneously present to activate the gate. Bouton (1993, 1994) proposed that (a) spontaneous recovery is the consequence of the CX–CS gate becoming inactive because the CX of extinction is different from the CX of testing as time changes; (b) renewal is the consequence of the CX–CS gate becoming inactive when the spatial CX of extinction is different from the spatial CX of testing; (c) reinstatement is explained because CX–US associations are part of the conditioning context, and presenting the US in the context is equivalent to returning the animal to the context of conditioning (Bouton *et al.*, 1993), or because presentation of the US in the context of testing constitutes a context change between extinction and testing (Garcia-Gutierrez & Rosas, 2003). According to Bouton (personal communication), these assumptions make reinstatement identical to an *ABA* or *ABC* renewal. Finally, according to the Bouton (1993, 1994) model, reacquisition might be faster than acquisition because CX changes with time, or because presentation of the US is equivalent to replacing the animal in the context of training (such as in an *ABA* renewal), thereby decreasing the inhibition during the second phase of conditioning. Reacquisition might be slower, however, if the conditions during reacquisition matched those of extinction (e.g. Bouton & Swartznentruber, 1989), which is equivalent to replacing the animal in the context of extinction. For Bouton (1993, 1994), no direct CX–US associations are present, because they would become inhibitory during extinction, an outcome at odds with his experimental results (Bouton & Bolles, 1979a; Bouton & King, 1983; Bouton & Swartzentruber, 1989). In contrast, excitatory CX–US associations have been inferred from experimental results showing a preference for nonreinforced contexts (Bouton & King, 1983; Bouton, 1984). This particular assumption would make the contextual stimuli different from other regular CSs.

Bouton's (1993, 1994) functional model cannot explain why (a) spontaneous recovery decreases with repeated presentations of the CS (e.g. Thomas *et al.*, 2003), or (b) when the acquisition–extinction interval increases (Rescorla, 2004b), and (c) renewal is eliminated with prolonged extinction (Denniston *et al.*, 2003) (because extinction depends only on the CX–CS gate-US association).

As mentioned in the subsection on "Predictions", when an adaptation of Bouton's (1993, 1994) CX–CS gate is incorporated into our model, the model generates results that are similar to those shown here. In this case, however, predictions for the proposed experiment on summation tests in the context of extinction show small differences in responding in both the control group (which receives presentations of a novel CS in a new context before testing the excitatory CS), and the experimental group (which receives presentations of a novel CS in the test context).

Summary

In this chapter, we showed that although associative mechanisms are sufficient to account for some of the reported results regarding extinction, additional novelty-driven attentional mechanisms, such as those included in the SLG model, are also required. Attentional processes are needed to explain extinction bursts, why contextual inhibition is not detectable in summation tests, the properties of spontaneous recovery, external disinhibition, magnitude of renewal with different procedures (*ABA* and *ABC* vs. *AAB*), the elimination of renewal by massive extinction and slow reacquisition.

However, the description of the properties of extinction cues seems to require supplementary configural mechanisms (see Chapter 16). Furthermore, CX–CS configural stimuli might be also needed to improve the description of some experimental results regarding the associative properties of the context. These configural mechanisms can be easily integrated with the simple associative and attentional mechanisms in the present version of the model, along the lines suggested by Buhusi and Schmajuk (1996; Schmajuk, 1997, page 132).

In addition to providing a theoretical framework able to analyze and explain a large number of experimental results, the model generates a series of testable predictions that can guide future experimental research. One of these predictions (a period of time interposed between acquisition and extinction decreases renewal) has been recently confirmed.

10

The neurobiology of extinction

Following the ideas proposed in Chapters 4 and 6, this chapter applies the "conceptual nervous system" provided by the SLG model in order to establish brain–behavior relationships during extinction. We applied the SLG model to the simulation of (a) Frohardt et al.'s (2000) data showing that neurotoxic hippocampal lesions eliminate reinstatement in rats, and (b) LaBar and Phelps's (2005) study showing that reinstatement is absent in patients with hypoxic damage to the hippocampus. As in Chapter 6, we assumed that neurotoxic hippocampal lesions, which injure the hippocampus proper (HPLs), impair the formation of (and changes in) CS–CS, CS–CX, CX–CX and CX–CS associations as defined in the SLG model. We will refer to these associations as between-CS associations. The same assumption was made for hypoxic hippocampal damage which results in human amnesia.

Effects of neurotoxic hippocampal lesions

A number of studies seem to indicate that selective excitotoxic hippocampal lesions (Talk, Gandhi & Matzel, 2002; but see Ward-Robinson et al., 2001), fimbrial lesions (Port & Patterson, 1984), and kainic lesions of hippocampal CA_1 (Port, Beggs & Patterson, 1987) impair the acquisition of between-CS associations. In the framework of the SLG model, Buhusi et al. (1998, see Equations [6.1a] and [6.1b] in Chapter 6) described the effect of these lesions by assuming that between-CS associations, presumably stored in cortical areas, remain zero. They also assumed that associations of the CS with itself, which produce habituation to the CS, are modified when the CS is perceived. Under these assumptions, the model is able to explain the apparently contradictory effects (i.e. impairment, preservation and facilitation) of hippocampal lesions on latent inhibition. Furthermore, Pothuizen et al. (2006) recently confirmed the predictions generated by the model regarding the facilitation of latent inhibition of conditioned taste aversion by selective lesions of a brain region that receives hippocampal input, i.e. the shell of the nucleus accumbens.

Also, in support of these assumptions, the marked deficits shown by amnesic patients in learning unrelated word pairs (Shimamura & Squire, 1984) can be represented by impairments in between-CS associations.

Simulation procedures

As in Chapter 9, in our simulations several tonic contextual stimuli were used to represent the contexts in which the animals are placed. As in Chapter 9, we used CXh = 0.5 and CXg = 0.4 in all simulations, but CXc varied according to the training apparatus used in the different experiments.

Simulated results

In this section, data from Frohardt et al. (2000, Experiment 1) were simulated to account for the consequences of neurotoxic hippocampal damage on fear reinstatement in rats. In addition, data from LaBar and Phelps (2005, Experiment 3) were simulated to explain the reported absence of reinstatement of skin conductance response in amnesic patients with hypoxic damage to the hippocampus.

Effect of hippocampal lesions on reinstatement after aversive conditioning
Experimental data

Frohardt et al. (2000) reported that neurotoxic hippocampal lesions eliminated reinstatement. In their experiment, rats received: (a) alternated training to criterion in two Skinner boxes (Contexts A and B) in a VI 90 schedule; (b) conditioning in CXA consisting of 8 conditioning trials during 2 days, with a 60-s light offset CS followed by a 0.5-s footshock alternated with exposure to CXB; (c) extinction in CXA for 4 days alternated with exposure to CXB; (d) reinstatement consisting of 8 US presentations in CXA (Same group) alternated with exposure to CXB; and (e) testing in CXA for 4 trials. A control group (Different group) received 8 US presentations in CXB. As shown in Figure 10.1 (upper panel), whereas sham-lesioned rats in the Same group (shocks delivered in the extinction context) showed small suppression ratios (strong responding) to the CS, HP-lesioned rats and sham-lesioned rats in the Different group (shocks delivered in another CX) show large suppression ratios (no responding).

236 Attentional and associative mechanisms

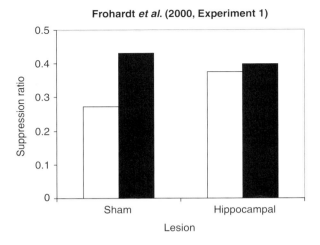

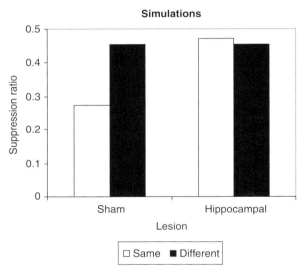

Figure 10.1 Effect of neurotoxic lesions on reinstatement in rats. *Upper panel*: Data from Frohardt *et al.* (2000, Experiment 1) indicate suppression ratios following US presentations in the Same (TC) or Different (CC) context in animals receiving Sham or Hippocampal lesions. *Lower panel*: Computer simulations. Average suppression ratio over 7 test trials.

Simulation results

Simulations assumed that between-CS associations are impaired by the excitoxic hippocampal lesions. That is, Equation [2.9b] for normal animals (see Chapter 2), becomes $\Delta V_{CS,CX} = 0$ after the lesions. As shown in Figure 10.1 (lower panel), simulations for the Sham Same group showed a smaller suppression ratio (stronger responding) than the Sham Different and both Hippocampal groups (weak conditioning).

According to the model, in both groups, extinction in CX*A* results in a decreased CS–US association and an inhibitory CX*A*–US association and, consequently, in the absence of a CR. However, the groups differ during the US administration in CX*A*. Because normal animals expect to find the CS in CX*A*, when the CS is absent, Novelty′ and attention to CX*A* increase. Consequently, the CX*A*–US association becomes excitatory. Instead, in the absence of CX*A*–CS associations, hippocampal lesion animals are insensitive to the absence of CS in CX*A*. Therefore, Novelty′ and attention to CX*A* do not increase. As a result, the CX*A*–US association remains inhibitory. Finally, during testing in CX*A*, because of the remaining CX*A*–US inhibitory association, hippocampal-lesioned animals do not show reinstatement. In sum, because they are insensitive to the absence of the CS in CX*A*, hippocampal-lesioned animals do not increase attention to CX*A* and the CX*A*–US association responsible for reinstatement remains inhibitory.

In addition to impairment of reinstatement, Frohardt *et al.* (2000, Experiment 2) also reported that neurotoxic hippocampal lesions spare renewal, i.e. the re-emergence of the CR following extinction in a different context. The model correctly describes this result. According to the model, renewal is the consequence of the formation of inhibitory associations between the context of extinction and the US, which are not active when the animal is removed from that context. Because neurotoxic lesions do not affect CX–US inhibitory associations, renewal is not expected to be impaired.

In sum, under the assumption that between-CS associations are absent after neurotoxic hippocampal lesions, the model correctly describes impaired reinstatement and preserved renewal. Interestingly, Wilson *et al.* (1995) reported that fornix lesions, in addition to having similar consequences on reinstatement and renewal, also preserve spontaneous recovery. This additional result is also described by the model under the assumption that between-CS associations are absent. In the normal case, the model describes spontaneous recovery because Novelty′ increases due to the unpredicted re-appearance of the CS, which had been absent for some time, and was therefore unpredicted by the context. This increase in Novelty′ increases attention to the CS, thereby allowing for the expression of the remaining CS–US association, and the generation of a CR (see Chapter 9). In the lesioned case, spontaneous recovery is also present because Novelty′ increases when the CS is presented even in the absence of CX–CS associations. Attention increases because attention to the CX decreases during the absence of the CS (as CX–CX associations increase), making the CX unable to activate the CX–US inhibitory association and counterbalance the CS–US excitatory association. Again, this increase in Novelty′ increases attention to the CS and results in the generation of a CR.

Comparison with alternative assumptions

In contrast to our assumption that excitoxic hippocampal lesions impair the formation of between-CS associations, Frohardt *et al.* (2000) suggested that neurotoxic hippocampal lesions impede the formation of CX–US excitatory associations. This suggestion has received recent support from Otto and Poon's (2006) report showing that excitotoxic lesions of the dorsal hippocampus impair contextual fear conditioning; presumably mediated by CX–US associations. Lack of a CX–US excitatory association would impair reinstatement by eliminating the effect of reinforced CX exposures, but not affect renewal, because it does not depend on those associations. In agreement with these notions, computer simulations with the SLG model also show that, under the assumption that CX–US excitatory associations are absent after selective hippocampal lesions, reinstatement is abolished.

Even if the simulations generated assuming an (a) impairment of between-CS associations, and (b) CX–US excitatory associations are similar for the spontaneous recovery, reinstatement and renewal cases, they differ in their predictions for fast reacquisition with few extinction trials, and for slow reacquisition with many extinction trials (Ricker & Bouton, 1996). Whereas impairment of CX–US association should not have a major effect on either fast or slow reacquisition, impairment of between-CS associations will have an impact on slow acquisition because attention to the CS and the CX will not change during the prolonged extinction. According to the model, presentation of the US during reacquisition causes the initially inhibitory CX–US association to become excitatory, thereby exposing the remaining CS–US excitatory association. When attention to the CX is small, the CX–US association will change slowly and reacquisition will be slow.

Finally, it is worth observing that, in contrast to the effect of neurotoxic hippocampal lesions and fornix lesions, muscimol infusions in the dorsal hippocampus (Corcoran & Maren, 2004), muscimol infusions in the ventral hippocampal (Hobin, Ji & Maren, 2005) or electrolytic lesions of the dorsal hippocampus (Ji & Maren, 2005) impair (at least some types of) renewal. Such results would call for assuming both the impairment of excitatory *and* inhibitory CX–US associations; an assumption that might simply imply the absence of context representations following hippocampal lesions (O'Keefe & Nadel, 1978). In the absence of inhibitory CX–US associations, CS–US associations are necessarily eliminated, and any form of recovery from extinction is impossible. To the extent that inhibitory CS–US associations are intact following neurotoxic hippocampal lesions (Chan, Jarrard & Davidson, 2003), this impairment would be limited to inhibitory CX–US associations.

In sum, it is possible that a combination of impaired (a) between-CS associations, and (b) excitatory and inhibitory CX–US associations is needed to correctly describe the effect of excitotoxic hippocampal lesions on both latent inhibition and post-extinction effects.

Reinstatement in amnesic patients

Experimental data

LaBar and Phelps (2005, Experiment 3) reported that reinstatement of skin conductance response was absent in two amnesic patients with hypoxic damage to the hippocampus. In this experiment, participants received: (a) 4 habituation trials of a visual CS presented alone (a blue square, 4-s duration), (b) 4 acquisition trials of the CS paired with a shock US delivered to the wrist (200-ms duration, coterminating with the CS, 100% reinforcement), (c) 8 extinction trials of the CS alone, (d) 4 trials of re-exposure to the US alone, (e) 8 CS-alone recovery test trials. Five-minute waiting periods separated the extinction and US re-exposure phases, as well as the US re-exposure and CS recovery test phases. The ITI duration was 16 (+ 2) s throughout all experimental phases, except for the US re-exposure phase, which was 50 (+ 1) s in duration. Amnesic patients and one group of matched healthy control subjects (N = 8) were placed in the same environmental context for all phases of the study. Another healthy control group (N = 8) underwent US re-exposure in a novel (irrelevant) context.

Recovery of extinguished fear (vertical solid line in Figure 10.2, upper panel) occurred only for control participants who underwent reinstatement in the same environmental context. Amnesics showed intact acquisition and extinction of fear during the initial training session, but they did not show fear recovery following reinstatement in the same environmental context.

Although renewal has been demonstrated recently in humans (Milad, Orr, Pitman & Rauch, 2005; Vansteenwegen *et al.*, 2005), its presence has not been yet assessed in amnesic patients.

Simulation results

We approximated the skin conductance response used in the experiment with the CR provided by the model. Therefore, habituation trials were not included in the simulations because, unlike the skin response, the CR starts at zero. Again, Equation [2.9b] for the normal case (see Chapter 6, Equation 6.1c), becomes $dV_{CS,CX}/dt = 0$ for the amnesic patients. We applied the same simulation parameters as before, with the exception of the number of extinction trials, which was reduced to only 8 in order to capture the incomplete extinction achieved in the LaBar and Phelps (2005) experiment.

240 Attentional and associative mechanisms

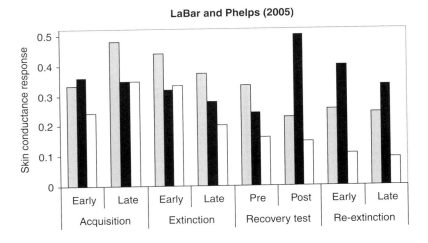

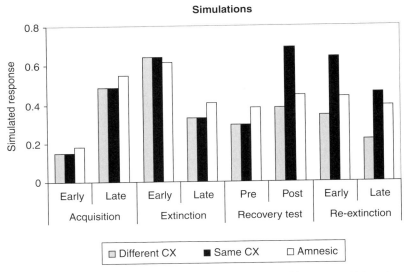

Figure 10.2 Reinstatement of conditioned fear in normal human participants and amnesic patients. *Upper panel:* Normalized skin conductance conditioned responses by group and experimental phase. Data from LaBar and Phelps (2005, Experiment 3). *Lower panel:* Computer simulations. Average conditioned responses over 2 trials.

According to the model, amnesic patients do not show reinstatement for the same reasons animals with neurotoxic hippocampal lesions fail to do it. Patients are insensitive to the absence of the CS during reinstatement trials and, therefore, do not increase attention to the CX when the US is presented in the absence of the CS, which is needed to make positive the CX–US association that is responsible for reinstatement.

Discussion

The present chapter shows that the SLG model provides a clear mechanistic account for how reinstatement is impaired by hippocampal dysfunction. As explained in Chapter 9, the model correctly describes that reinstatement (a) follows US presentations in the testing context (Bouton & Bolles, 1979a, Experiment 1; Bouton & King, 1983), (b) follows US presentations in the context of extinction with testing conducted in a different context (Westbrook *et al.*, 2002), (c) is related to CX–US excitatory associations formed after reinforcement trials (Bouton & King, 1983), (d) is eliminated by exposure to the context (Bouton & Bolles, 1979a, Experiment 2), and (e) is eliminated by exposure to an independently reinforced CS (Rescorla & Cunningham, 1977). In addition, according to the model, (f) the inhibitory CX-US associations formed during extinction are difficult to detect in summation tests (Bouton & King, 1983) either because the (a) CX is not attended; or (b) because CX and CS form a configural stimulus which is only partially activated and poorly attended, when the inhibitory power of the context is evaluated in the absence of the CS.

By assuming that there is a deficit in the formation of CS–CX and CX–CS associations, the SLG model describes the absence of reinstatement in (a) rats with hippocampal neurotoxic lesions (Frohardt *et al.*, 2000, Experiment 1), and (b) amnesic patients (LaBar & Phelps, 2005). According to the model, animals with hippocampal lesions, and amnesic patients fail to show reinstatement because they are insensitive to the absence of the CS during US presentation trials and, therefore, do not increase attention to CX and the CX–US association responsible for reinstatement.

Other extinction paradigms

In addition to the experimental results described above, the SLG model can be applied to the description of neurophysiological manipulations on other extinction paradigms. For example, studies demonstrating that NMDA receptors participate in acquisition (Miserendino *et al.*, 1990), extinction (Falls *et al.*, 1992) and reinstatement (Johnson *et al.*, 2000), are well described by the model by assuming that those receptors participate in the coding of the US (as suggested by Delamater, 2004). Data showing that lesions of dopamine neurons in the medial prefrontal cortex retard extinction (Morrow *et al.*, 1999) and that electrolytic lesions of the ventromedial prefrontal cortex impair spontaneous recovery but not reinstatement (Gewirtz *et al.*, 1997), can be explained under the assumption that Novelty´, coded by dopamine, and attention to the CS decrease following those lesions (see Chapter 6). Finally, Harris and Westbrook (1998) reported that GABA antagonists eliminate the inhibition of responding by contextual cues and that this effect does not add to renewal in a context different from that of extinction. These results are captured by the model under the assumption that GABA receptors are part of the biological substrate of the CX–US inhibitory associations formed during extinction and eliminated during renewal. In addition, data showing that GABA antagonists have no effect on latent inhibition, even though latent inhibition was affected by a change in the context (Harris & Westbrook, 1998), are well replicated by the model under the assumption that latent inhibition is controlled by attention, which is proportional to Novelty′ coded by dopamine (see Chapter 6), and not by GABA.

Summary

The present chapter shows that the effect of both neurotoxic and hypoxic lesions of the hippocampus on reinstatement is well described by the SLG model under the assumption that between-CS associations are impaired. A similar assumption correctly describes reinstatement in amnesic patients. The assumption has been successful at describing the effect of hippocampal lesions and predicting the effect of accumbal lesions on latent inhibition (see Chapter 6). In addition to impaired reinstatement, the assumption describes preservation of spontaneous recovery and renewal. Although some data seem to support these predictions, other experimental results suggest that renewal might be impaired. Therefore, it is possible that a combination of impaired (a) between-CS associations, and (b) excitatory and inhibitory CX–US associations is needed to correctly describe the effect of excitotoxic hippocampal lesions on both latent inhibition and post-extinction effects.

Part III CONFIGURAL MECHANISMS

11

A configural model of conditioning

This chapter introduces a configural neural network model of classical conditioning offered by Schmajuk and DiCarlo (1992) and modified by Schmajuk, Lamoureux and Holland (1998) in order to account for data on occasion setting. In Chapter 12, we evaluate the model by applying it to the experimental results in which occasion setting is observed.

Skinner (1938) suggested that a discriminative stimulus in an operant conditioning paradigm does not elicit a response, but simply sets the occasion for the response to occur. More recently, many investigators (e.g. Bouton & Nelson, 1994; Holland, 1983, 1992; Rescorla, 1985, 1992) have claimed that Pavlovian conditioning procedures can also endow stimuli with occasion setting functions, as well as simple associative functions. Whereas a simple conditioned stimulus (CS) may elicit conditioned responses (CRs) because it signals the occurrence of an unconditioned stimulus (US), an "occasion setter" (Holland, 1983, 1992) or "facilitator" (Rescorla, 1985) may instead modulate responding generated by another CS by indicating the relation between that CS and the US. Rather than signaling the delivery of the US, an occasion setter indicates whether another cue is to be reinforced (or not reinforced), setting the occasion for its reinforcement (or nonreinforcement). Extensive evidence seems to support this distinction between simple conditioning and occasion setting, and suggests that these two functions are acquired under different circumstances, manifest many different behavioral properties and often act quite independently of each other (see Holland, 1992; Swartzentruber, 1995, for comprehensive reviews.)

Schmajuk, Lamoureux and Holland (SLH) (1998) extended a successful, real-time network model of conditioning presented by Schmajuk and DiCarlo (SD) (1992) in an attempt to precisely define simple associative and occasion setting functions of a CS, and account for the existing data base. Like many network models (e.g. Sutton & Barto, 1981; Kehoe, 1988), the heritage of the SD model can be traced to the Rescorla and Wagner (1972) competitive rule, in that learning is related to discrepancies between the values of reinforcers actually received, and aggregate predictions of those reinforcers based on the excitatory and inhibitory associations of conditioned stimuli. However, the SD model diverges from that of Rescorla and Wagner (1972) and related network models in many ways. First, it describes behavior in real time, i.e. changes in the variables are described as a moment-to-moment (instead of trial-to-trial) phenomena. Second, it incorporates a layer of "hidden" units between the input CSs and output CR units, which internally code configural stimuli. Third, CS inputs are connected to the CR outputs both directly and indirectly through the hidden-unit layer. This last feature enables the SD model to describe the distinction between simple conditioning and occasion setting: a CS acts as a simple stimulus through its direct connections with the output units and as an occasion setter through its indirect configural connections through the hidden units. Therefore, within the model many previous distinctions between "occasion setting" and "configural" processes (e.g. Holland, 1983, 1992; Pearce, 1987, 1994; Wilson & Pearce, 1990) simply disappear.

This chapter presents a configural model of classical conditioning capable of describing many of the properties of occasion setting.

A configural model of classical conditioning

As mentioned in Chapter 2, competition between CSs to gain association with the US is well described by the Rescorla and Wagner (1972) rule (and Equation [2.9a] in the SLG model). Several models (e.g. Sutton & Barto, 1981; Kehoe, 1988; the SLG model) utilize this "delta rule" (see Chapter 1) to change the connection strengths between their two layers, the CS-input layer and the US-output layer. In these models, the strength of CS–US associations or connections are changed until the difference between the US and its prediction is zero. Two-layer networks are successful at describing a number of compound conditioning phenomena based on the linear combination (summation and competition) of response tendencies among individual CS elements, for example, blocking, overshadowing and conditioned inhibition.

Two-layer models (including the SLG model), however, are unable to deal with conditioning phenomena, specifically those cases that involve nonlinear discriminations between compound stimuli and their elements, such as the logical "exclusive-OR" or negative patterning problem (Woodbury, 1943). In this problem, each of two CSs is separately paired with the US, but a compound of those two CSs is presented without the US (A+, B+, AB−). Animals learn to respond less on compound AB− trials than on either A+ or B+ trials. Within two-layer models, such a solution is impossible, because the joint activation of two inputs will always produce more activity in the output layer than the activation of only one input (but see Harris, 2006, for an alternative solution).

Rumelhart et al. (1986) showed that exclusive-OR problems can be solved by networks that incorporate a layer of "hidden" units positioned between input and output units. In such multilayer networks, the information coming from the input units is recoded by the hidden units into an internal representation. The exclusive-OR problem can be solved if this internal representation, active only with the simultaneous presentation of both inputs, acquires an inhibitory association with the output (see Kehoe, 1988). Because the output is proportional to the sum of the excitatory effects of the individual inputs and the inhibitory force of the internal representation, the output will be large when each input is presented separately, and small when both inputs are presented together. Importantly, the approach avoids the combinatorial explosion, which might result from the presentation of multiple possible combinations of inputs, by limiting the number of hidden units that can code the simultaneous presentation of inputs (Yadav et al., 2006). Unfortunately, having a limited number of hidden units leads to the problem of catastrophic interference, i.e. the undesirable change in established internal representations when new representations are established (McCloskey & Cohen, 1989).

The SD model for a single response system

In 1992, Schmajuk and DiCarlo (SD1992) proposed a real-time network that described both the effects of temporal parameters on classical conditioning, as well as nonlinear classical conditioning paradigms. Like the Rescorla–Wagner (1972) model, one central notion of the theory is that organisms only learn when the actual value of the US differs from its expected value (error-correction rule). A second notion is that CSs generate short-term memory traces which can become associated with the US, thereby describing trace conditioning. A third notion is that the expected value of the US is computed on a linear combination of the associative strength of all active CSs plus the associative strength of configurations of those active CSs.

Figure 11.1 offers a simplified diagram of the network presented by Schmajuk and DiCarlo (1992). It consists of one input layer, one hidden-unit layer and

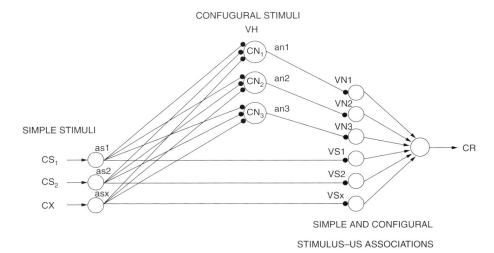

Figure 11.1 The SD model for a single response system. Diagram of a network that incorporates a layer of hidden units capable of describing stimulus configuration. CS: conditioned stimulus, as: short-term memory of a CS, CN: configural stimulus, an: short-term memory of a CN, VS: CS–US association, VN: CN–US associations, VH: CS–CN association, CR: conditioned response. Arrows represent fixed synapses. Solid circles represent variable synapses.

two output layers. This section offers a formal description of the SD network in terms of differential equations that characterize the dynamics of the activity of nodes in the different layers, as well as the changes in their connectivity.

Short-term memory (STM) traces

Input units are activated by conditioned stimuli, CS_1, CS_2 and CS_3 (the context, CX). In order to capture the dynamics of stimulus presentation (CS and US duration and intensity, interstimulus and intertrial intervals) and describe the topography of behavior in real time, it is assumed that CS_i activates a short-term memory (STM) trace, as_i. The change in as_i over one time unit is defined by

$$d(as_i)/dt = -K_1 as_i + K_2(K_3 - as_i)(CS_i + f(as_i)), \quad [11.1]$$

where $-K_1 as_i$ represents the passive decay of the STM of CS_i and K_2 represents the rate of increase of as_i. Function $f(as_i)$ is given by $f(as_i) = K_4 as_i^m/(\beta^m + as_i^m)$, which with adequate parameters is relatively large even for relatively small values of as_i. According to Equation [11.1], as_i is active after CS_i onset and kept active by $f(as_i)$ even after CS_i offset. As as_i decreases, $f(as_i)$ remains active for sometime and then decays to zero with as_i.

Figure 11.1 shows that CS_i acquires a *direct* association with the output units and becomes *configured* with other CS_is in a hidden unit that, in turn, acquires associations with the output units. Direct input–output connections, relaying stimulus components, and indirect connections, relaying configured stimuli from the hidden-unit layer, compete to establish associations with the output units under the regulation provided by a delta rule. Schmajuk and DiCarlo (1992) demonstrated that this parallel input–output architecture is necessary for (a) a description of compound conditioning in which simple and configural stimuli have separate representations that compete to establish associations with the US (Kehoe, 1986), and (b) a plausible mapping of the model over established cortical and subcortical parallel brain circuits.

Input–hidden unit associations

Stimulus configuration is achieved by adjusting the strength of the association between input i and hidden unit j, VH_{ij}. Changes in VH_{ij} are given by

$$d(VH_{ij})/dt = K_5 as_i(1 - |VH_{ij}|)EH_j, \quad [11.2]$$

where term $(1 - |VH_{ij}|)$ bounds VH_{ij} between -1 and 1. By Equation [11.2], VH_{ij} changes only when as_i is active and the hidden-unit error, EH_j, is not zero.

Rumelhart et al. (1986; Werbos, 1987) proposed a method to "backpropagate" the output error to the hidden units. They assumed that EH_j is proportional to $d(an_j)VN_jEO$, where $d(an_j)$ is the derivative of the output of hidden unit j with respect to its total input, VN_j is the association of hidden unit j with the output, and EO is the output error defined below. Because in most three-layer networks an_j is not readily available and $d(an_j)$ is not easily calculated, it is not clear how a neurophysiological mechanism can compute EH_j. By contrast, in the network illustrated in Figure 11.1, VN_j is stored in a separate output layer and the activity of each of the elements in this layer is proportional to an_jVN_j. Therefore, EH_j is assumed to be proportional to the product an_jVN_jEO, where the derivative of the output of the hidden unit with respect to its total input, $d(an_j)$, is replaced by the output of the hidden unit, an_j. Accordingly, hidden-unit error, EH_j, is given by

$$EH_j = 1/1 + \exp(-K_6 an_j VN_j EO) - 0.5, \quad [11.3a]$$

where EO is given by Equation [11.7].

By Equations [11.2] and [11.3a], when the US is present ($EO > 0$), EH_j and as_i–H_j associations increase in proportion to as_i and an_jVN_jEO. Therefore, as_i–H_j associations increase faster when the output of the hidden units an_j is active (Equations [11.4] and [11.5]). In the case of serial feature discriminations (see Chapter 12), this activation is provided by the target CS, whose trace is strong when the US is presented.

Because Equation [11.3a] approximates the result of the interaction between medial septal and entorhinal cortex inputs to dentate gyrus, CA_3 and CA_1 hippocampal fields (Schmajuk & DiCarlo, 1992, Appendix C), the neural architecture shown in Figure 11.1 implements a biologically plausible version of backpropagation consistent with accepted neurobiological principles.

Configural stimuli

The output activities of the hidden-unit layer, an_1, an_2 and an_3, are assumed to code configural stimuli denoted by CN_1, CN_2 and CN_3. Hidden unit j is activated by the STM of different CS_is in proportion to their connections with the hidden unit

$$sum_j = \Sigma_i VH_{ij} as_i, \quad [11.4]$$

where VH_{ij} represents the association between the STM trace as_i and hidden unit j.

The output of the hidden units is a sigmoid given by

$$an_j = (sum_j^n / \beta^n + sum_j^n), \quad [11.5]$$

where β is the value of sum_j for which an_j equals 0.5, and the exponent n determines the slope of an_j. Whereas Rumelhart et al. (1986) assumed that the output of neural units was active even in the absence of any input, by Equation [11.5] the hidden units are active only when inputs are present, i.e. $an_j = 0$ if $sum_j = 0$.

Equations [11.4] and [11.5] specify how the output of the hidden units, an_j is computed. Traditionally, a configural stimulus has been defined as a stimulus that is active when its component stimuli are active together (e.g. CS_1 and CS_2) and which can acquire excitatory or inhibitory associations with the US (e.g. Rescorla, 1973; Rescorla & Wagner, 1972, page 86; Pearce, 1987; Kehoe, 1988). By contrast, in the SD model, configuring is accomplished by training hidden units that may respond to various combinations of stimuli (e.g. CS_1 and not CS_2), and each one may acquire excitatory or inhibitory associations with the output units. In other words, while hidden units may come to respond to the concurrent activation of multiple stimuli, as in the classic view of configuration (Rescorla, 1973), VH_{ij} associations may also result that are inhibitory in nature. In this case, a hidden unit may be activated by presentation of one or more CSs, but the addition of another CS will inhibit or attenuate the activation of that node.

Because some hidden units might be activated by different CSs, initial values of VH_{ij} define the amount of generalization between CSs. We assume that generalization is stronger between stimuli in the same modality than between stimuli in different modalities.

Input–output associations

Input units form direct associations, VS_1, VS_2 and VS_3, with the output layer. Changes in the associations between input i and the output layer, VS_i, are given by a delta rule

A configural model of conditioning 251

$$d(VS_i)/dt = K_7 as_i(1 - |VS_i|)EO. \quad [11.6a]$$

By Equation [11.6a], VS_i changes only when as_i is active and the output error EO_i is not zero. The term $(1 - |VS_i|)$ bounds VS_i ($-1 \leq VS_i \leq 1$).

Output error EO is given by

$$EO = US - B_{US}. \quad [11.7]$$

The aggregate prediction of the US is given by

$$B_{US} = \Sigma_i as_i VS_i + \Sigma_j an_j VN_j. \quad [11.8]$$

We assume that when $B_{US} < 0$ then $B_{US} = 0$. This assumption implies that, in agreement with experimental data (Zimmer-Hart & Rescorla, 1974), the network characterizes conditioned inhibition as not extinguishable by nonreinforced presentations of a conditioned inhibitor. Also in agreement with experimental data (Baker, 1974), the assumption implies that the network does not generate an excitatory CS when a neutral CS is presented with an inhibitory CS.

Differences between input–output and input–hidden associations

Although Equation [11.2] is similar to Equation [11.6a], learning input–output and input–hidden units associations show important differences. According to Equation [11.6a], input–output associations, VS_i, increase proportionally to the magnitude of as_i when the US is present. Therefore, when as_i is relatively weak at the time of the US presentation, VS_i will be small. As explained in Chapter 12, as_i is weak in the case of a CS in that precedes the US by a long time interval (trace conditioning), such as a feature CS in a serial feature discrimination. Also explained in Chapter 12, direct CX–US associations are weak in the case of contextual discriminations, because the association undergoes extinction during the duration of the trial. Because direct feature-US and CX–USs are relatively weak, their effect on the CR, proportional to $as_i VS_i$ (Equation [11.8]), will be weak as well.

By contrast, input–hidden units associations follow different rules. According to Equations [11.2], [11.3a] and [11.5], input–hidden units associations VH_{ij} change only when the output of the hidden unit j, an_j, is active. That implies that the initial (random) input–hidden unit association VH_{ij} of that hidden unit with input as_i, should be sufficiently strong to activate an_j and allow changes in the connections of that hidden unit. Therefore, a CS with a strong as_i, such that of a target in the case of a serial feature discrimination or a contextual discrimination, is first needed to activate a hidden units that can become associated with the weak as_i of a feature or the context. Once a hidden unit is active, a feature CS or the CX will be able to change their own input–hidden unit association VH_{ij} with that hidden unit. Even if, by

Equation 11.2], changes in the VH_{ij} association are still relatively small when the as_i is weak, the effect of the feature or the CX on the output is amplified by a highly nonlinear sigmoid function (with a relatively high threshold) that controls the output of the hidden unit (Equation [11.5]).

In sum, CSs with weak as_i can form only weak associations with both hidden and output units. However, whereas the activity of the output units is a linear function of those associations, the activity of hidden units is a highly nonlinear function that amplifies the small effect that a weak as_i might have. Therefore, even though a feature or the CX might not be able to establish strong direct as_i–US associations (by Equation [11.6]), they still can control the output of the hidden units through as_i–H_j associations (by Equation [11.2]), formed when those hidden units are activated by the target CS.

Hidden-unit output associations

Hidden units form associations, VN_1, VN_2, VN_3 with the output layer. Changes in the association between hidden unit j and the output layer, VN_j, are given by a delta rule

$$d(VN_j)/dt = K_7 an_j (1 - |VN_j|)EO, \qquad [11.9a]$$

By Equation [11.9a], VN_j changes only when an_j is active and EO is not zero. $K_7 = K_7'$ when EO = 0, and $K_7 = K_7''$ when EO < 0. The term $(1 - |VN_j|)$ bounds VN_j ($-1 \leq V_j \leq 1$). Equation [11.9] forces as_i and an_j to compete to gain associations with the US: once any as_i or an_j fully predicts the US, EO becomes zero thereby preventing other as_i and an_j from forming associations with the US.

CR generation

The output of the network is trained by the US to generate a CR when the right combination of CSs is presented. In the single response system model, the CR output of the system is given by

$$CR = R_1[\Sigma_i as_i VS_i + \Sigma_j an_j VN_j], \qquad [11.10]$$

where R_1 is a response function, as_i represents the direct inputs, VS_i represents their associations (excitatory or inhibitory) with the US, an_j represents the output of the hidden units, and VN_j their associations (excitatory or inhibitory) with the US. Notice that, because the aggregate prediction B_{US} is also proportional to $\Sigma_i as_i VS_i + \Sigma_j an_j VN_j$ (Equation [11.8]), B_{US} is an efference copy of the CR.

Simple CSs vs. occasion setters

In the framework of Equation [11.10], a stimulus performs as a simple CS when it acts on the CR through direct associations, VS_i, and as an occasion setter when it acts through indirect hidden-units output associations, VN_j. Furthermore, although direct and indirect associations compete in acquisition through a delta rule, a stimulus can act both as a simple CS and as an occasion setter at the same time. Importantly, connections between the hidden units and the output units are not qualitatively different from direct connections between input and output units. Notice that responding to a CS_1-CS_2 compound will equal the linear combination (summation) of the responding to the separate presentation of elements CS_1 and CS_2, when CS_1 and CS_2 act as simple CSs, through their direct connections VS_1 and VS_2. In contrast, responding to a CS_1-CS_2 compound will differ from the linear combination of responding to the elements CS_1 and CS_2, when either CS_1 or CS_2 act as an occasion setter, through indirect connections VN_j.

Paradigms described by the SD model

Schmajuk and DiCarlo (1992; see also Schmajuk, 1997) demonstrated that their configural model described the following classical conditioning paradigms: (a) acquisition of delay and trace conditioning, (b) extinction, (c) acquisition–extinction series of delay conditioning, (d) blocking, (e) overshadowing, (f) discrimination acquisition, (g) discrimination reversal, (h) feature-positive discrimination, (i) conditioned inhibition, (j) negative patterning, (k) positive patterning, and (l) generalization. In addition, Schmajuk and Blair (1993) showed that the model is able to describe place learning, that is, learning the location of a place in space using only distal landmarks.

Recent computer simulations with the SD model show that the model describes the results reported by Brandon, Vogel and Wagner (2000) regarding the participation of configural cues in generalization and discrimination. In the Brandon *et al.* (2000) study, one group of rabbits was conditioned to an *ABC* compound, a second group was conditioned to an *AB* compound, and a third one to stimulus *A* alone. As expected, the group trained to *ABC* showed increasingly weak responding when tested with *AB* or *A* alone. Instead, the group trained to *A* showed increasingly decreased responding when tested with *AB* and *ABC*. Finally, the group trained with *AB* showed maximal responding to *AB*, and smaller responding to *A* than to *ABC*. According to the SD model, the first result is the effect of the model's competitive rule (overshadowing). The second result is the consequence of the inhibition exerted by stimuli *B* and *C* over excitatory hidden units that predict the US: as the number of CSs increases, so does the inhibition, and consequently, the contribution of the hidden units to the CR decreases. The third result was a combination of both the stronger competitive and the weaker inhibitory mechanisms. Therefore, the SD model offers – as an emergent property – an alternative explanation for results that, until now, were explained only by the Brandon *et al.* (2000) "replaced elements" model. In contrast, the Pearce (1987, 1994) configural theory wrongly predicts a symmetrical decrement when the *AB*-trained group is tested with either *A* or *ABC*.

The SD model for multiple response systems

The model presented above (as well as all other extant conditioning models), assumes that all inputs activate essentially the same CR, that is, that the form of the CR is determined by the choice of US. As discussed before, much investigation of occasion setting has exploited the fact that the form of the CR is often determined not only by the US, but also by the nature of the CS. As mentioned above, Holland (1977) found that rats exhibited very different CR forms during visual (e.g. rear behavior) and auditory (e.g. head-jerk behavior) signals for food. Consequently, both to more completely describe conditioned behavior in general, and to address the outcomes of experiments in occasion setting in particular, the present section extends the SD model to describe multiple response systems.

It is important to notice that this extension does not modify the model's learning rules for VS_i, VN_j, and VH_{ij}, the computation of the aggregate prediction, B_{US}, or the computation of the output error, EO, but only the computation of the responses (performance rules) for the different systems. The simplicity of this extension is based on the fact, noted before, that the associations of hidden units with the US, VN_j, and the associations of input units with the US, VS_i, are stored in a separate layer of neural elements. Therefore, the activity of each of these neural elements is available for the computation of CR_1 and CR_2.

Figure 11.2 shows a diagram of the SD/SLH neural network for multiple response systems. Figure 11.2 illustrates how the associations computed by the network shown in Figure 11.1 are organized to generate not only aggregate prediction B_{US} and the CR, but also CR_1 and CR_2. Figure 11.2 shows a multilayer network with one input layer, one hidden-unit layer, and two sets of output layers; one controlling CR_1 and another one controlling CR_2. Input units are activated by conditioned stimuli *A* and *B*, assumed to be in one, sensory modality (e.g. auditory), and by conditioned stimuli *X* and *Y*, assumed to be in a different sensory modality (e.g. visual).

Briefly, in this model a stimulus (a) accrues a *direct* excitatory association with the output units of a specific response system, (b) accrues a *direct* inhibitory association with the output units of every response system, and (c) becomes *configured* with other CSs in hidden units which in turn acquire associations with the output units of every response system. In Figure 11.2, inputs *A* and B_{US} (e.g. auditory CSs) activate CR_1 (e.g. headjerk) through their direct *excitatory* associations with the US (VS_A, VS_B). Symmetrically, inputs *X* and *Y* (e.g. visual CSs) activate CR_2 (e.g. rearing) through their direct *excitatory* associations with

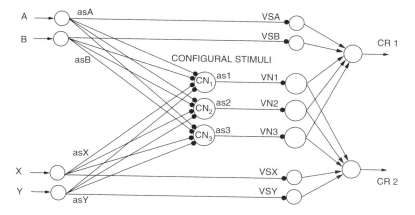

Figure 11.2 The SD/SLH model for multiple response systems. Diagram of a network that incorporates a layer of hidden units capable of describing stimulus configuration in classical conditioning. A and B: inputs to response system CR_1, as_A and as_B: short-term memory of A and B, X and Y: inputs to response system CR_2, an_X and an_Y: short-term memory of X and Y, CN: configural stimulus. CR_1 and CR_2: conditioned responses, VS_A, VS_B, VS_X and VS_Y: CS–US associations, VN_1, VN_2 and VN_3: CN–US associations. Arrows represent fixed synapses. Solid circles represent variable synapses.

the US (VS_X, VS_Y). The output activities of the hidden-unit layer (assumed to code configural stimuli CN_1, CN_2 and CN_3) excite or inhibit both CR_1 and CR_2, through their associations with the US (VN_1, VN_2 and VN_3). Although not shown in Figure 11.2, inputs A, B, X and Y inhibit both CR_1 and CR_2 through their direct inhibitory associations with the US ($-VS_A$, $-VS_B$, $-VS_X$ and $-VS_Y$).

Therefore, input A (a) excites or inhibits CR_1 through direct associations; (b) excites or inhibits CR_1 and CR_2 indirectly through associations with configural, hidden units; but (c) only inhibits CR_2 through direct associations.

In simple mathematical terms, response system 1 generates

$$CR_1 = as_A VS_A + as_B VS_B + \Sigma_j an_j VN_j + I, \qquad [11.11]$$

whereas response system 2 generates

$$CR_2 = as_X VS_X + as_Y VS_Y + \Sigma_j an_j VN_j + I. \qquad [11.12]$$

In Equations [11.11] and [11.12], the term I is the output of a system that provides inhibition to both response systems, and given by

$$I = as_A(-VS_A) + as_B(-VS_B) + as_X(-VS_X) + as_Y(-VS_Y). \qquad [11.13]$$

In short, CR_1 and CR_2 result from simply reorganizing the values of as_A, as_B, an_j, VS_A, VS_B and VN_j computed for the single response system. Because CR_1 is activated by $CS(A)$ we refer to it as CR_A, similarly because CR_2 is activated by X we refer to it as CR_X. In this framework, a stimulus performs as a simple CS when it acts on a response system through its direct excitatory or inhibitory associations, and as an occasion setter when it acts on a response system through its configural associations. Although a CS's simple and configural representations compete to gain association with the US, a CS can act as both a simple CS and an occasion setter at the same time. Furthermore, as illustrated in the next chapter (e.g. Figure 12.14), these roles can be completely antithetical: a CS can simultaneously behave as an excitatory simple CS and as an inhibitory occasion setter.

Summary

This chapter describes the SD model, a real-time, single-response neural network that successfully describes many classical conditioning paradigms. In the network, a CS can establish direct, simple CS–US associations and indirect, configural H–US associations, all formed in accordance with a competitive rule. It is worth noticing that CS–H associations follow a slightly different rule, which explains why even though a CS might not be able to establish strong direct CS–US associations, it can still establish strong CS–H associations with the hidden units.

The chapter also presents an extension of the SD model, the SLH model, which includes (a) multiple response systems that establish associations with simple and configural stimuli to control different responses (e.g. eyeblink, headjerk, rearing), (b) a system that provides both stimulus configuration and generalization to the different response systems, and (c) a system that provides inhibition to the different response systems. The SLH model offers a precise description of the different roles a CS can play in classical conditioning. A CS acts as a simple CS through its direct, simple associations with the US, or as an occasion setter through indirect, configural associations with the US. As a result of their own excitatory or inhibitory associations with the output (US) units, hidden units corresponding to stimulus configurations join the action of direct excitatory and inhibitory associations of simple CSs with the US. Interestingly, hidden-unit action does not reflect a special process or function. Connections between the hidden units and the output units are not qualitatively different from direct connections between input and output units, and are both controlled by a competitive rule.

12

Occasion setting

This chapter illustrates how the SD/SLH model (presented in Chapter 11) describes situations in which a CS behaves as a simple CS or as an occasion setter. We will analyze its performance in (a) a simultaneous FP discrimination with a strong feature and a weak target, (b) a simultaneous FP discrimination with a weak feature and a strong target, (c) a simultaneous FN discrimination, (d) a serial FP discrimination, (e) a serial FN discrimination, and (f) a contextual discrimination. As in previous chapters, in each case, we first present sample empirical data from test sessions administered after training, then show simulations of those data from the model, and finally detail the mechanisms by which the model acquires those discriminations.

It will be shown that a CS acts as a simple CS (through its direct associations with the US) or an occasion setter (through its indirect associations with the US via the hidden units), depending on the strength of its direct association with the US (a function of its intensity, duration and the CS–US interval) and the requirements of the task at hand.

Distinctions between occasion setting and simple conditioning

Investigations of the nature of learning in feature-positive (FP) and feature-negative (FN) conditional discriminations (Jenkins & Sainsbury, 1969) were the starting point for much of the research in occasion setting. In an FP discrimination, presentations of the "feature" cue, X, and the "target" cue, A, are accompanied by the US, and presentations of A alone are not reinforced (XA+/A−), whereas in an FN discrimination the element alone is reinforced and the compound nonreinforced (A+/XA−). According to most conditioning theories (e.g. Pearce & Hall, 1980; Rescorla & Wagner, 1972; see also Jenkins & Sainsbury, 1969, for an earlier statement), in FP discriminations, X should acquire conditioned excitatory associations with a representation of the US, and in FN discriminations, X should acquire conditioned inhibitory associations with the US representation, directly suppressing its activation. However, Ross and Holland (1981) suggested that, under some circumstances, notably when X preceded A on compound trials, X would instead acquire the ability to depress or enhance the action of associations between representations of A and the US.

Three operational sets of criteria were offered (Holland, 1983, 1992) to distinguish between X's simple conditioned tendencies (excitation or inhibition) and its occasion setting powers: (a) response form, (b) extinction/counterconditioning effects, and (c) summation/transfer effects.

Response form

A simple conditioning account for the solution of XA+/A− FP discriminations attributes CRs during the XA compound to X–US associations, whereas an occasion setting account attributes them to A–US associations, which are enhanced or enabled by X's occasion setting powers. To determine which of these associations controlled responding in FP discriminations, Ross and Holland (1981) exploited a conditioning preparation in which the form of the CR is substantially determined by the nature of the CS associated with the US. In that preparation, auditory and localized visual cues paired with food produce easily distinguishable CRs in food-deprived rats (Holland, 1977). Rats reared on their hind legs ("rear" behavior) during the presentation of visual stimuli associated with food, and generated rapid head movements ("headjerk" behavior) during the presentation of auditory stimuli associated with food. Consequently, the form of the CR evoked by a light + tone compound exposes its associative origins: if the CR is dominated by behaviors characteristic of visual cues (rear behavior), then that responding is the consequence of light–food associations, but if it comprises behaviors characteristic of auditory cues (headjerk behavior), it reflects tone–food associations. Thus, in an FP discrimination, the form of responding observed during the reinforced XA compound stimulus provides information about whether that CR is the consequence of X–US or A–US associations. Ross and Holland (1981) monitored the CRs of rats in FP discriminations with simultaneous (XA+/A−) and serial (X→A+/A−) training procedures. With both procedures, the rats quickly learned to respond on reinforced compound trials and withhold responding on A− trials. However, with the simultaneous training procedure, the form of the CR during the compound CS was characteristic of X–US associations, whereas with the serial procedure, the rats displayed CRs characteristic of A–US associations. Thus, these observations of response form suggested that the simultaneous procedure encouraged solution of the

FP discrimination by simple conditioning of X, but with the serial procedure, X came to set the occasion for responding based on an association between A and the US.

Extinction and counterconditioning

If performance on FP and FN discriminations is based on simple associations between X and the US, then elimination of those associations after discrimination training should abolish discriminative performance. Thus, to the extent that rats solve XA+/A− discriminations by forming simple excitatory associations between X and the US, then repeated presentation of X alone after training should extinguish those associations, and hence eradicate responding to the XA compound. Similarly, to the extent that rats solve XA−/A+ discriminations by forming simple inhibitory associations between X and the US, then repeated reinforced presentations of X after training should establish excitatory X–US associations, and hence eliminate X's simple inhibitory powers. Indeed, with simultaneous training compounds, Holland (1989b, 1989d) found that posttraining extinction of X produced severe decrements in performance on FP discriminations, and posttraining counterconditioning of X destroyed performance on FN discriminations. In contrast, when training involved serial (XA) compounds, thought to establish occasion setting, posttraining extinction or counterconditioning of X had minimal effects (Holland, 1989b, 1989d; Rescorla, 1985, 1986). Thus, X's ability to elicit conditioned responding when presented alone was largely irrelevant to the production of responding during the XA serial compound. Apparently, X's ability to modulate the action of an A–US association (occasion setting) is independent of X's own simple associations with the US.

Transfer

To the extent that performance on FP and FN discriminations is based on simple associations between X and the US, the identity of A should be unimportant for the display of X's powers. In FP discriminations, X should elicit responding regardless of whether it is accompanied by A, some other cue, or even no cue. Similarly, in FN discriminations, X should inhibit responding to any CS associated with the same US. On the other hand, if X sets the occasion for responding based on an A–US association, then its powers might be more limited to its original target (A). Several early experiments from Holland's laboratory examined transfer of X's powers to another target cue, which was trained separately from the FP or FN discrimination (e.g. Holland & Lamarre, 1984; Holland, 1986b). When simultaneous XA training trials were used, X's powers transferred broadly to other target cues after both FP and FN discriminations. However, after training with serial X→A compounds, X's occasion setting powers were much more restricted to its original training target A, as might be anticipated if X had acquired a modulatory link to a specific A–US association. Although later data (e.g. Davidson & Rescorla, 1986; Holland, 1989b; Rescorla, 1985), discussed later, forced modifications of this simple claim (occasion setters have been found to modulate responding to targets of other occasion setters, and trained-and-extinguished cues, for example), it is fair to say that in many circumstances, transfer of occasion setting is more limited than transfer of simple conditioning (Holland, 1992).

Simulated results

In the following sections, we will apply the SD/SLH model to the empirical evidence that distinguishes occasion setting from simple association, including data on response form, extinction and counterconditioning, transfer effects, as well as other factors that influence occasion setting. All computer simulations were carried out with the same set of parameter values (presented in Appendix 12.1).

Response form

Simultaneous FP discrimination

Experimental results

Ross and Holland (1981, Experiment 1) observed the responding of rats in simultaneous (XA+/A−) FP discriminations. The form of the CR acquired to the XA compound was characteristic of the predictive X feature (CR_X). When a 5-s light + tone compound was reinforced and the 5-s tone alone nonreinforced, the tone alone elicited no behavior, and the compound evoked high levels of rear behavior characteristic of the light feature (CR_X) and minimal headjerk behavior, characteristic of the tone target (CR_A) (upper panel of Figure 12.1). Similarly, if the light + tone compound was reinforced and the light alone was nonreinforced, the light alone elicited no behavior, and the compound evoked headjerk behavior. Furthermore, in both cases, when the feature was presented separately in a test session, it controlled the same behavior as the compound, CR_X (not shown in Figure 12.1). Consequently, Ross and Holland (1981) concluded that responding in simultaneous FP discrimination procedures was entirely the result of feature–US associations.

Simulation results

The lower panel of Figure 12.1 shows a computer simulation of responding to the XA compound stimulus in the test phase in Ross and Holland's (1981, Experiment 1) study. As in the empirical data (upper panel), the SD/SLH

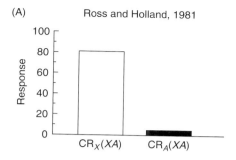

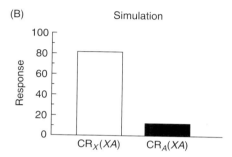

Figure 12.1 Response form (CR_A and CR_X) during simultaneous feature-positive discriminations. *Upper panel*: (Data from Ross & Holland, 1981). CR_X and CR_A responding of rats trained in simultaneous feature positive (FP) discriminations. The bars indicate the percentage of behavior that was CR_X (open bar) or CR_A (solid bar) on XA compound trials. *Lower panel*: Simulation. Peak CR_X (open bar) and CR_A (solid bar) on XA trials after 30 training trials (15 XA+ trials alternated with 15 A– trials).

model generated mostly CR_X during XA presentations, that is, responding of a form appropriate to the X feature CS.

Figure 12.2 illustrates the real-time values of X, A and the US during a reinforced trial in a simultaneous FP discrimination. The feature X and target A generate short-term memory traces that grow after X and A onset and decay to zero after X and A termination. According to the SD/SLH model (see Equation [11.6] in Chapter 11), in a reinforced trial, X–US and A–US associations increase in proportion to the shaded area under each trace, and decrease in proportion to the clear area under the curve. Because during a nonreinforced trial, A–US associations decrease in proportion to the whole area under the curve, X–US associations are strong and A–US associations are weak. Furthermore, when X and A are presented together on a reinforced trial, the X–US associations block the formation of A–US associations. Thus, FP discrimination is achieved because XA presentations result in strong CRs based on X–US associations, but A presentations result in weak CRs.

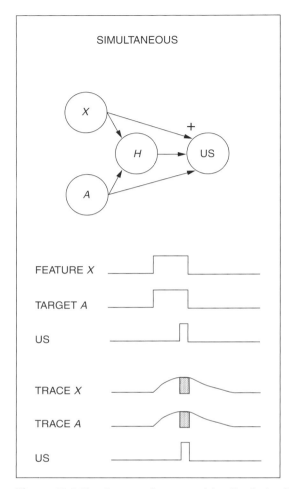

Figure 12.2 Simultaneous feature-positive discrimination. *Upper panel*: Simplified diagram of the circuit in Figure 11.2. *Lower panel*: Real-time representations of X and A, their traces, and of the US on an XA+ trial.

Figure 12.3 shows a computer simulation of the acquisition of a simultaneous FP discrimination when X and A have similar intensities. The upper panel of Figure 12.3 shows CR_X and CR_A responding to XA and A. At the beginning of training, CR_X responding to the XA compound and CR_A responding to A are similar. With increasing number of trials, CR_X responding to the XA compound increases and CR_A responding to A decreases, and the discrimination is achieved. The lower panel of Figure 12.3 shows that the US prediction generated by X increases (as $_X VS_X$, see Chapter 11), as the US prediction generated by A (as $_A VS_A$, see Chapter 11) decreases, over trials. The US prediction generated by the collection of all hidden units H ($\Sigma_h an_h VN_h$, see Chapter 11) remains close to zero.

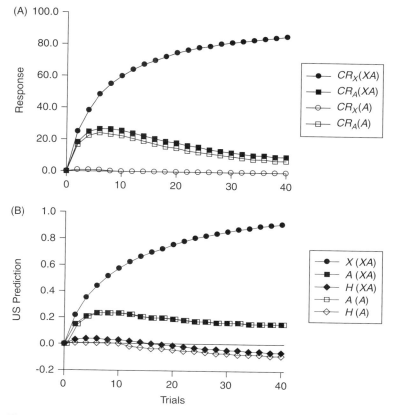

Figure 12.3 Acquisition of simultaneous feature-positive discrimination (simulation). *Upper panel*: CR$_X$(XA): feature-appropriate responses during XA trials, CR$_A$(XA): target-appropriate responses during XA trials, CR$_X$(A): feature-appropriate responses during A trials, CR$_A$(A): target-appropriate responses during A trials. *Lower panel*: X(XA): US prediction by X during XA trials, A(XA): US prediction by A during XA trials, H(XA): US prediction by all the hidden units during XA trials, A(A): US prediction by A during A trials, H(A): US prediction by all the hidden units during A trials.

Thus, in a simultaneous FP discrimination with X and A of comparable salience, X acts as a simple CS because it accrues strong direct associations with the US, and the hidden units play little role.

Simultaneous FP discrimination and target A intensity

Experimental results

Holland (1989c) and Holland and Haas (1993) examined the acquisition of simultaneous FP (XA+/A−) discriminations in Pavlovian and operant (respectively) procedures, as a function of the intensity of the A target cue. The data indicated that when A was of relatively low intensity (Group LO), X–US

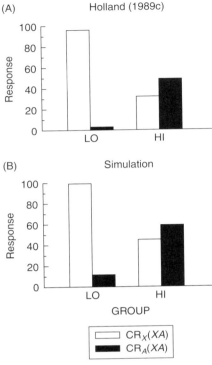

Figure 12.4 Simultaneous feature-positive discrimination. A intensity. *Upper panel*: (Data from Holland, 1989c). $CR_A(XA)$ (solid bars) and $CR_X(XA)$ (open bars) responding after simultaneous feature-positive (FP) training with low- (Group Lo) or high- (Group Hi) intensity auditory A cue. *Lower panel*: Simulation. Peak CR(XA) and $CR_X(XA)$ after 30 training trials (15 XA+ trials alternated with 15 A− trials) for Group LO (X salience 0.95, and A salience 0.3), and 250 training trials (125 XA+ trials alternated with 125 A− trials) for Group HI (X salience 0.3 and A salience 0.95).

associations were formed, but if A was of high intensity (Group HI), X came to modulate the action of A. These conclusions were based on observations of response form, the extent of transfer and the effects of X extinction. For example, the top portion of Figure 12.4 shows the incidence of responding of a form appropriate to X (CR_X) and to A (CR_A) during XA compound presentations in Holland's (1989c, Experiment 1) study. In Group LO, in which A was relatively weak, responding to XA was dominated by CR_X, but in Group HI, in which A was relatively strong, responding to XA was predominantly CR_A.

Simulated results

The lower portion of Figure 12.4 shows that, in agreement with experimental data, CR_A is stronger than CR_X when a relatively strong A is used (Group HI), but CR_A is weaker than CR_X when a relatively weak A is used (Group LO). In

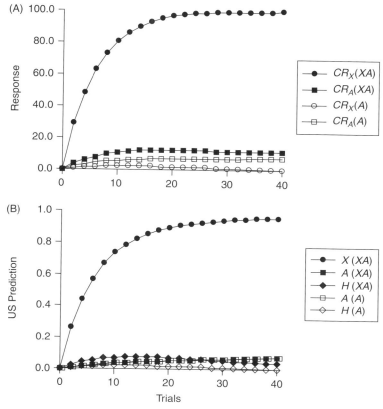

Figure 12.5 Acquisition of simultaneous feature-positive discrimination with a low-intensity target (simulation). *Upper panel*: CR_X (XA): feature-appropriate responses during XA trials, CR_A(XA): target-appropriate responses during XA trials, CR_X(A): feature-appropriate responses during A trials, CR_A(A): target-appropriate responses during A trials. *Lower panel*: X(XA): US prediction by X during XA trials, A(XA): US prediction by A during XA trials, H(XA): US prediction by all the hidden units during XA trials, A(A): US prediction by A during A trials, H(A): US prediction by all the hidden units during A trials.

the framework of the SD/SLH model, X–US and A–US associations increase with increasing intensities of A and X (see Equations 11.6 in Chapter 11). When X is more intense than A, the X–US associations block A–US associations and the FP discrimination is solved by the simple X–US associations. However, when X is less intense than A, the A–US associations block X–US associations and the FP discrimination is solved by occasion setting. That is, A–US associations activate CR_A, but CR_A is inhibited by the hidden units in the absence of the XA compound.

Figure 12.5 shows a computer simulation of the acquisition of a simultaneous FP discrimination when A is less salient than X. Simulated results resemble

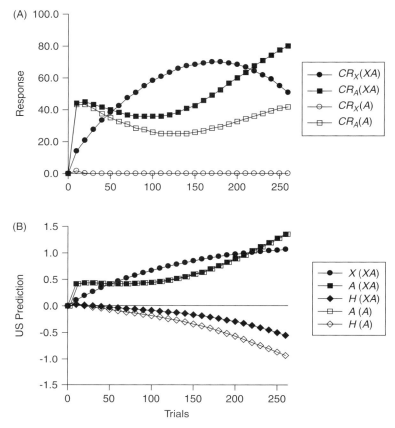

Figure 12.6 Acquisition of simultaneous feature-positive discrimination with a high-intensity target (simulation). *Upper panel*: $CR_X(XA)$: feature-appropriate responses during *XA* trials, $CR_A(XA)$: target-appropriate responses during *XA* trials, $CR_X(A)$: feature-appropriate responses during *A* trials, $CR_A(A)$: target-appropriate responses during *A* trials. *Lower panel*: $X(XA)$: US prediction by X during *XA* trials, $A(XA)$: US prediction by A during *XA* trials, $H(XA)$: US prediction by all the hidden units during *XA* trials, $A(A)$: US prediction by A during *A* trials, $H(A)$: US prediction by all the hidden units during *A* trials.

those shown in Figure 12.3, although the lower salience of *A* leads to less responding on *A*-alone trials, and more on *XA* trials.

A computer simulation of the acquisition of a simultaneous FP discrimination when *A* is more salient than *X*, shown in Figure 12.6, presents quite a different picture. The upper panel of Figure 12.6 shows CR_X and CR_A responding to *XA* and *A*. At the beginning of training, CR_A responding to the *XA* compound and CR_A responding to *A* are similar. With increasing number of trials, CR_X and CR_A responding to the *XA* compound increases and CR_A responding to *A* decreases, and the discrimination is achieved. However, in contrast to the simulations shown in Figure 12.3, CR_X responding to the XA compound first increases and then decreases, and by Trial 250, the dominant response to the XA compound is CR_A. Interestingly, the experimental data reported by Holland (1989c, Figure 1) show all of these patterns, although the SD/SLH model slightly overpredicts CR_A responding on A- alone trials.

The lower panel of Figure 12.6 shows that the US prediction generated by X (as $_X VS_X$, see Chapter 11) and the US prediction generated by A (as $_A VS_A$, see Chapter 11) increase over trials. Importantly, the US prediction generated by the collection of all hidden units H ($\Sigma_h an_h VN_h$, see Chapter 11) becomes more inhibitory on A trials than on XA trials. Therefore, the discrimination is achieved (stronger CR_A responding on XA trials than on A trials) because for the same US prediction generated by A on XA and A trials, the hidden units generate a stronger inhibition on A than XA trials. Thus, feature X acts as an occasion setter because it acts on the A response system through its configural associations.

Figure 12.7 summarizes the SD/SLH model's solutions for simultaneous FP discriminations in a simplified depiction of the network where X and A represent, respectively, feature and target, and H represents the hidden-unit layer. Direct associations of A and X with the US determine the response form, CR_A or CR_X. In the case of simultaneous FP discriminations with X more salient than or equally salient to A (left portion of Figure 12.7), X acts as a simple CS because it acquires strong direct associations with the US, and A acquires weak ones. In the case of simultaneous FP discriminations with X less salient than A (right portion of Figure 12.7), X and A acquire strong direct associations with the US. On A-alone trials, A also activates the hidden units, which then inhibit the US prediction, preventing the display of responding based on the A–US association. But on XA trials, X weakens the inhibition normally exerted by the hidden units (lower panel of Figure 12.6), enabling the performance of both CR_A and CR_X. Thus, X acts both as a simple CS because of its strong direct excitatory association with the US, and as an occasion setter, because of its inhibitory associations with the hidden units.

Notice that, in the case of a single response system, Figure 12.7 can be interpreted as representing the predictions of the US by X, A and H. In the case of simultaneous FP discriminations with X more salient than or equally salient to A (left portion of Figure 12.7), X acts as a simple CS to elicit the CR. In the case of simultaneous FP discriminations with X less salient than A (right portion of Figure 12.7), X and A acquire strong direct associations with the US. On A-alone trials, A also activates the hidden units, which then inhibit the US prediction, preventing the display of the CR based on the A–US association. But on XA

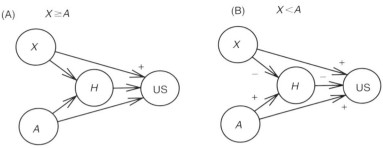

Figure 12.7 Simultaneous feature-positive discrimination. Simplified diagram of the circuit represented in Figure 11.2. *Left panel*: When A is a low intensity target, X acquires a direct excitatory association with the US. *Right panel:* When A is a high intensity target, X acquires a direct excitatory association with the US and an inhibitory association with the hidden units (H). A acquires an excitatory association with the US and an excitatory association with the hidden units (H). The hidden units acquire an inhibitory association with the US. Associative strengths are represented by plus or minus signs placed next to the arrows connecting the nodes. Only significant associative strengths are indicated.

trials, X weakens the inhibition normally exerted by the hidden units, enabling the performance of the CR.

Simultaneous FN discrimination

Experimental results

Holland (1989d, Experiment 1) studied simultaneous FN (A+/XA−) discrimination learning in rats when the feature (X) and target (A) were of comparable salience. The rats learned to perform target-appropriate CR_A on A+ trials and to withhold responding on XA− trials. The upper panel of Figure 12.8 shows CR_A responding on XA, A and X-alone trials in a test session; there was no substantial CR_X (feature-appropriate) responding.

Simulation results

The lower panel of Figure 12.8 shows the simulated results of Holland's (1989d, Experiment 1) study; the simulated and empirical (upper panel) results are comparable. Figure 12.9 shows a computer simulation of the acquisition of the simultaneous FN discrimination when X and A have similar saliences. The upper panel of Figure 12.9 shows CR_X and CR_A responding to XA and A. At the beginning of training, CR_A responding to the XA compound and to A alone differ only slightly. The discrimination is achieved as CR_A responding to A increases and CR_A responding to the XA compound decreases over trials. The lower panel of Figure 12.9 shows that the US prediction generated by A becomes increasingly excitatory (as $_AVS_A$, see Chapter 11), as the US prediction generated by X

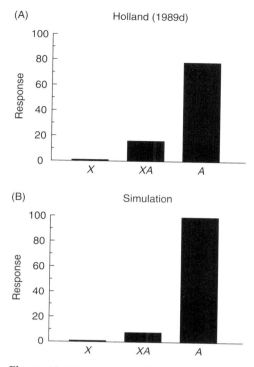

Figure 12.8 Simultaneous feature-negative discrimination. *Upper panel*: (Data from Holland, 1989d). CR_A responding on X, XA and A trials after simultaneous feature-negative training. *Lower panel*: Simulation. $CR_A(X)$, $CR_A(A)$ and $CR_A(XA)$ after 30 training trials (15 $XA-$ trials alternated with 15 $A+$ trials).

(as $_X VS_X$, see Chapter 11) becomes inhibitory, over trials. The US predictions generated by the collection of all hidden units H ($\Sigma_h an_h VN_h$, see Chapter 11) on both XA and A trials remain close to zero. Thus, in a simultaneous FN discrimination with X and A of comparable salience, X acts as a simple inhibitory CS because it accrues strong inhibitory X–US direct associations, so that the system generates a response to XA that is weaker than the response to A.

Figure 12.10 summarizes the results for simultaneous FN discrimination in a simplified depiction of the network where X and A represent, respectively, feature and target, and H represents the hidden-unit layer. Direct associations of A with the US determine the response form, CR_A. In the case of simultaneous FN discriminations with X more salient than or equally salient to A (left portion of Figure 12.10), X acts as a simple inhibitory CS and A acts as a simple excitatory CS.

In the case of simultaneous FN discriminations with X less salient than A (right portion of Figure 12.10), X cannot generate an inhibitory prediction of the US strong enough to counteract the excitatory prediction of A. Therefore, X

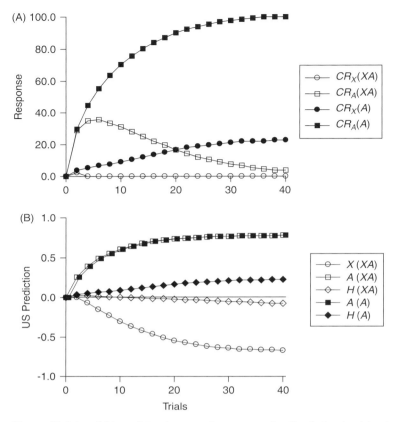

Figure 12.9 Acquisition of simultaneous feature-negative discrimination (simulation). *Upper panel*: $CR_X(XA)$: feature-appropriate responses during *XA* trials, $CR_A(XA)$: target-appropriate responses during *XA* trials, $CR_X(A)$: feature-appropriate responses during *A* trials, $CR_A(A)$: target-appropriate responses during *A* trials. *Lower panel*: $X(XA)$: US prediction by X during *XA* trials, $A(XA)$: US prediction by A during *XA* trials, $H(XA)$: US prediction by all the hidden units during *XA* trials, $A(A)$: US prediction by A during *A* trials, $H(A)$: US prediction by all the hidden units during *A* trials.

acts as an occasion setter, acquiring an excitatory association with the hidden units H, while H acquires an inhibitory association with the US. On *A* trials, A strongly activates the representation of the US through its direct connections, and a large CR is generated. On *XA* trials, A and X activate H which strongly inhibits the representation of the US, and a small CR is generated.

Although we have not explicitly examined the effects of X's and A's salience on the nature of learning in simultaneous FN discriminations, some existing data might be interpreted as consistent with this claim. Rescorla (e.g. 1989) consistently finds evidence for occasion setting (e.g. resistance to counter-conditioning) in pigeon autoshaping, regardless of whether feature and target are presented serially or simultaneously. It is worth noting that in those

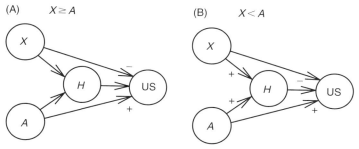

Figure 12.10 Simultaneous feature-negative discrimination. Simplified diagram of the circuit represented in Figure 11.2. *Left panel*: When A is a low-intensity target, X acquires a direct inhibitory association and A a direct excitatory association with the US. *Right panel*: When A is a high-intensity target, X acquires an excitatory association with the hidden units (H). A acquires an excitatory association with the US and an excitatory association with the hidden units (H). The hidden units acquire an inhibitory association with the US. Associative strengths are represented by plus or minus signs placed next to the arrows connecting the nodes. Only significant associative strengths are indicated.

experiments with simultaneous compounds, the features are diffuse visual or auditory stimuli, and the targets are discrete keylight cues. Other investigators (e.g. LoLordo, 1979) have argued that the latter cues are far more salient than the former in pigeon autoshaping. In contrast, the stimuli used in Holland's (1989d, Experiment 1) simultaneous FN discriminations were of more equal salience, and the results were more characteristic of simple conditioned inhibition (e.g. more susceptibility to counterconditioning).

Serial FP discrimination

Experimental results

Compared to their findings with simultaneous compounds in FP discriminations, Ross and Holland (1981) observed quite different patterns of behavior when the X feature preceded the A target on compound trials. When a 5-s light, 5-s empty interval, and 5-s tone (XA) serial compound was reinforced and separate presentations of the tone were nonreinforced, the rats exhibited substantial headjerk behavior (characteristic of auditory CSs) during the tone on compound trials (CR_A, upper panel of Figure 12.11), but not on tone-alone trials. Similarly, when a tone and empty interval and light compound was reinforced and the light alone nonreinforced, the rats acquired rear behavior during the light, but only on compound trials. Thus, in both cases, the target cues controlled behavior characteristic of target–US associations (CR_A). Because responding to the target occurred only on serial compound trials, Ross and Holland (1981) suggested that the feature set the occasion for the occurrence of responding based on target–US associations.

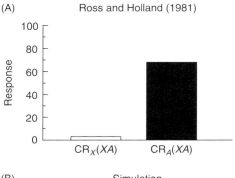
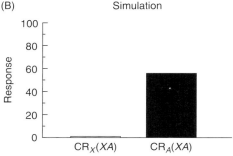

Figure 12.11 Response form (CR_A and CR_X) during serial feature-positive discriminations. *Upper panel*: (Data from Ross & Holland, 1981). CR_X and CR_A responding of rats trained in serial feature-positive (FP) discriminations. The bars indicate the percentage of behavior that was CR_X (open bars) or CR_A (solid bars) on X→A compound trials. *Lower panel*: Simulation. Peak CR_X (open bars) and CR_A (solid bars) on X→A trials after 60 training trials (30 X→A+ trials alternated with 30 A− trials).

Simulation results

The lower panel of Figure 12.11 shows a simulation of the test results of Ross and Holland's (1981, Experiment 2) study. In agreement with the experimental data (upper panel), the SD/SLH model generated mostly CR_A responding during X→A trials. Thus, the model captures the essence of Ross and Holland's (1981) response form distinction: whereas with simultaneous presentation of the feature and target cues in training, response form was appropriate to the feature (see Figure 12.1), with serial feature–target arrangements, response form was appropriate to the target.

Figure 12.12 illustrates the real-time values of X, A and the US during a reinforced trial in a serial FP discrimination. According to the SD/SLH model (see Equation [11.6] in Chapter 11), on a reinforced trial, X–US and A–US associations increase in proportion to the shaded area under each trace, and decrease

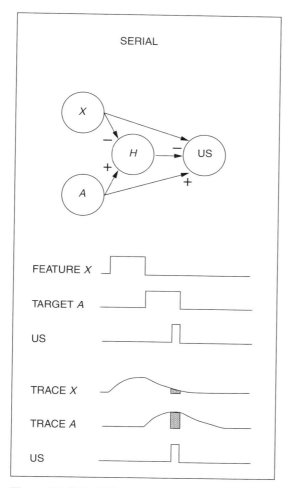

Figure 12.12 Serial feature-positive discrimination. *Upper panel*: Simplified diagram of the circuit in Figure 11.2. *Lower panel*: Real-time representations of X and A, their traces, and of the US on an X→A+ trial.

in proportion to the clear area under the curve. Therefore, during reinforced X→A trials X–US associations grow less than A–US associations.

Figure 12.13 shows a computer simulation of the acquisition of a serial FP discrimination. The upper panel of Figure 12.13 shows CR_X and CR_A responding to XA and A. As the number of trials increases, CR_A responding to the XA compound increases and CR_A responding to A decreases, and the discrimination is achieved. The lower panel of Figure 12.13 shows that the US prediction generated by X ($an_X VS_X$, see Chapter 11) increases only slightly because it is temporally remote from the US, and the US prediction generated by A ($an_A VS_A$, see Chapter 11) increases substantially because it is contiguous with the US. Because the US is overpredicted by the A–US associations on A's nonreinforced

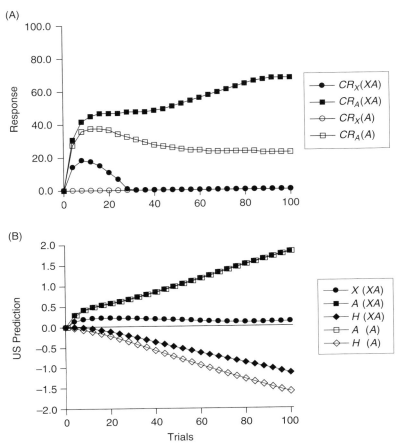

Figure 12.13 Acquisition of serial feature-positive discrimination (simulation). *Upper panel*: $CR_X(XA)$: feature-appropriate responses during $X \to A$ trials, $CR_A(XA)$: target-appropriate responses during $X \to A$ trials, $CR_X(A)$: feature-appropriate responses during A trials, $CR_A(A)$: target-appropriate responses during A trials. *Lower panel*: $X(XA)$: US prediction by X during $X \to A$ trials, $A(XA)$: US prediction by A during $X \to A$ trials, $H(XA)$: US prediction by all the hidden units during $X \to A$ trials, $A(A)$: US prediction by A during A trials, $H(A)$: US prediction by all the hidden units during A trials.

trials, the system trains its hidden units (see section on "Differences between input–output and input–hidden associations" in Chapter 11) in order to reduce the output error by generating a strong inhibition on the output on A trials and a weak inhibition on $X \to A$ trials. Thus, the US predictions generated by the collection of all hidden units H ($\Sigma_h an_h VN_h$, see Chapter 11) become increasingly negative, being more inhibitory on A trials than on XA trials.

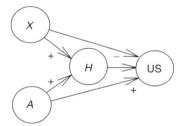

Figure 12.14 Serial feature-positive discrimination. Simplified diagram of the circuit represented in Figure 11.2. X acquires an inhibitory association with the hidden units (H). A acquires an excitatory association with the US and an excitatory association with the hidden units (H). The hidden units acquire an inhibitory association with the US. Associative strengths are represented by plus or minus signs placed next to the arrows connecting the nodes. Only significant associative strengths are indicated.

Figure 12.14 shows a simplified depiction of the network after 100 training trials where X and A represent, respectively, feature and target, and H represents the hidden-unit layer. Because of their different temporal relations with the US (see Figure 12.12), A accrues strong A–US associations but X accrues only weak X–US associations. Since the form of responding is determined by the direct associations of X and A with the US, responding on X→A presentations comprises mostly CR_A (see Figure 12.11). Furthermore, responding on A-alone trials is small because it is strongly inhibited by the hidden units (which are activated by A), and responding on X→A presentations is large because the inhibitory influence of the hidden units is ameliorated by the inhibitory action of X on the hidden units.

Notice that, in the case of a single response system, Figure 12.14 can be interpreted as representing the predictions of the US by X, A and H. In the case of serial FP, A accrues strong A–US associations but X accrues only weak X–US associations. The CR on A-alone trials is small because it is strongly inhibited by the hidden units (which are activated by A and the context), and the CR on X→A presentations is large because the inhibitory influence of the hidden units is inhibited by the action of X. In a way, A acts simultaneously as an excitor and an inhibitor.

Serial FN discrimination

Experimental results

Holland (1989d, Experiment 1) trained rats with a serial (A+/X→A−) FN discrimination procedure, in which a 5-s tone CS was reinforced with food when it was presented alone, but not reinforced when it was preceded, 10 s earlier, by a 5-s light cue. The upper panel of Figure 12.15 shows test session CR_A

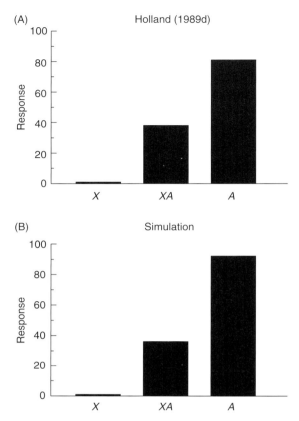

Figure 12.15 Serial feature-negative discrimination. *Upper panel*: (Data from Holland, 1989d). Feature-appropriate CR_X responding on X-alone trials and target-appropriate CR_A responding on XA and A trials after serial feature-negative (FN) training. *Lower panel*: Simulation. Peak $CR_X(X)$, $CR_A(A)$, and $CR_A(XA)$ after 200 training trials (100 X→A− trials alternated with 100 A+ trials).

responding during A on A-alone and X→A trials, and responding on X-alone trials during the interval in which A would have been presented. The substantial CR_A responding observed on A-alone trials was suppressed on X→A trials.

Simulation results

The lower panel of Figure 12.15 shows a computer simulation of Holland's (1989d, Experiment 1) results. The SD/SLH model correctly generates mostly CR_A responding, appropriately discriminated on reinforced and nonreinforced trials.

Figure 12.16 shows a computer simulation of the acquisition of a serial FN discrimination. The upper panel of Figure 12.16 shows CR_X and CR_A responding to XA and A. At the beginning of training, CR_A responding to the XA compound

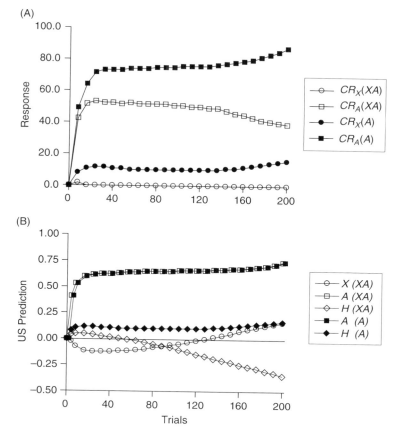

Figure 12.16 Acquisition of serial feature-negative discrimination (simulation). Upper panel: $CR_X(XA)$: feature-appropriate responses during X→A trials, $CR_A(XA)$: target-appropriate responses during X→A trials, $CR_X(A)$: feature-appropriate responses during A trials, $CR_A(A)$: target-appropriate responses during A trials. Lower panel: X(XA): US prediction by X during X→A trials, A(XA): US prediction by A during X→A trials, H(X→A): US prediction by all the hidden units during XA trials, A(A): US prediction by A during A trials, H(A): US prediction by all the hidden units during A trials.

is similar to CR_A responding to A. With increasing number of trials, CR_A responding to the XA compound decreases and CR_A responding to A increases, and the discrimination is achieved. The lower panel of Figure 12.16 shows that the US prediction generated by X shows little change (as$_X$VS$_X$, see Chapter 11) and the US prediction generated by A (as$_A$VS$_A$, see Chapter 11) increases over trials. The US predictions generated by the collection of all hidden units H ($\Sigma_h$an$_h$VN$_h$, see Chapter 11) first increase slightly and then decrease, being inhibitory on XA trials and slightly excitatory on A trials.

278   Configural mechanisms

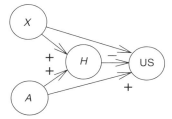

Figure 12.17 Serial feature-negative discrimination. Simplified diagram of the circuit represented in Figure 11.2. Both *A* and *X* acquire excitatory associations with the hidden units (*H*). *A* acquires an excitatory association with the US. The hidden units acquire an inhibitory association with the US. Associative strengths are represented by plus or minus signs placed next to the arrows connecting the nodes. Only significant associative strengths are indicated.

Figure 12.17 shows a simplified depiction of the network after 200 training trials where *X* and *A* represent, respectively, feature and target, and *H* represents the hidden-unit layer. Direct associations of *A* and *X* with the US determine the response form, CR_A or CR_X. *A* accrues excitatory *A*–US associations as well as excitatory *A*–*H* associations. The hidden units (*H*) exert inhibition on the US prediction, somewhat moderating responding. Because the trace of *X* is small at the time of the *A* presentation on *X*→*A* trials (see Figure 12.12), *X* does not acquire inhibitory *X*–US associations. Instead, *X*–*H* excitatory associations are formed and *X* acts as an occasion setter (see section on "Differences between input–output and input–hidden associations" in Chapter 11). Consequently, the discrimination is acquired because hidden units *H* exert a stronger inhibition on the US prediction on *XA* trials than on *A* trials, due to the combined excitation of *H* by *X* and *A*.

Contextual discrimination

Just as in the case of simultaneous FP or FN discriminations with the target more salient than the feature or serial FP or FN discriminations, in contextual discriminations, in which a CS is reinforced in one context and non-reinforced in another, the contexts seem to act as occasion setters. That is, the contexts control responding through their configural associations, but not through their direct associations.

Experimental data

Bouton and Swartzentruber (1986) studied contextual discrimination by administering a tone CS and a shock US during sessions in Context 1 and the CS alone in Context 2. After the discrimination was learned, neither context showed any reliable evidence of direct excitatory or inhibitory association with the US, and nonreinforced exposure to Context 1 alone did not reduce

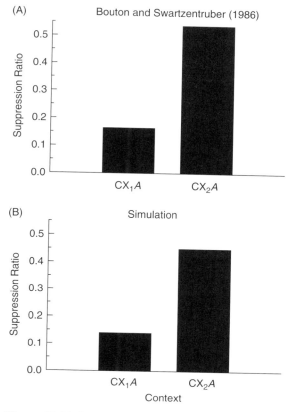

Figure 12.18 Contextual discrimination. *Upper panel*: (Data from Bouton & Swartzentruber, 1986). Suppression ratios to the CS in Context 1 (CX_1A) and in Context 2 (CX_2A). *Lower left*: Simulation. Suppression ratios to the CS in Context 1 (CX_1A) and in Context 2 (CX_2A), after 300 training trials (150 A+ trials in Context 1 alternated with 150 A− trials in Context 2). Suppression ratio was computed by $(L - CR_A)/(L + (L - CR_A))$ where L ($L = 64$) represents the baseline rate of lever pressing as a percentage of the maximum CR_A.

responding to the CS in that context. Figure 12.18 (upper panel) shows more conditioned suppression to the CS in Context 1 than in Context 2 (Bouton & Swartzentruber, 1986, Experiment 1).

Simulation results

The lower panel of Figure 12.18 shows that suppression to A is large in Context 1 and small in Context 2. These simulated results are in agreement with Bouton and Swartzentruber's (1986) data (upper panel).

Figure 12.19 shows the computer simulation of the acquisition of a contextual discrimination. The upper panel of Figure 12.19 shows CR_A responding to CS_A in Context 1 and Context 2. At the beginning of training, CR_A responding

280 Configural mechanisms

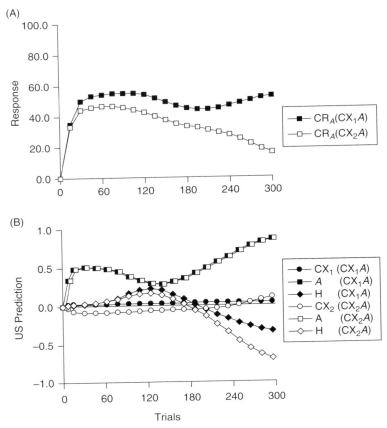

Figure 12.19 Acquisition of a contextual discrimination (simulation). *Upper panel*: $CR_A(CX_1A)$: responding in Context 1 during A+ trials, $CR_A(CX_2A)$: responding in Context 2 during A− trials. *Lower panel*: $CX_1(CX_1A)$: US prediction by CX_1 during A trials, $A(CX_1A)$: US prediction by A during A trials in Context 1, $H(CX_1A)$: US prediction by all the hidden units during A trials in Context 1, $CX_2(CX_2A)$: US prediction by CX_2 during A trials in Context 2, $A(CX_2A)$: US prediction by A during A trials in Context 2, $H(CX_2A)$: US prediction by all the hidden units during A trials in Context 2.

is similar in both contexts. With increasing number of trials, CR_A responding decreases in Context 2 and increases in Context 1, and the discrimination is achieved. The lower panel of Figure 12.19 shows that the US predictions generated by the contexts show little change ($as_{CX}VS_{CX}$, see Chapter 11) and the US prediction generated by A (as $as_A VS_A$, see Chapter 11) increases over trials. The US predictions generated by the collection of all hidden units H ($\Sigma_h an_h VN_h$, see Chapter 11) first increase and then decrease, being more inhibitory on Context 2 trials than on Context 1 trials.

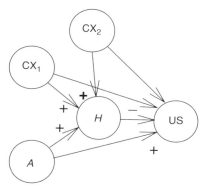

Figure 12.20 Contextual discrimination. Simplified diagram of the circuit represented in Figure 11.2. CX_1 acquires an excitatory association with the hidden units (H), CX_2 acquires a stronger excitatory association with the hidden units (H), A acquires an excitatory association with the US and an excitatory association with the hidden units (H). The hidden units acquire an inhibitory association with the US. Associative strengths are represented by plus or minus signs placed next to the arrows connecting the nodes. Only significant associative strengths are indicated.

Figure 12.20 shows a simplified depiction of the network where CX_1 and CX_2 represent, respectively, Contexts 1 and 2, A represents the CS, and H represents the hidden-unit layer. Whereas A accrues strong A–US associations due to its temporal proximity to the US, the contexts accrue only weak direct associations with the US because they are present for most of the time in the absence of the US. Therefore, the discrimination is acquired because hidden units H (see section on "Differences between input–output and input–hidden associations" in Chapter 11) exert a stronger inhibition on the US prediction during A presentations in Context 2 than during A presentations in Context 1. Both contexts acquire an almost negligible association with the US.

Summary

In summary, a CS acts as a simple CS (through its direct associations with the US) or an occasion setter (through its indirect associations with the US via the hidden units) depending on the strength of its direct association with the US, as determined by its intensity (in the case of simultaneous discriminations), ISI (in the case of serial discriminations), and duration (in the case of contextual discriminations). In other words, a CS acts as an occasion setter when, its direct associations with the US being too weak to control behavior, it becomes configured with the target CS to control the nonlinear outputs of the hidden units (see section on "Differences between input–output and input–hidden associations", in Chapter 11). This section demonstrates that, when direct and indirect CS–US associations are combined in a real-time network, the same competitive rule that governs phenomena entailing compound stimuli – such as blocking, overshadowing or conditioned inhibition – dictates the "role" of a CS (a simple CS or an occasion setter), as indicated by response form data.

Furthermore, the SD/SLH model correctly describes many aspects of the dynamics of discrimination learning, and the relative ease of the solution in a variety of discrimination procedures. That is, simultaneous discriminations are learned more rapidly than serial discriminations, and serial FP discriminations are learned more quickly than serial FN discriminations. On the other hand, with the parameter set used here, context discrimination is learned relatively slowly, compared to many reports (e.g. Bouton & Swartzentruber, 1986). Unfortunately, we have no data on context conditioning comparable to those described for the FP and FN discriminations.

Effects of extinction and counterconditioning on an occasion setter

The previous section illustrated the conditions under which a CS may act as a simple CS (through its direct associations with the US), as an occasion setter (through its indirect associations with the US), or a combination of both. According to the SD/SLH model, manipulations that change only CS–US direct connections should affect the simple properties of the CS, but not its occasion setting properties. This section explores the effects of (a) feature extinction after FP discrimination training; and (b) feature counterconditioning after FN training, when either simultaneous or serial compound stimuli were used, that is, as a function of the simple conditioning or occasion setting faculties of the features.

Simultaneous discriminations

FP discrimination and extinction

Experimental results

To the extent that responding to the XA compound in a simultaneous FP discrimination with a relatively salient X is the consequence of simple conditioning of X, repeated presentations of X alone after training should extinguish that conditioning, and hence abolish any responding to the compound that reflected X–US associations. Holland (1989b) reported that nonreinforced presentations of X substantially reduced responding to both X and the XA compound after simultaneous FP discrimination training (upper left panel of Figure 12.21).

Simulation results

The lower left panel of Figure 12.21 shows, that in agreement with the experimental data, nonreinforced presentations of X substantially reduced

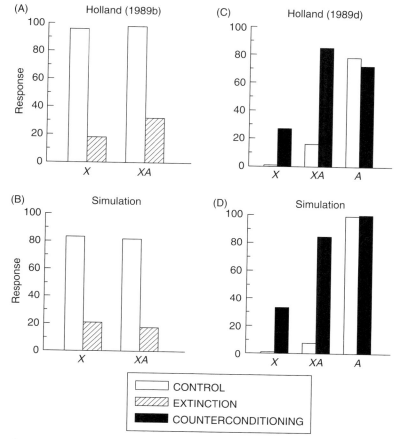

Figure 12.21 Simultaneous feature-positive and feature-negative discriminations. Effects of feature extinction and counterconditioning. *Upper left*: (Data from Holland, 1989b). CR_X responding during *XA* compound and *A*-alone trials after simultaneous feature-positive (FP) training followed by *X* extinction (filled bars) or not (open bars). *Lower left*: Simulation. Peak $CR_X(X)$ and $CR_X(XA)$ after (open bars) 15 *XA+* trials alternated with 15 *A−* trials, or (filled bars) 15 *XA+* trials, alternated with 15 *A−* trials followed by 5 *X−* trials. *Upper right*: (Data from Holland, 1989d). CR_X responding on *X*-alone trials and CR_A responding on *XA* and *A* trials after simultaneous feature-negative (FN) training followed by *X*→US counterconditioning (solid bars) or not (open bars). *Lower right*: Simulation. Peak $CR_X(X)$, $CR_A(A)$, and $CR_A(XA)$ after (open bars) 15 *XA−* trials alternated with 15 *A+* trials, or (solid bars) 15 *XA−* trials alternated with 15 *A+* trials, followed by 3 *X+* trials.

responding, not only to *X* alone, but also to the *XA* compound after simultaneous FP discrimination training.

According to the SD/SLH model (see Figure 12.3 and left panel in Figure 12.7), during simultaneous FP discrimination with a relatively salient *X*, *X* acquires strong excitatory associations with the US. Nonreinforced presentations of *X* decrease those *X*–US associations, thereby decreasing responding to the *XA* compound.

FN discrimination and counterconditioning

Experimental results

Effects analogous to those obtained in FP discriminations were reported by Holland (1989d) in FN discrimination learning (see upper right panel of Figure 12.21). Rats received simultaneous *A+/ XA–* training, followed by *reinforced* presentations of *X* (filled bars), or no such counterconditioning (open bars). With simultaneous training, counterconditioning of *X* abolished its ability to inhibit responding to *A* on *XA* trials, consistent with the claim that simultaneous training establishes inhibitory associations between *X* and the US.

Simulation results

The lower right panel of Figure 12.21 shows that, in agreement with experimental data, reinforced presentations of *X* substantially increased responding to both *X*-alone and the *XA* compound after simultaneous FN training. According to the SD/SLH model (see Figure 12.9 and left panel of Figure 12.10), during simultaneous FN discrimination, *X* acquires strong inhibitory associations with the US, with minimal contribution of the hidden units. Subsequent reinforced presentations of *X* yield excitatory *X*–US associations, increasing responding to the *XA* compound during testing.

Serial discriminations

FP discrimination and extinction

Experimental results

Holland (1989b) reported the effects of *X* extinction after serial FP training. Although responding controlled directly by *X* was significantly reduced, *X*'s ability to modulate behavior controlled by the *A* target was unaffected (upper left panel of Figure 12.22). Thus, after serial training, *X*'s ability to evoke simple CRs, and its ability to set the occasion for responding to the *X→A* compound were largely independent.

Simulation results

The lower left panel of Figure 12.22 shows that, in agreement with experimental data, nonreinforced presentations of *X* did not reduce responding

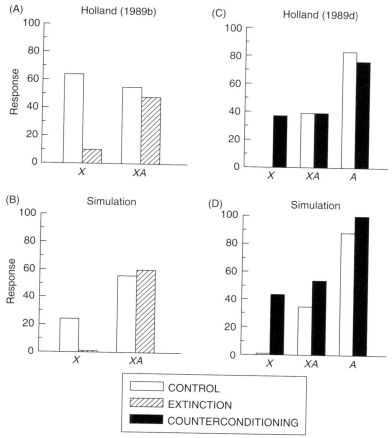

Figure 12.22 Serial feature-positive and feature-negative discriminations. Effects of feature extinction and counterconditioning. *Upper left*: (Data from Holland, 1989b). CR_A responding during A on $X \to A$ trials and CR_X responding on X-alone trials after serial FP training followed by X extinction (filled bars) or no X extinction (open bars). *Lower left*: Simulation. $CR_X(X)$ and $CR_A(X \to A)$ after (open bars) 30 $X \to A+$ trials, alternated with 30 $A-$ trials, or (filled bars) 30 $X \to A+$ trials, alternated with 30 $A-$ trials followed by 5 $X-$ trials. *Upper right*: (Data from Holland, 1989d). CR_X responding on X-alone trials and CR_A responding on $X \to A$ and A-alone trials after serial FN training followed by $X \to US$ counterconditioning (solid bars) or not (open bars). *Lower right*: Simulation. Peak $CR_X(X)$, $CR_A(A)$, and $CR_A(X \to A)$ after 110 $X \to A-$ trials alternated with 110 $A+$ trials (open bars), or 110 $X \to A-$ trials alternated with 110 $A+$ trials followed by 1 $X+$ trial (solid bars).

to the $X{\rightarrow}A$ compound after serial FP training, although responding to X was substantially reduced. Notice, however, that the SD/SLH model underpredicts the magnitude of CR_X responding to X.

According to the SD/SLH model, in the case of serial FP discrimination (see Figures 12.14 and 12.16), A acquires strong excitatory associations and X weak excitatory associations with the US. Since the hidden units provide a strong inhibition on A trials and a weak inhibition on $X{\rightarrow}A$ trials, $X{\rightarrow}A$ presentations result in strong responding and A presentations result in weak responding. Subsequent nonreinforced presentations of X decrease the weak X-US association so that CR_X responding to X is reduced. But, because nonreinforced presentations of X do not modify its associations with the hidden units, CR_A responding to the $X{\rightarrow}A$ compound remains mostly unchanged.

FN discrimination and counterconditioning

Experimental results

Holland (1989d) reported that after serial FN learning, establishment of excitation to X through $X{\rightarrow}US$ pairings had only a small effect on X's ability to inhibit responding to A. This finding suggests that serial training endowed X with the ability to modulate the effectiveness of the A-US association (upper right panel of Figure 12.22).

Simulation results

The lower right panel of Figure 12.22 shows that, in agreement with experimental data, reinforced presentations (counter conditioning) of X did not substantially increase responding to the $X{\rightarrow}A$ compound after serial FN training.

According to the SD/SLH model, in the case of serial FN discrimination, A acquires strong excitatory associations with the US (see Figures 12.17 and 12.19). Since the hidden units provide strong inhibition on $X{\rightarrow}A$ trials, A presentations result in strong CR_A responding and $X{\rightarrow}A$ presentations result in weak CR_A responding. Subsequent reinforced presentations of X yield strong X-US associations, but these associations mainly increase the strength of CR_X but only slightly the strength of CR_A.

Summary

The SD/SLH model generates simulations that were in accord with all reported empirical differences concerning the effects of extinction and counterconditioning on simple conditioning, obtained in simultaneous FP and FN discriminations, and occasion setting, observed after serial FP and FN discrimination training. According to the model, whereas extinction and counterconditioning procedures alter X-US associations involved in simple conditioning, they do not modify X-H or H-US associations involved in occasion setting.

Occasion setting

Transfer effects

The previous sections demonstrated that the SD/SLH model captures the heart of several empirical distinctions between simple conditioning and occasion setting. This section explores data regarding the transfer of the occasion setting properties of a feature from the original target to another target. According to the model, whereas simple CS–US associations can be readily combined with the CS–US associations of other CSs, configural associations follow more complicated rules. Therefore, this section examines transfer effects after simultaneous and serial FP or FN discriminations.

Simultaneous discriminations

FP discrimination

Experimental results

Holland (1986a) performed transfer tests after simultaneous (XA+/A−) FP discrimination training. Training trials with another target (B) were intermixed with the FP discrimination training trials. Initially, B was reinforced until conditioned responding occurred at a high level, and then B was presented without the food US until responding to B extinguished to a low level. Finally, the X feature was presented in compound with either its original target (A), or with the transfer target (B). Figure 12.23 (upper left panel) shows that after simultaneous training, transfer of responding to the XB compound was substantial.

Simulation results

The left bars in the lower panel of Figure 12.23 show that, in agreement with experimental data, the difference between peak CR amplitudes evoked by XB and B (TRAN) is similar to the difference between peak CR amplitudes evoked by XA and A (ORIG) after simultaneous FP discrimination. Stimuli A and B are assumed to be in the same sensory modality.

According to the SD/SLH model, transfer is nearly complete after simultaneous FP training because responding to the XA compound is based on X–US associations (see Figure 12.3 and left panel in Figure 12.7) and, therefore, replacing A for B does not change the CR substantially.

FN discrimination

Experimental results

Holland (1989d) conducted transfer tests after simultaneous (A+/XA−) FN discrimination training. In those tests, the feature was presented in compound with either its original target (A), or with a transfer target (B) that had been separately paired with the US. As with FP training, there was substantial transfer after simultaneous training (upper right panel of Figure 12.23).

288 Configural mechanisms

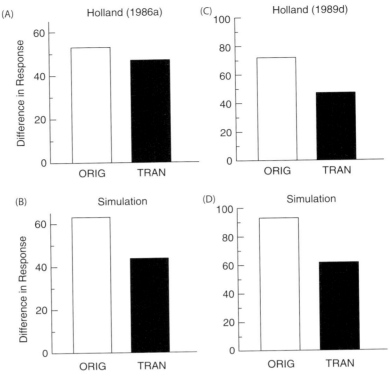

Figure 12.23 Simultaneous feature-positive and feature-negative discriminations. Transfer effects. *Upper left:* (Data from Holland, 1986a). Difference in CR_X responding between XA and A trials (the original discrimination, open bars) and between XB and B trials (transfer discrimination, solid bars) when B is trained and extinguished after simultaneous feature-positive (FP) training. *Lower Left*: Simulation. Difference between peak $CR_X(XA)$ and $CR_X(A)$ after 10 XA+ trials, alternated with 10 A− trials (open bars) and difference between peak $CR_X(XB)$ and $CR_X(B)$ after 10 XA+ trials alternated with 10 A− trials and followed by 8 B+ trials and 2 B− trials (filled bars). *Upper right:* (Data from Holland, 1989d). Difference in CR_A responding on A and XA trials (the original discrimination, open bars) and between B and XB trials (transfer discrimination, filled bars) after simultaneous feature-negative training. *Lower right*: Simulation. Difference between peak $CR_A(A)$ and $CR_A(XA)$ after 15 XA− trials alternated with 15 A+ trials (open bars), difference between peak $CR_A(B)$ and $CR_A(XB)$ after 15 XA− trials alternated with 15 A+ trials and followed by 8 B+ trials (solid bars).

Simulation results

The right bars in the lower panel of Figure 12.23 show that, in agreement with experimental data, after simultaneous FN discrimination training, the difference between peak CR amplitudes evoked by B and XB (TRAN) is similar to the difference between peak CR amplitudes evoked by A and XA (ORIG). Stimuli A and B are assumed to be in the same sensory modality.

According to the SD/SLH model, transfer is nearly complete after simultaneous FN training because the weak responding to the XA compound is based on the combination of inhibitory X–US associations and excitatory A–US associations (see Figure 12.9 and left panel in Figure 12.10). Therefore, replacing the excitatory A target for the excitatory B target does not change the CR substantially, and the discrimination is preserved. The modest decrease in the magnitude of the discrimination is due to the small inhibitory action of the hidden units (see Figure 12.9), which respond preferentially to A.

Serial discriminations

FP discrimination

Transfer to a trained and extinguished cue

Conflicting data have been collected on transfer to new targets in serial discriminations. Whereas some studies of rat appetitive conditioning (e.g. Holland, 1986a) reported little transfer, other studies (e.g. Davidson & Rescorla, 1986; Jarrard & Davidson, 1991) found substantial transfer after training and extinguishing the new target.

Experimental results

Holland (1986a) examined transfer in serial (X→A+/A−) FP discriminations. After discrimination was complete, another target (B) was first paired with the US, and then extinguished. When the X feature was presented in compound with the transfer target (B), transfer to the X→B compound was minimal (see Figure 12.24, left bars in upper panel). In contrast to Holland's (1986a, 1989a) results, Jarrard and Davidson (1991) reported strong transfer when the training trials with another target (B) were intermixed with the original FP discrimination training trials and extinction succeeded the discrimination trials (see Figure 12.24, right bars in upper panel).

Simulation results

The lower panel of Figure 12.24 shows simulated test phase data for the original discrimination (ORIG) and the discrimination with the trained and extinguished target (T/E) for different context saliences. In the SD/SLH model, context salience not only represents the salience of the contextual cues, but also controls the degree of generalization between CSs (see section on "Feature–target (X–A) similarity"). According to the model, whereas the original discrimination increases with decreasing context salience, transfer to the trained and extinguished cue increases with increasing context salience. Consequently, depending on the salience of contextual cues, the SD/SLH model can capture the essence of both Holland's (1986a) transfer results (large discrimination and small transfer to a trained and extinguished cue) and Jarrard and Davidson's (1991) results (smaller discrimination and large transfer to a

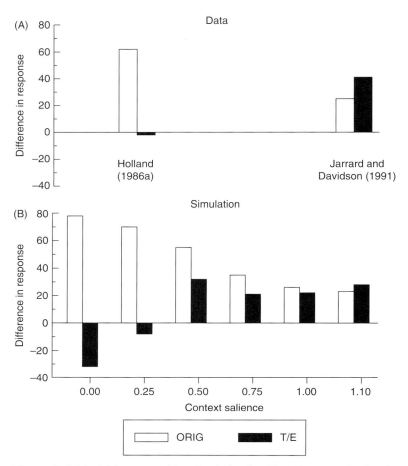

Figure 12.24 Serial feature-positive discrimination. Transfer to a trained and extinguished CS in different contexts. *Upper panel*: (Data from Holland, 1986a and Jarrard & Davidson, 1991). Differences in CR_A responding during A between X→A and A trials (open bars), and during B between X→B and B trials (solid bars) after serial FP training and training and extinguishing B. *Lower panel*: Simulations. Differences between peak $CR_A(X{\to}A)$ and $CR_A(A)$ (open bars) and between peak $CR_A(X{\to}B)$ and $CR_A(B)$ (solid bars) after 38 X→A+ trials alternated with 38 A− trials and followed by 6 B+ trials and 3 B− trials for different values of context salience.

trained and extinguished cue). Unfortunately, there is no independent evidence for context salience or generalization between CSs in either experiment.

Figure 12.25 shows a computer simulation of the training and extinction phases of a transfer experiment using a context with low salience. The upper panel of Figure 12.25 shows CR_A (CR_1 in Figure 11.2) responding to B during six B+ trials followed by five B− trials. Bars indicate responding during X→A, A, X→B and B

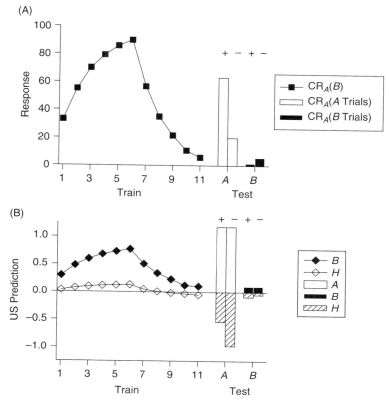

Figure 12.25 Serial feature-positive discrimination. Training and extinction of a cue in a weak context. *Upper panel*: Line: CR$_A$(B): target-appropriate responses during B acquisition and extinction trials. Bars: CR$_A$ (A Trials): target-appropriate responses during X→A+ or A− test trials, CR$_A$ (B Trials): target-appropriate responses during X→B+ or B− test trials. *Lower panel*: Lines: B: US prediction by B during B trials, H: US prediction by all the hidden units during B trials. Bars: A: US prediction by A during X→A+ or A-alone test trials (−), B: US prediction by B during X→B+ or B-alone test trials (−), H: US prediction by H during X→A+, A-alone (−), X→B+, or B-alone (−) test trials.

test trials. The lower panel of Figure 12.25 shows that the US prediction generated by B increases over reinforced, and decreases over nonreinforced trials. The US predictions generated by the collection of all hidden units, H, show little change. That is, extinction of responding to B is mediated through decreases in the B–US simple associations. Because B–US associations are small, little responding to X→B is observed on transfer test trials (see bars in the lower panel of Figure 12.25).

Figure 12.26 shows a computer simulation of the acquisition and extinction phases of transfer using a salient context. The upper panel of Figure 12.26 shows CR$_A$ responding to B during six B+ trials followed by five B− trials. Notice that

292   Configural mechanisms

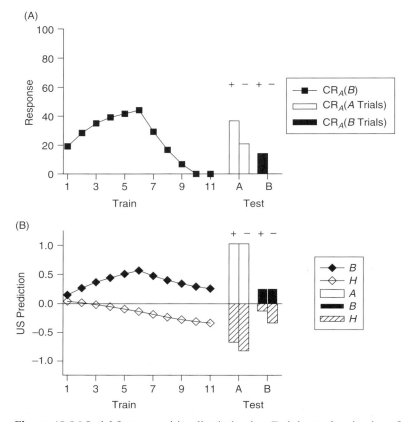

Figure 12.26 Serial feature-positive discrimination. Training and extinction of a cue in a salient context. *Upper panel*: Line: CR$_A$(B): target-appropriate responses during B acquisition and extinction trials. Bars: CR$_A$ (A Trials): target-appropriate responses during X→A+ or A− test trials, CR$_A$ (B Trials): target-appropriate responses during X→B+ or B− test trials. *Lower panel*: Lines: B: US prediction by B during B trials, H: US prediction by all the hidden units during B trials. Bars: A: US prediction by A during X→A+ or A-alone test trials (−), B: US prediction by B during X→B+ or B-alone test trials (−), H: US prediction by H during X→A+, A-alone (−), X→B+ or B-alone (−) test trials.

extinction of CR(B) is more complete with a salient than with a weak context. Bars indicate responding during X-A, A, X-B and B test trials. The lower panel of Figure 12.26 shows that the US prediction generated by B increases over reinforced trials, but barely decreases over nonreinforced trials. In contrast, the US predictions generated by the collection of all hidden units, H, become increasingly inhibitory during extinction. That is, extinction of responding to B is mediated through increasingly inhibitory H–US associations. Because simple B–US associations remain relatively large, strong responding to X→B is

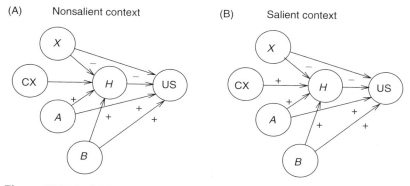

Figure 12.27 Serial feature-positive discrimination. Transfer to a trained and extinguished cue. Simplified diagrams of the circuit represented in Figure 11.2 including the context (CX). *Left panel*: X acquires an inhibitory association with the hidden units (H), CX acquires a weak excitatory association with the hidden units (H), A acquires an excitatory association with the US and an excitatory association with the hidden units (H), B acquires a weak excitatory association with the US. The hidden units acquire an inhibitory association with the US. No transfer is observed. *Right panel*: CX acquires a strong excitatory association with the hidden units (H), B acquires a strong excitatory association with the US. Transfer is observed. Associative strengths are represented by plus or minus signs placed next to the arrows connecting the nodes. Only significant associative strengths are indicated.

observed on transfer test trials. Presentation of X inhibits the inhibitory H–US associations, thereby increasing responding on X–B trials (see bars in the lower panel of Figure 12.26).

Interestingly, our account of extinction as the consequence of B's configural associations cancelling the action of B's direct associations, but leaving them intact, is similar to Bouton's (1994) suggestion that acquisition and extinction are mediated by separate, context-controlled mechanisms. The interaction between B's direct and configural associations also allows the SD/SLH model to describe the renewal phenomenon (see Chapter 9), in which the CS generates a CR when, after undergoing extinction in one context, it is tested in another context. In the case of renewal, excitatory CS–US associations are preserved, as inhibitory contextual configural associations are formed. Return to the original training context reduces the contribution of those inhibitory configural associations, resulting in the reemergence of responding.

Figure 12.27 summarizes the results for transfer to a trained and extinguished cue in a serial FP discrimination using a simplified depiction of the network where X and A represent, respectively, feature and target, H represents the hidden-unit layer, and B represents the trained and extinguished cue. Direct associations of A or B with the US determine response form CR_A and direct associations of X with the US determine the response form CR_X. In the case of a weak context (left portion of Figure 12.27), hidden units are not activated by the context, and transfer is small because the B–US association is not preserved during extinction, and there is little generalization between A and B. In the case of a salient context (right portion of Figure 12.27), hidden units are activated by the context and transfer is large because (a) the B–US association is preserved during extinction by the hidden unit–US associations becoming inhibitory; and (b) there is generalization, through the context-activated hidden units, between A and B. In other words, whereas in a salient context, extinction is mediated by the occasion setting properties of the context, in a weak context, extinction is mediated by B's simple stimulus properties. Because B's simple CS and occasion setting properties are needed to solve the discrimination, transfer is observed in salient, but not in weak contexts.

In general then, the SD/SLH model predicts greater transfer under any condition that preserves B–US simple associations during extinction or, equivalently, any condition in which extinction of responding to B is mediated through inhibitory H–US associations. Other preliminary simulations indicate that the relative salience of the cues, the particular trial sequences used, whether training of the new target is carried out during or after the acquisition of the original discrimination, and the extent of extinction training all may affect the preservation of B–US associations over extinction, and hence the amount of transfer. However, as with context salience and stimulus generalization, in the absence of parametric variation of these variables within the same experiment, it is difficult to attribute the discrepancies in the available transfer literature to these factors.

Within-category transfer

Although Holland's (1986a, 1989a) results suggest that, under certain conditions, occasion setters act only on their original targets and do not transfer to other trained and extinguished cues, Holland (1989a) also reported that there is substantial transfer to a cue that was trained as a target in another, comparable discrimination involving occasion setting.

Experimental data

The top portion of Figure 12.28 shows the results of transfer testing after training on two serial FP discriminations, X→A+/A− and Y→B+/B−, and intermixed training and extinction of another CS, C (Holland, 1989a). There was reliable transfer to the target of the other feature, but transfer to cue C was minimal.

Simulation results

The lower panel of Figure 12.28 shows that, in agreement with experimental data, transfer is larger when the transfer cue was a target in another

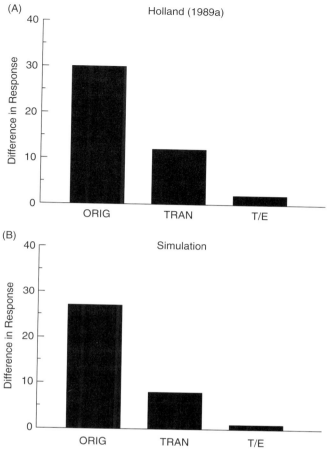

Figure 12.28 Serial feature-positive discrimination. Within-category transfer. *Upper panel*: (Data from Holland, 1989a). Serial feature-positive (FP) discrimination performance (difference in CR_A responding) with the originally trained (ORIG) target cues (average of X→A minus A and Y→B minus B), the target cues from another (TRAN) serial FP discrimination (average of X→B minus B and Y→A minus A), and a separately trained and extinguished (T/E) target cue (average of X→C and Y→C minus C). *Lower panel*: Simulation. Average of the difference between peak $CR_A(X→A)$ and $CR_A(A)$ and the difference between peak $CR_A(Y→B)$ and $CR_A(B)$ after alternated 37 Y→B+ trials, 37 B− trials, 37 X→A+ trials, and 37 A− trials [ORIG], average of the difference between peak $CR_A(Y→A)$ and $CR_A(A)$ and the difference between peak $CR_A(X→B)$ and $CR_A(B)$ after alternated 37 Y→B+ trials, 37 B− trials, 37 X→A+ trials and 37 A− trials [TRAN], and the difference between peak $CR_A(B)$ and $CR_A(X→B)$ after alternated 37 X→A+ trials, 37 A− trials and 37 alternated B+ trials followed by 37 B− trials [T/E].

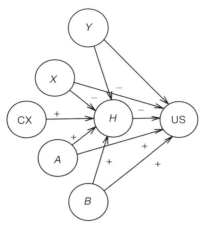

Figure 12.29 Serial feature-positive discrimination. Within-category transfer. Simplified diagram of the circuit represented in Figure 11.2. X and Y acquire inhibitory associations with the hidden units (H), CX acquires an excitatory association with the hidden units (H), A and B acquire excitatory associations with the US and excitatory associations with the hidden units (H). The hidden units acquire an inhibitory association with the US. Transfer is observed. Associative strengths are represented by plus or minus signs placed next to the arrows connecting the nodes. Only significant associative strengths are indicated.

FP discrimination (TRAN), than when the transfer cue was a trained and extinguished cue (T/E). Target stimuli are in the same sensory modality.

Figure 12.29 shows that, during serial FP discrimination learning, the system acquires (a) excitatory A–US and B–US associations, (b) A–hidden units and B–hidden units excitatory associations, (c) X–hidden units and Y–hidden units inhibitory associations, and (d) hidden units–US inhibitory associations. Therefore, strong responding to the X→A compound is the result of the combination of the A–US excitation and the H–US inhibition decreased by X. Similarly, strong responding to the Y→B compound is the result of the combination of the B–US excitation and the H–US inhibition decreased by Y. During transfer, using the associations acquired during the initial discriminations, strong responding to the X→B compound is the result of the combination of the B–US excitation and the H–US inhibition decreased by X (which inhibits the same hidden units as the X→A original compound). Similarly, strong responding to the Y→A compound is the result of the combination of the A–US excitation and the H–US inhibition decreased by Y (which inhibits the same hidden units as the Y→B original compound). In sum, FP discriminations transfer to each other because (a) both targets have excitatory associations with the US, (b) both targets have excitatory associations with the hidden units (generalization), (c) both features have inhibitory associations with the hidden units (generalization), and (d) hidden units have inhibitory associations with the US. Generalization between X→B and X→A compounds and between Y→A and Y→B compounds, present during within-category transfer, increases as the salience of the context increases. However, as shown in Figure 12.28, within-category transfer is larger than transfer to a trained and extinguished cue, because the same context salience capable of supporting generalization between compounds (necessary for within-category transfer) may be incapable of preventing the extinction of B–US associations (needed for transfer to a trained and extinguished cue). As described in the previous section, the use of more salient contexts would produce greater transfer to both another target and a trained and extinguished cue, but preserve the superiority of transfer to another target.

FN discrimination

Experimental results

Holland (1989d) conducted transfer tests after training on two serial FN discriminations, A+/X→A– and B+/Y→B–, and separate training of C (Holland, 1989d, Experiments 1 and 2; upper panel of Figure 12.30). Although transfer to the target of the other feature (Y→A and X→B) was substantial (TRAN), transfer to the separately trained C cue was minimal (EXC).

Simulation results

Figure 12.30 shows that, in agreement with experimental data, transfer is large when the transfer target was trained as a target in another FN discrimination (TRAN), but not when it was simply a trained cue (EXC). The SD/SLH model overpredicts, however, the transfer to the trained cue.

During the acquisition of two serial FN discriminations, A+/X→A– and B+/Y→B–, the system acquires (a) excitatory A–US and B–US associations, (b) excitatory A–hidden units and B–hidden units associations, (c) excitatory X–hidden units and Y–hidden units associations, and (d) inhibitory hidden units–US excitatory associations. Therefore, responding to A is the result of the A–US excitatory association. Decreased responding to X→A is the result of the A–US excitatory association and the hidden unit–US inhibition activated by X and A. Likewise, responding to B is the result of the B–US excitation; responding on Y→B trials is reduced because Y and B excite the inhibitory output of the hidden units. Because either X or Y excites the inhibitory output of the hidden units, transfer to a target trained within the other serial FN discrimination is substantial. However, a separately trained excitor (C) is much less likely to gain access to the hidden units. As a result, X and Y, which act by exciting the inhibitory output of the hidden units when presented with A or B, have much less inhibitory effect on responding to the separately trained excitor, C.

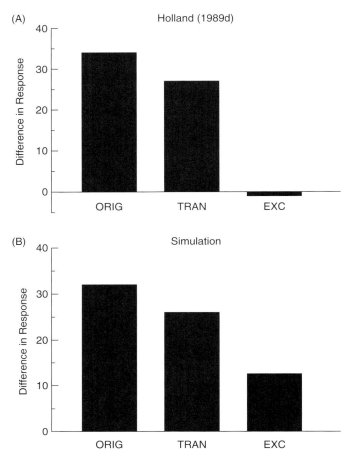

Figure 12.30 Feature-negative discrimination. Within-category transfer. *Upper right:* (Data from Holland, 1989d). Serial feature-negative (FN) discrimination performance (difference in CR_A responding) with the originally trained (ORIG) target cues (average of A minus X→A and B minus Y→B), the target cues from another (TRAN) serial FN discrimination (average of B minus X→B and A minus Y→A), and a separately trained excitor (EXC) (C minus average of X→C and Y→C). *Lower right*: Simulation. Average of the difference between peak $CR_A(A)$ and $CR_A(X→A)$ and the difference between peak $CR_A(B)$ and $CR_A(Y→B)$ after alternated 70 Y→B− trials, 70 B+ trials, 70 X→A− trials, and 70 A+ trials (ORIG), average of the difference between peak $CR_A(A)$ and $CR_A(Y→A)$ and the difference between peak $CR_A(B)$ and $CR_A(X→B)$ after alternated 70 Y→B− trials, 70 B+ trials, 70 X→A− trials, and 70 A+ trials (TRAN), and the difference between peak $CR_A(X→B)$ and $CR_A(B)$ after 70 X→A− trials alternated with 70 A+ trials and followed by 70 B+ trials (EXC).

Summary

The SD/SLH model accounts for the observation, in both serial FP and FN discriminations, of substantial transfer to a target of another occasion setter, despite little transfer to separately trained cues in similar conditions. In addition, the model suggests that the final strength of the simple association of the trained and extinguished cue with the US may account for the conflicting findings about transfer to those cues after serial FP discrimination training. Weak simple associations, and therefore weak transfer, should be obtained when training and extinction of the cue occur in a weak context, one that does not prevent the extinction of the simple cue–US associations. Strong simple associations and, therefore, strong transfer should be observed when training and extinction of the cue occur in a salient context, one that prevents the extinction of the cue–US associations. Since it is difficult to evaluate the actual salience of the contexts used in different experiments, a parametric study is needed to assess the accuracy of this account.

Other factors that influence occasion setting

Feature-target (X–A) similarity

Experimental results

In several experiments, the use of similar-modality X and A cues slowed the acquisition of serial FP (Holland, 1989a) and FN (Lamarre & Holland, 1987) discriminations, but not of simultaneous discriminations. In Holland's (1989a) experiment, rats in Group SER received training with two serial FP discriminations, with 5-s feature cues followed, after 5-s empty intervals, by 5-s target cues, in each of the two types of reinforced compound trials. In one subgroup of rats (SAME), one compound comprised a visual feature and a visual target, and the other compound comprised an auditory feature and an auditory target. In another subgroup of rats (DIFFERENT), one compound included a visual feature and an auditory target, and the other included an auditory feature and a visual target. Thus, both subgroups received the same events as features and targets, and differed only in the similarity relations present within the compound trials. The rats in Group SIM received trials identical to those just described, except that the 5-s feature and target cues were presented simultaneously on compound trials. Learning was more rapid when the features and targets were in different modalities. The top portion of Figure 12.31 (right bars) shows the number of trials to criterial discrimination performance of rats trained with serial procedures in Holland's (1989a) study. In contrast, the acquisition of simultaneous FP discriminations was not reliably affected by feature–target similarity (Holland, 1989a; top portion of Figure 12.31, left bars), as would be anticipated if the solution of simultaneous FP discriminations involved simple X–US associations rather than configural associations.

Simulation results

The lower panel of Figure 12.31 shows that, in agreement with Holland's (1989a) data, acquisition of the serial FP discrimination was faster (fewer trials to criterion) when X and A were in different modalities than when they were in the same modality.

The success of the SD/SLH model in dealing with the effects of stimulus similarity hinges on its treatment of generalization. The model accounts for the generalization between any two cues, X and A, in terms of X and A having initial

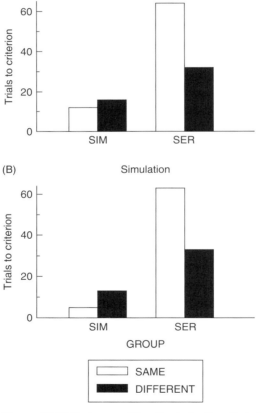

Figure 12.31 Feature-positive discrimination. X–A similarity. Number of trials to discrimination criterion in simultaneous (SIM) and serial (SER) feature-positive (FP) discrimination training when X and A were of the SAME (both auditory or both visual) or DIFFERENT (visual–auditory or auditory–visual) modalities. *Upper panel:* (Data from Holland, 1989a). *Lower panel:* Simulation.

excitatory connections to the same hidden units. Generalization was evaluated by measuring the CR elicited by A in the same and different modalities following conditioning to X. The selected set of initial random weights yielded greater generalization when X and A were in the same modality (36%) than when they were in different modalities (4%).

During serial FP discriminations, when X and A are in the same modality, CR_A on A trials increases due to the generalization of X–US associations, thereby hindering the acquisition of the discrimination. At the network level, the retardation in acquiring the serial FP discrimination in Group SAME is due to the time needed for the hidden units responsible for the generalization between X and A to change their initial excitatory connections with X to inhibitory connections.

In contrast to the serial FP simulation (lower panel of Figure 12.31), simultaneous FP discrimination was slightly faster when X and A were in the same modality than when they were in the different modalities. Simultaneous FP discrimination acquisition is facilitated when X and A are in the same modality because A–US and X–US associations both contribute to generation of the CR on X-A trials. Furthermore, simultaneous FP discrimination acquisition is hindered when X and A are in different modalities because the X–US association needs to block the A–US association to yield stronger responding on X-A trials. Although Holland (1989a) reported no statistically significant effect of similarity of modality in the simultaneous case, it is interesting that the model shows a clear interaction between similarity and ISI, which was indeed observed by Holland (1989a) when serial and simultaneous positive patterning (e.g. XA+/A−/X−), rather than FP discriminations were trained.

Pretraining of X and A

Within the SD/SLH model, the occurrence of outcomes described as reflecting simple conditioning or occasion setting relate to the differential influence of direct associations between CSs and USs, and indirect associations that involve configural hidden units. Thus, it is likely that prior training that influences the mix of these classes of associations would influence occasion setting. Several studies examined the effects of establishing conditioning to the individual X, A, or both stimuli prior to the administration of procedures designed to establish occasion setting. These studies were originally performed primarily to illuminate the relation between simple conditioning and occasion setting strategies in solving conditional discrimination tasks. For example, to the extent that the two solution strategies were competitive, prior establishment of X–US associations might be anticipated to interfere with subsequent use of an occasion setting strategy to solve an FP discrimination. In contrast, to the extent that an occasion setting solution demanded the formation of *A*–US associations, prior training of *A* might be expected to facilitate the use of that strategy.

Experimental results

Ross (1983), using Pavlovian appetitive conditioning procedures with rats, and Rescorla (1986), using the pigeon autoshaping procedure, both found substantial enhancement of *X*'s occasion setting power in FP training after prior reinforced training of *A* (see upper left panel of Figure 12.32). Interestingly, Rescorla (1988) also reported that prior training of *A* as a conditioned *inhibitor* in a *B*+/*BA*– procedure, enhanced acquisition of occasion setting in an *XA*+/*A*– procedure in pigeon autoshaping (upper right panel of Figure 12.32).

In contrast, Ross (1983) and Rescorla (1986) showed that prior *X*–US pairings interfered with the acquisition of occasion setting to that *X* (upper panel of Figure 12.33). Finally, Rescorla (1986) found that prior excitatory training of *A* ameliorated the decremental effects of prior training of *X* alone, that is, prior training of *both X* and *A* did not significantly slow the subsequent acquisition of occasion setting (upper panel of Figure 12.33).

Simulated results

The lower left panel of Figure 12.32 shows that, in agreement with the experimental data, pretraining of *A* facilitates FP discrimination by increasing responding to the *XA* compound. In the framework of the SD/SLH model, excitatory pretraining of *A* increases *A*–US associations, thereby increasing the output error on *A*– trials and the rate of learning of the discrimination. Similar results are obtained for the case of prior inhibitory *A* training. The lower right panel of Figure 12.32 shows that, in agreement with the experimental data, inhibitory pretraining of *A* facilitates FP discrimination acquisition by increasing responding to the *XA* compound.

Finally, the lower panel of Figure 12.33 shows that, in agreement with the experimental data, prior *X*–US pairings hindered the acquisition of the FP discrimination (bars labeled *X*+), but the addition of *A*+ pretraining ameliorated that deficit (bars labeled *X*+*A*+). In the framework of the SD/SLH model, increased *X*–US associations tend to block *A*–US associations, thereby decreasing CR_A responding to *XA*. When *A*–US associations are increased as well (that is, by prior training of both *A* and *X*), increased *X*–US associations are less able to block *A*–US associations, and therefore, CR_A responses to *XA* will increase.

Discussion

Schmajuk and DiCarlo (1992) offered a real-time, single-response neural network that successfully describes many classical conditioning paradigms.

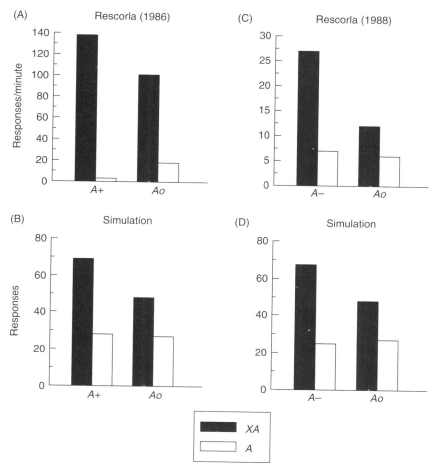

Figure 12.32 Feature positive discrimination. Pretraining of A. *Upper left panel*: (Data from Rescorla, 1986). Rate of key-pecking on XA (solid bars) and A trials (open bars) at the end of FP discrimination training preceded by reinforced (bars labeled A+) or nonreinforced (bars labeled Ao) pretraining of A. *Upper right panel*: (Data from Rescorla, 1988). Rate of key-pecking on XA (solid bars) and A trials (open bars) at the end of FP discrimination training preceded by inhibitory, B+/AB− (bars labeled A−), or nonreinforced (bars labeled Ao) pretraining of A. *Lower left panel*: Simulation. Peak $CR_A(X \to A)$ and $CR_A(A)$ after 25 XA+ trials alternated with 25 A− trials (bars labeled Ao), or 25 XA+ trials alternated with 25 A− trials preceded by 30 A+ trials (bars labeled A+). *Lower right panel*: Simulation. Peak $CR_A(X \to A)$ and $CR_A(A)$ after 25 XA+ trials alternated with 25 A− trials (bars labeled Ao), or 25 XA+ trials alternated with 25 A− trials preceded by 75 B+ trials alternated with 75 AB− trials (bars labeled A−).

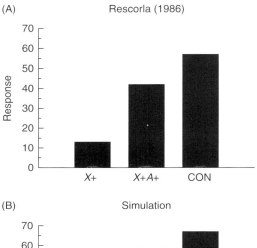
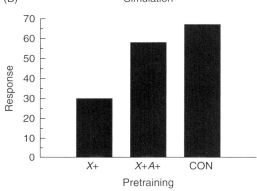

Figure 12.33 Feature-positive discrimination. Pretraining of X and A. *Upper panel*: (Data from Rescorla, 1986). Key pecks per minute on XA trials at the end of feature-positive discrimination training preceded by reinforced X (bar labeled X+), reinforced X and reinforced A (bar labeled X+A+), or nonreinforced (bar labeled CONT) pretraining. *Lower panel*: Simulation. Peak $CR_A(X \rightarrow A)$ after 31 XA+ trials alternated with 31 A− trials (bar labeled CONT), 31 XA+ trials alternated with 31 A− trials preceded by 20 X+ trials (bar labeled X+), and 31 XA+ trials alternated with 31 A− trials preceded by 20 X+ trials and 20 A+ trials (bar labeled X+A+).

In the network, a CS can establish direct, simple associations and indirect, configural associations with the US, all of which operate in accordance with a modified version of the Rescorla–Wagner (1972) rule. The present chapter extends this network theory to include (a) multiple response systems that establish associations with simple and configural stimuli to control different responses (e.g. eyeblink, headjerk, rearing), (b) a system that provides both stimulus configuration and generalization to the different response systems, and (c) a system that provides inhibition to the different response systems.

The SD/SLH model offers a precise description of the different roles a CS can play in classical conditioning. A CS acts as a simple CS through its direct, simple

Table 12.1. *Occasion setting: simulations obtained with the SD/SLH model.*

Paradigms
When does a CS behave as an occasion setter?
Simultaneous FP discrimination (response form)
Simultaneous FP discrimination with a strong target
Simultaneous FN discrimination
Serial FP discrimination (response form)
Serial FN discrimination
Contextual discrimination
Extinction and counterconditioning
Simultaneous FP discriminations and extinction
Simultaneous FN discrimination and counterconditioning
Serial FP discriminations and extinction
Serial FN discrimination and counterconditioning
Transfer effects
Simultaneous FP discrimination
Simultaneous FN discrimination
Serial FP transfer to a trained and extinguished cue
Serial FP within-category transfer
Serial FN within-category transfer
Other factors
X–A similarity
Pretraining of A
Pretraining of X and A

associations with the US, or as an occasion setter through indirect, configural associations with the US. As a result of their own excitatory or inhibitory associations with the output (US) units, hidden units corresponding to stimulus configurations join the action of direct excitatory and inhibitory associations of simple CSs with the US. Importantly, hidden-unit action does not reflect a special process or function. Connections between the hidden units and the output units are not qualitatively different from direct connections between input and output units, and are both controlled by a modified Rescorla–Wagner (1972) rule.

The SD/SLH model simulated the principal differences in response form, extinction and counterconditioning effects, and breadth of transfer found after serial and simultaneous FP and FN discrimination training. Furthermore, it anticipated the importance of a variety of temporal and physical variables in determining the acquisition of occasion setting, and it simulated more subtle aspects of those data, such as the importance of the nature of A training on transfer that have been difficult to interpret within other frameworks. Table 12.1 summarizes these findings.

In addition, computer simulations not presented in this chapter showed that the SD/SLH model qualitatively addresses how temporal factors, such as (a) feature-target (X–A) and X–US intervals; (b) X–A, X–US and A–US intervals; (c) termination asynchrony; and (d) relation of within- and between-trial time intervals, affect the nature of FP discriminations. Some of these issues are addressed in Chapter 14.

Across a range of experimental conditions, the SD/SLH model does an excellent job of specifying when a CS would control responding primarily as the consequence of its direct associations with the US, and hence display characteristics of a simple stimulus, and when a CS would modulate responding as the consequence of its action on the configural units, and hence show characteristics of an occasion setter. Unlike the casual hypotheses that often guided investigations of the conditions for acquiring an occasion setting role, the model precisely defines the necessary conditions in which a CS will act as an occasion setter. For instance, in a serial FP discrimination, X becomes an occasion setter because A–US excitatory associations are stronger than X–US excitatory associations, and the system is required to generate a response to XA stronger than the response to A. In a serial FN discrimination, X becomes an occasion setter because X–US associations are not strong enough to inhibit excitatory A–US associations sufficiently for the system to generate a response to A stronger than the response to XA. In a simultaneous FP discrimination with a strong A, X becomes an occasion setter because A–US excitatory associations are stronger than X–US excitatory associations, and the system is required to generate a response to XA stronger than the response to A. In a contextual discrimination, the context in which A is reinforced becomes an occasion setter because A–US excitatory associations are stronger than context–US excitatory associations, and the system is required to generate a response to A stronger in that context than in a different context.

Alternative theories of occasion setting

An adequate theoretical description of occasion setting should account for the distinctions between the outcomes characterized as reflecting simple conditioning and those indicating occasion setting, and address the two basic concerns of any comprehensive learning theory: the nature or content of the learning in question (i.e. what is learned) and the conditions under which that learning occurs (e.g. Rescorla, 1975). Three approaches to the issue of what is learned in occasion setting can be identified. The first posits a modulatory process, the second a configural process, and the third one a combination of configural and elementary processes.

Modulatory accounts

These accounts posit that occasion setters modify the effectiveness of associations between representations of individual CSs and USs. For instance, Holland (1983; Ross & Holland, 1981) proposed that occasion setters acted by either increasing or decreasing the effectiveness of associations between specific CSs and USs. Rescorla (1985, 1986) proposed that occasion setters (which he termed facilitators and inhibitors) modulate the threshold of activation of the US representation. Wagner (1992; Wagner & Brandon, 1989) suggested that occasion setters engage motivational systems that modulate CRs produced by more discrete response systems engaged by shorter-interval CSs.

Configural accounts

These accounts suggest that unitary representations of nominally compound stimuli are associated with USs separately from associations between elemental CSs and USs. For instance, Pearce (1987) proposed a simple configural model of conditioning, which he later extended to deal with occasion setting phenomena (Wilson & Pearce, 1989, 1990), and more recently reformulated as a connectionist model (Pearce, 1994). According to Pearce's (1987) model, a compound stimulus is always treated as a unique stimulus and does not include its elementary stimuli. Thus, compound stimulus–US pairings would produce CRs to an element of that compound presented alone only to the extent of the generalization between the compound and that element. Within Pearce's (1987) model, the extent of the generalization between a compound and its various elements is related to the relative salience of those elements. Wilson and Pearce (1989, 1990) and Holland (1992) described how this simple model, coupled with assumptions about how compound–element generalization might be affected by various experimental variables such as CS–CS time intervals, can account for the three basic characteristics of occasion setting described earlier.

Gluck and Myers (1993) presented a computational theory of classical conditioning. The model includes two or more three-layer networks working in parallel. The output and hidden layers of the first network are trained to associate CSs with those same CSs and the US. The output of the hidden layers of the first network is used to train the hidden units of the second network. The output layer of the second network is also trained by the US.

Combined configural/elementary accounts

These accounts propose a combination of configural and elementary processes, whereby unitary representations of compound stimuli are associated with USs jointly with associations between elemental CSs and USs. For instance, Bouton and his colleagues (e.g. Bouton & Swartzentruber, 1986; Bouton &

Nelson, 1994) have pointed out analogies between the action of contextual cues and occasion setters. Bouton and Nelson (1994) suggested that discrimination learning produces a combination of direct and indirect associative links among the CSs, the context and the US. A control element, which gates the action of the associations formed between the CSs and the US, is established. Both the CSs and the contextual cues form associations with that control element. Consequently, some associations are only effective when the control element is jointly activated by the CSs and the contextual cues. Bouton and Nelson (1994) suggest that different conditioning procedures might encourage the encoding of one type of link over the other (e.g. direct connections with the US versus indirect connections with the control element), although they have not proposed explicit rules.

Kehoe (1988) presented a three-layer network model in which representations of the individual elements of compounds are maintained in an input layer at the same time that a configural representation of the compound is synthesized in an intermediate, hidden-unit layer. Depending on the relative contributions of the input (element) and hidden (configural) units, responding to compound cues might be highly sensitive to, or resistant to variations in the strengths of their individual elements. Morell and Holland (1993) noted that, at least in the case of serial FN discrimination and serial negative patterning, Kehoe's (1988) model can be made to handle much of the data of transfer of occasion setting, and the independence of the feature's simple and occasion setting powers, with only a few additional assumptions.

Wagner (1992) introduced his version of his real-time theory of classical conditioning (SOP; Wagner, 1981), which assumes that CS units have both direct and indirect (through hidden units) modifiable connections with the adaptive unit of the US. Because competition between hidden units and CS units for associative strength with the US depend on the temporal relationship between the CSs, the model can describe differences between simultaneous versus serial FP discriminations. This theory also incorporates Wagner and Brandon's (1989) modulatory mechanism for interactions between emotional and discrete CR systems. Later, Brandon and Wagner (1998) applied a combination of the Wagner and Brandon (1989) model with their "replaced elements view" (see Brandon, Vogel & Wagner, 2000) to occasion setting.

Although not applied to occasion setting, Sutherland and Rudy (1989) suggested a model in which individual elements and configurations, rather than competing to gain association with the US as in the SD model, become independently associated with the US in two separate association systems. When output of the two systems conflict (as when discriminations are trained between compounds and their elements), that of the configural system takes precedence.

Comparison of the SD/SLH model with modulatory approaches to occasion setting

As noted in the section on "Alternative theories of occasion setting", several aspects of occasion setting data are equally consistent with modulatory accounts, in which X comes to enhance or suppress the action of an A–US association, and configural accounts, in which an X–A unit acquires associations with the US. For example, both the immunity to extinction or counterconditioning of X and the specificity of responding to the original X–A compound are consistent with either approach. But several kinds of evidence led Holland (e.g. 1983, 1992) to favor a modulatory account of occasion setting data.

Much of this evidence against configural approaches was based on assumptions about the nature of configural processes that were not systematically derived, but rather based loosely on Gestalt principles of perceptual organization. For example, it seemed plausible that an X–A configuration would be more likely when those two elements were presented simultaneously, rather than when they were separated in time. But occasion setting was enhanced by increased separation of X and A in time, both in terms of onset asynchrony, and in the presence of a gap between X termination and A onset (Holland, 1986a, 1992).

However, configural models need not follow these Gestalt principles exclusively. Within the present model, because the configural units are in competition with simple associations, they gain important control over behavior whenever the temporal (or other) parameters discourage the formation of simple feature–US associations. Separating X from A also ensures greater separation of X from the US, while maintaining A's contiguity with the US. Thus, within the SD/SLH model, configural units will have greater influence when serial, rather than simultaneous compounds are used (regardless of how readily such units may be formed or engaged in the first place).

Similarly, Holland (1989c, 1992) and Ross (1983) argued that it seemed logical to assume that separate training of A–US associations or of both A–US and X–US associations would be detrimental to the establishment of an independent X + A configuration in FP discrimination learning. Nevertheless, the data showed those manipulations to have no debilitating effect on the acquisition of occasion setting; indeed, the former procedure apparently enhanced occasion setting. However, as described in the section on "Pretraining of X and A", these outcomes are consistent with the present model, in which prior A–US pairings force the network to acquire stronger CR_A associations to XA, which in turn results in a greater contribution of associations of X and A with configural hidden units. Hence, within the SD/SLH model, training of individual element associations can enhance the contribution of configural units.

Finally, Holland (1983, 1989a, 1992; Lamarre & Holland, 1987) argued that within Gestalt traditions it was reasonable to assume that two cues of similar modalities would be more likely to be configured than two of different modalities. Thus, occasion setting based on configural processes should be more likely with similar than with dissimilar X and A cues. However, the data showed more rapid and substantial acquisition of occasion setting when X and A were of different modalities in both serial FP (Holland, 1989a) and serial FN (Lamarre & Holland, 1987) discriminations. Nonetheless, as we showed in the section on "Feature–target (X–A) similarity", the SD/SLH model also predicts more rapid acquisition of serial FP discriminations when X and A are of different modalities than when they are of the same modality. At the network level, generalization between same-modality X and A cues slows acquisition of the inhibitory X–hidden unit associations needed to solve the serial FP discrimination (recall that this discrimination involves acquisition of excitatory A–hidden unit associations).

Thus, although the timing, pretraining and similarity data cited by Holland (1992) in support of modulatory processes are inconsistent with a set of plausible, but casual assumptions about how configural learning is affected by those variables, they are easily embraced by the configural processes described in the SD/SLH model. At the same time, the SD/SLH model correctly accounts for target training effects on the amount of transfer, permits real-time descriptions of performance and predicts pretraining effects that are not compatible with Holland's (1983) hypotheses (e.g. enhancement of serial FP learning by prior inhibitory training of A).

In a similar manner, the SD/SLH model captures many aspects of Rescorla's (1986) modulatory model. For example, within Rescorla's (1986) account, X's acquisition of occasion setting is favored by manipulations that ensure that X is reinforced in the presence of an A that possesses a strong inhibitory component. As noted in previous sections, within the SD/SLH model these kinds of manipulations all guarantee an important role for configural associations by increasing output error after individual event presentations. The model adds to these hypotheses by accounting for patterns of transfer that are not easily dealt with by Rescorla's (1986) ideas, and providing more specific, real-time descriptions of how temporal and other variables (e.g. similarity) contribute to occasion setting.

Wagner and Brandon's (1989) modulatory model presumes that with the use of long CS–US intervals emotional conditioning is engaged, which can modulate performance of more discrete responses conditioned to CSs trained with short CS–US intervals. This modulatory function is dependent solely on the establishment of simple associations between the long-interval CS and the

US, and does not demand special training procedures for either the modulating or target stimuli. Indeed, Brandon and Wagner (1991) reported that the modulatory cue was more effective when it was simply paired with the US than when it was trained in compound with a short-interval target cue. In contrast, experiments with rat appetitive conditioning (e.g. Holland, 1986b) and pigeon autoshaping (Rescorla, 1986) serial conditioning procedures show that, in those preparations, occasion setters must be trained within discrimination procedures: simple trace or long-delay conditioning, or training within a serial compound without separate nonreinforced presentations of the target cue do not endow cues with occasion setting properties. The model correctly predicts these latter data. It remains to be seen whether variations in parameter sets, or more likely, the context salience, might yield other predictions. In this regard it is worth noting that, unlike Holland's (1986b) data, Bonardi and Hall (1984b) report that nondiscriminative appetitive conditioning procedures may generate cues with occasion setting properties.

Thus, although the SD/SLH model avoids hypothesizing a new and separate modulatory function, it retains many of the advantages of those accounts. The SD/SLH model provides a single, consistent theoretical framework for dealing with a large body of literature in occasion setting, without demanding additional ad hoc assumptions for accounting for each new class of experimental data.

Comparison of the SD/SLH model with other configural theories of occasion setting

In the SD/SLH model, whereas simple CSs control CRs primarily through their simple associations with the US, occasion setters act primarily indirectly, through their associations with configural hidden units. For example, in an $X \rightarrow A+$, $A-$ serial FP discrimination, A acquires strong direct excitatory associations with the US, and an excitatory association with a hidden unit that has inhibitory associations with the US, while X acquires weak (or no) excitatory associations with the US and inhibitory associations with the hidden units (Figure 12.14). Thus, on A-alone trials, the strong A–US associations are countermanded by A's activation of inhibitory hidden units, whereas on $X \rightarrow A$ trials, the inhibitory influence of the hidden units is itself countermanded by X–hidden unit associations. The network acquires this combination of strengths in circumstances in which A–US associations are stronger than X–US associations, and the system is nevertheless required to generate a strong output in the presence of XA, and weak output in the presence of A alone.

In contrast to the SD/SLH model, in Pearce's (1987, 1994) model, individual element–US associations play no direct role in the display of behavior during the presentation of compound cues. Compounds and their putative elements are viewed as completely separate cues, sharing responding and competing for associations only to the extent that a compound and one of its so-called elements generalize. The conditions that favor the development of occasion setting phenomena in Pearce's model, including serial arrangement of X and A, or the use of more intense A cues, are those that increase the salience of A relative to that of X, resulting in more generalization of XA's strength to A than to X. Within the SD/SLH model, these conditions also result in a greater contribution of configural hidden units to simple conditioned responding. Thus, despite substantial differences in the roles ascribed to configural and elemental associations within Pearce's model and ours, in many cases, both suggest comparable conditions for learning of occasion setting.

However, in Pearce's (1987) theory, the mechanisms by which experimental variables alter the salience of cues, and hence the generalization between compounds and elements, are not specified by the theory itself, but are contained in ad hoc assumptions. For example, because it is not a real-time theory, even temporal variables are assumed to have their effects indirectly, by affecting the salience of individual cues, rather than directly affecting association formation. Elsewhere, Holland (1989c) argued that the assumptions needed by Pearce's theory to generate occasion setting response form data forced incorrect predictions about the relative rates of learning in various feature positive discriminations and about response form in other training conditions. Similarly, Pearce's theory deals with the observation that target training affects transfer only by additional assumptions about how generalization among features and targets is altered by various training experiences.

As does the SD/SLH model, Kehoe's (1988) model adopts a unique-stimulus hypothesis that assumes maintenance of representations of both compound stimuli and their individual elements. In contrast to the backpropagation procedure employed in the SD/SLH model, in Kehoe's model, all configural units are trained with the same US signal and, therefore, they all learn the same configuration; a procedure that might subtract power from the system. One major distinction, however, between the the SD/SLH model and the Kehoe models resides in the way they configure stimuli. Whereas in the Kehoe model, configuring is equivalent to creating compound stimuli that are active when the component stimuli are presented together with the US, in the SD/SLH model, configuring is a more general process accomplished by training hidden units that may respond to various combinations of stimuli either in the presence, or the absence of the US. A second major difference is that the SD model is a real-time model capable of describing the effects of the temporal arrangements of features and targets on occasion setting.

Although Gluck and Myers's (1993) computational theory of classical conditioning is similar in some aspects to the SD/SLH model, it lacks some critical components that are unique to the SD/SLH model. For instance, because the model is not a real-time system, it cannot describe occasion setting in serial FP or FN discriminations. Furthermore, because the model does not have CS-US direct connections as the SD model has, the model neither distinguishes between simple association and occasion setting properties of a CS, nor describes other properties of occasion setting such as response form.

Gluck and Myers (1994) have recently shown that their model is able to describe the occasion setting properties of context. They applied their model to Bouton and Swartzentruber's (1986) contextual discrimination protocol in which a CS is reinforced in Context 1 and not reinforced in Context 2, and showed that the model learns to respond to the CS in Context 1 but not in Context 2. Because the model does not show responding to Context 1 alone, they conclude that Context 1 has not acquired a direct association with the US. In addition, because the model demonstrates transfer to another CS when it is trained in Context 1, they conclude that it is not a Context 1-CS configuration that has become associated with the US. Therefore, without any mechanistic explanation, Gluck and Myers assume that Context 1 operates as an occasion setter. However, we have shown that there is no contradiction between transfer and configural associations, and therefore it is likely that their model accounts for contextual discrimination as a consequence of a Context 1-CS configuration associated to the US. More recently, Zackheim, Myers and Gluck (1998) developed a recurrent real-time version of the Gluck and Myers (1993) model that captures how (a) the context, and (b) phasic CSs act as occasion setters.

Brandon and Wagner (1998) combined the AESOP model (Wagner and Brandon, 1989), which incorporates both emotional and sensory representations of the US, with the replaced elements view (described in Brandon *et al.*, 2000). This view assumes that cues A and X are represented, respectively, by context independent elements a and x (A alone and X alone) and by context-dependent elements $a_{\sim x}$ and $x_{\sim a}$ (A in the absence of X and X in the absence of A). When A and X are presented together, the context-dependent representations $a_{\sim x}$ and/or $x_{\sim a}$ are replaced by context-dependent elements a_x and x_a. Brandon and Wagner (1998) indicated that the model addresses the temporal aspects of occasion setting, and offers the unique advantage of describing the emotional modulation of the phenomenon; without further assumptions it is unable to describe the CS specificity of the responses (response form).

Summary

This chapter shows that the SD/SHL model can describe situations in which a CS behaves as a simple CS, or as an occasion setter: a CS acts as an occasion setter when, its direct associations with the US being too weak to control behavior, it becomes configured with the target CS to control the nonlinear outputs of the hidden units. Direct and indirect CS–US associations compete, through the same rule that governs blocking, overshadowing or conditioned inhibition, to decide the "role" of the CS as a simple CS, or as an occasion setter.

Many of the basic aspects of the data seem compatible with a simple–configural association interaction view without ad hoc assumptions. Furthermore, the use of a model that represents quantitatively the temporal, intensity and similarity relations among cues, might permit us to account for differences in experimental results due to quantitative differences in otherwise identical empirical procedures. Due to the direct CS–US associations, the SD/SLH model seems the only model able, at this point, to describe the CS-specificity of the CRs (response form).

Potentially, the model might contribute to our understanding of a variety of phenomena related to hierarchical organizations of associations, including the discriminative control of operant behavior (Colwill & Rescorla, 1990), stimulus equivalence (Sidman, 1986), context effects (Bouton & Swartzenruber, 1986; see Chapter 15) and the modulation of learned behavior by motivational events (Davidson, 1993; Holland, 1991). Finally, as shown in Chapter 13, the model can also forward our comprehension of some of the neurophysiological aspects of occasion setting.

Appendix 12.1   Simulation parameters

Simulations with the SD/SLH model assume that one simulated time step is equivalent to 250 ms. Each trial consists of 400 time units, equivalent to 100 s. Unless specified, the simulations assume 5-s X and A. X and A intensities are 0.95. US intensity is 1. X and A onsets are at 5 s in simultaneous discriminations. X onset is at 5 s and A onset is at 15 s in serial discriminations. That is, in the case of simultaneous FP discriminations, both feature and target CSs overlapped, and in the case of serial FP discriminations, the feature CS preceded the target CS by 5 s. In both cases, the last 1.25 s of A overlap the US. Context trace is 0.5 in all simulations, except those of within-category transfer where it is 0.4, and contextual discrimination, where it is 0.3.

Parameter values are $K_1 = 0.028$, $K_2 = 0.2$, $K_3 = 0.95$, $K_4 = 2$, $K_5 = 0.0225$, $K_6 = 5$, $K_7' = 0.0075$, $K_7'' = 0.00675$, $\beta = 0.925$, $m = 8$, $n = 1.5$. Six hidden units were used in all simulations.

Input–hidden unit association weights, VH_{ij}, were randomly assigned using a uniform distribution ranging between ±0.25 in all simulations. Although simulation results were very robust and discriminations (CR(*A*) vs. CR(*XA*)) can be reached with many different parameter values, all simulations were carried out with identical parameter values. The type of solution (*X*–US, *A*–US and *H*–US associative values), however, depended on the initial values of the input–hidden unit associations.

Simulated peak CR amplitudes are expressed as the percentage of the asymptotic peak CR amplitude for simple conditioning (100 CR/CR_{MAX}). The number of simulated trials are those necessary to reach a discrimination criterion similar to that achieved in the experimental data.

13

The neurobiology of occasion setting

As described in Chapter 12, Schmajuk, Lamoreux and Holland (SLH) (1998) showed that an extension of a neural network model introduced by Schmajuk and DiCarlo (SD) (1992) characterizes many of the differences that distinguish simple conditioning from occasion setting.

In the framework of their model, Schmajuk and DiCarlo (1992) proposed that the hippocampus modulates (a) the competition among simple and complex stimuli to establish associations with the unconditioned stimulus, and (b) the configuration of simple stimuli into complex stimuli. Furthermore, Schmajuk and Blair (1993, 1995) suggested that (a) nonselective aspiration lesions of the hippocampal formation impair both configuration and competition, and (b) ibotenic acid selective lesions of the hippocampus proper impair only stimulus configuration.

These assumptions are somewhat similar to the ones made in Chapter 6 for the SLG model, in which we assumed that CS_1-CS_2, CS-CX, CX-CX and CX-CS associations, instead of the configural associations, were impaired by the selective hippocampal lesions (HPLs). In both models, we assume that competition is eliminated by nonselective lesions of the hippocampal areas.

In order to provide a theoretical account of hippocampal participation in occasion setting, the present chapter combines (a) Schmajuk, Lamoureux and Holland's (1998) account of occasion setting (see Chapter 12) with (b) Schmajuk and DiCarlo's (1992) and Schmajuk and Blair's (1993, 1995) assumptions regarding the effect of HPLs and HFLs on classical conditioning mechanisms. The results have implications for both the behavioral and neurophysiological aspects of occasion setting.

Correspondence between the neural network architecture and the organization of the brain

Figure 13.1 shows how nodes and connections in the SD/SLH model are mapped onto different brain regions. In the brain-mapped SD/SLH model, (a) neural elements represent specific neural populations in the brain, (b) connections between neural elements represent known anatomical connections, (c) the activity of neural elements represents activity of specific neural populations, (d) learning sites represent known locations of plasticity in the brain, and (e) neurophysiological manipulations (e.g. lesions of neural elements) represent equivalent manipulations of specific neural populations in the brain. The resulting brain-mapped model is applied to the description of the effects of lesions of the hippocampus and neocortex.

Schmajuk and DiCarlo (1992) suggested that the hippocampus computes (a) the aggregate prediction of the US (B_{US}), and (b) the error for the hidden units (EH_j). These assumptions are respectively consistent with data showing that the neural activity of some pyramidal cells in the dorsal hippocampus is positively correlated with the CR topography during classical conditioning (Berger & Thompson, 1978a; Berger, Rinaldi, Weisz & Thompson, 1983; Berger & Thompson, 1982; Weiss, Kronforst-Collins & Disterhoft, 1996), and with results indicating that the neural activity of some hippocampal cells first increase and then decrease during classical conditioning (Sears & Steinmetz, 1990). The aggregate prediction, $B_{US} = \Sigma_i as_i VS_i + \Sigma_j an_j VN_j$, represents the prediction of the intensity of the US based on all the stimuli present at a given time. Aggregate prediction B is used to determine the output error signal, $EO = US - B_{US}$, that controls the associations of different as's and an's with the US. Whereas the aggregate prediction of the US regulates the *competition* among direct and configural stimuli to establish associations with the US, hidden unit error signals control changes in the *configuration* of simple stimuli as complex stimuli. Notice that, because the CR is also proportional to $\Sigma_i as_i VS_i + \Sigma_j an_j VN_j$, B_{US} is an efference copy of the CR.

The output of CA_1 pyramidal neurons, proportional to $EH_j = an_j VN_j EO$, regulates learning in cortical hidden units, thereby controlling stimulus configuration. This suggestion is in accord with evidence showing that the hippocampus receives information ($an_j VN_j$) from, and sends information back ($an_j VN_j EO$) to different polysensory associative cortices through the entorhinal cortex (Squire, Shimamura & Amaral, 1989), and with results demonstrating that lesions of the hippocampus result in a severe impairment to acquire some types of new knowledge (anterograde amnesia) (Hamann & Squire, 1995) without significantly affecting old memories (limited retrograde amnesia) (Milner, 1966).

318 Configural mechanisms

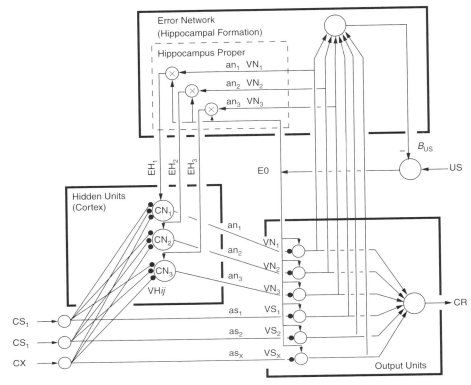

Figure 13.1 The SD model for a single response system. Diagram of a network that incorporates a layer of hidden units capable of describing stimulus configuration. CS: conditioned stimulus, as: short-term memory of a CS, CN: configural stimulus, an: short-term memory of a CN, VS: CS–US association, VN: CN–US associations, VH: CS–CN association, US: unconditioned stimulus, B: aggregate prediction, CR: conditioned response, EH: error signal for hidden units, EO: error signal for output units. Anatomical areas indicated in parentheses refer to the mapping of different nodes in the network onto various brain regions. The hippocampus block includes the hippocampus proper (CA_1 and CA_3 regions) and the dentate gyrus. The hippocampal formation block includes the hippocampus proper, dentate gyrus, subiculum, presubiculum and the entorhinal cortex. Arrows represent fixed synapses. Solid circles represent variable synapses.

Weinberger (1995) has recently reviewed data demonstrating cortical learning during classical conditioning. An alternative locus for the storage of configural stimuli is the superior colliculus, which combines auditory and visual spatial information into a shared neurophysiological representation (Knudsen & Brainard, 1995).

Schmajuk and DiCarlo (1992) suggested that the theta rhythm in the medial septum codes the output error signal EO = US − B_{US}; a hypothesis consistent with data showing that medial septal activity decreases during acquisition of classical conditioning (Berger & Thompson, 1978b). The hippocampus uses theta information (proportional to EO) to modulate the entorhinal input (proportional to an_jVN_j) and sends the result (proportional to an_jVN_jEO) back to (the hidden units in) polysensory associative cortices where an_jVN_j information had originated.

Schmajuk and DiCarlo (1992) proposed that, in the case of the rabbit's nictitating membrane (NM) preparation, as_i–US and an_j–US associations are stored in cerebellar areas, thereby controlling the generation of NM CRs. Simple stimuli activity, as_i, and configural stimuli activity, an_i, reach the sensory input to the pontine nuclei. This assumption is compatible with histological information showing that the pontine nuclei receive direct input from sensory inputs (see Rosenfield & Moore, 1995) and indirect connections from the cortex (Brodal, 1992); results indicating that lesions of cerebellar areas impair the acquisition and retention of classical conditioning of the NM response (Desmond & Moore, 1982; Thompson, 1986), and data showing that the neural activity of interpositus (Berthier & Moore, 1990) and red nucleus (Desmond & Moore, 1991) are positively correlated with the CR Upperography during classical conditioning.

Schmajuk, Thieme and Blair (1993) proposed that a circuit storing as_i–US and an_j–US associations that control food-cup approach (the CR) in rats includes the caudate. This idea is supported by the fact that the caudate receives direct input from the thalamus, and indirect connections from the cortex (Brodal, 1992), and that lesions of the caudate impair the ability of rats to approach a rewarded cue (Packard & McGaugh, 1992).

Finally, we speculate that CR_1 and CR_2 described in Figure 11.2 are controlled by a circuit that includes the central nucleus (CN) of the amygdala. Gallagher, Graham and Holland (1990) reported that neurotoxic lesions of CN impair the acquisition of conditioned orienting responses to the light (rearing) and conditioned orienting responses to the tone (startle) in a simple discrimination paradigm. CN lesions did not impair, however, either the habituation of the unconditioned orienting responses, or the conditioned response. Importantly, LeDoux (1992) has described how information about auditory CSs reaches the amygdala. The CS is first relayed through auditory pathways to the medial geniculate body (MGB) of the thalamus, and from there to the lateral nucleus (LN) of the amygdala through two parallel pathways. One pathway involves a direct projection from the medial division of the MGB and the posterior intralaminar nucleus (PIN) to the lateral amygdala; the other pathway includes a projection from the ventral and medial divisions of the MGB and the PIN to the auditory cortex, to the perirhinal cortex, to the lateral amygdala. Whereas the direct pathway is capable of establishing "quick and dirty" connections, the cortical system is needed for discriminatory functions. Therefore, we conjecture that direct and indirect pathways controlling CR_1 and CR_2 in Figure 12.2 correspond to direct thalamic–amygdalar and indirect thalamic–cortical pathways. Ouputs from the amygdala that reach the caudate and putamen (striatum) through the substantia nigra might be involved in controlling conditioned orienting movements.

It is worth emphasizing that, as required by the model, putative loci of associations with the US in cerebellum, striatum and amygdala receive both direct sensory inputs and indirect cortical inputs.

Effects of brain lesions

Based on the mapping onto the brain circuitry of nodes and connections in the SD/SLH model (illustrated in Figure 13.1), the effect of hippocampal and cortical lesions are described below. These descriptions are in line with those presented in Chapters 6 and 10. Parameter values for the normal case have been described in Chapter 12, Appendix 12.1.

Effects of hippocampal lesions

Effect of nonselective hippocampal lesions

Schmajuk and DiCarlo (1992) proposed that HFLs, produced by aspiration, electrolysis, or colchicine-kainic acid, can be simulated by removing the Hippocampal Formation block in Figure 13.1. HFLs produce important changes. One consequence of HFLs is that hidden unit error signals are no longer computed, and therefore no new associations are formed in the association cortex, thereby impairing paradigms that require readjustment of the configural stimuli.

After HFLs, the hidden-unit error EH_j is given by

$$EH_j = 0, \qquad [11.3b]$$

and, therefore, $d(VH_{ij})/dt = 0$. Equation [11.3b] replaces Equation [11.3a] presented in Chapter 11.

Another outcome of HFLs is that the aggregate prediction of the US, $B = \Sigma_i as_i VS_i + \Sigma_j an_j VN_j$, is no longer computed. In the absence of B, as_i–US and an_j–US associations become independent of the associations accrued by the rest of the as's and an's with the US, i.e. they no longer compete to establish associations with the US. After HFLs, changes in input–output associations, VS_i, are given by

$$d(VS_i)/dt = K_7 as_i (1 - |VS_i|)(US - as_i VS_i). \qquad [11.6b]$$

Equation [11.6b] replaces Equation [11.6a] presented in Chapter 11.

Similarly, changes in hidden unit–output associations, VN_j, are given by

The neurobiology of occasion setting 321

$$d(VN_j)/dt = K_7 an_j(1 - |VN_j|)(US - an_j VN_j). \qquad [11.9b]$$

Equation [11.9b] replaces Equation [11.9a] presented in Chapter 11.

Effect of selective hippocampal lesions

Extending Schmajuk and DiCarlo's (1992) view, Schmajuk and Blair (1993) suggested that the effect of HPLs can be described by removing or disconnecting the block labeled Hippocampus Proper in Figure 13.1. Equivalently, the effect of selective HPLs can be described by assuming that hidden unit error signals, EH_j, are equal to zero, that is

$$EH_j = 0, \qquad [11.3b]$$

and, therefore, $d(VH_{ij})/dt = 0$. Equation [11.3b] replaces Equation [11.3a] presented in Chapter 11.

Effects of cortical lesions

Schmajuk and DiCarlo (1992) suggested that the effect of complete cortical lesions (CLs) can be described by removing or disconnecting the block labeled Hidden Units in Figure 13.1. Equivalently, the effect of CLs can be described by assuming that input–hidden unit associations, VH_{ij}, are equal to zero, that is

$$VH_{ij} = 0. \qquad [11.6b]$$

Equation [11.6b] replaces Equation [11.6a] presented in Chapter 11.

Computer simulations

As mentioned before, according to the model, a CS acts as a simple CS (through its direct associations with the US) or an occasion setter (through its indirect associations with the US via the hidden units) depending on the strength of its direct association with the US, as determined by its intensity (in the case of simultaneous discriminations), interstimulus interval (in the case of serial discriminations) and duration (in the case of conditional contextual discriminations).

This section analyzes the effects of nonselective and selective lesions of the hippocampus on learning paradigms in which stimuli act as occasion setters: (a) simultaneous FP discriminations with a salient target, (b) serial FP discriminations, and (c) simple and conditional contextual discriminations. Identical parameter values, listed in Appendix 12.1, were used in all simulations.

Simultaneous FP discriminations with a salient target

Holland (1989a) examined the acquisition of simultaneous FP (XA+/A−) discriminations, as a function of the intensity of the *A* target cue. The data indicated that when *A* was of relatively low intensity, *X* acted as simple CS and *X*–US associations were formed. That is, the discrimination was solved in a linear fashion (responding to the *XA* compound equals the sum of responding to *X* [strong] plus responding to *A* [weak]). In contrast, if *A* was of high intensity, *X* acted also as an occasion setter and came to control the responding to *A*. That is, the discrimination was solved both in a linear fashion (*X* responding to the *XA* compound equals the sum of responding to *X* [strong] plus responding to *A* [weak]) and a nonlinear fashion (*A* responding to *XA* exceeds the sum of responding to *X* [weak] plus responding to *A* [weak]).

Experimental results

Loechner and Weisz (1987) investigated the effects of HFLs on a simultaneous FP discrimination using the rabbit NM preparation. The compound consisted of a tone (*T*) and a light (*L*). In this preparation, both light and tone elicit the same conditioned responses, namely NM CR. The light and the tone were used both as the feature or the target. Because when trained in a simple discrimination the tone elicited conditioned responses faster than the light, Loechner and Weisz (1987) concluded that the tone was more salient than the light.

The upper panel of Figure 13.2 summarizes Loechner and Weisz's (1987) experimental results. When the tone was used as the feature and the light as the target (*TL*+/*L*– case), both control and HFL animals show similar differences between responding to the *TL* compound and responding to *L* (left upper panel of Figure 13.2). When the light was used as the feature and the tone as the target (*TL*+/*T*– case), HFL animals were impaired in the discrimination (right upper panel of Figure 13.2). Notice that the rabbit NM preparation generates only one CR and, therefore, the data does not provide enough information to decide whether the light acts as a simple CS, an occasion setter, or both when the salient tone is used as a target.

Simulation results

The lower panel of Figure 13.2 shows simulated results. The left lower panel of Figure 13.2 shows percent NM CR amplitude to *LT* and *L*, following 10 *TL*+ trials alternated with 10 *L*– trials. As in the experiment, simulations assume that the tone CS was more salient than the light CS. In agreement with Loechner and Weisz's (1987) data, normal and HFL simulated animals exhibit simultaneous FP discrimination when the light is used as the target. The right lower panel of Figure 13.2 shows percent NM CR amplitude to *LT* and *T*, following 60 *LT*+ trials alternated with 60 *T*– trials. In agreement with Loechner and Weisz's (1987) data, normal, but not HFL animals, exhibit simultaneous FP discrimination when the tone is used as the target.

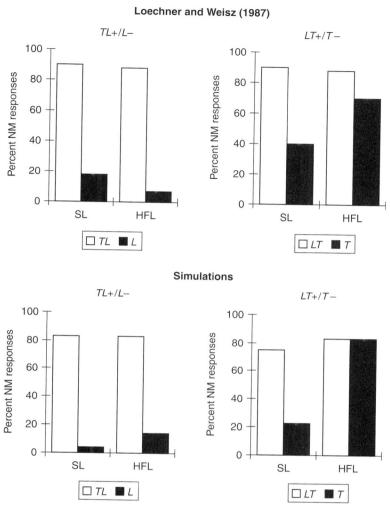

Figure 13.2 Simultaneous feature-positive discrimination with weak and strong targets: effects of hippocampal formation lesions. *Upper panel*: (Data from Loechner and Weisz, 1987). Percentage NM CR responding of rabbits trained in simultaneous feature-positive (FP) discrimination after sham lesions (SLs) and hippocampal formation lesions (HFLs). Percentage of CRs on *TL* (or *LT*) compound trials (open bars) and on *L*-alone (or *T*-alone) trials (solid bars) after *LT+/L−* (or *LT+/T−*) simultaneous FP discrimination training. *Lower panel*: Simulation. Percent CR amplitude on *TL* compound trials (or *LT*) (open bars) and on *L*-alone (or *T*-alone) trials (solid bars) after 10 *LT+* trials alternated with 10 *L−* trials (or 60 *LT+* trials alternated with 60 *T−* trials).

324 Configural mechanisms

Figure 13.3 (upper left panel) shows a simplified depiction of the network where *T* and *L* represent, respectively, feature and target, and *H* represents the hidden-unit layer. Figure 13.3 (upper right panel) shows a simplified depiction of the network where *L* and *T* represent, respectively, feature and target. Figure 13.3 (lower panels) illustrates the real-time values of *L*, *T* and the US during a reinforced trial in a simultaneous FP discrimination. Both *T* and *L* generate short-term memory traces that grow after *T* and *L* onset, and decay to zero after *T* and *L* termination. According to the SD/SLH model (see Equation [11.6]), on a reinforced trial, *T*–US and *L*–US associations increase in proportion to the shaded area under each trace, and decrease in proportion to the clear area under the curve. Therefore, *T*–US associations will be stronger than *L*–US associations.

In the framework of the model, normal and HFL animals exhibit simultaneous FP discrimination when the light is used as the target (*TL*+/*L*–) because in all cases *L*–US associations are small when compared to the *T*–US associations and animals can easily discriminate between *TL* and *L* trials (left panel in Figure 13.3).

According to the model, normal animals exhibit simultaneous FP discrimination when the tone is used as the target (*LT*+/*T*– case) because the less salient light competes with the more salient tone for association with the US during *LT*+ trials. On *LT*+ trials, because *T* is more salient than *L*, *T*–US associations grow faster than *L*–US associations (right panel in Figure 13.3). At the beginning of training, when aggregate prediction *B* is small compared to the US (see Equation [11.6]), increments in *T*–US associations on *LT*+ trials are larger than decrements in *T*–US associations on *T*– trials. As training progresses and the value of the aggregate prediction *B* approaches the value of the US, increments in *T*–US associations on *LT*+ trials become smaller than decrements in *T*–US associations on *T*– trials. As the net value of the *T*–US associations decrease, *L*–US associations slowly but relentlessly increase on *LT*+ trials and, therefore, discrimination is acquired based on the properties of *L* as a simple CS. According to the model, this solution changes when, with additional trials, *T* acquires strong direct associations with the US. On *T*-alone trials, *T* also activates the hidden units, which then inhibit the US prediction, preventing the display of responding based on the *T*–US association. But on *LT* trials, *L* weakens the inhibition normally exerted by the hidden units, enabling the performance of both CR_T and CR_L. Thus, with increasing number of trials, *L* acts both as a simple CS because of its strong direct excitatory associations with the US, and as an occasion setter, because of its inhibitory associations with the hidden units (right panel in Figure 13.3).

Because in the HFL case the model lacks the aggregate prediction B_{US}, competition between *T*–US and *L*–US associations is absent, and *T*–US associations do not decrease (reaching an asymptotic value characteristic of partial

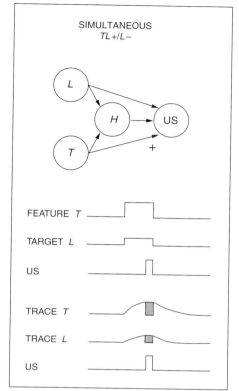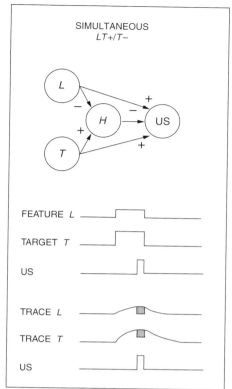

Figure 13.3 Simultaneous feature-positive discrimination. *Left panels*: Weak target TL+/L− discrimination. *Upper*: Simplified diagram of the circuit depicted in Figure 12.2. T is more salient than L and acquires a stronger excitatory association with the unconditioned stimulus (US). *Lower*: Real-time representations of T and L, their traces, and of the US on a TL+ trial. *Right panels*: Strong target LT+/T− discrimination. *Upper*: Both L and T acquire excitatory associations with the unconditioned stimulus (US). The hidden units H, excited by T and inhibited by L, acquire an inhibitory association with the US. *Lower*: Real-time representations of L and T, their traces, and of the US on a LT+ trial.

reinforcement), thereby impairing simultaneous FP discrimination. In other words, the failure to learn the simultaneous FP discrimination with a salient target shown by HFL animals is due to the lack of competition between the feature and the target to gain association with the US.

Computer simulations not shown here demonstrate that animals with HPLs can solve simultaneous FP discriminations with both salient and nonsalient targets. HPL-simulated animals solve the FP discrimination with a salient target because, even if unable to train the hidden units, competition between the

326　Configural mechanisms

feature and the target to gain association with the US is intact. HPL animals show discrimination based on the properties of *L* as a simple CS. Computer simulations also show that animals with CLs can solve simultaneous FP discriminations with a salient and nonsalient targets, because, even in the absence of hidden units, competition between the feature and the target to gain association with the US is intact.

Serial FP discriminations

Compared to their findings with simultaneous compounds in FP discriminations with a nonsalient target (see Chapter 12), Ross and Holland (1981) observed quite different patterns of behavior when the *X* feature preceded the *A* target on compound trials. In this case, *A* controlled behavior characteristic of target–US associations (CR_T). Because responding to the target occurred only on serial compound trials, Ross and Holland (1981) suggested that the feature set the occasion for the occurrence of responding based on target–US associations. That is, the serial FP discrimination was solved in a nonlinear fashion (*A* responding to the *XA* compound is stronger than the sum of responding to *X* [weak] plus responding to *A* [weak]).

Experimental results

Ross *et al.* (1984) examined the effects of HFLs on simple and conditional (serial FP) discriminations in rats. HFL animals were unable to acquire and retain a conditional (serial FP) discrimination (upper panels of Figure 13.4). In contrast, in agreement with previous results by Buchanan and Powell (1982), Berger and Orr (1983) and Orr and Berger (1985), acquisition and retention of a simple discrimination was preserved after HFLs.

Simulation results

The left lower panel of Figure 13.4 (Acquisition) shows percentage peak amplitude of headjerk $CR_T(LT)$ and $CR_T(T)$ following 52 alternated *LT*+ and *T*– trials. In agreement with Ross *et al.*'s (1984) results, HFL simulated animals are impaired in the serial FP discrimination. Also in agreement with the experimental results, computer simulations (not shown here) demonstrate that acquisition of a simple discrimination is preserved after HFLs. Furthermore, in agreement with experimental data (Oakley & Russell, 1975), computer simulations show that animals with CLs can solve a simple discrimination, in which stimulus configuration is not required.

The right lower panel of Figure 13.4 (Retention) shows percentage peak amplitude of headjerk $CR_T(LT)$ and $CR_T(T)$ following 104 alternated *LT*+ and *T*– trials for the normal case. Retention for the HFL case was simulated by 52 alternated *LT*+ and *T*– trials, before and after the simulated lesion. Figure 13.4 (lower

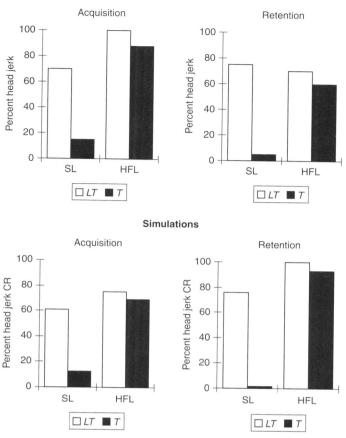

Figure 13.4 Serial feature-positive discrimination. *Left upper panel (Acquisition)*: (Data from Ross et al., 1984). Percentage of headjerk CR_T responding on *LT* compound (open bars) and on *T*-alone trials (solid bars) after *LT+/T−* serial FP discrimination training administered after sham lesions (SLs) and hippocampal formation lesions (HFLs). *Right upper panel (Retention)*: (Data from Ross et al., 1984). Percentage of headjerk CR_T responding on *LT* compound trials (open bars) and on *T*-alone trials (solid bars) after *LT+/T−* serial FP discrimination training administered before and after sham lesions (SLs) and hippocampal formation lesions (HFLs). *Left lower panel (Acquisition)*: Simulation. Percentage headjerk CR_T amplitude on *LT* compound trials (open bars) and on *T*-alone trials (solid bars) after serial FP discrimination training consisting of 26 *LT+* trials alternated with 26 *T−* administered after sham lesions (SLs) and hippocampal formation lesions (HFLs). *Right lower panel (Retention)*: Simulation. Percent headjerk CR_T amplitude on *LT* compound trials (open bars) and on *T*-alone trials (solid bars) after serial FP discrimination training consisting of 26 *LT+* trials alternated with 26 *T−* administered before HFLs and additional 26 *LT+* trials alternated with 26 *T−* training administered after HFLs.

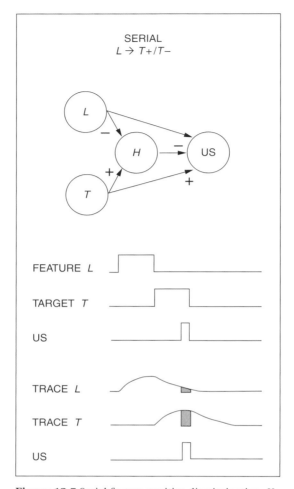

Figure 13.5 Serial feature-positive discrimination. *Upper panel*: T acquires an excitatory association with the unconditioned stimulus (US), whereas the hidden units (H), excited by T and inhibited by L, acquire an inhibitory association with the US. *Lower panel*: Real-time representations of L and T, their traces, and of the US on a LT+ trial.

panels) shows that, in agreement with the experimental data, HFL animals are impaired in both the acquisition and retention of the serial FP discrimination. Also in agreement with the experimental results, computer simulations (not shown here) demonstrate that retention of a simple discrimination is preserved after HFLs.

Figure 13.5 (upper panel) shows a simplified depiction of the network where L and T represent, respectively, feature and target, and H represents the hidden-unit layer. Direct associations of T and L with the US determine the response form, CR_T or CR_L.

Figure 13.5 (lower panel) illustrates the real-time values of L, T and the US during a reinforced trial in a serial FP discrimination. According to the SD/SLH model (see Equation [11.6]), in a reinforced trial, L–US and T–US associations increase in proportion to the shaded area under each trace, and decrease in proportion to the clear area under the curve. Therefore, during reinforced LT trials, T–US associations grow more than L–US associations. During non-reinforced T trials, T–US associations tend to decrease, but this decrement is partially prevented by inhibitory associations formed between hidden units (that respond to joint presentations of the CX and T) with the US. Therefore, although during T trials the system generates a weak response due to the inhibitory hidden unit–US associations, T–US associations are still strong. Moreover, because on LT trials the system needs to generate a strong output, input–hidden associations are modified in order to decrease the activity of the hidden units when the trace of L is active with CX and T, thereby increasing responding. Figure 13.5 illustrates that the hidden units exert a strong inhibition on the output on T trials and a weak inhibition on LT trials. In this case, during LT trials the form of responding is still determined by the unextinguished direct associations of T with the US and, therefore, responding comprises mostly CR_Ts (headjerk).

In the case of HFLs, the absence of the aggregate prediction B, and therefore of competition, prevents the system from generating the H–US inhibitory associations needed on T trials to counteract the T–US excitatory associations, and the system cannot acquire the serial FP discrimination. The HFL case does not show retention of serial FP discrimination because, although the previously formed configural stimuli are not changed after HFLs, in the absence of competition through the aggregate prediction B, the system cannot sustain the adequate connections between the hidden units with US to generate a larger response during XA trials than during A-alone trials (see Figure 13.5).

Finally, computer simulations not shown here demonstrate that CLs, assumed to eliminate the hidden-unit layer in the network, impair the acquisition of conditional discriminations; a result in agreement with data reported by Daum, Channon, Polkey and Gray (1991).

Transfer of occasion setting

Conflicting data have been collected on transfer to a new trained and extinguished target B in serial discriminations. Whereas some studies (e.g.

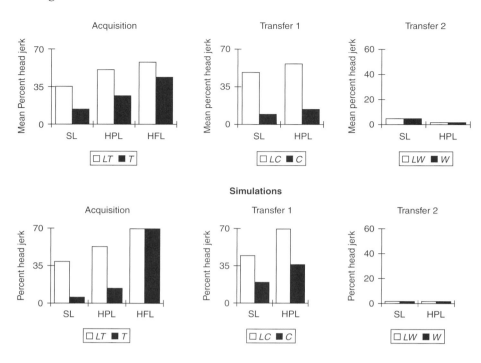

Figure 13.6 Transfer of occasion setting. *Left upper panel (Acquisition)*: **(**Data from Jarrard & Davidson, 1991). Percentage of headjerk CR_T responding on *LT* trials (open bars) and on *T*-alone trials (solid bars) after serial FP discrimination training administered after sham lesions (SLs), hippocampus proper lesions (HPLs) and hippocampal formation lesions (HFLs). *Middle upper panel (Transfer to a reinforced and extinguished cue C)*: (Data from Jarrard & Davidson, 1991). Percentage of headjerk CR_T responding on *LC* trials (open bars) and on *C*-alone trials (solid bars) after serial FP discrimination training and reinforcement and extinction of a clicker (*C*) cue administered after sham lesions (SLs) and hippocampus proper lesions (HFLs). *Right upper panel (Transfer to a nonreinforced cue W)*: (Data from Jarrard & Davidson, 1991). Percentage of headjerk CR_T responding on *LW* trials (open bars) and on *W*-alone trials (solid bars) after serial FP discrimination training and nonreinforced presentations of a white noise (*W*) cue administered after HPLs. *Left lower panel (Acquisition)*: Simulation. Percentage of headjerk CR_T amplitude on *LT* trials (open bars) and on *T*-alone trials (solid bars) after serial FP discrimination training consisting of 64 *LT*+, *T*− , *C*+ (clicker) and *C*− alternated trials followed by 6 *C*+ and 2 *C*− trials, administered after sham lesions (SLs), hippocampus proper lesions (HPLs) and hippocampal formation lesions (HFLs). *Middle lower panel (Transfer to a reinforced and extinguished cue C)*: Simulation. Percentage of headjerk CR_T amplitude on *LC* trials (open bars) and on *C*-alone trials (solid bars) after serial FP discrimination training consisting of 64 *LT*+, *T*− , *C*+ and *C*− alternated trials, followed by 6 *C*+ and 2 *C*− trials, administered after sham lesions (SLs) and hippocampus proper lesions (HPLs). *Right lower panel (Transfer to a nonreinforced cue W)*: Simulation. Percentage

Holland, 1986a, 1986b) reported little transfer, other studies (e.g. Jarrard & Davidson, 1991) found substantial transfer after training and extinguishing the new target. In Chapter 12 we showed that, depending on the salience of contextual cues, the model can capture the essence of both Holland's (1986a, 1986b) transfer results (large discrimination and small transfer to a trained and extinguished cue) and Jarrard and Davidson's (1991) results (smaller discrimination and large transfer to a trained and extinguished cue).

According to Jarrard and Davidson (1991), transfer testing is important to examine the hypothesis proposed by Rescorla (1973) that serial FP discriminations are solved by combining the STM trace of X with the presentation of A to form a configural cue that is directly associated to the US. They reasoned that, if serial FP was dependent upon a configural cue unique to XA, transfer would not be possible. However, our model suggests that this configural cue might generalize to XB and, if B–US simple associations are still present after extinction of B, transfer is feasible.

Experimental results

Using a preparation and experimental design similar to that used by Ross *et al.* (1984), Jarrard and Davidson (1991; Davidson & Jarrard, 1989) studied the effect of HFLs and HPLs on the acquisition of conditional and simple discriminations. As Ross *et al.* (1984), they found that HFL animals were unable to acquire a serial FP (conditional) discrimination (left upper panel of Figure 13.6). In addition, they determined that HPLs did not impair the acquisition of the serial FP discrimination or its transfer to a trained and extinguished cue (left and middle upper panels of Figure 13.6). They also reported that neither normals nor HPL animals exhibited transfer of occasion setting to a nonreinforced cue (right upper panel of Figure 13.6). Finally, also in agreement with Ross *et al.*'s (1984) data, they reported that neither HFL nor HPL animals were impaired in the acquisition of the simple C+ versus W− discrimination.

Simulated results

The left lower panel of Figure 13.6 (Acquisition) shows percentage peak amplitude of headjerk $CR_T(LT)$ and $CR_T(T)$ after 16 sessions consisting of 1 $LT+$ trial, alternated with 1 $L-$ trial, 1 (clicker) $C+$ trial and 1 $C-$ trial, followed by 6

Caption for fig. 13.6 (cont.)

of headjerk CR_T amplitude on *LW* trials (open bars) and on *W*-alone trials (solid bars) after serial FP discrimination training consisting of 48 $LT+$, $T-$ and $W-$ (white noise) alternated trials, administered after sham lesions (SLs) and hippocampus proper lesions (HPLs).

$C+$ and 2 $C-$ trials. In agreement with Jarrard and Davidson's (1991) results, HFL, but not HPL simulated animals are impaired in the acquisition of the serial FP discrimination.

The middle lower panel of Figure 13.6 (Transfer 1) shows percentage peak amplitude of headjerk $CR_T(LC)$ and $CR_T(C)$ following training. Also in agreement with Jarrard and Davidson's (1991) results, HPL simulated animals are not impaired in the transfer of occasion setting to a reinforced and extinguished cue.

Finally, the right lower panel of Figure 13.6 (Transfer 2) shows percentage peak amplitude of headjerk $CR_T(LW)$ and $CR_T(W)$ following 16 sessions consisting of 1 $LT+$ trial, alternated with 1 $L-$ trial, and 1 (white noise) $W-$ trial. In agreement with Jarrard and Davidson's (1991) results, SL and HPL simulated animals show little transfer of occasion setting to the nonreinforced cue W.

Figure 13.7 shows a simplified depiction of the network where L represents the feature, CX represents the context, T represents the original target, C represents the conditioned and extinguished cue, W the nonreinforced cue, and H represents the hidden-unit layer. According to the model, acquisition of the serial FP discrimination by HPL animals can be explained in the following terms. As in other backpropagation procedures, the hidden-unit layer in the model is initialized with random input–hidden unit weights, VH_{ij}. In the normal case, these initial input–hidden unit weights are modified during training to generate the configural stimuli needed to solve the task. Figure 13.7 (see also Figure 13.5) illustrates the case in which hidden units exert a strong inhibition on the output on T trials and a weak inhibition on LT trials. Although the HPL model cannot train input–hidden unit weights, the network can still solve the discrimination if the untrained hidden units provide stimulus configurations which become inhibitorily associated to the output, thereby allowing the network to solve the problem. In the case of a serial FP discrimination, a hidden unit that responds in the presence of T and the absence of L (see Figure 13.7), or one that responds in the presence of both T and L are sufficient to solve the problem.

In the normal case, the SD/SLH model describes the transfer of occasion setting to a reinforced and extinguished cue as the result of two basic processes. First, when C is trained and then extinguished, C–US connections are preserved because hidden units responding to the context and C become inhibitorily associated to the US, thereby decreasing behavioral responding to C without extinguishing the C–US associations (see Figure 13.7). Second, C activates the hidden units energized by T and inhibited by T which are responsible for the increased responding during LT presentations in the original discrimination. Therefore, the model will generate strong responses on LC trials

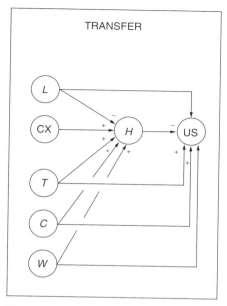

Figure 13.7 Transfer of occasion setting. *T* (a tone) acquires an excitatory association with the unconditioned stimulus (US), whereas the hidden units (*H*), excited by *T* and inhibited by *L*, acquire an inhibitory association with the US. This combination produces a strong response on *LT* trials and a weak response on *T*-alone trials. *C* (a clicker) maintains its excitatory association with the US even after extinction. *W* (a white noise) never becomes excitatory.

and weak responses on *C*-alone trials. As mentioned above, although the HPL model cannot train input–hidden unit weights, the network can still display transfer if untrained hidden units provide the configural stimuli needed to solve the task. Finally, according to the model, occasion setting shows little transfer to a nonreinforced cue because the cue does not have *W*–US direct connections (see Figure 13.7).

Contextual discriminations

Bouton and Swartzentruber (1986) studied conditional contextual discriminations by administering a tone CS and a shock US during sessions in Context 1 and the CS alone in Context 2 (see Chapter 12). After the discrimination was learned, neither context showed any reliable evidence of direct excitatory or inhibitory association with the US, and nonreinforced exposure to Context 1 alone did not reduce responding to the CS in that context. Therefore, contexts seem to act as occasion setters. That is, the conditional contextual discrimination was solved in a nonlinear fashion (*A* responding in Context 1 is

334 Configural mechanisms

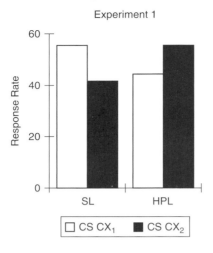

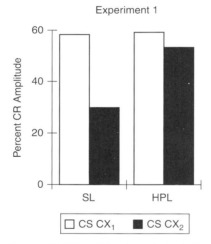

Figure 13.8 Conditional contextual discrimination. *Upper panel*: (Data from Honey & Good, 1993, Experiment 1). Percentage CR responding on CS CX_1 (open bars) and on CS CX_2 trials (solid bars) after contextual discrimination training, administered after sham lesions (SLs) and hippocampus proper lesions (HPLs). *Lower panel*: Simulation. Percentage CR responding on CS CX_1 (open bars) and on CS CX_2 trials (solid bars) after 200 CS CX_1+ trials followed by 100 CS CX_1+ and CS CX_2− alternated trials administered after sham lesions (SLs) and hippocampus proper lesions (HPLs).

stronger than the sum of responding to Context 1 [weak] plus responding to A [weak]).

Experimental results

Honey and Good (1993, Experiment 1) employed rats in an autoshaping procedure to study the effect of HPLs on a conditional contextual discrimination. They reported that, whereas normal animals showed conditional contextual discrimination (they were more likely to respond when the CS was presented in the context in which it was reinforced), this phenomenon was absent in HPL animals (upper panel of Figure 13.8). Using a similar procedure, Good and Honey (1991, Experiment 2) studied the effect of HFLs on a conditional contextual discrimination, and reported that this phenomenon was also absent in HFL animals (left upper panel of Figure 13.9). Good and Honey (1991, Experiment 3) also studied the effect of HPLs on a simple contextual discrimination in which, following Experiment 2, animals received alternated reinforced presentations in Context 1 and nonreinforced presentations in Context 2. Both SL and HPL animals showed higher responding in the reinforced than in the nonreinforced context (right upper panel of Figure 13.9).

Interestingly, Holland (1991) and Davidson (1993) suggested that, like contexts, motivations can act as occasion setters. Like conditional contextual discriminations, conditional motivational discriminations are also impaired either by HPLs (Davidson & Jarrard, 1993) or HFLs (Hirsh, Leber & Gillman, 1978).

Simulation results

Figure 13.8 (lower panel) shows simulated results for Honey and Good's (1993, Experiment 1) study. Simulated SL and HPL animals first received 200 reinforced presentations of CS_1 in Context 1, followed by 100 alternated reinforced presentations of CS_1 in Context 1 and nonreinforced presentations of CS_1 in Context 2. In agreement with experimental results, whereas SL animals showed conditional contextual discrimination, this phenomenon is absent in HPL animals.

Figure 13.9 (left lower panel) shows simulated results for Good and Honey's (1991, Experiment 2) study. Simulated SL and HFL animals first received 200 reinforced presentations of CS_1 in Context 1, followed by 100 alternated reinforced presentations of CS_1 in Context 1 and nonreinforced presentations of CS_1 in Context 2. In agreement with experimental results, whereas SL animals showed conditional contextual discrimination, this phenomenon is eliminated in HFL animals. According to the model, whereas SL animals are able to solve the conditional contextual discrimination problem by generating CS_1–CX configurations that are more inhibitory in Context 1 than in Context 2, thereby

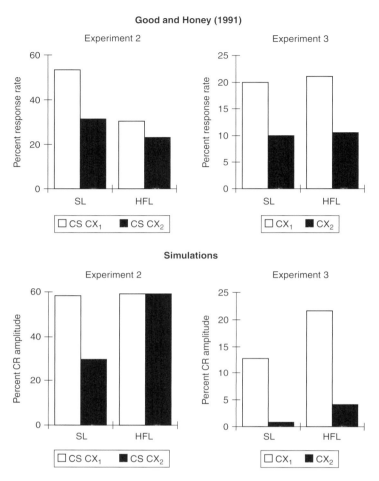

Figure 13.9 Conditional and simple contextual discriminations. *Upper left panel*: (Data from Good & Honey, 1991, Experiment 2). Percentage CR responding on CS CX_1 (open bars) and on CS CX_2 trials (solid bars) after contextual discrimination training, administered after sham lesions (SLs) and hippocampal formation lesions (HFLs). *Upper right panel*: (Data from Good & Honey, 1991, Experiment 3). Percentage CR responding on CX_1 (open bars) and on CX_2 trials (solid bars) after CX_1+ and CX_2- trials following discrimination training and administered following sham lesions (SLs) and hippocampal formation lesions (HFLs). *Lower left panel*: Simulation. Percentage CR responding on CS CX_1 (open bars) and on CS CX_2 trials (solid bars) after 200 CS CX_1+ trials and 100 CS CX_1+ and CS CX_2- alternated trials, administered after sham lesions (SLs) and hippocampal formation lesions (HFLs). *Lower right panel*: Simulation. Percentage CR responding on CX_1 (open bars) and on CX_2 trials (solid bars) after 200 CS CX_1+ trials, 100 CS CX_1+ and CS CX_2- alternated trials, and 50 CX_1+ and CX_2- alternated trials, administered after sham lesions (SLs) and hippocampal formation lesions (HFLs).

achieving discrimination, HFL simulated animals can neither generate CS_1–CX configurations, nor develop inhibitory associations between these configurations and the US. However, in contrast to the experimental data, simulated HFL animals show a higher level of responding than SL animals; a result compatible with most data on the affect of HFLs on the acquisition of classical conditioning. This discrepancy might be due to a retardation in learning caused by the thalamic lesions suffered by HFL animals (see Figure 2 in Honey and Good, 1991), or to the use of an autoshaping procedure.

Figure 13.9 (right lower panel) shows simulated results for Good and Honey (1991, Experiment 3). Following training in the conditional contextual discrimination, simulated SL and HFL animals received 25 reinforced presentations in Context 1 alternated with 25 nonreinforced presentations in Context 2. In agreement with Good and Honey's data, normal and HPL animals showed simple contextual discrimination, that is, higher responding in the reinforced context. According to the model, SL and HPL animals are able to build CX_1–US associations, and therefore, to discriminate between contexts. Furthermore, in agreement with Winocur, Rawlins and Gray (1987), because HFL animals lack the competition provided by the aggregate prediction they show abnormally strong conditioning to the context. However, as explained in the preceding paragraph, CS_1–CX configurations and not CX–US associations are essential to achieve conditional contextual discriminations.

Figure 13.10 shows a simplified depiction of the network where CS represents the target, CX_1 represents Context 1, CX_2 represents Context 2, and H represents the hidden-unit layer. In the model, contextual cues are treated as tonic CS, that generally do not gain strong associations with the US. Therefore, in order to control responding in the case of conditional contextual discriminations, contexts act as occasion setters in normal animals (Bouton & Swartzentruber, 1986) by activating the inhibitory action of the hidden units on the prediction of the US. In the HPL case, existing input–hidden unit associations are not adequate to solve the problem, therefore, contextual discriminations are impaired. In the HFL case, hidden unit–US associations are absent, and therefore the animals cannot achieve conditional contextual discriminations. However, HFL animals are able to build CX–US associations, and therefore, to respond more strongly in the reinforced than in the nonreinforced context. Furthermore, because HFL animals lack competition provided by the aggregate prediction, HFL animals might show abnormally strong conditioning to the context, a result in agreement with Winocur, Rawlins and Gray (1987).

Finally, computer simulations not shown here demonstrate that CLs, assumed to eliminate the hidden-unit layer in the network, impair the acquisition of conditional, but not simple contextual discriminations.

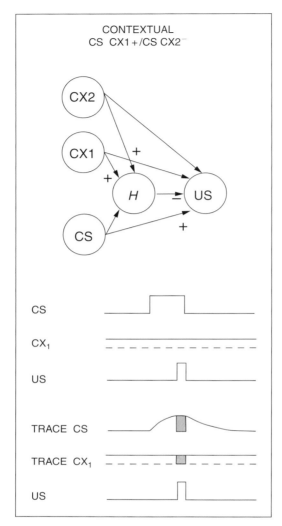

Figure 13.10 Conditional contextual discrimination. *Upper panel*: The CS acquires an excitatory association with the unconditioned stimulus (US), whereas the hidden units (H), excited more strongly in Context 2 than in Context 1, acquire an inhibitory association with the US. *Lower panel*: Real-time representations of the CS and Context, their traces, and of the US on a CS CX_1+ trial.

Discussion

Chapter 11 presented the SD/SLH model, a configural theory of classical conditioning that describes conditioning in multiple response systems (e.g. eyeblink, headjerk, rearing). In the framework of this model, a stimulus performs as a simple CS when it acts on a response system through its direct excitatory associations, and as an occasion setter when it acts on a response system through its indirect associations. Schmajuk and DiCarlo (1992) suggested that whereas direct CS–US and indirect CN–US associations were stored in cerebellar areas which control the (single) NM response system, stimulus configuration occurred in neocortical regions of the brain (see Figure 13.1). It is tempting to speculate that simple and configural representations are respectively computed in subcortical and neocortical regions of the brain. Whereas the subcortical areas employ simple representations capable of solving relatively elementary problems, neocortical areas build the configural representations needed to solve more difficult tasks.

Schmajuk and DiCarlo (1992) suggested that (a) nonselective aspiration lesions of the hippocampus disrupt the computation of the aggregate prediction of the US (thereby impairing competition) and the hidden-unit error signals (thereby impairing configuration), and (b) cortical lesions eliminate the hidden-unit layer (the configural system) in the network. In addition, Schmajuk and Blair (1993) proposed that (c) selective ibotenic acid lesions of the hippocampus only disrupt hidden-unit error signals (thereby only impairing the configuration). Under these assumptions, the SD/SLH model seems to describe most of the effects of hippocampal aspiration and ibotenic acid lesions on paradigms in which stimuli act as occasion setters.

Specifically, we applied the model to Loechner and Weisz's (1987) study of the effects of HFLs on a simultaneous FP discrimination, Ross *et al.*'s (1984) experiments on the effects of HFLs on simple and serial FP discriminations, Jarrard and Davidson's (1991) studies of the effect of HFLs and HPLs on the acquisition of conditional and simple discriminations, and Honey and Good's (1993) examination of the effect of HPL on a conditional contextual discrimination.

In addition to the experimental results analyzed in this chapter, the model seems able to account for the results of more recent studies. For instance, Holland *et al.* (1999) reported that hippocampal lesions interfere with negative occasion setting. Adding to the body of data on occasion setting, Campolattaro and Freeman (2006a, 2006b) found that lesions of the perirhinal cortex, which can influence CS input to the cerebellum, impair simultaneous FP and FN, but not serial FP discriminations in the rabbit nictitating membrane.

Discrimination difficulty and the effect of hippocampal and cortical lesions

Table 13.1 summarizes experimental and simulated results regarding the effect of HFLs, HFLs and CLs on different types of discriminations. Discriminations are divided into linear, i.e. discriminations in which experimental data (and the model) show that stimuli act as simple CSs and responding to a compound results from the addition of the responses to the individual elements, and nonlinear, i.e. discriminations in which experimental data (and the model) show that stimuli act as occasion setters and responding to a compound differs from the addition of the responses to the individual elements.

Experimental data suggest that discrimination difficulty increases with increasing similarity between the patterns for reward and nonreward. Similarity increases as (a) the number and salience of irrelevant (common) elements in the input patterns increase, (b) the number and the salience of discriminative relevant elements decreases, and (c) the difference between discriminative elements along a continuum decreases. For example, Redhead and Pearce (1995a) reported that the number of trials to criterion increases with increasing number of irrelevant elements in the patterns for reward and nonreward. Redhead and Pearce (1995b), using negative patterning, and Loechner and Weisz (1987), using simultaneous FP patterning, showed that normal animals take more trials to criterion when the irrelevant cue is salient than when it is not. Ross and Holland (1981) reported that serial FP discriminations, in which the feature (the discriminative element) is presented before the US, require more trials to criterion than simultaneous FP discriminations, in which the feature is presented simultaneously with the US. Sutherland, Mackintosh and Mackintosh (1963) showed that a discrimination between two stimuli that are more analogous (a straight square and a square tilted 15°) requires more trials than a discrimination between more disparate stimuli (a straight square and a square tilted 45°) along a continuum.

Shepard (1991) suggested that a measure of similarity between patterns X and Y is given by the Euclidian distance $D(X,Y) = [\Sigma_i(X_i - Y_i)^2]^{1/2}$ (see Appendix 13.2). Therefore, an estimation of discrimination difficulty can be obtained by computing the distance between reinforced and nonreinforced patterns. As shown in Table 13.1, distance D decreases, and therefore similarity and difficulty increase, from simple contextual discriminations which share no common elements, simple discriminations which share a weak common context, simultaneous FP discriminations with a nonsalient target which share the weak common context and the nonsalient target and capitalize in a salient discriminative feature, simultaneous FP discriminations with a salient target which share the weak common context and the salient target and capitalize

Table 13.1. Simulated results for HFL, HPL and CL cases compared with experimental results in several discrimination tasks.

Paradigms and references	Solution		Euclidian distance between patterns	HFL		HPL		CL	
	Data	Model		Data	Model	Data	Model	Data	Model
			Spared by HFL, HPL and CL						
Simple discrimination									
Good and Honey (1991)	Linear	Linear	1.41	0	0	?	0	?	0
Simple discrimination									
Ross, Orr, Holland and Berger (1984)	Linear	Linear	1.41	0	0	0	0	0	0
Jarrard and Davidson (1991)									
Oakley and Russell (1975)									
Simultaneous FP discrimination with a nonsalient target (TL+/L−)	Linear	Linear	1.25	0	0	?	0	?	0
Loechner and Weisz (1987)									

Table 13.1. (Cont.)

Paradigms and references	Solution		Euclidian distance between patterns	HFL		HPL		CL	
	Data	Model		Data	Model	Data	Model	Data	Model
Spared by HPL and CL; impaired by HFL									
Simultaneous FP discrimination with a salient target (LT+/T−) Loechner and Weisz (1987) Holland (1989a)	Linear/ Nonlinear	Linear/ Nonlinear	0.21	−	−	?	0	?	0
Spared by HPL; impaired by HFL and CL									
Serial FP discrimination (L→T+/T−) Acquisition Ross, Orr, Holland and Berger (1984) Jarrard and Davidson (1991) Daum, Channon, Polkey and Gray (1991)	Nonlinear	Nonlinear	0.21	−	−	0	0	−	−
Retention Ross, Orr, Holland and Berger (1984)	Nonlinear	Nonlinear	0.21	−	*	?	0	?	−
Transfer to a trained and extinguished cue				*	*	0	0	*	*

Jarrard and Davidson (1991)

	Impaired by HFL, HPL and CL						
Conditional contextual discrimination	Nonlinear	0.07	–	–H	–	?	–
Good and Honey (1991)							
Honey and Good (1993)							
Conditional motivational discrimination	Nonlinear	0.07	–	–	–	?	–
Hirsh, Leber and Gillman (1978)							
Davidson and Jarrard (1993)							
Place discrimination	Nonlinear	0.004	–	–	–	–	–
Morris, Garrud, Rawlins and O'Keefe (1982)							
Morris, Schenk, Tweedie and Jarrard (1990)							
Gallagher and Holland (1992)							
DiMattia and Kesner (1988)							

Note: – = deficit, 0 = no effect, ? = no available data, * = the original discrimination is not learned. *H*: responding is stronger than in the experimental data.

in just a nonsalient discriminative feature, serial FP discriminations which share the weak common context and a salient target and capitalize in just a nonsalient trace of the feature, to conditional contextual and motivational discriminations which share a strong common stimulus element and capitalize in just two weak discriminative contexts. In the case of place discrimination in a Morris tank, because the reinforced location of the platform and nonreinforced locations slightly removed from the platform share a large number of common elements, and the discriminative stimuli along the spatial continuum are very analogous, the distance between reinforced and nonreinforced locations is extremely small. Examination of the distances listed in Table 13.1 suggest that, whereas normal animals are able to learn difficult nonlinear discriminations, HFL, HPL and CL animals are increasingly impaired as the difficulty of the discriminations increases (see Schmajuk & Blair, 1993).

Effects of HFLs

According to the model, the effects of HFLs can mainly be derived from the absence of competition between CSs and CNs to gain associations with the US as a consequence of the lack of the aggregate prediction B. Competition limits the excitatory associations accrued by CSs and CNs presented together in the presence of the US, and permits the formation of inhibitory associations. In the case of a discrimination between patterns that share irrelevant common elements, competition will result in the elimination of the associations of the irrelevant common elements with the US, thereby contributing to the acquisition of the discrimination. Although in the absence of competition difficult linear discriminations cannot be achieved, simple linear discriminations can still be accomplished.

In agreement with Ross *et al.*'s (1984) and Jarrard and Davidson's (1991) results, HFL simulated animals show normal acquisition and retention of a simple discrimination ($D = 1.41$) because, even in the absence of competition, the irrelevant common element (context) does not gain a strong association with the US. In agreement with Loechner and Weisz's (1987) data, HFL simulated animals exhibit simultaneous FP discrimination when a nonsalient light is used as the target ($D = 1.41$) because, even in the absence of competition, the irrelevant common element (nonsalient target) does not gain a strong association with the US. However, when the irrelevant common element (salient target) accrues a strong association with the US in the absence of competition, in agreement with Loechner and Weisz's (1987) data, HFL animals do not exhibit simultaneous FP discrimination when a salient tone is used as the target ($D = 0.21$).

In agreement with Ross et al.'s (1984) and Jarrard and Davidson's (1991) results, HFL simulated animals are impaired in both the acquisition and retention of a serial FP discrimination ($D = 0.21$) because competition is also needed to establish inhibitory hidden unit–US associations. For a similar reason, in agreement with Good and Honey (1991, Experiment 2) Honey and Good (1993, Experiment 1), the model shows that HFL animals respond strongly to the presentation of a reinforced CS without showing conditional contextual discrimination ($D = 0.07$). Similarly, in agreement with Morris, Garrud, Rawlins and O'Keefe (1982) and DiMattia and Kesner (1988), the model shows that HFL animals are impaired in a place discrimination task ($D = 0.004$) because there are many irrelevant common elements (distal landmarks) shared by rewarded and nonrewarded places. However, because two different contexts share few irrelevant common elements, competition is not needed to establish simple contextual discriminations ($D = 1.41$). Therefore, in accord with Good and Honey (1991, Experiment 3), the model shows that HFL animals, although incapable of contextual conditional discriminations, are able to discriminate between reinforced and nonreinforced contexts.

In sum, HFLs impair the acquisition of simultaneous FP discriminations when a salient CS is used as the target ($D = 0.21$), serial FP discriminations ($D = 0.21$), conditional contextual discriminations ($D = 0.07$), conditional motivational discriminations ($D = 0.07$) and place learning ($D = 0.004$). HFL animals are not impaired, however, in the acquisition of relatively easy discriminations that do not require competition, such as simultaneous FP discriminations when a nonsalient CS is used as the target ($D = 1.25$), simple discriminations ($D = 1.41$) or contextual discriminations ($D = 1.41$).

Effects of HPL

According to the model, the effects of HPLs can be derived from the absence of training of the hidden units. In contrast to normal simulated animals, HPL-simulated animals are limited to use a pool of unmodifiable hidden units, and hence unmodifiable CNs, to meet the demands of the discriminations. Although in the absence of training of the hidden units difficult nonlinear discriminations cannot be achieved, easy nonlinear discriminations can still be accomplished.

As summarized in Table 13.1, according to the model, as the difficulty of the discriminations increases, HPL animals are increasingly impaired. HPL animals are not impaired in the acquisition of simple contextual discriminations ($D = 1.41$), simple discriminations ($D = 1.41$) or serial FP discriminations ($D = 0.21$). However, HPLs impair the acquisition of the more difficult conditional contextual discriminations ($D = 0.07$), conditional motivational discriminations ($D = 0.07$) and place learning discrimination ($D = 0.004$) (Morris, Schenk, Tweedie & Jarrard, 1990; Gallagher & Holland, 1992).

In agreement with Jarrard and Davidson's (1991) results, HPL-simulated animals are not impaired either in a serial FP discrimination ($D = 0.21$), because just one hidden unit is needed in order to respond more strongly to A-alone presentations than to XA presentations, and therefore, HPL animals perform approximately in the same manner as normal animals. In contrast, in agreement with Honey and Good's (1993, Experiment 1) results, HPL-simulated animals do not show conditional contextual discrimination. According to the model, a more difficult ($D = 0.07$) conditional contextual discrimination imposes the more stringent requirement that at least one hidden unit respond more strongly to CX_1A than to CX_2A, where CX_1 is precisely the context in which A is reinforced, to respond more strongly to CX_1A than to CX_2A. Therefore, HPL animals performance is impaired relative to that of normal animals.

In addition to its mostly correct descriptions of existing data, question marks in Table 13.1 indicate several predictions that await experimental testing. For instance, the model predicts that HPLs will not impair either the relatively easy simple contextual discriminations and simultaneous FP discriminations with a nonsalient target, or the more difficult simultaneous FP discriminations with a salient target. The model also predicts that HPLs should not affect the retention of a serial FP discrimination because, once the hidden units have been trained during the acquisition of the discrimination, nothing will change after the lesion.

Effects of CLs

According to the model, the effects of CLs can be derived from the absence of hidden units. Although in the absence of hidden units nonlinear discriminations cannot be achieved, even difficult linear discriminations can still be accomplished. Table 13.1 shows that CLs spare the acquisition of simple linear discriminations ($D = 0.21$) (Oakley & Russell, 1973, 1975) but impair the acquisition of serial FP discriminations ($D = 0.21$) (Daum *et al.*, 1991) and place learning ($D = 0.004$) (DiMattia & Kesner, 1988).

In addition, question marks in Table 13.1 indicate several predictions that await experimental testing. For example, the model predicts that CLs will not impair linearly solvable discriminations, from the relatively easy simple contextual discriminations and simultaneous FP discriminations with a nonsalient target, to the more difficult simultaneous FP discriminations with a salient target. The model predicts, however, that CLs will impair the acquisition of nonlinear serial FP discriminations (of difficulty comparable to that of simultaneous FP discriminations with a salient target), as well as conditional contextual and motivational discriminations. The model also predicts that CLs should impair the retention of a serial FP discrimination because, once the hidden units have been trained during the acquisition of the discrimination, they are eliminated by the lesion.

Other effects of hippocampal and cortical lesions

Besides the paradigms listed in Table 13.1, Table 13.2 shows that the model correctly predicts deficits shown by HFL animals in overshadowing (Rickert, Lorden, Dawson, Smyly & Callahan, 1979; Schmajuk, Spear & Isaacson, 1983), blocking (Rickert, Bent, Lane & French, 1978; Solomon, 1977; Gallo & Candido, 1995; but see Garrud, Rawlins, Mackintosh, Goodal, Cotton & Feldon, 1984) and inhibitory conditioning (Micco & Schwartz, 1972; but see Solomon, 1977), as well as negative patterning (NP) and positive patterning (PP) (Rudy & Sutherland, 1989; Whishaw & Tomie, 1991; but see Davidson, McKernan & Jarrard, 1993).

Table 13.2 also shows that the model correctly describes the effects of HPLs on negative and positive patterning (Gallagher & Holland, 1992) and acquisition and retention of place learning (Morris *et al.*, 1990; Gallagher & Holland, 1992). The model is also able to describe the effect of CLs on delay conditioning (Oakley & Russell, 1972), trace conditioning (Yeo, Hardiman, Moore & Russell, 1984), extinction (Oakley & Russell, 1972), blocking (Moore, personal communication, 1990), conditioned inhibition (Moore, Yeo, Oakley & Russell, 1980) and discrimination acquisition (Oakley & Russell, 1973, 1975). Yet, the model has difficulties describing the effects of CLs on discrimination reversal (Oakley & Russell, 1975). In addition, as indicated with question marks in Table 13.2, the model makes numerous novel predictions about the effects of HFLs, HPLs and CLs on many learning paradigms.

Although the present rendering of the SD/SLH model does not describe latent inhibition, Buhusi and Schmajuk (1996, see Chapter 15) combined the SD/SLH model with the SLG attentional model introduced in Chapter 2, which describes many of the features of LI. In the resulting attentional–configural model, LI is the result of the reduced novelty that follows the prediction of the CS by other CSs and the Context. According to the model, and in agreement with experimental data, whereas absence of the aggregate prediction of the US following HFLs produces deficits in LI (e.g. Schmajuk, Lam & Christiansen, 1994), absence of training to the hidden units following HPLs only affects the contextual dependency of LI (Honey & Good, 1993).

Table 13.2. *Simulated results for HFL, HPL and CL cases compared with experimental results in different learning paradigms.*

Paradigm	HFL Data	HFL Model	HPL Data	HPL Model	CL Data	CL Model
Delay conditioning	+, 0	+	?	+	0	0
Trace conditioning	+, 0, −	+	?	0	0	0
Extinction	0, −	−	?	−	0	0
Explicitly unpaired extinction	0	−[a]	?	−	?	−
Acquisition series	−	−	?	−	?	−
Extinction series	−	−	?	−	?	−
Blocking	0, −	−	?	0	0	0
Overshadowing	0, −	−	?	0	?	0
Simple discrimination						
Acquisition	0	0	0	0	0	0
Reversal	−	−	?	0	+	0[a]
Conditioned inhibition	0	−*	?	0	0	0
Differential conditioning	−	−	?	0	?	0
Simultaneous feature-positive discrimination (TL+/T−)	−	−	0	0	?	−
Simultaneous feature-positive discrimination (TL+/L−)	0	0	0	0	?	−
Serial feature-positive discrimination						
Acquisition	−	−	0	0	0	0
Retention	−	−	0	0	0	0
Negative patterning						
Acquisition	0, −	−	0	0	?	−
Retention	−	−	?	0	?	−
Positive patterning						
Acquisition	?	−	0	0	?	−
Retention	?	−	?	0	?	−
Context switching	−	−	?	0	?	−
Contextual discrimination	−	−[†]	−	−	?	−
Motivational discrimination	−	−	?	−	?	−
Discrimination between contexts	0	0	?	0	?	0
Place learning						0
Acquisition	−	−	−	−	0, −	
Retention	−	−	0	0	−	−

Note: HFL = hippocampal formation lesions; HPL = hippocampus proper lesions; CL = cortex lesions; T = tone; L = light. − = Deficit; + = facilitation, 0 = no effect, ? = no available data, †: responding is stronger than in the data, [a] = the model fails to describe accurately the experimental data.

Brain neural activity during associative learning

In addition to simulating the behavioral effects of different types of brain lesions, Schmajuk and DiCarlo (1992) showed that the model describes neural activity in hippocampus, lateral septum, medial septum and dorsal accessory olive of normal animals. Table 13.3 summarizes the descriptions of neural activity provided by the SD/SLH model. The model correctly describes neural activity of hippocampal pyramidal cells, medial septum and lateral septum during acquisition and extinction of classical conditioning. Only pyramidal cell activity during the CS period in the course of extinction was not correctly described. In addition, the model also describes the activity of some pyramidal cells during place learning, as well as decrements in medial septal activity during acquisition of classical conditioning.

The model also describes the neural activity of hippocampal pyramidal cells which preferentially fire at the spatial location of the US during place learning (Breese, Hampson & Deadwyler, 1989).

Related theories of hippocampal function

Hirsh (1974) proposed that two memory systems (associative and contextual) participate in learning; the contextual system depending on a working hippocampus. Similar to occasion setters, contextual cues (including motivational states) determine which stimuli control behavior. Associative and contextual systems in the Hirsh scheme roughly correspond to the direct–US and hidden unit–US associations in the SD/SLH model. Hirsh suggested that in intact animals the associative system is suppressed and behavior is controlled exclusively by the contextual system. Because the contextual system is needed to solve conditional logic tasks, animals with hippocampal lesions are impaired in conditional discriminations.

Related to the notion of contextual retrieval, Wickelgren (1979) suggested that the hippocampus participates in configural learning. According to Wickelgren, the organization of various stimuli into an associative group (chunking) is the basis of configuring in conditioning. Wickelgren (1979) proposed that the hippocampus plays a critical role in cortical chunking. The hippocampus partially activates free, as opposed to bound cortical neurons. Bound neurons reduce their connections to the hippocampus, consolidating memory by protecting the neurons from hippocampal input. HFLs produce amnesia due to a disruption in chunking. According to the backpropagation procedure used in the SD/SLH model, cortical neurons associate various components into a configural stimulus when (a) the output of that neuron has an effect on the CR output, and (b) there is a mismatch between the actual and predicted US. In contrast, in Wickelgren's model, cortical neurons associate different components into a

Table 13.3. *Simulations of neural activity compared with experimental results in different learning paradigms.*

Brain region	Paradigm	Data	Model
Hippocampal pyramidal cells	*Classical conditioning*		
	Acquisition	Increases	Increases
	CS period	Increases–decreases	Increases–decreases
		Precedes behavior	Precedes behavior
	Extinction	Decreases	Decreases
	CS period	Precedes behavior	Succeeds behavior*
	Acquisition	Increases	Increases
	US period	Increases–decreases	Increases-decreases
		Precedes behavior	Precedes behavior
	Extinction	Decreases	Decreases
	US period	Precedes behavior	Precedes behavior
	Place learning		
	Reinforced locations	Increases	Increases
	Classical conditioning		
Lateral septum	Acquisition	Increases	Increases
Medial septum	Acquisition	Decreases	Decreases
Dorsal accessory olive	Acquisition	Decreases	Decreases

Note: ? = no available data, * = the model fails to describe accurately the experimental data.

new chunk when they are not yet committed to the representation of a compound stimulus. Whereas the SD/SLH model specifies that simple and configural stimuli compete to gain association with the US according to a simple delta rule, Wickelgren's model does not indicate how component and compound stimuli interact with the US.

More recently, Rudy and Sutherland (1989) proposed that the hippocampus participates in the acquisition and storage of configural associations. According to Rudy and Sutherland (1989), while two memory systems (simple and configural) subserve learning, only the configural association system depends critically on the integrity of the hippocampal formation. They assume that (a) memory is simultaneously stored in both the simple and configural systems; and (b) that if the configural association has a greater predictive accuracy than a simple association involving one of the relevant elements, then the simple association's output is suppressed. Although the SD/SLH model and Rudy and Sutherland's theory share several basic assumptions, such as the inclusion of simple and configural associations, some differences should be noted. For example, while they assume that if the configural association has a greater predictive accuracy than a simple association then the simple association's output is suppressed, the SD/SLH model assumes that simple and configural associations compete in equal terms to gain association with the US. Also, whereas Rudy and Sutherland's theory asserts that HFLs impair only configural learning and simple learning rules remain intact, the SD/SLH model contends that HFLs impair configural learning *and* change the rules for simple associations. The SD/SLH model also diverges from the Rudy and Sutherland theory in that it assumes that configural associations are stored in the association cortex and not in the hippocampus. Recently, Rudy and Sutherland (1995) proposed two simple modifications that make their model closer to the SD/SLH model. They suggested that (a) the critical system for configural associations is cortical circuitry outside the hippocampus, and (b) the output of the hippocampal formation contributes to configural processing by selectively enhancing cortical units representing stimulus conjunctions. This enhancement has two important consequences: (a) it decreases the similarity between configural and simple systems, and (b) it increases the rate at which configural units acquire associative strength.

Using an approach similar, in some respects, to that previously introduced by Schmajuk and DiCarlo (1992), Gluck and Myers (1993) presented a computational theory that, as in the SD/SLH model, also models hippocampal function in terms of a network that trains a hidden layer through a backpropagation procedure. According to Gluck and Myers (1993), the hippocampal region develops new stimulus representations that enhance the discriminability of differentially predictive cues while compressing the representation of redundant cues. Hippocampal representations are assumed to recode sensory representations in cortical and cerebellar regions. The authors assume three three-layer networks that work in parallel. One of the networks represents the hippocampal region, another one the cortex and a third one the cerebellum. The output and hidden layers of the hippocampal network are trained to associate stimulus inputs with those same stimulus inputs and the US. The output layers of both cortical and cerebellar networks are also trained by the US. However, hidden units of both cortical and cerebellar networks are trained by the hidden units of the hippocampal network. According to Gluck and Myers, lesions of the hippocampal region eliminate training of the cortical and cerebellar hidden units, and therefore, cortical and cerebellar regions use previously established fixed representations. However, Schmajuk and Dragoi (unpublished results) demonstrated that a mathematical analysis of the model's equations reveals that, when training of the cortical and cerebellar hidden units is eliminated, previously established representations in cortical and cerebellar regions are extinguished, and the model wrongly predicts complete retrograde amnesia after hippocampal lesions.

Neurophysiological evidence suggests that the distinction between Schmajuk and DiCarlo's (1992) assumption (hippocampal lesions cause the error signals for cortical hidden units to be zero), and Gluck and Myers' (1993) assumption (hippocampal lesions cause the teaching signal for cortical hidden units to be zero), is an important one. McCormick, Steinmetz and Thompson (1985) reported that lesions of the rostromedial portions of the inferior olive, which presumably serves as a pathway for information from the US to reach the cerebellum, causes a previously learned classically conditioned response to extinguish even with US presentations. That is, disruption of the US teaching signal for the cerebellum causes the extinction of previously acquired associations. Similarly, in the Gluck and Myers' (1993) model, disruption of the hippocampal teaching signal for cortical hidden units causes the extinction of previously acquired associations. Therefore, because hippocampal lesions do not produce devastating retrograde amnesia, hippocampal output probably does not act as a teacher for the cortex; as suggested by Gluck and Myers.

Gluck and Myers (1994) have recently shown that their model is able to describe the occasion setting properties of context. They applied their model to Bouton and Swartzentruber's (1986) contextual discrimination protocol in which a CS is reinforced in Context 1 and not reinforced in Context 2, and showed that the model learns to respond to the CS in Context 1 but not in Context 2. Because the model does not show responding to Context 1 alone, they conclude that Context 1 has not acquired a direct association with the US. In addition, because the model demonstrates transfer to another CS trained in Context 1, they conclude that its is not a Context 1–CS configuration that has become associated with the US. Therefore, without any mechanistic explanation, Gluck and Myers assume that Context 1 operates as an occasion setter. However, Schmajuk and Holland (1995) have shown that there is no contradiction between transfer and configural associations, and therefore it is likely that the model accounts for contextual discrimination as a consequence of Context 1–CS configuration associated to the US.

Although Gluck and Myers' (1994, Zackheim et al., 1998) model is able to describe the occasion setting properties of context and serial FP, the model neither distinguishes between simple association and occasion setting properties of a CS, nor describes other properties of occasion setting such as response form. In contrast to our model, Myers, Gluck and Granger (1995) indicated that their model is unable to describe the effect of HPLs in negative patterning (Gallagher & Holland, 1992) or conditional discriminations (Jarrard & Davidson, 1991) such as those described in the present chapter. Finally, because the Gluck and Myers' model does not assume that HFLs affect competition, it cannot describe Loechner and Weisz's (1987) results. It is worth mentioning that Meeter, Myers and Gluck (2005) integrated Myers and Gluck's previous computational models of corticohippocampal function, and developed a unified theory of hippocampal participation in episodic memory and incremental learning.

Summary

The present chapter addresses the question of hippocampal participation in occasion setting by using the SD/SLH model described in Chapter 12. The model describes classical conditioning in multiple response systems and, therefore, is able to characterize many of the properties of occasion setting. As indicated, the parallel input–output architecture of the model is critical to (a) the description of the conditioning of compound stimuli, (b) the description of multiple response systems, and (c) the correct mapping of the model over parallel brain circuits.

In this chapter, nodes and connections in the SLH network were mapped onto cortical, subcortical and hippocampal circuits. We hypothesized that (a) CS–US and CN–US associations are stored in subcortical regions, and (b) stimulus configuration takes place in the association cortex. Regarding hippocampal function, we theorized that the hippocampus computes and broadcasts (a) a signal proportional to the aggregate prediction of the US to subcortical areas (e.g. cerebellum, caudate) in order to control the associations formed with the US (competition), and (b) error signals to the association cortex in order to modulate stimulus configuration.

Under the assumptions that HFLs impair both competition and configuration and HPLs impair only configuration, we demonstrated through computer simulations that the model correctly describes most experimental results regarding the effect of these lesions on tasks in which stimuli act as occasion setters. In addition to the experimental results analyzed in this chapter, the model seems able to account for the results of more recent studies.

Also, we showed that whereas normal animals are able to learn difficult discriminations, HFL, CL and HPL animals (in that order) become increasingly impaired as the difficulty of the discriminations, measured by the similarity between reinforced and nonreinforced patterns, increases.

354   Configural mechanisms

In sum, the results shown in this chapter contribute to a large inventory of learning paradigms, including classical conditioning and spatial learning (see Schmajuk & Blair, 1993), for which the SD/SLH model provides adequate characterizations of (a) normal behavior and its associated neural activity, and (b) the effect of different types of brain lesions.

Appendix 13.1   Simulation parameters

Unless indicated, parameters are identical to those used in Appendix 12.1. Tones and lights were represented by CSs of intensity 0.95. The nonsalient light used by Loechner and Weisz (1987) was represented by a CS of intensity 0.2. In the cases of the restrained rabbit NM response preparation used by Loechner and Weisz (1987), the free rat preparation used by Ross *et al.* (1984), and the rat autoshaping preparation employed by Honey and Good (1993) and Good and Honey (1991), the intensity of the context was assumed to be 0.2. In the case of the free rat preparation used by Jarrard and Davidson (1991) the intensity of the context was assumed to be 0.8.

Appendix 13.2   Computation of similarity

The similarity between reinforced pattern X and nonreinforced pattern Y is roughly estimated by the Euclidian distance $D(X,Y) = [\Sigma_i(X_i - Y_i)^2]^{1/2}$ (Shepard, 1991). We have considered that X_i are normalized values computed by $X_i = x_i / [\Sigma_i x_i^2]^{1/2}$, where x_i is the raw value of the element. Tones and lights were represented by xs of intensity 0.95. The nonsalient light used by Loechner and Weisz (1987) was represented by an x of intensity 0.2. In the cases of the restrained rabbit NM response preparation used by Loechner and Weisz (1987), the free rat preparation used by Ross *et al.* (1984), and the rat autoshaping preparation employed by Honey and Good (1993) and Good and Honey (1991), the intensity of the context was assumed to be 0.05. The trace of the feature in the Ross *et al.*'s (1984) experiment was represented by an x of intensity 0.2.

Similarity for simple contextual discriminations computed as the distance (D = 1.41) between the reinforced pattern (CX_1 = 0.05) and the nonreinforced pattern (CX_2 = 0.05), for simple discriminations as the distance (D = 1.41) between the reinforced pattern (CX = 0.05, CS(A) = 0.95) and the nonreinforced pattern (CX = 0.05, CS(B) = 0.95), for simultaneous FP discrimination with a nonsalient target as the distance (D = 1.25) between the reinforced pattern (CX = 0.05, T = 0.95, L = 0.2) and the nonreinforced pattern (CX = 0.05, L = 0.2), and for simultaneous FP discrimination with a salient target (D = 0.21) as the distance between the reinforced pattern (CX = 0.05, L = 0.2, T = 0.95) and the nonreinforced pattern (CX = 0.05, T = 0.95). Similarity for serial FP discriminations is computed as the distance (D = 0.21) between the reinforced pattern (CX = 0.05, X = 0.2, A = 0.95) and the nonreinforced pattern (CX = 0.05, A = 0.95), for conditional contextual and motivational discriminations (D = 0.07) as the distance between the reinforced pattern (CX_1 = 0.05,

CS[A] = 0.95) and the nonreinforced pattern (CX_2 = 0.05, CS[A] = 0.95), and for place learning in a Morris tank as the distance (D = 0.004) between the reinforced pattern (CX = 0.05, A = 0.95, B = 0.95, C = 0.95) where A, B and C are the distances from the hidden platform to three different distal landmarks, and the nonreinforced pattern (A = 0.85, B = 0.85, C = 0.85) removed a distance from the location of the center of platform that exceeds the diameter of the platform.

Part IV ATTENTIONAL, ASSOCIATIVE, CONFIGURAL AND TIMING MECHANISMS

14

Configuration and timing: timing and occasion setting

Buhusi and Schmajuk (1999) presented a neural network model of conditioning that combines the SD/SLH configural model presented in Chapter 11, with a timing model offered by Grossberg and Schmajuk (1989). The Grossberg and Schmajuk (1989) timing model incorporates a mechanism by which a CS can predict the time when the US is presented. In this timing model, stimuli evoke multiple traces of different duration and amplitude, that peak at different times after CS presentation. In the "configural–timing" model offered by Buhusi and Schmajuk (1999), these traces compete to become associated directly and indirectly (through hidden units) with the US, as described in Chapter 11. The output of the system predicts the value, moment and duration of presentation of reinforcement.

Most interestingly, and in contrast to the SD/SLH model, in the configural–timing neural network described in this chapter, a stimulus may assume different roles (simple CS, occasion setter, or both) at different time moments. Moreover, while in the SD/SLH model competition between CSs is purely associative, in the configural–timing model competition between CSs is both associative and temporal. CSs compete to predict not only the presence and the intensity of the US, but also its temporal characteristics: time of presentation and duration. The configural–timing model is able to address both the temporal and associative properties of simple conditioning, compound conditioning and occasion setting.

The content of learning in Pavlovian conditioning

In simple Pavlovian conditioning, animals learn not only that the presentation of the CS precedes the US, but also the time at which the US follows the CS. For example, Smith (1968) reported that for different CS–US interstimulus intervals (ISIs), the conditioned response (CR) shows a peak at the time of the US presentation and the duration of the CR increases with increasing ISIs. Similar results were reported in operant conditioning with fixed interval schedules in rats (Meck & Church, 1982, 1987) and pigeons (Roberts, Cheng & Cohen, 1989).

Temporal information seems to also play a role in paradigms involving multiple CSs, such as blocking. Reduced conditioning (blocking) is observed not simply when a new stimulus is reinforced in the presence of a good predictor of the US (Kamin, 1969a,1969b), but only when the blocked stimulus, B, is paired in the same temporal relationship with the US as the blocking stimulus, A (Schreurs & Westbrook, 1982; Goddard & Jenkins, 1988; Barnet, Grahame & Miller, 1993). In sum, in simple and compound Pavlovian conditioning, when a stimulus acts as a simple CS, the content of learning seems to be sensitive to both the presence of the US and the timing of the US presentation.

In the above-mentioned paradigms, stimuli are considered to act as simple CSs when they elicit CRs by signaling the occurrence of the US. Alternatively, a stimulus can play the role of an "occasion setter" when it controls the responses generated by another CS by indicating the relation between that second CS and the US (see Chapter 12). When a stimulus acts as an occasion setter, the content of learning is sensitive to the timing of the target presentation. For example, in a feature-positive (FP) discrimination, animals have to respond when feature stimulus X is paired with target stimulus A, and to refrain from responding to the target stimulus A when presented alone. Holland (1992) demonstrated that the content of learning in this paradigm is different when feature X and target A are presented simultaneously ($XA+/A-$) or serially ($X{\rightarrow}A+/A-$). Moreover, recent data (Holland, Hamlin & Parsons, 1997) show that training with a specific feature–target interval (FTI) in a serial FP discrimination ($X{\rightarrow}A+/A-$), results in better discrimination when the X–A delay matches the training interval than at other shorter or longer X–A intervals. Thus, temporal information seems to be acquired not only in simple and compound conditioning, but also in occasion setting.

Although a number of hypotheses have been advanced to address the properties of conditioning in paradigms such as those described above, possibly due to the complexity of the phenomena, these hypotheses tend to separately address their temporal and associative aspects. Classical learning theories develop around notions such as "associative strength" and "surprise", largely ignoring the timing of the CR. For example, in order to deal with blocking, Kamin (1969a,1969b) proposed that the failure of B to be associated with the US is due to the US being already signaled by A, and thus being unsurprising. This (and other) phenomena prompted Rescorla and Wagner (1972) to propose that association is gained in proportion to the difference between the actual and the predicted value of the US. In their model, Rescorla and Wagner assume that the content of learning includes only predicted values of reinforcement. Therefore, although the Rescorla and Wagner (1972) model correctly describes many Pavlovian conditioning phenomena, it does not address their temporal properties.

Some associative and temporal characteristics of conditioning can be jointly described by associative models that depict conditioning in "real-time". Such models (e.g. Moore & Stickney, 1980; Wagner, 1981; Sutton & Barto, 1990; Schmajuk & Moore, 1988; Schmajuk & DiCarlo, 1992) are able to capture temporal aspects of conditioning such as the effect of CS duration, US duration and ISI interval. However, they do not speak to the temporal specificity of conditioning.

Among these real-time models, the SD/SLH model describes not only simple and compound conditioning, but also some of the associative and temporal properties of occasion setting (see Chapters 11 and 12). In the model, each CS activates a single short-term memory trace which can become associated directly and indirectly (through hidden units) with the US. This feature enables the model to describe the distinction between simple conditioning and occasion setting: a CS acts as a simple stimulus through its direct connections with the output units, and as an occasion setter through its indirect configural connections through the hidden units. Although the SD/SLH model is able to describe some of the effects of temporal parameters on occasion setting (Chapter 12) the model still cannot describe the timing of the CR.

On the other hand, a number of timing models have been proposed in classical (e.g. Grossberg & Schmajuk, 1989; Grossberg & Merrill, 1992; Moore & Choi, 1998) and operant conditioning (e.g. Gibbon, 1977; Church & Broadbent 1990; Killeen & Fetterman, 1988; Machado, 1997; Staddon & Higa, 1999). These theories concentrate on the idea that peak time CR and CR duration increase with increasing ISIs, and propose various timing mechanisms by which the temporal properties of these phenomena can be explained: a pacemaker/accumulator process (Gibbon, 1977; Meck & Church, 1987), a set of oscillators (Church & Broadbent, 1990), a sequence of behaviors (Killen & Fetterman, 1988), or a set of memory traces (Grossberg & Schmajuk, 1989; Schmajuk, 1990; Machado, 1997; Staddon & Higa, 1999).

For example, Grossberg and Schmajuk's (1989) spectral timing model assumes that a CS activates multiple memory traces. Those traces active at the time of the US presentation become associated with the US in proportion to their activity. The outputs generated by all traces are added in order to determine the magnitude of the CR. The CR shows a peak at the time when the traces that have been active simultaneously with the US are active again. Although the model is able to describe CR timing with single and multiple USs, and a Weber law for temporal generalization, the model cannot describe important associative phenomena such as compound conditioning and occasion setting.

Finally, a few models aim at describing both temporal and associative property conditioning (Grossberg & Merrill, 1992; Desmond & Moore, 1988; Moore & Choi, 1998). Although Grossberg and Merrill's (1992) and Desmond and Moore's (1988; Moore & Choi, 1998) models address temporal specificity in simple and compound conditioning, they do not describe the complex properties of occasion setting (see Chapter 12).

In order to describe temporal specificity in simple conditioning, compound conditioning and occasion setting, this chapter presents a configural–timing model of Pavlovian conditioning in which stimuli compete to predict the value, moment and duration of presentation of reinforcement. The model incorporates three notions. First, it is a real-time model, therefore being able to describe various temporal relations between stimuli, and between stimuli and the reinforcement. The real-time properties of the model follow the hypothesis proposed by Grossberg and Schmajuk (1989), which assumes that stimuli evoke multiple traces with different temporal properties. Second, we assume that these traces become associated directly and indirectly (through hidden units) with a representation of the reinforcement (SD/SLH model described in Chapter 11), thus allowing for the description of the difference between simple CSs and occasion setters. In contrast to the SD/SLH model, the present model assumes that a stimulus may adopt the roles of a simple CS and/or an occasion setter at different moments in time, thus providing support for the description of the temporal specificity of the action of a CS. Third, while in the SD/SLH model, competition between CSs is purely associative (CSs compete to predict the presence and the value of the US), in the present model, competition between CSs is both associative and temporal. CSs compete to predict the presence and the value of the US, as well as the moment and duration of US presentation. Thus, the model can address both the associative and temporal characteristics of the competition between CSs in conditioning paradigms such as blocking.

This chapter demonstrates that the model is able to address temporal specificity and associative properties characterizing conditioning paradigms of different levels of complexity. We detail computer simulations of CR topography in simple conditioning (Smith, 1968), temporal competition between simple CSs in paradigms like blocking (Barnet, Grahame & Miller, 1993), and temporal specificity of serial FP discriminations (Holland, Hamlin & Parsons, 1997). The chapter suggests that a necessary characteristic of any putative timing mechanism is to co-operate (operate in conjunction) with associative mechanisms in order to describe both the temporal and associative properties of conditioning.

A timing version of the SD/SHL model

In order to describe temporal specificity in simple conditioning and occasion setting, we introduced a real-time neural network model of conditioning in which stimuli compete in predicting the value, time of presentation and duration of the reinforcement. The model incorporates three notions, which refer to timing, associations and stimulus competition.

The first notion, derived from Grossberg and Schmajuk's (1989) spectral timing hypothesis, is that a stimulus CS_i, evokes k multiple memory traces, denoted τ_{ik}. These traces have three (related) properties. They peak at different moments in time, have different peak amplitudes and are active over different time intervals. Fast τ_{ik} traces peak closer to the onset of CS_i, have large amplitudes and are active for a relatively short period of time. Slow τ_{ik} traces peak far away in time from the onset of CS_i, and are active for a relatively long period of time, but have smaller peak amplitudes than fast traces. The topography of the traces is a key issue in the model, in that it determines the moment of the peak, the amplitude and the duration of the CR.

The second notion refers to the type of associations that are assumed to be part of the content of learning. In a manner consistent with Schmajuk and DiCarlo's (1992) model, we assume that stimuli are associated directly and indirectly (through "hidden" units) with a representation of the US. Each trace τ_{ik} is independently associated through direct connections V_{ik} with the US, and through indirect connections VH_{ikj} with hidden unit H_j. Thus, each trace is capable of separately exciting or inhibiting a representation of the US, and to excite or inhibit hidden units. On the other hand, each hidden unit H_j activates a trace τ_j which becomes associated with the US through connection VN_j. A CS acts as a simple CS through the direct connections of its traces with the US, and as an occasion setter through the indirect connections of its traces with the US via the hidden units H_j.

However, while in the Schmajuk and DiCarlo (1992) model a CS can adopt the role of a simple CS and/or an occasion setter, throughout the duration of the trial, in the present model, a CS can have different roles at different moments during the trial. The CS can act as a simple CS at one moment in time, and as an occasion setter at another moment in time, and as both at other time moments, by the activation of the different traces evoked by the stimulus. Therefore, in the present model, an occasion setter is a stimulus that controls both the associative and temporal properties of the relation between another CS and the US.

The third notion is that the multiple traces of the CSs and of the hidden units compete to predict the value of the US. We assume that the prediction of the US

364   Attentional, associative, configural and timing mechanisms

is the sum of the direct and indirect associations (Schmajuk & DiCarlo, 1992). However, while in the Schmajuk and DiCarlo (1992) model the competition between CSs is purely associative (CSs compete to predict the presence and the value of the US), in the present model the competition between CSs is both associative and temporal. CSs compete to predict not only the presence and intensity of the US, but also its duration and time of presentation. Importantly, in the model, timing is mapped onto trace-specific associations, i.e. the US is predicted at specific time moments after CS onset by the differential activation of specific CS–US and hidden–US associations. As mentioned, traces τ_{ik} are most active (peak) at different moments in time. Therefore, the fact that a specific trace τ_{ik} is associated with the US signifies that the US representation is more active (or inhibited) at a specific moment in time. Similarly, hidden units are more active (or inhibited) at a specific moment in time. The fact that hidden units are co-activated by the (multiple) traces of all the CSs present, allows them to be active (or inhibited) when specific temporal relations among CSs hold. In sum, the competition between traces for the prediction of the value of the US is also a temporal competition for the prediction of the time and duration of US presentation.

Figure 14.1 summarizes the main features of the model. Each conditioned stimulus, CS_1 and CS_2, activates k multiple memory traces, τ_{1k} and τ_{2k}. The context CX is assumed to activate constant traces τ_{CXk}. Note that CS_1 and CS_2 can be presented with different interstimulus intervals. Their traces are associated directly with a representation of the US through direct connections V_{1k} and V_{2k} (although not shown in the figure, the context CX can also activate the representation of the US). Spectral traces τ_{1k}, τ_{2k} and τ_{CXk}, jointly activate the hidden units

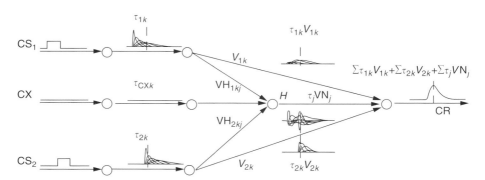

Figure 14.1 The main features of the model. CS_1, CS_2: conditioned stimuli; CX: experimental context; τ_{1k}, τ_{2k}: spectral traces; V_{1k}, V_{2k}: direct CS–US connections; V_{1kj}, V_{2kj}: indirect CS–hidden unit H_j connections; VN_j: direct hidden unit Hj–US connections; τ_j: activation trace of the hidden unit H_j; $\tau_{1k}V_{1k}$, $\tau_{2k}V_{2k}$: US predictions by direct traces; $\tau_j VN_j$: US prediction by hidden units; CR: conditioned stimulus, proportional to the sum of US predictions based on direct and indirect connections.

H_j, through indirect connections VH_{1kj}, VH_{2kj} and VH_{CXkj}. In order to illustrate the idea that hidden units are active when stimuli CS_1 and CS_2 are in a specific temporal relation, Figure 14.1 depicts the hypothetical activation of the hidden units, τ_j. Each hidden unit H_j becomes associated with the representation of the US, through indirect connection VN_j. The US prediction is computed as the sum of US predictions based on direct connections, $\tau_{1k}V_{1k}$, $\tau_{2k}V_{2k}$ and indirect connections $\tau_j VN_j$. As in the Rescorla–Wagner (1972) rule, the difference between the actual and predicted value of the US is used to adapt direct connections V_{ik}. The same error is used in a variation of the back-propagation algorithm (Werbos, 1974; Rumelhart, Hinton & Williams, 1986) introduced by Schmajuk and DiCarlo (1992) to train the hidden connections VH_{1kj}, VH_{2kj} and VH_{CXkj}. The differential equations that characterize the model are presented in Appendix 14.1. Parameter values used in the simulations are shown in Appendix 14.2.

In the following sections we present computer simulations obtained with the configural–timing model for different learning paradigms. We demonstrate that the model is able to address not only the associative, but also the temporal properties of conditioning paradigms of different complexity. We present computer simulations of the CR topography in simple conditioning (Smith, 1968), the effect of changing the temporal CS–US interval between the phases of a blocking paradigm (Barnet, Grahame & Miller, 1993) and the temporal specificity of the action of a CS acting as an occasion setter in a serial feature-positive discrimination (Holland, Hamlin & Parsons, 1997).

CR topography in simple conditioning

An important property of the CR topography in simple conditioning is that CR peak time and CR duration are proportional to the CS–US interval. Smith (1968) recorded the nictitating membrane response (NMR) in rabbits that received a 50-ms tone paired with a 50-ms periocular shock of intensity 1, 2 and 4 mA, with an ISI of 125, 250, 500 and 1000 ms. Smith (1968) found that for different CS–US intervals, the NMR is centered around the time of US presentation, and CR duration increases for longer CS–US intervals. On the other hand, US intensity determines the strength of the response, but not its timing.

The present model is able to describe these phenomena due to the topography of the traces. Figure 14.2 illustrates the way traces contribute to the peak and duration of the CR. Each panel presents the value of a specific variable of the model during a test trial after simple conditioning. Panel A shows conditioned stimulus CS_i. The dotted line shows the moment of presentation of the US. Panels B and C show that stimulus CS_i activates potential x_{ik}, which in turn activates potential y_{ik}. When combined (panel D), x_{ik} and y_{ik} form the spectral traces, τ_{ik} (see Appendix 14.1 for details).

The topography of the traces determines the CR to peak at the moment of US presentation. Traces τ_{ik} peak at different moments in time. They increase their associations V_{ik} in proportion to their overlap (co-activation) with the US, and decrease (extinguish) their associations at moments when they are active in the absence of the US. Therefore, traces τ_{ik} that peak close to the moment of US presentation will be reinforced more (will gain a larger V_{ik}) than those that peak at other moments. Also, due to the associative competition, strong traces (of higher peak amplitude) will tend to overshadow the weak ones. The predictions of the US based on direct associations V_{ik}, $\tau_{ik} V_{ik}$, are presented in panel E of Figure 14.2, which shows that the combination of predictions is governed by the traces whose peak time is close to the moment of US presentation. Therefore, the CR, which is proportional to the sum of the predictions $\Sigma_i \tau_{ik} V_{ik}$ (panel F), will peak approximately at the moment of US presentation.

The topography of the traces also determines the duration of the CR to be proportional to the ISI (scalar timing). By Equation [14.1] in Appendix 14.1, spectral traces that peak further away in time vary more slowly in time and last longer (panel D). Since the CR peaks at the moment of US presentation, its duration will be proportional to the ISI (panel F). Should the ISI increase, correspondingly slower traces would be closer to the moment of US presentation, would be reinforced more and the CR would be accordingly of increased duration.

The model correctly describes the CR topography when the ISI and US intensity are varied. The upper panels of Figure 14.3 show the topography of the nictitating membrane response (NMR) as reported by Smith (1968). The left upper panel shows that the CR peaks approximately at the moment of US presentation and that the duration of the response increases with the ISI. The upper right panel shows that an increase in US intensity does not greatly affect either the peak time, or the duration of the CR. Lower panels present simulations with the model showing that the model correctly addresses the experimental data. The CR peaks approximately at the moment of US presentation (left lower panel), and an increase in US intensity (right lower panel) minimally affects the peak time and the duration of the CR.

These results demonstrate that spectral traces are able to correctly describe CR topography (see also Grossberg & Schmajuk, 1989, and Grossberg & Merrill, 1992). The next section examines the competition between simple stimuli for the prediction of the value, moment and duration of US presentation.

Temporal competition in simple conditioning

Barnet *et al.* (1993) examined the temporal specificity of blocking. In the first phase of the experiments, stimulus A was paired with the US (shock)

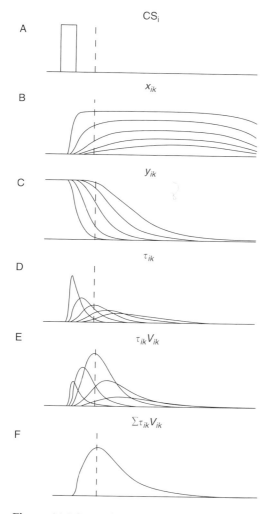

Figure 14.2 Spectral traces and CR topography. Presentation of conditioned stimulus CSi (panel A) determines two family of potentials, x_{ik} (panel B) and y_{ik} (panel C), which combine to form the memory traces τ_{ik} (panel D). Traces τ_{ik} that peak closer to the moment of US presentation (dotted line) are reinforced more (panel E). Therefore, traces that peak closer to the moment of US presentation dominate the prediction of the US, $\tau_{ik}V_{ik}$, and thus control the peak and the duration of the CR (panel F).

simultaneously (A+, common onset and offset) or serially (A→+, US onset at CS termination). In the second phase, stimuli A and B were paired with the shock, either simultaneously or serially. In the test phase, animals were required to complete 5 cumulative seconds of licking in the presence of the CSs. A high latency in the presence of the stimulus was taken as evidence for conditioning, while a reduced latency was taken as evidence for blocking.

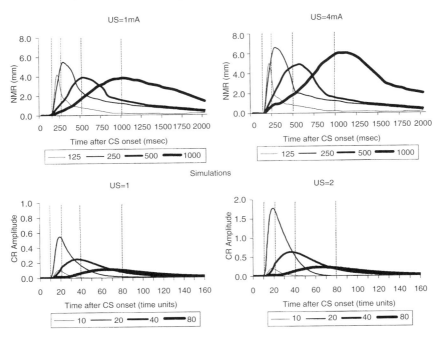

Figure 14.3 CR topography in simple conditioning. *Upper panels*: Mean CR topography of the nictitating membrane response (NMR) after 10 conditioning days in which a 50-ms CS was paired with a periocular shock US, at four different CS–US intervals: 125 ms, 250 ms, 500 ms and 1000 ms (Smith, 1968). *Left upper panel*: CR topography for 1-mA US shock (Smith, 1968). *Right upper panel*: CR topography for 4-mA US shock (Smith, 1968). *Lower panels*: Simulated CR amplitude after 50 CS–US pairings at four different CS–US intervals: 10, 20, 40 and 80 time units. *Left lower panel*: Simulated CR amplitude for US = 1. *Right lower panel*: Simulated CR amplitude for US = 2.

In their setting, Barnet et al. (1993) observed a significantly reduced conditioning for simultaneous pairing than for serial pairing. When B is reinforced in simultaneous position this phenomenon could partly obscure unblocking of B. Therefore, Barnet et al. (1993) further subjected animals for which B was reinforced in the second phase in simultaneous position to second-order conditioning to B. In a third phase of the paradigm, a new stimulus, C, was paired serially with the blocked stimulus B ($C{\rightarrow}B$). A high latency in the presence of the stimulus C was taken as evidence for conditioning to B, while a reduced latency was taken as evidence for B blocking.

Table 14.1 shows details of the experimental procedures used by Barnet et al. (1993). Following Barnet et al.'s (1993) notation, we use the letters "S" or "s" for simultaneous CS–US pairings, CS+, and the letters "F" or "f" for serial ("forward") CS–US pairings, CS+. The first letter of the name of the group refers to the A–US interval in phase 1, the second refers to the B–US interval in phase 2, while the third refers to the A–US interval in phase 2. Each group consists of experimental animals that receive all phases, and control animals that do no receive phase 1 of the paradigm. For example, in group S-F(f), while experimental animals receive simultaneous A+ pairings in phase 1, and serial AB→+ pairings in phase 2, control animals receive only serial AB→+ pairings (phase 2).

In our simulations we assume that latency is proportional to CR. The larger the CR, the longer it takes to complete the required licking period (see Appendix 14.2 for details). Since in the model the difference in conditioning between simultaneous and serial pairing is not extremely large, and since the model does not address second-order conditioning, simulations comprise only the first two phases of the Barnet et al. (1993) experiments.

In sum, in the test phase, animals were required to complete 5 cumulative seconds of licking in the presence of the blocking stimulus A, the blocked stimulus B, or the second-order CS, C. Barnet et al. (1993) reported a reduced latency (blocking) in the presence of stimuli B or C in the groups that receive A training in phase 1 in the same temporal relation to the US as B in the second phase. In the following subsections we demonstrate that the model correctly accounts for these results.

Competition to predict the moment of US presentation

Part of the temporal specificity of blocking can be accounted for by assuming that the "surprise" that drives learning (Kamin, 1969a,1969b) includes a temporal component (see e.g. Schreurs & Westbrook, 1982; Goddard & Jenkins, 1988). According to this account, which we refer to as "temporal surprise", in the first phase of the blocking experiment, animals learn not only that A predicts the US, but also the moment of US presentation. In the second phase of the paradigm, when stimuli A and B are paired with the US, B is blocked if the moment of US presentation is rendered unsurprising by presenting the US in the same temporal relation with A as in the first phase. In line with this idea, Barnet et al. (1993, Experiments 1 and 3) showed that when the temporal A–US interval in both phase 1 and 2 is the same, latency to respond in the presence of stimuli B or C is reduced (blocking). On the other hand, a change in the temporal A–US interval from phase 1 to phase 2, renders the US unpredicted in time and unblocks B (increased latency to respond in the presence of stimuli B or C). As shown in the upper panels of Figure 14.4, blocking (reduced latency) is present in experimental animals (open bars) relative to controls (closed bars) in groups F-F(f) and S-S(s), in which the temporal A–US interval in phases 1 and

Attentional, associative, configural and timing mechanisms

Table 14.1. *Temporal specificity of blocking as examined by Barnet et al. (1993).*

Group		Phase 1	Phase 2	Phase 3
F–F(f)	Experimental	A→+	AB→+	
	Control		AB→+	
S–F(f)	Experimental	A+	AB→+	
	Control		AB→+	
S–S(s)	Experimental	A+	AB+	C→B
	Control		AB+	C→B
F–S(s)	Experimental	A→+	AB+	C→B
	Control		AB+	C→B
S–F(s)	Experimental	A+	B→A+	
	Control		B→A+	
F–S(f)	Experimental	A→+	A→B+	C→B
	Control		A B+	C B

Note: In the simulations presented, we considered only the first two phases of the paradigm.

2 is the same, but not in groups S–F(f) and F–S(s), in which the temporal A–US interval is changed from phase 1 to phase 2.

The lower panels of Figure 14.4 show the simulated conditioned response amplitude to the blocked stimulus B for Barnet *et al.*'s (1993) Experiments 1 and 3. Details of simulations are given in Appendix 14.2. As shown in the lower panels of Figure 14.4, in general agreement with Barnet *et al.*'s (1993) results, blocking is present in simulated experimental animals (open bars) relative to controls (closed bars) in F–F(f) and S–S(s) groups, but not in S–F(f) and F–S(s) groups.

In the model, stimuli compete to predict the moment of US presentation. Thus, blocking is observed in experimental animals (open bars) of both groups F–F(f) and S–S(s) relative to controls (closed bars) because in the second phase A is a good predictor of the moment of US presentation. Since the value and moment of US presentation do not change from phase 1 and 2, when A and B are paired with the US, the reinforcement is "unsurprising", the error in prediction is small and B gains little associative strength, as shown in the lower panels of Figure 14.4.

However, a change in A–US interval from phase 1 to phase 2 renders the US "temporally" surprising and unblocks B. When A predicts the reinforcement at a

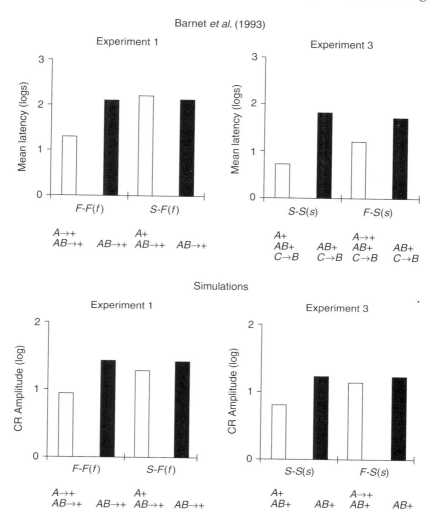

Figure 14.4 Temporal specificity of blocking. *Upper panels*: Mean latency to complete 5 consecutive seconds of licking (in log scale). Open bars: experimental; closed bars: controls. *Left upper panel*: Latency in the presence of the blocked stimulus B in groups F–F(f) and S–F(f) (Barnet et al., 1993, Experiment 1). *Right upper panel*: Latency in the presence of a second-order CS paired with the blocked stimulus B in groups S–S(s) and F-S(s) (Barnet et al., 1993, Experiment 3). *Lower panels*: Simulated conditioned response amplitude (in log scale) after 100 trials of phase 1 and 100 trials of phase 2 of the blocking paradigm (Table 14.1). Open bars: experimental; closed bars: controls. *Left lower panel*: Simulated CR to the blocked stimulus B in groups F–F(f), S–F(f). *Right lower panel*: Simulated CR to the blocked stimulus B in groups S–S(s) and F–S(s).

specific moment, but animals receive the reinforcement at a different moment, two types of surprise (error) occur. At the moment predicted by A the US is expected but is not delivered, and the error in prediction is negative. Therefore, the A traces that predict the US will extinguish. By contrast, at the moment when the US is presented in the second phase, the US is unexpected, the error in prediction is positive, and both A and B traces that are most active at that moment will gain associative strength. In sum, blocking is absent in the experimental animals (open bars) of both groups $S-F(f)$ and $F-S(s)$, relative to their controls (closed bars), because in the second phase A is a good predictor of the presence of the US, but a rather imprecise predictor of the moment the US will be delivered.

Competition to predict the intensity and the duration of the US

Although the "temporal surprise" notion of is able to account for the unblocking of B when the A–US interval is different in the two phases of the paradigm, this concept can not explain the unblocking of B in groups $S-F(s)$ and $F-S(f)$, as reported by Barnet *et al.* (1993, Experiments 2 and 4). In these groups, latency to respond to stimuli B or C is increased although the A–US interval is the same in the two phases of the blocking paradigm, and thus A is presumably a good predictor of the presence and timing of the US.

Competition to predict the intensity of the US

In the $S-F(s)$ experimental group, animals receive simultaneous A–US pairings, $A+$, in phase 1 and 2, but B is paired serially with the US, $B\rightarrow+$, in phase 2. The "temporal surprise" notion implies that B would be blocked, because the presence and timing of the US are predicted by A. However, the left panel of Figure 14.5 shows that blocking is absent in experimental animals (open bars) relative to controls (closed bars) in group $S-F(s)$. According to Barnet *et al.* (1993) these results suggest that animals associate a specific time-to-US-presentation with each CS. Only if this interval is the same for A in phase 1 and for B in phase 2, B-to-US timing is superfluous, and blocking occurs. As an alternative, here we demonstrate that the result can also be explained if one assumes that each CS evokes spectral traces, which compete not only to predict the time of US presentation, but also the US value.

In the model, the shape of the traces imply that in the associative competition between traces, serial conditioning is favored over simultaneous conditioning (in Figure 14.3, the CR for 10 t.u. is much smaller than the CR for 20 t.u.). During phase 1, the simulated experimental $S-F(s)$ animals (open bars) receive simultaneous $A+$ pairings, so that the spectral traces of stimulus A gain VA_k

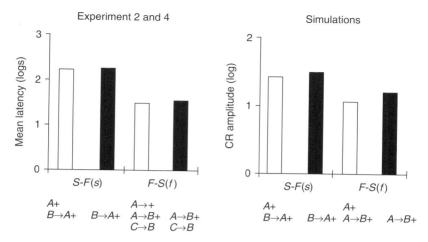

Figure 14.5 Temporal specificity of blocking. *Left panel*: Mean latency to complete 5 consecutive seconds of licking (in log scale) in the presence of the blocked stimulus B in group S–F(s) (Barnet *et al.*, 1993, Experiment 2), and in the presence of a second-order CS paired with the blocked stimulus B in group F–S(f) (Barnet *et al.*, 1993, Experiment 4). Open bars: experimental; closed bars: controls. *Right panel*: Simulated conditioned response amplitude (in log scale) to the blocked stimulus B in groups S–F(s), and F–S(f) after 100 trials of phase 1 and 100 trials of phase 2 of the blocking paradigm (Table 14.1). Open bars: experimental; closed bars: controls.

associative strength to predict the reinforcement simultaneously with A. However, according to the ISI curves presented in Figure 14.3, after similar number of A–US pairings, more conditioning is gained if the A is paired serially with the US, A+, that simultaneously, A+.

Therefore, slow B traces successfully compete with fast A traces to predict the intensity of the US. In phase 2 of the experiment, in the simulated experimental S–F(s) animals, A is paired simultaneously with the US and B serially, B→A+. Since in phase 1 fast A traces gain relatively low associative strength with the US, the error in prediction is still positive, so that slow B traces successfully compete and gain associative strength. Thus, latency to respond to B is increased in both experimental (open bars) and control animals (closed bars) in the simulated S–F(s) group (right panel of Figure 14.5). In sum, in the model, the lack of blocking in the S–F(s) group can be explained by the competition for the prediction of the value of the US. Details of the simulations are given in Appendix 14.2.

Competition to predict the time and duration of the US

In the previous section, the central notion is that CSs compete to predict the US. Here we refine this notion by showing that the unblocking observed by Barnet *et al.* (1993, Experiment 4) in group *F–S(f)* can be attributed to the competition to predict the duration of the US. This competition is largely due to the scalar property provided by the traces.

This notion can explain the lack of blocking in the *F–S(f)* group (Barnet *et al.*, 1993, Experiment 4). In the *F–S(f)* experimental animals, *A* is paired serially with the US in phase 1 and 2, *A*→+, while *B* is paired in phase 2 simultaneously with the US, *B*+. Since *B* is reinforced in the second phase in simultaneous position, both experimental and control animals acquire little conditioning. In order to reveal the blocking effect, Barnet *et al.* (1993) further subjected the animals to second-order conditioning to *B*. In a third phase of the paradigm, a new stimulus, *C*, was paired serially with the blocked stimulus *B* (*C*→*B*), thus allowing *C* to reveal the associative strength acquired by *B*. A reduced latency to respond to *C* shows that *B* was blocked, while an increased latency shows that *B* acquired associative strength (unblocked). The left panel of Figure 14.5 shows latency to respond to *C*. The results suggest that *B* blocking is absent in experimental animals (open bars) relative to control animals (closed bars) in group *F–S(f)* (Barnet *et al.*, 1993, Experiment 4).

In the model, traces that better match the ISI are promoted over faster or slower traces. A trace gains associative strength when the trace is active in the presence of the US (overlaps with the US), and loses associative strength when the trace is activated in the absence of the US. Slow traces are long in duration, and hence less precise in predicting the moment and the duration of the US. They are co-active with the US for a longer duration, but also lose more association when they are active in the absence of the US. Fast traces overlap less with the US, but are also less active in the absence of the US. Therefore, fast traces, although short in duration, when simultaneously paired with the US (short ISI), predict the moment and duration of US presentation better than slow traces. In this case, their balance is better: they gain little associative strength, but lose less.

In sum, when competing to predict the US in simultaneous position, fast traces, which predict the duration and the moment of the US presentation more accurately, are favored over slow traces. During phase 1, the experimental *F–S(f)* animals (open bars) receive serial *A*→+ pairings, so that the slow spectral traces of stimulus *A* gain V$_{Ak}$ associative strength to predict the reinforcement at *A* offset. However, as explained above, fast *B* traces are favored against slow *A* traces, because they predict more accurately the moment and the duration of US presentation. The right panel of Figure 14.5 shows that in general agreement with Barnet *et al.*'s (1993, Experiment 4) results, little (if any) blocking is observed in the *F–S(f)* experimental group, relative to its controls. Details of the simulations are given in Appendix 14.2.

Summary of blocking simulations

In a blocking paradigm, A and B traces compete to accurately predict the value, the time and the duration of US presentation. Blocking is observed only if the A–US interval in phase 1 matches the B–US interval in phase 2, because the US is already accurately predicted both in value, in time and in duration. By contrast, if the A–US interval is changed between the 2 phases of the experiment, the error in predicting the moment of reinforcement renders the B unblocked (Barnet *et al.*, 1993, Experiments 1 and 3).

In the model, traces that better match the ISI are promoted over faster or slower traces. For example, even if the A–US interval does not change (i.e. the moment of US presentation is correctly predicted by A), if B is able to predict the time and duration of US presentation more accurately than A, B is unblocked (Barnet *et al.*, 1993, Experiment 4).

Interestingly, in the Barnet *et al.* (1993) setting, and in the model, the scalar property is counterbalanced by an asymmetry between serial and simultaneous CS–US pairings in gaining associative strength. Simultaneous A+ pairings are not capable of completely blocking serial $B{\rightarrow}$+ pairings even if the US is predicted more accurately after simultaneous pairings (Barnet *et al.*, 1993, Experiment 2), and this effect reflects the competition between traces to better predict the intensity of the US.

The simulations presented above demonstrate that competition between traces is a viable mechanism for explaining temporal specificity in a class of paradigms in which stimuli act as simple CSs. In the next section we explore the temporal properties of a class of paradigms in which stimuli act as occasion setters.

Temporal specificity of the action of an occasion setter

Recently, Holland *et al.* (1997) investigated temporal specificity in a serial FP discrimination ($X{\rightarrow}A+/A-$); a paradigm in which feature X acts as an occasion setter, and showed that temporal information not only determines the type of solution, but is also part of the content of learning. Here we show that by allowing the CSs to evoke multiple spectral traces, an associative structure in which stimuli may act as simple CSs or as occasion setters (see Chapter 12) is able to correctly address the temporal and associative characteristics of serial FP discriminations.

Importantly, because in the Schmajuk and DiCarlo (1992) model a CS evokes a single trace, it can adopt the role of a simple CS and/or an occasion setter throughout the duration of the trial. If in a serial FP discrimination feature X were an occasion setter throughout the $X{\rightarrow}A$+ trial (as in the Schmajuk and DiCarlo, 1992, model) the solution would lack temporal specificity. In contrast, since in the present model a CS evokes multiple traces, it can have different roles at different moments in time by the differential activation of its traces. While one of its traces might behave as a simple CS (through a direct trace–US association), another trace can assume the role of an occasion setter (through indirect trace–hidden unit associations). By allowing a CS to have different roles at different moments in time, the present model is able to address the temporal specificity of occasion setting. Therefore, the present model extends the notion of occasion setter to describe a stimulus that controls both the temporal and associative relations between another CS and the US.

Temporal specificity of an occasion setter

Holland et al. (1997) shaped rats to chain pull in order to receive food reward, and placed chain pulling under the control of to-be-target CSs. Afterwards, two groups of rats were trained in two serial FP discriminations with 5- or 25-s X–A feature–target intervals (FTIs). After training, the rats were tested at different FTIs. Both groups were tested at short FTIs ranging from 0 to 30 s (Test 2). The 25 s group was also tested at long FTIs ranging from 25 to 55 s (Test 3). A short latency to pull was taken as evidence of conditioning.

Holland et al.'s (1997) experimental results are shown in the left panel of Figure 14.6, which depicts the difference in latency to chain pull to A-alone or X→A for different feature–target (X–A) intervals. Three points should be noted. First, when tested at both short (Test 2) and long (Test 3) FTIs, both groups seem to discriminate better when tested at FTIs similar to that used in training, than at shorter or longer FTIs. Second, discrimination in the 5-s X–A group is sharper than in the 25-s X–A group, which suggests that the temporal specificity of serial FP discriminations follows a scalar property (a shallower temporal gradient for longer FTIs). Third, better discrimination (larger peak) is observed in the 5-s X–A group than in the 25-s X–A group, although the 25-s X–A group received more training (40 training sessions) than the 5-s group (only 15 training sessions). In sum, the above observations favor the suggestion (Holland et al., 1997) that the content of learning includes the X–A temporal interval, and that the data is consistent with scalar timing.

The right panel of Figure 14.6 presents simulations of the Holland et al. (1997) results. The model was trained (separately) in two serial FP discriminations (X→A+/A−) with 15 or 75 time unit FTIs. After the training stage, the model was tested for different FTIs. Latency to respond was assumed to decrease linearly with the CR. Therefore, difference in latency is proportional to difference in CR amplitude (see Appendix 14.2).

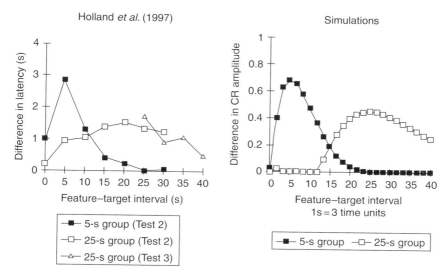

Figure 14.6 Temporal specificity of serial feature-positive discrimination. *Left panel*: Difference in latency to chain pull in the presence of *A* alone or *X→A* for different feature–target (*X–A*) intervals (Holland *et al.*, 1997), in two groups of animals that were trained for serial FP discriminations with 5-s feature–target interval (5-s group), or 25-s feature–target interval (25-s group) (see text). *Right panel*: Difference in simulated conditioned response to the presentation of *X→A*, or *A* alone, for different feature–target (*X–A*) intervals, after 300 alternated *X→A+/A−* trials with 15 time units *X–A* interval (5-s group), or 800 alternated *X→A+/A−* trials with 75 time units FTI (25-s group). Latency to respond is assumed to decrease linearly with the CR: Latency = 1 − CR. Therefore, difference in latency Latency$_A$ − Latency$_{X→A}$ = (1 − CR$_A$) − (1 − CR$_{X→A}$) = CR$_{X→A}$ − CR$_A$.

The right panel of Figure 14.6 demonstrates that in the model, serial FP discrimination qualitatively follows the properties reported by Holland *et al.* (1997). First, the difference in the simulated conditioned response to the presentation of *X→A* and *A* alone is larger at the training FTI than at other shorter or longer intervals. Second, the difference in CR shows the scalar property, i.e. its duration is proportional to the FTI. Finally, the 25-s *X–A* group shows less discrimination after 800 alternated *X→A+/A−* trials than the 5-s *X–A* group after only 300 alternated *X→A+/A−* trials.

Although simulations generally agree with data, differences exist. For example, the discrimination gradient in the 25-s group is shallower in the data (left panel) than in the simulations (right panel). Difference in latency is relatively large in the 25-s group animals tested with 5- and 10-s FTIs, but difference in CR is null in the model for these intervals. Possible explanations of this discrepancy are worth discussing. First, in the present model, a CS can have different roles (simple CS and/or occasion setter) at different moments in time. For example, in a serial FP discrimination with a long FTI, slow X traces are occasion setters (that act through hidden units) and fast X traces become direct inhibitors of the US, and sharpen the discrimination gradient in the 25-s group. If feature X were an occasion setter throughout the X→A+ trial, the solution would lack temporal specificity and the discrimination gradient would be shallower.

Second, the simulated gradient depends on the assumed relation between the CR magnitude and latency to respond. In the present simulations we simply assumed that latency in response decreases linearly with the CR. Therefore, difference in latency is assumed proportional to difference in CR magnitude (see Appendix 14.2).

Third, in the model, the shape of the gradient varies with the number of training trials and the number of traces evoked by a stimulus. By increasing the number of training trials the discrimination increases, and so does temporal specificity. In our simulations we used a number of training trials proportional to the number of experimental training sessions. Also, by increasing the number of traces evoked by one stimulus, the discrimination gradient becomes more symmetrical. However, an increase in the number of traces evoked by one stimulus would also increase the computational complexity of the model. Therefore, in the present simulation we used a relatively low number of traces, $k = 5$.

Fourth, although the model describes classical conditioning, we applied it to the description of operant conditioning, given that in operant conditioning discriminative stimuli become classically conditioned to the US (Mackintosh, 1983). Therefore, the discrepancy between data and simulations might be due to the differences between the experimental (operant) procedure and the simulated (classical) one. Indeed, a much steeper gradient was recently reported by Holland (1998) by using a classical conditioning procedure. While temporal specificity of occasion setting seems to be procedure-independent, the discrimination gradient might be procedure-dependent.

In sum, the model qualitatively describes the temporal specificity of serial FP discrimination. Importantly, besides describing its temporal characteristics, the model is also able to describe the type of associations learned by animals.

The feature controls both temporal and associative mechanisms

Figures 14.7 and 14.8 show that at both short and long FTIs the feature controls both the temporal and associative properties of the solution learned by the animals. Figure 14.7 shows that for short FTIs the discrimination is solved mainly through direct X–US, and indirect H–US connections. Indeed, the prediction of the US by the target, $B(A,US)$ is very small. When presented alone (right panels), the small prediction of the US by the target, $B(A,US)$ is cancelled

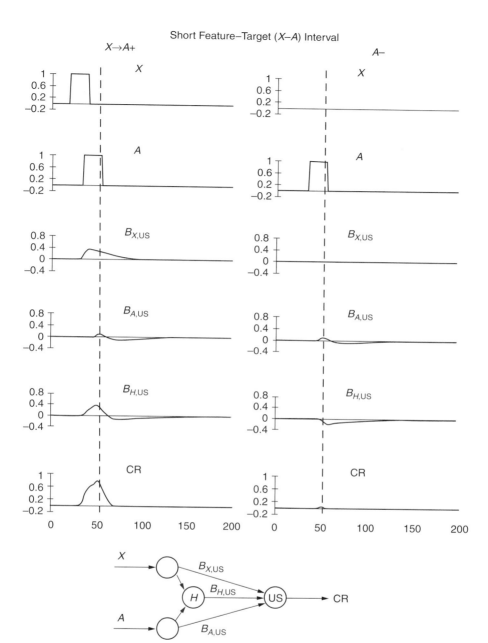

Figure 14.7 Feature directly controls the temporal and associative mechanisms of a serial FP discrimination with short FTI. Two test trials after 300 alternated X→A+/A− trials with a 15 time unit FTI. *Left panels*: Temporal map of the activation of the model during the first 200 time units of an X→A test trial. *Right panels*: Temporal map of the activation of the model during the first 200 time units of an A test trial. *Lower panel*: A schematic of the associations involved in solving the discrimination. X: feature; A: target; B(X,US): US prediction by feature X; B(A,US): US prediction by target A; B(H,US): US prediction by hidden units; CR: conditioned response. Dotted line: moment of US presentation during training.

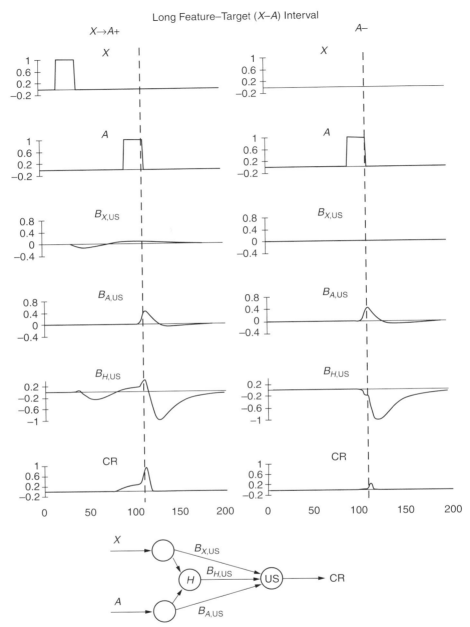

Figure 14.8 Feature indirectly (through hidden units) controls the temporal and associative mechanisms of a serial FP discrimination with long FTI. Two test trials after 800 alternated X→A+/A− trials with a 75 time unit FTI. *Left panels*: Temporal map of the activation of the model during the first 200 time units of an X→A+ test trial. *Right panels*: Temporal map of the activation of the model during the first 200 time units of an A− test trial. *Lower panel*: A schematic of the associations involved in solving the discrimination. X: feature; A: target; B(X,US): US prediction by feature X; B(A,US): US prediction by target A; B(H,US): US prediction by hidden units; CR: conditioned response. Dotted line: moment of US presentation during training.

by a small negative prediction of the US by the hidden units, B(H,US), so that A alone does not evoke any response. By contrast, on X→A presentation (left panels) the US is predicted by the feature and by the hidden units, so that the sum of these predictions (CR) is positive. Importantly, since feature directly controls the temporal and associative properties of the solution (the CS acts as a simple CS), these properties do not survive after feature extinction (see next section).

Figure 14.8 shows that at long FTIs the feature indirectly controls (through hidden units) both the temporal and associative properties of the solution learned by the animals. Figure 14.8 shows that in this case the discrimination is solved mainly through direct target A–US, and indirect H–US connections controlled by the feature. Indeed, both panels show that the prediction of the US by the feature, B(X,US), is very small. When the target is presented alone (right panels), the large prediction of the US by the target, B(A,US) is cancelled by a large negative prediction of the US by the hidden units, B(H,US), so that A alone evokes a response small in amplitude and duration. By contrast, on X A presentations (left panels of Figure 14.8) the US is predicted mainly by the target and by the hidden units (controlled by X) so that the sum of these predictions (CR) is positive and large, both in amplitude and duration.

Importantly, note the control of the discrimination by the feature through hidden units. The US prediction by hidden units changes sign (at the moment of US presentation marked by the dotted line) from positive (X→A, left panels) to negative (A alone, right panels). As will be shown in the next section, since the feature controls indirectly (through hidden units) the temporal and associative properties of the solution (the CS acts as an occasion setter), these properties survive after feature extinction.

Temporal specificity of serial FP discrimination after feature extinction

Whereas feature X extinction eliminates a simultaneous FP discrimination in which X acts as a simple CS, it does not affect a serial FP discrimination in which X acts as an occasion setter (Holland, 1992). Therefore, in order to further characterize the solution found by the animals (simple conditioning or occasion setting), Holland *et al.* (1997) tested the two groups of animals after the feature cue was extinguished. Importantly, Holland *et al.* (1997) also examined the effect of feature extinction on the temporal specificity of serial FP discrimination.

Figure 14.9 shows that after feature extinction, discrimination decreases dramatically in the 5-s X–A group but not in the 25-s X–A group, suggesting that in the two cases animals use different solutions to solve the discrimination (Holland *et al.*, 1997). The upper panels of Figure 14.9 show the results of the test before (left upper panel) and after (right upper panel) feature extinction (Holland *et al.*, 1997), in two groups of animals trained with 5-s and 25-s X–A intervals. The bars represent difference in latency to chain pull in the presence

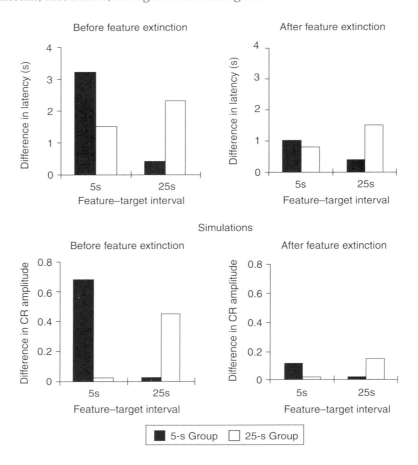

Figure 14.9 Temporal specificity of serial FP discrimination before and after feature extinction. *Upper panels*: Difference in latency to chain pull in the presence of A alone, or X→A, for two feature–target intervals, 5 s and 25 s (Holland et al., 1997). *Left upper panel*: Difference in latency before feature X extinction (Holland et al., 1997). *Right upper panel*: Difference in latency after feature X extinction (Holland et al., 1997). *Lower panels*: Simulated difference in CR to X→A, or A alone, before and after 40 trials of feature X extinction, in two simulated groups that received 300 alternated X→A+/A− trials, with 15 time units FTI (5-s group), or 800 alternated X→A+/A− trials, with 75 time unit X–A interval (25-s group). *Left lower panel*: Difference in latency before feature X extinction. *Right lower panel*: Difference in latency after feature X extinction. Latency to respond is assumed to decrease linearly with the CR: Latency = 1 − CR. Therefore, difference in latency is $\text{Latency}_A - \text{Latency}_{X \to A} = (1 - CR_A) - (1 - CR_{X \to A}) = CR_{X \to A} - CR_A$.

of X→A or A alone. The left upper panel of Figure 14.9 shows that animals discriminate better at the training FTI than at the other interval. The right upper panel shows the effect of feature extinction. In the 5-s X–A group feature extinction almost eliminates discriminations, suggesting that the solution of the discrimination relies on direct feature–US associations. In the 25-s X–A group, feature extinction has less effect on the discrimination, suggesting that the solution does not depend on direct feature–US associations.

The lower panel of Figure 14.9 shows that the model is able to correctly describe the effect of feature extinction on serial FP discrimination. The left lower panel of Figure 14.9 demonstrates that before feature extinction the model discriminates better at the training FTI than at the other interval. The right lower panel of Figure 14.9 shows the effect of 40 feature extinction trials on serial FP discrimination. Discrimination decreases dramatically in the 5-s X–A group (closed bars), but only moderately in the 25-s X–A group (open bars).

In sum, the model is not only able to describe the temporal specificity of serial FP discrimination, but also the associative relations that would allow feature extinction to eliminate FP discrimination in the 5-s group, but not in the 25-s group. The simulated 5-s X–A group solves the discrimination by engaging direct X–US connections and positive H–US connections activated by the presentation of X in the experimental context (Figure 14.7). When these associations extinguish, discrimination is abolished. However, as shown in Figure 14.8, in the simulated 25-s X–A group, discrimination is solved by engaging direct A–US connections, and inhibitory H–US connections inhibited by the presentation of X→A in the experimental context. Extinguishing the direct X–US associations, and the few H–US associations activated by X in the experimental context, has only a moderate effect on the discrimination.

Summary of serial FP discrimination simulations

Simulations show that by allowing the CSs to evoke multiple spectral traces, the model is able to correctly address both the temporal specificity and the associative relations involved in a serial FP discrimination. The model is qualitatively in accord with Holland *et al.*'s (1997) data showing that animals discriminate better when tested at the training FTI than at shorter or longer FTIs, the discrimination gradient is consistent with scalar timing, and animals discriminate better when trained with short, but not long FTIs.

Importantly, the model incorporates the associative structures that would allow feature extinction to eliminate FP discrimination in the short, but not long FTI group. Further research remains to explore if other properties of occasion setting mentioned in Chapter 12 are preserved in the model.

Discussion

Perhaps due to the complexity of conditioning, it is common to develop theories and models which deal with very specific properties of learning. For example, theories tend to separately address the associative and temporal aspects of Pavlovian conditioning.

Associative models largely ignore temporal properties and concentrate on the type of associations that seem to be involved in conditioning. For example, a popular associative model, the Rescorla and Wagner (1972) model, assumes that animals learn to predict the value of the US, but ignore its timing. Although "real-time" models of associative learning (e.g. Moore & Stickney, 1980; Wagner, 1981; Schmajuk & Moore, 1988; Schmajuk & DiCarlo, 1992) are able to address some temporal aspects of conditioning, they do not speak to the temporal specificity of these phenomena. Other theories (e.g. Grossberg & Merrill, 1992; Moore & Choi, 1998), although addressing timing in simple and compound conditioning, do not apply to the more complex characteristics of occasion setting.

As mentioned, a wide range of models describe the temporal properties of classical (Grossberg & Schmajuk, 1989; Grossberg & Merrill, 1992) and operant conditioning (Gibbon, 1977; Church & Broadbent, 1990; Killen & Fetterman, 1988; Machado, 1997; Staddon & Higa, 1999), and the neurophysiological aspects of timing (see e.g. Meck & Church, 1987), but ignore all other aspects of the paradigms. Various mechanisms were proposed to underly timing: a pacemaker/accumulator process (Gibbon, 1977; Meck & Church, 1987), a set of oscillator processes (Church & Broadbent, 1990), a sequence of behaviors (Killen & Fetterman, 1988), or a set of memory traces (Grossberg & Schmajuk, 1989; Machado, 1997; Staddon & Higa, 1999).

A central problem from the perspective of timing theories is to differentiate between "clock"-based (pacemaker- or oscillator-based) timing mechanisms, and "clock-free" timing mechanisms (see e.g. Staddon & Higa, 1999). Although this approach is useful to investigate particular timing mechanisms, it says little about how these mechanisms are used in conjunction with other learning mechanisms, given that animals seem to learn not only *when* the US is presented, but also other aspects of the conditioning paradigm, such as *what* are the best predictors of the US. In contrast, this chapter suggests that efforts towards addressing timing questions should be complemented by a combined analysis of the temporal and associative characteristics of these paradigms.

The present model describes timing in classical conditioning by assuming that CSs activate a set of processes that develop at different time scales. These processes and their configurations compete to become associated with the US. Based on these properties the model is able to describe ISI curves in simple conditioning, the temporal specificity of blocking and the temporal specificity of the action of occasion setters. The results presented here suggest that the spectral timing hypothesis has an important characteristic: it can work in conjunction with associative mechanisms to explain timing in associative learning.

For example, we suggest that in a blocking paradigm, stimuli compete to accurately predict the value, the time and the duration of US presentation. In line with Barnet et al. (1993) more blocking is observed if the A–US interval in phase 1 matches the B–US interval in the second phase of the blocking paradigm, because the US is already accurately predicted. According to the model, the accuracy of US prediction implies not only a precise moment, but also an appropriate CR duration. If these criteria are not satisfied, unblocking occurs. In particular, if B is able to predict the value, time and duration of the US more accurately than A, B is able to gain associative strength and predict the appropriate properties of the US.

In order to deal with conditioning paradigms that involve multiple CSs, Grossberg and Merrill (1992) extended the Grossberg and Schmajuk (1989) model by assuming that (a) internal representations of the CSs compete among them thereby modifying their magnitude, (b) spectral traces controlled by CS internal representations establish associations with the US in proportion to their magnitude, and (c) trace–US associations change in proportion to the moment-to-moment change in the US internal representation. Therefore, their model is able to describe some aspects of timing in simple and compound conditioning, including second-order conditioning. By contrast, the model presented here follows Moore and Stickney's (1980; Schmajuk & Moore, 1988; Schmajuk, 1997, page 39) assumption that the unchanged internal representations (traces) of the CSs compete to gain association with the US.

A model that is also able to address some temporal aspects of blocking and second-order conditioning was presented by Desmond and Moore (1988; Moore & Choi, 1998). In their model, each CS evokes a cascade of activations that compete in predicting the value and the time of US presentation through a time-derivative rule (Sutton & Barto, 1990). Although in the original formulation the model does not describe the scalar property of CR topography (increased CR duration for larger CS–US intervals), the model might address this issue by allowing parameter delta to progressively vary with elapsed time after CS onset (John W. Moore, personal communication, April 1998).

Importantly, the present model and the above competing models differ in their predictions relative to the blocking paradigm. The Grossberg and Merrill (1992) and the Desmond and Moore (1988; Moore & Choi, 1998) models are able to account for the unblocking due to a change in the A–US interval between the two phases of the blocking paradigm. Also, as an outcome of their description of second-order conditioning, they might be able to account for unblocking when A–US duration is the same in both phases ($A+$), but B is paired serially with A and the US, $B{\rightarrow}A+$, as in group S-F(s) (Barnet et al., 1993, Experiment 2). However, while Grossberg and Merrill (1992) and the Desmond and Moore (1988; Moore & Choi, 1998) models would presumably predict that in this case animals learn the B–A interval, the present model predicts that animals learn the B–US interval. Unlike the present model, these models also have problems accounting for unblocking when A–US duration is the same in both phases ($A{\rightarrow}+$) and B is paired simultaneously with the US, $A{\rightarrow}B+$, as in group F-S(f) (Barnet et al., 1993, Experiment 4). The present model describes this result due to competition between spectral traces to predict the US duration. Thus, the present model also predicts that in this case animals learn the B–US interval.

Furthermore, in contrast to the Grossberg and Merrill (1992) and the Desmond and Moore (1988; Moore & Choi, 1998) models, the present model is also able to address temporal specificity in a different class of phenomena that seems to require, besides competition between simple CS–US associations, the involvement of hidden H–US associations. More precisely, we showed that when activated by spectral traces, a structure consistent with Schmajuk and DiCarlo's (1992) model is able to correctly address temporal specificity of serial FP discriminations. At short feature–target intervals, the associative solution is similar to simple conditioning, and therefore is abolished by feature extinction. When trained with large feature–target intervals, the associative solution is similar to occasion setting, and therefore is not abolished by feature extinction. In sum, for both short and long FTIs, the model describes both the temporal and associative properties of serial FP discrimination

However, while in the Schmajuk and DiCarlo (1992) model the competition between CSs is purely associative, in the present model the competition between CSs is both associative and temporal. In short, CSs compete to predict not only the presence and intensity, but also the duration and time of presentation of the US. For example, blocking is observed only when specific temporal relationships occur. Similarly, in a serial FP discrimination, feature X simultaneously controls both the associative and temporal properties of the solution found by animals.

Most importantly, while in the Schmajuk and DiCarlo (1992) model a CS assumes the same role (simple CS, and/or occasion setter) throughout a trial duration, in the present model, a CS can assume different roles at different time moments. Thus, a CS can act as a simple CS at one moment in time, and as an occasion setter at other moments in time, by the activation of its different traces. In the Schmajuk and DiCarlo (1992) model an $X{\rightarrow}A+/A-$ discrimination with a long FTI is solved indirectly (through hidden units), while direct X–US associations are null. At no moment during the trial does the feature directly activate the US representation. Therefore, the solution is not time specific. In contrast, in the present model, in an X→A+/A− discrimination with a long FTI, slow X traces are occasion setters (they act through hidden units) and fast X traces become direct inhibitors of the US. Feature X directly inhibits the US representation shortly after X presentation, but indirectly activates it (through hidden units) later on in time. Feature X assumes different roles during the trial, thus allowing for temporal specificity of the solution. Therefore, in the present model, temporal specificity is incorporated into the solution found by the system by allowing any CS to have different roles at different moments in time. The present model extends the notion of occasion setter to describe a stimulus that controls both the temporal and associative relations between another CS and the US.

Further research is nonetheless necessary to resolve some conflicting assumptions incorporated in the present model; assumptions that were favored in the past by the separate analysis of conditioning from the timing, or associative, perspective. For example, the temporal assumption adopted in the present model that CSs evoke spectral traces that are independent of CS duration (Grossberg & Schmajuk, 1989) precludes the description of the effect of CS duration on conditioning; a feature well described by other "real-time" associative models (Moore & Stickney, 1980; Schmajuk & Moore, 1988; Schmajuk & DiCarlo, 1992). On the other hand, since in the present model a CS can have different roles (simple CS or occasion setter) at different time moments, the description of the multiple associative features of occasion setting (see Chapter 12 and e.g. Schmajuk & Buhusi, 1997; Schmajuk, Lamoureux & Holland, 1998; Lamoureux, Buhusi & Schmajuk, 1998) is partly obscured by the temporal interactions between multiple traces.

Despite the shortcomings discussed above, the present model gives an adequate description of both temporal and associative aspects of some conditioning protocols, and can guide the independent study of both timing and associative mechanisms engaged in animal learning. Nonetheless, this chapter suggests that a combined analysis of timing and associative learning is needed for a better understanding of conditioning phenomena.

Application of the model to serial order

Terrace and McGonigle (1994) studied the abilities of animals and children to learn and construct sequences. They presented evidence that traditional chaining models, such as the associative system in the SLG model (see Figure 2.1) fail to "describe the formation of templates of item order and the ability to assign items from novel lists to particular ordinal positions." It is conceivable, however, that an "ordinal" form of the configural–timing model described in this chapter would be able to learn particular ordinal positions by associating a particular trace with a particular item on the lists.

Summary

In order to describe the temporal specificity in simple conditioning and occasion setting, we present a computational model of Pavlovian conditioning in which stimuli compete to predict the value, the duration and the time of presentation of the reinforcement. The model develops three basic notions. First, it is a real-time model, and therefore able to describe various temporal relations between stimuli, and between stimuli and the reinforcement. The real-time properties of the model follow from the hypothesis that stimuli evoke multiple traces with different temporal properties (Grossberg & Schmajuk, 1989). Second, spectral traces become associated directly and indirectly (through hidden units) with a representation of the reinforcement, in a manner consistent with Schmajuk and DiCarlo's (1992) model. Therefore, the model is able to describe the difference between simple CSs and occasion setters. However, in contrast to Schmajuk and DiCarlo's (1992) model, in the present model a stimulus may assume different roles (simple CS, occasion setter, or both) at different time moments, thus providing support for the description of the temporal specificity of occasion setting. Therefore, the present model extends the notion of occasion setter to describe a stimulus that controls both the temporal and associative relations between another CS and the US. Third, the direct and indirect (hidden unit) traces compete to predict the presence and characteristics of the US. Importantly, while in the Schmajuk and DiCarlo (1992) model competition between CSs is purely associative (CSs compete for the prediction of the presence and value of the US), since in the present model timing is mapped onto trace-specific associations, competition between CSs is both associative and temporal. The associative competition for the prediction of presence and value of the US is also a temporal competition for the prediction of the temporal properties (time of presentation and duration) of the US. Thus, the model can address associative and temporal competition between CSs, as in the blocking paradigm.

Computer simulations demonstrate that the model is able to address temporal specificity in conditioning paradigms in which stimuli act as simple CSs or as occasion setters. We show that the model describes the CR topography in simple conditioning (Smith, 1968), the temporal competition between simple CSs in paradigms like blocking (Barnet, Grahame & Miller, 1993) and the temporal specificity in serial feature-positive discriminations (Holland, Hamlin & Parsons, 1997).

This chapter suggests that efforts towards characterizing animal timing should include a joint analysis of the temporal and associative mechanisms that allow animals to simultaneously learn both *what* and *when* significant events occur.

In the simulations presented in Figures 15.4 and 15.5 we considered only the first two phases of the paradigm.

Appendix 14.1 The configural–timing model

Conditioned stimulus CS_i, evokes k spectral traces, with different temporal properties. These traces are computed using activities x_{ik} and y_{ik}. Activity x_{ik} is given by

$$dx_{ik}/dt = (k_1/k)(1 - x_{ik})(CS(i) + k_2 f_1(i)) - k_3 x_{ik} \qquad [14.1]$$

where $CS(i)$ is 1 when CS_i is present and 0 otherwise, k_1 is a growth rate parameter, k_3 is a decay parameter, and k_2 and function f_1 denote the combined feedback contribution of the traces corresponding to stimulus CS_i. According to Equation [14.1A], each activity x_{ik} increases at a different rate, inversely related with index k of the trace. Activities x_{ik} are initialized with null values. The feedback function f_1 is given by Equation [14.5].

Activity y_{ik} is given by

$$dy_{ik}/dt = k_4(1 - y_{ik}) - k_5 f_2(x_{ik}) y_{ik} \qquad [14.2]$$

where f_2 is the sigmoid function

$$f_2(a) = a^\alpha/(a^\alpha + k^\alpha). \qquad [14.3]$$

According to Equation [14.2A], for each CS_i, the set of k activities y_{ik} increases with growth rate k_4, and decays in proportion to activity x_{ik}. While x_{ik} increase with a rate inversely related to index k of the trace, y_{ik} decrease with a rate inversely related to index k of the trace. Activities y_{ik} are initialized with values 1.

The k spectral traces of stimulus CS_i are given by

$$\tau_{ik} = y_{ik} f_2(x_{ik}) \qquad [14.4]$$

The spectral traces peak at different moments in time, and rise and decay in proportion to index k of the trace. For each CS_i, traces τ_{ik} are fed back into activities x_{ik}, through function f_1 given by

$$f_1(i) = f_3(\Sigma_k x_{ik}) \qquad [14.5]$$

where f_3 is the sigmoid function

$$f_3(a) = a^\beta/(a^\beta + k_7^\beta). \qquad [14.6]$$

The context, CX, is assumed to activate k traces, τ_{CXk}, of constant intensity. The spectral traces, τ_{ik}, of each CS_i are configured in the hidden unit H_j through indirect connections VH_{ikj}. Activity of hidden unit H_j is given by

$$\text{act}_j = \Sigma_i \Sigma_k \tau_{ik} VH_{ikj}. \quad [14.7]$$

Each hidden unit H_j evokes a memory trace τ_j proportional with its activity:

$$\tau_j = k_8 f_4(\text{act}_j) \quad [14.8]$$

where f_4 is the sigmoid function

$$f_4(a) = a^\gamma / (a^\gamma + k_9^\gamma). \quad [14.9]$$

Hidden units H_j are associated with the US through connections VN_j. The prediction of the US by the hidden unit H_j is given by

$$B_{Hj,US} = \tau_j VN_j. \quad [14.10]$$

Spectral traces, τ_{ik}, of each CS_i are directly associated with the US through direct connections V_{ik}. The direct prediction of the US by stimulus CS_i is given by

$$B_{CSi,US} = \Sigma_k \tau_{ik} V_{ik}. \quad [14.11]$$

The aggregate prediction of the US is given by the sum of the direct and indirect contributions of all the stimuli, CS_i, and of all hidden units H_j:

$$B_{US} = \Sigma_i B_{CSi,US} + \Sigma_j B_{Hj,US}. \quad [14.12]$$

Importantly, in order to prevent the extinction of conditioned inhibition by the presentation of CS alone (Zimmer-Hart & Rescorla, 1974), B_{US} is considered non-negative. If $B_{US} < 0$ then $B_{US} = 0$.

The conditioned response, CR, is considered proportional with the aggregate prediction of the US:

$$CR = k_{10} B_{US}. \quad [14.13]$$

Direct associations V_{ik} are adapted according to

$$dV_{ik}/dt = k_{11} \tau_{ik}(US - B_{US})(1 - |V_{ik}|) \quad [14.14]$$

where the growth rate k_{11} is equal to k_{11}' if $US \geq B_{US}$, or equal to k_{11}'' if $US < B_{US}$. Similarly, the direct hidden–US associations VN_j, are adapted according to

$$dVN_j/dt = k_{12} \tau_j (US - B_{US})(1 - |VN_j|) \quad [14.15]$$

where the growth rate k_{12} is equal to k_{12}' if $US \geq B_{US}$, or equal to k_{12}'' if $US < B_{US}$. The indirect, CS_i–US associations, VH_{ikj}, are adapted by

$$dVH_{ikj}/dt = k_{13} \tau_{ik} \tau_j (US - B_{US}) VN_j. \quad [14.16]$$

Appendix 14.2 Simulation parameters

Simulations assume 20 hidden units and $k = 5$ traces per stimulus, including the context CX. Context traces, τ_{CXk} are 0.01 in intensity. The initial values of V_{ik} and VN_j

are null. Input–hidden unit association weights, VH_{ijk}, are randomly assigned using a uniform distribution ranging between ±0.25 in all simulations.

Each simulation trial consists of 400 time steps. The simulations assume 20-time unit CSs, and a 3-time unit US of intensity US = 1 if not otherwise specified.

All simulations use the same set of parameters: $\alpha = 8$, $\beta = 2$, $\gamma = 1.5$, $k_1 = 0.05$, $k_2 = 2$, $k_3 = 0.01$, $k_4 = 0.0001$, $k_5 = 0.125$, $k_6 = 0.8$, $k_7 = 0.005$, $k_8 = 1.5$, $k_9\gamma = 0.35$, $k_{10} = 5$, $k_{11}' = 0.005$, $k_{11}'' = 0.0045$, $k_{12}' = 0.015$, $k_{12}'' = 0.018$, and $k_{13} = 0.5$.

In simulations presented in Figure 14.3, the CS is presented between time units 20–40. The CS–US intervals are 10, 20, 40 or 80. US intensity is US = 1 or US = 2. In simulations presented in Figure 14.3 the nictitating membrane response is assumed to be proportional to the CR, NMR = CR.

Figures 14.4 and 14.5 present simulations of the first two phases of the paradigms described in Table 14.1. The CS–US interval is 14 time units for simultaneous CS–US pairing, CS+, and 34 time units for serial ("forward") CS–US pairings, CS→+. In simulations presented in Figures 14.4 and 14.5 latency to complete the licking period is assumed to be proportional to the CR, Latency = CR.

In simulations presented in Figures 14.6–14.9, feature X is presented between time units 20–40. The target is presented between time units 35–55 for group 5 s, and between time units 95–115 for group 25 s. The target–US interval is 13 time units. In simulations presented in Figures 14.6–14.9 latency to respond is assumed to decrease linearly with the CR: Latency = 1 – CR. Therefore, difference in latency is proportional to difference in CR:

$$\text{Latency}_A - \text{Latency}_{X \to A} = (1 - CR_A) - (1 - CR_{X \to A}) = CR_{X \to A} - CR_A.$$

15

Attention and configuration: extinction cues

In order to jointly explain latent inhibition and occasion setting, Buhusi and Schmajuk (1996) combined the attentional and associative systems of the SLG neural network (Chapter 2) with the configuration mechanisms of the SD neural network (Chapter 11). As mentioned, the SLG model describes the multiple properties of latent inhibition (Chapter 5), overshadowing and blocking (Chapter 8) and extinction (Chapter 9). On the other hand, the SD/SLH model correctly describes negative and positive patterning, as well as most of the features of occasion setting (Chapter 12). Buhusi and Schmajuk (1996) demonstrated that the attentional–configural model describes a broad range of experimental results, including: (a) acquisition and extinction series, (b) overshadowing and blocking, (c) discrimination acquisition and reversal, (d) conditioned inhibition and inhibitory conditioning, (e) simultaneous feature-positive discrimination, (f) serial feature-positive discrimination (occasion setting), (g) negative patterning, (h) positive patterning, and (i) reduced responding when context is switched. In addition, the model describes (j) sensory preconditioning, (k) latent inhibition, (l) contextual effects on LI, and (m) place and maze learning. This chapter presents a simplified version of the attentional–configural model described Buhusi and Schmajuk (1996) to address the properties of extinction cues.

Attentional and configural mechanisms in extinction

Some experiments have studied the effect of extinction cues (ECs), i.e. temporally discrete cues preceding the presentation of the CS during extinction. During testing, the presence of ECs tends to decrease responding. As mentioned in Chapter 9, the SLG model is unable to describe the effects of an EC on spontaneous recovery (Brooks & Bowker, 2001) or renewal (Brooks & Bouton, 1994). Moreover, Brooks and Bouton (1993, Experiment 2) showed that, even when an EC could have decreased renewal by becoming a conditioned inhibitor, ECs show no evidence of having become inhibitory in summation and retardation tests. Therefore, the EC might act as a negative occasion setter, setting the occasion for the CS not to generate a CR.

As indicated in Chapter 12, occasion setting is well described by the SD/SLH model. When ECs are present during extinction, the SD model (a) forms weakly inhibitory direct EC–US connections; (b) configures H units by adjusting their connections with the EC, CS and CX inputs; and (c) establishes strongly inhibitory H–US associations. Because activation of H units is maximal at the time of the presentation of the CS, H units strongly inhibit responses at the time when the CS is present. Consequently, ECs are able to decrease both spontaneous recovery and renewal. As mentioned, Brooks and Bouton (1993) reported that an EC does not appear to be inhibitory in summation and retardation tests. The model correctly describes these results because the direct EC–US inhibitory associations are weak and the H units are only partially activated in the absence of the CS.

Configuration and attention

According to the SD model, during extinction the CS–US association somewhat decreases and the H–US association becomes inhibitory. In this condition, the CS simultaneously maintains both direct excitatory and indirect inhibitory associations with the US; a condition postulated by other models of extinction (e.g. Konorski, 1948; Pearce & Hall, 1980). The model predicts that when the CX is weak, H is not activated and the CS–US association is completely eliminated (see Chapter 12, Figure 12.25). However, as in the case of the SLG model, when H is active during extinction, the H–US inhibitory association protects the CS–US from complete extinction (see Chapter 12, Figure 12.26). Notice that, if H is partially active in the absence of the CS, H will be inhibitory and the model cannot explain, as described in Chapter 9 (section on "Extinction"), the reported results of summation tests. According to the SD model, renewal is the result of the inhibition being eliminated when the context is changed and H is partially deactivated, thereby allowing the nonextinguished CS–US association to activate the CR. In terms of the model, reinstatement is the result of the formation of simple CX–US associations following US presentation in the CX.

Notice that, even if configural mechanisms can explain the effects of ECs, renewal and reinstatement, attentional processes are still needed to explain the experimental results listed in Chapter 9 (e.g. contextual inhibition not detectable in summation tests, spontaneous recovery, external disinhibition, magnitude of renewal with different procedures, elimination of renewal by massive extinction and slow reacquisition). Therefore, this chapter combines the attentional-associative properties of the SLG model with the configural properties of the SD model, along the lines suggested by Buhusi and Schmajuk (1996).

The attentional-configural form of the SLG model

A simplified version of the model described Buhusi and Schmajuk (1996) is presented in Figure 15.1. This simplified version consists of adding a configural element, similar to a hidden unit (H) in the SD model, to the SLG attentional-associative model. In Figure 15.1, the EC is represented both as a simple stimulus (EC) and a configural stimulus (H) activated by the CS, the EC and CXc (the intensity of contextual cues in the training cage, not shown in the figure). As in the original SLG model (see Chapter 2), attention to the CS, EC and H varies with Novelty'. As in the SLG model (Chapter 2), the configural stimulus H can be associated with all other simple and configural CSs. Notice that configural stimulus H is equivalent to the "unique stimulus hypothesis" proposed by Rescorla and Wagner (1972) to explain, for example, negative patterning.

Although one possible way for an EC to decrease responding is by forming an EC-US inhibitory association during extinction, as mentioned, the data do not support this option (Brooks & Bouton, 1993). In terms of the attentional-configural form of the SLG model, the trace of the EC is too weak at the time of the CS presentation to inhibit responding. Instead, in the model, the EC acts as a simple stimulus and establishes a weak direct inhibitory association with the US representation, and forms a configuration H, which establishes a strong inhibitory association with the US. We assume that H is active at the time when the CS is present simultaneously with the EC trace. When the CS is absent, we assume that H is only a part (2/3) of its original value. When the CS is absent and the animal is placed in a different context (CXc = 0), H is zero and the EC is only represented as a simple CS.

Effects of presenting an extinction cue on spontaneous recovery
Experimental results
Presentation during testing of an EC attenuates spontaneous recovery. In Brooks and Bowker's (2001) study, rats received 84 CS-US presentations. The CS duration was 30 s, with a mean ITI of 270 s, the US consisted of food delivery coinciding with the CS termination, and the CR was measured as the number of food magazine entries. During each of the two extinction sessions, there

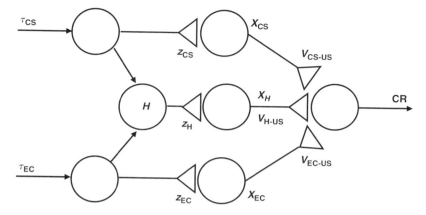

Figure 15.1 A simplified diagram of the attentional–configural form of the SLG model. CS: conditioned stimulus; US: unconditioned stimulus; t: traces of the CS or EC; z: attention to the CS or EC; X: representations of the CS, EC or H; V: CS–US, EC–US or H–US association; CR: conditioned response. Triangles: variable connections between nodes. Arrows: fixed connections between nodes. H represents a configuration of the CS with the EC. The configural variable H takes a larger value when the CS and the EX are present in a trial than when one of them is absent.

were 12 nonreinforced CS presentations. Three out of four CS presentations were preceded by the EC, which terminated 40 s before the onset of the CS. After extinction, animals in groups EC + NC and EC + C received 48 EC–US presentations with the US onset coinciding with the CS termination, whereas animals in groups C and NC spent an equivalent amount of time without any stimuli (retention interval). Finally, rats were tested for spontaneous recovery in the presence (C) and the absence (NC) of the EC.

As shown in Figure 15.2 (upper panel), Brooks and Bowker (2001) reported that a group presented with an EC during testing (Group EC) showed less spontaneous recovery than one not presented with the cue (Group NC). In addition, they showed that spontaneous recovery is attenuated even more if a US is presented immediately following the EC before the testing session (Group EC + EC versus Group EC + NC). Although not shown in Figure 15.2 (upper panel), Brooks and Bouton (1993, Experiment 3) had demonstrated that a cue presented during conditioning does not decrease spontaneous recovery (Group Conditioning C).

Simulations

Simulations consisted of 10 CS–US trials, 10 cued extinction trials, 20 retention-interval trials, and 2 test trials. Simulations of 5 home-cage trials (CXh =

396 Attentional, associative, configural and timing mechanisms

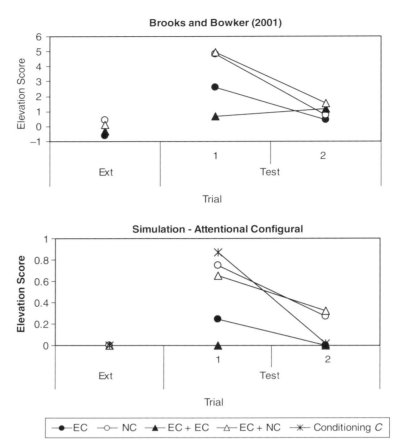

Figure 15.2 Spontaneous recovery is attenuated by presentation of a nonreinforced or reinforced extinction cue. *Upper panel*: Data from Brooks and Bowker (2001). *Lower panel*: Simulations with the attentional–configural form of the model. EC: testing with a nonreinforced extinction cue, NC: testing with no extinction cue, EC + EC: testing with a reinforced extinction cue, EC + NC: reinforced cue not present during testing. Conditioning C: testing with a conditioning cue. Compare with Figure 9.8.

0.5, CXg = 0.4) were included between the different sessions. The salience of the training context was set to CXc = 0.7 and CXg = 0.4. The CS had a 20 t.u. duration and salience 2, and the EC was 20 t.u. long with salience 1.5. In cued-CS trials (extinction and testing), the configural stimulus H was presented overlapping with the CS (20 t.u. duration), and the offset of the EC coincided with the onset of the CS. The ITI was 100 t.u. The configural stimulus H had salience 2. The conditioning cue was identical to the EC, but was presented during conditioning and testing.

Figure 15.2 (lower panel) shows that the attentional–configural version of the SLG model is able to replicate the experimental results.

Figure 15.3 shows the values of variables, just before the time of presentation of the US on reinforced trials, in the attentional–configural version of the SLG model during the treatment period that follows extinction. From top to bottom, the left panels correspond to groups NC, EC, and EC + EC and show that the CR is large in Group NC, relatively small in Group EC, and nil in Group EC + EC. By the end of extinction (Trial 30), in every group CS–US associations are excitatory, CXc–US associations are inhibitory, EC–US associations are barely inhibitory ($V_{EC-US}= -0.01$), and H–US associations are inhibitory. The EC gains only a weak inhibitory association with the US because its relatively weak trace has to compete with the traces of the CX and the configural stimulus H; both strongly active at the time when the absent US is maximally predicted by the CS. In Group EC + EC (left bottom panel), EC–US pairings cause EC–US associations to become excitatory during the treatment period (Trials 31 to 50). Importantly, given the temporal proximity of the EC and the US, the EC–US association blocks the formation of a CX–US association that could augment responding. The CX–US association, $V_{CXc,US}$, first increases and then decreases, but always stays inhibitory. As shown in Figure 15.3 (right panels), the main consequence of EC–US pairings is that, by the end of the treatment period (Trials 46 to 50), attention to CXc is relatively high in Group EC + EC, but low in groups NC and EC.

In Group NC (see right upper panel), spontaneous recovery is the consequence of a faster increase in the representation of the excitatory CS, X_{CS}, than that of the inhibitory CXc during test, allowing for the expression of the $V_{CS,US}$ association that survived extinction. In Group EC (see right middle panel), the presence of the EC also results in a fast activation of the representation of the inhibitory H representation X(H) overlapping with X(CS), thereby resulting in a relatively small CR. Group EC + EC (right lower panel) also shows attenuation of recovery because, although its simple representation is excitatory, the presence of the EC produces a fast activation of the inhibitory H configural stimulus and, in addition, of the inhibitory CXc representation. In this way, the EC simultaneously maintains both direct excitatory, and indirect inhibitory associations with the US. Comparison of the right middle and lower panels in Figure 15.3 reveals that Group EC + EC shows more attenuation than Group EC because, as described above, attention to the inhibitory CXc does not decrease during the treatment period, thereby increasing inhibition during testing. Notice that even when the simple representation of EC has become excitatory during treatment, it does not contribute to the CR because its trace is barely active at the time of the CS presentation (X_{EC} is very small).

Retardation and summation tests of an extinction cue

Importantly, Brooks and Bouton (1993, Experiments 2 and 4) reported that the EC did not show evidence of having become inhibitory in a retardation test or a summation test.

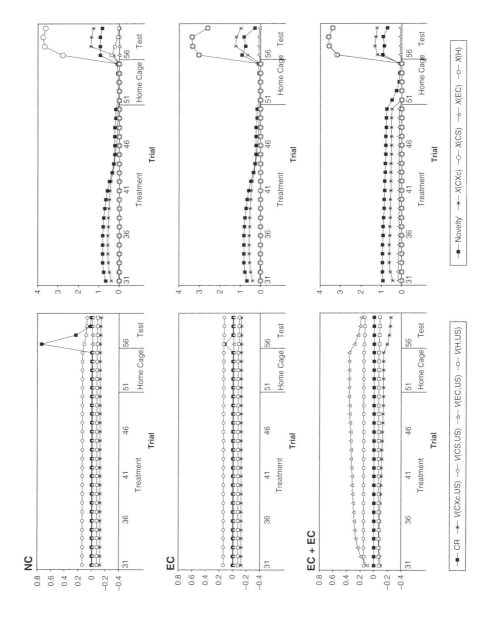

Experimental results

In Brooks and Bouton (1993, Experiment 2), rats received 104 CS–US presentations. The CS (T) was a 30-s, 80-dB pure tone, presented with a mean ITI of 190 s, and the US consisted of two pellets of food delivered immediately following the CS. During the extinction session in Group L–T (L as an Extinction Cue), there were 24 nonreinforced T presentations. Three out of four T presentations were preceded by the extinction cue L (30-s light off), which terminated 15 s before the onset of T. Group L– underwent a similar procedure (preexposure), with the exception that no tone CSs were presented (L as a Control Cue). A retardation test, with 16 L–US trials, compared the conditioning rates of L in groups L– and LT–.

Brooks and Bouton (1993, Experiment 4) also conducted a summation test for conditioned inhibition. Initially, conditioning of an excitor CS (N) was conducted, by delivering 64 N–US pairings. This CS was a 30-s, 65-dB white noise, presented with a mean ITI of 190 s. Then, acquisition and extinction of a pure-tone CS (T) was conducted as described above. After extinction, 2 sessions of summation test were carried out, each consisting of 6 alternated presentations of N–US and nonreinforced LN trials in groups L– and LT–.

Brooks and Bouton (1993, Experiments 2 and 4) reported that the EC did not become inhibitory in a retardation test (Figure 15.4a, upper left panel) or a summation test (Figure 15.4b, upper right panel).

Simulations

The parameters used were identical to those described for Bouton and Brooks' (2001) experiment (see Chapter 9), excluding the retention interval phase. Simulations for the summation test included additional 10 N-US trials at the beginning of the experiment, and N-US trials intermixed with the LN-trials during testing. Figures 15.4a and 15.4b show that simulations with the attentional-configural version of the SLG model are able to approximate the experimental results.

Caption for fig. 15.3 (cont.)

Figure 15.3 Variables in the attentional–configural form of the model during attenuation of spontaneous recovery by presentation of a nonreinforced or reinforced extinction cue. *Left panels*: Values of the conditioned response, CR; the association between the context and the unconditioned stimulus, $V_{CXc,US}$; the EC–US association, $V_{EC,US}$; the CS–US association, $V_{CS,US}$; as a function of trials. *Right panels:* Values of Novelty'; the internal representation of CXc, X_{CXc}; EC internal representation X_{EC}; and CS internal representation X_{CS} and H–US association, V_{H-US}; as a function of trials. *Upper panels*: Group NC, no extinction cue is presented, *Middle panels*: Group EC, extinction cue is presented, *Lower panels*: Group EC + EC, a reinforced extinction cue is presented. Treatment Trials 31–50, Home-cage Trials 51–55, Test Trials 56–59.

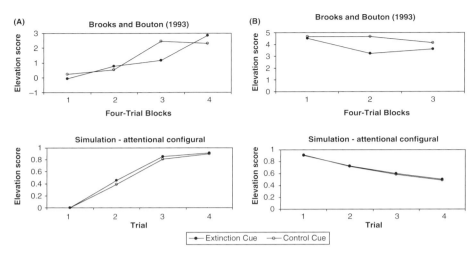

Figures 15.4A and 15.4B Retardation and Summation tests with the extinction cue. *Upper left panel*: Retardation data from Brooks and Bouton (1993, Experiment 2). *Upper right panel:* Summation data from Brooks and Bouton (1993, Experiment 4). *Lower panels*: Simulations with the attentional–configural form of the SLG model.

Effects of presenting an extinction cue during renewal testing

Experimental

Brooks and Bouton (1994) reported that presentation of an EC during testing attenuates renewal. Figure 15.5 (upper panel) shows that renewal is stronger with a neutral CS than with an EC. In Brooks and Bouton's (1994) Experiment 2, rats received five conditioning sessions of a tone-CS. Each session consisted of 16 CS–US presentations in Context A, with a mean ITI of 310 s. The CS was a 30-s, 80-dB pure-tone stimulus, and the US consisted in the delivery of two 45-mg food pellets. Two extinction sessions took place in Context B, each consisting of 24 nonreinforced CS presentations. An EC preceded the nonreinforced presentation of the CS on three of every four extinction trials. For half of the animals, the cue was the offset of the house light; for the rest, it was a 30-s, 65-dB intermittent white noise. Test sessions were conducted in Context A, where the CS was preceded either by the extinction cue (Group Extinction Cue) or the opposite cue (Group Neutral Cue).

Simulations

The parameters used were identical to those described for the Brooks and Bowker (2001) study on spontaneous recovery (see Chapter 9), without the retention interval phase. The salience of H was lowered from 2 to 1.4 during test trials because only two (CS, EC) out of three (CS, EC, CX_c) input stimuli

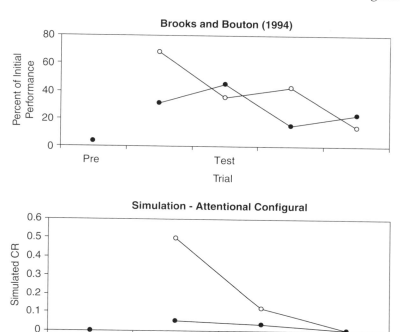

Figure 15.5 Renewal is attenuated by presentation of extinction cue. *Upper panel*: Data from Brooks and Bouton (1994, Figure 3). *Lower panel*: Simulations with the attentional–configural form of the SLG model. Compare with Figure 9.13.

were present. Computer simulations with the attentional–configural version of the SLG model are able to reproduce the data (Figure 15.5, lower panel). Because the acquisition and extinction procedures used to simulate renewal in this section are identical to those used for Brooks and Bowker (2001) in the section for spontaneous recovery, both retardation and summation tests shown in Figures 15.4a and 15.4b also apply to the renewal case shown here.

Summary

An attentional–configural form of the SLG model shows that spontaneous recovery and renewal are attenuated by an EC (Brooks & Bowker, 2001; Brooks & Bouton, 1994), and that the EC shows no evidence of becoming inhibitory in summation and retardation tests (Brooks & Bouton, 1993). According to the model, when ECs are present during extinction, the model (a) forms weakly inhibitory direct EC–US connections; (b) activates an H unit in proportion to the salience of the CS, EC and CX_c inputs; and (c) establishes strongly inhibitory H–US associations. Because activation of H units is maximal at the time of the presentation of the CS, H units strongly inhibit responses at the time when the CS is present. Consequently, ECs are able to decrease both spontaneous recovery and renewal. According to the model, an EC does not appear to be inhibitory in summation and retardation tests because the direct EC–US inhibitory associations are weak and the H units are only partially activated in the absence of the CS.

Notice that, in addition to EC effects, the attentional–configural form of the SLG model might also be able to capture the occasion setting properties of the context, which are needed to explain why the context does not appear to be inhibitory following a contextual discrimination (see Chapter 12). However, although the addition of H units might confer greater power to the SLG attentional–associative model, it might result in a combinatorial explosion when multiple CSs, ECs and CXs are considered. Therefore, a full integration between the SLG and the SD model as described by Buhusi and Schmajuk (1996) would be preferable to the present attentional–configural version of the SLG model.

16

Attention, association and configuration: causal learning and inferential reasoning

In this chapter, we apply the attentional–configural form of the SLG neural network model (see Chapter 15), which combines attentional, configural and associative mechanisms to the description of causal learning and inferential reasoning.

Causal learning

As described in Chapter 8, and according to associative theories (e.g. Rescorla & Wagner, 1972; the SLG model; and the SD/SLH model), blocking is the consequence of stimulus A winning the competition with X to predict the US, because the US is already predicted by A at the time of A–X–US presentations. In contrast, according to the inferential process view (see Beckers *et al.*, 2006), blocking is the consequence of an inferential process based on the assumptions of additivity and maximality. Maximality refers to the evidence that the Outcome produced by each potential cause has not reached the maximal possible value. Additivity denotes the fact that two causes, each one independently producing a given Outcome, produce a stronger Outcome when presented together.

Supporting the inferential explanation, recent experimental results have shown that, both in humans (Lovibond *et al.*, 2003; Beckers *et al.*, 2005) and rats (Beckers *et al.*, 2006), blocking was stronger if the maximal premise (the outcome of each cause does not reach the maximum possible value) and the additivity premise (the outcomes of effective causes can be added) are satisfied. In contrast, Beckers *et al.* (2005) showed that the Rescorla–Wagner (1972) model incorrectly predicts more blocking with an intense (maximal) than with a moderate (submaximal) outcome.

Livesey and Boakes (2004) suggested that the decreased blocking in the subadditive case could be explained by associative theories in terms of configurations of *A* and *X*. According to this view, when the combined effect of two stimuli is less than the sum of their individual effects, those stimuli are combined into a single stimulus. Blocking would not be present in the subadditive case because the *A*–Outcome association is not activated by the *A*–*X* configural stimulus, which gains a strong association with the Outcome. As Livesey and Boakes (2004, page 376) point out, further assumptions are needed to explain how a response to *X* is produced, when the strong association is formed between the *A*–*X* configuration, and not *X*, with the Outcome.

In this chapter, we show that the attentional–configural version of the SLG model presented in Chapter 15, can describe maximality and additivity effects on blocking and backward blocking. Notice that the inclusion of a configural mechanism is in line with the above-mentioned Livesey and Boakes (2004) suggestion.

The attentional–configural form of the SLG model

A simplified diagram of this attentional–configural version of the SLG model as applied to maximality and additivity experiments in blocking is presented in Figure 16.1. According to Figure 16.1, simultaneously active τ_A and τ_X activate configural stimulus *C*, and simultaneously active τ_G and τ_H activate configural stimulus *C'*. Attention to τ_A, τ_X, *C*, τ_G, τ_H, and *C'* varies with Novelty'. Configural stimuli *C* and *C'* can be associated with all other simple and configural CSs. As shown by the dashed double arrow in Figure 16.1, we assume generalization between compounds *C'* (*GH*) and *C* (*AX*) to be strong based in Young and Wasserman's (2002) experimental data showing that generalization between simple stimuli is much smaller than generalization between compounds. In addition, the model implements generalization among elements, and between elements and compounds, through the presence of a common contextual stimulus that is always active. Young and Wasserman (2002) also reported relatively strong generalization between common stimuli; an assumption also contemplated in the model.

It is important to remember that Node *X*, can be activated either by τ_X or by its prediction by τ_A (or, as explained below, by $\tau_{Outcome}$). This means that z_X can still be modified by Novelty' even if τ_X is absent. This is an important feature of the model that explains backward blocking (see Chapter 8) and additivity post-training, as described below.

In our simulations, the rating of the different stimuli was given by the sigmoid Rating = Predicted Outcome^6/(Predicted Outcome^6 + β^6), where β = Average of the predicted outcomes for all *A*, *X*, *K* and *L*.

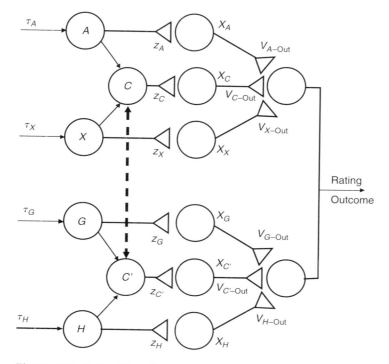

Figure 16.1 A simplified diagram of the attentional–configural form of the SLG model. τ: traces of A, X, G and H; z: attention to A, X, G and H; X: representations of A, X, G, H and C; Out: Outcome; V: A-Outcome, X-Outcome, C-Outcome or C'-Outcome associations. When simultaneously active τ_A and t_X activate configural stimulus C, and when simultaneously active τ_G and τ_H activate configural stimulus C'. The dashed arrow connecting C and C' represents the generalization between them. Through that generalization, τ_A and τ_X can activate the $V_{C\text{-Out}}$ association, and τ_G and τ_H can activate the $V_{C\text{-Out}}$ association. Triangles: variable connections between nodes, z and V. Arrows: fixed connections between nodes.

Additivity training preceding blocking

Experimental data

In Beckers et al.'s (2005, Experiment 2) additivity pretraining study, one additive and one subadditive group were used. The additive group received G+/H+/GH++/I+/Z− alternated trials (prior training), followed by alternated A+/Z− trials (elemental training), and finally by AX+/KL+/Z− alternated trials (compound training). Whereas G, H, I, Z, A, X, K and L denote stimuli, symbols + and ++ indicate the same outcome with different intensities. Stimuli were pictures of different foods on a computer screen, and the outcome (an allergic reaction) was represented by a red bar of different lengths on the screen. The probability, assigned by each group, of *X* and *K* (and *L*) to cause the outcome (ratings) were then compared. Subadditive groups received *GH*+ instead of *GH*++ presentations and *I*++ instead of *I*+ presentations. As shown in Figure 16.2 (upper left panel), although blocking was present in both groups, it was stronger in the additive than in the subadditive case.

Simulated results

Simulations for Experiment 2 included 90 additive or subadditive pretraining, 20 *A*+, 40 *AX*+ trials and 40 *KL*+ trials. Stimulus duration was 10 time units, stimulus intensity was 0.6, C intensity was 1, the Outcome was 2 for additivity and 1 for subadditive training. The Outcome was simulated as another CS (equivalent to the red bar), able to form both Outcome–Outcome and Outcome–CS associations. Notice that this is similar to the approach (sensory preconditioning between the CSs and a surrogate US) used in Chapter 8 to simulate backward blocking.

In line with Livesey and Boakes' (2004) view, the attentional–configural form of the SLG model is able to explain how additivity pretraining influences subsequent blocking. Like the Rescorla–Wagner (1972) model, the SLG model explains blocking because, at the time of the presentation of *X*, *A* already predicts the Outcome (or US, see Chapter 8). As shown in Figure 16.1, compound stimulus *C'*, activated by τ_G and τ_H and associated with the Outcome during pretraining, is fully activated when compound stimulus *C* is activated by τ_A and τ_X. The $V_{C'\text{-Out}}$ association, together with the blocker stimulus *A*, contributes to predict the Outcome, thereby increasing blocking. Because the $V_{C'\text{-Out}}$ association acquired during pre-training is stronger in the additive than in the subadditive case, blocking is stronger in the former than in the latter case (Figure 16.2, left lower panel).

Additivity training following blocking

Experimental data

In Beckers *et al.*'s (2005, Experiment 4) additivity posttraining study, the additive group received *G*+/*H*+/*GH*++/*I*+/*Z*– alternated trials, and the subadditive group *G*+/*H*+/*GH*+/*I*+/*Z*– alternated trials, following blocking training. As shown in Figure 16.2 (upper right panel), although blocking was present in both groups, it was stronger in the additive than in the subadditive case

Simulated results

Simulations for Experiment 4 included 20 *A*+, 20 *AX*+, 20 *KL*+ and 90 additive or subadditive post-training trials. Other parameters were identical to those used for Experiment 2.

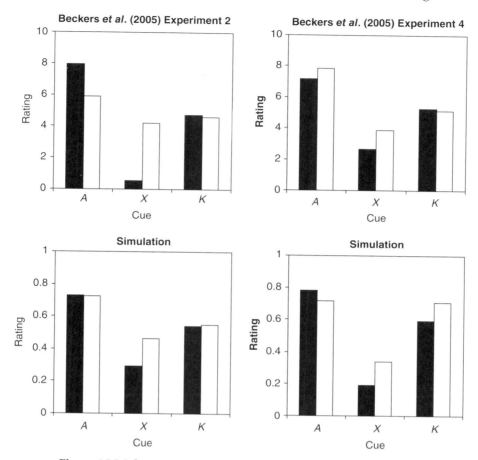

Figure 16.2 *Left upper panel*: Mean causal ratings for cue A, X and K in a blocking experiment following additivity pre-training. Data from Beckers *et al.* (2005) Experiment 2. *Right upper panel*: Same blocking experiment followed by additivity post-training; data from Beckers *et al.* (2005) Experiment 4. *Lower panels*: Corresponding simulations with the attentional–configural form of the SLG model.

As Mitchell *et al.* (2005) correctly observed, in the absence of pretraining, the C compound is already associated with the Outcome during post-training and, therefore, C–Outcome associations cannot be used to explain increased blocking. However, the attentional–configural form of the SLG model provides an attentional interpretation for the result. In terms of the model, during the AX+ phase of blocking, Outcome–X and C–X associations are formed. During the subsequent additivity post-training, Outcome–X and C–X associations predict X, but X is not there. In the additive case, the stronger Outcome extinguishes its Outcome–X association faster than the weaker non-additive Outcome does. During additivity post-training, Novelty′ increases because of the presentation of the novel G and H stimuli and absence of the older A and X stimuli. Thus, because the representation of X is weaker in the additive case, attention to X increases less in the additive than in the non-additive case. As shown in Chapter 2 (Equation [2.10a]), X predicts the Outcome in proportion to the product $X_X V_{X\text{-Out}}$. Therefore, the Rating is weaker (and blocking is stronger) in the additive than the non-additive case (see Figure 16.2, right lower panel).

Additivity training preceding backward blocking

In Beckers *et al.*'s (2005, Experiment 3) additivity pretraining study, one additive and one subadditive group were used. The additive group received $G+/H+/GH++/I+/Z-$ alternated trials (prior training), followed by $AX+/KL+/Z-$ alternated trials (compound training), and finally by alternated $A+/Z-$ trials (elemental training). Subadditive groups received $GH+$ instead of $GH++$ presentations and $I++$ instead of $I+$ presentations. Backward blocking was stronger in the additive than in the subadditive case.

Computer simulations (results not shown here) show that the attentional–configural form of the SLG model is also able to explain the effect of additivity information preceding backward blocking ($AX+$ followed by $A+$ training). According to the SLG model (see Chapter 8), backward blocking is the result of the Outcome–X associations activating the representation of X during AX presentations. During this period, Novelty′ in the experimental group first increases and then decreases because X, predicted both by the Outcome and by A, is absent. Therefore, the representation of X becomes associated with Novelty′ and attention to X slightly increases. In the control group, Novelty′ increases even more because X and A, both predicted by the Outcome, are absent. Therefore, the representation of X becomes more strongly associated with Novelty′ and attention to X is stronger in the control than in the experimental group. Correspondingly, during testing, responding to X is stronger in the control than in the experimental group, that is, backward blocking is present.

According to the model, during additivity pretraining, Novelty′ and attention to the Outcome are larger in the nonadditive than in the additive case. Therefore, the representation of the Outcome is larger during the $AX+$ phase of backward blocking, and the Outcome–X association is stronger in the nonadditive case than in the additive case. Because a stronger Outcome–X association results in a stronger representation of X during the $A+$ phase, attention to X increases more in the nonadditive case than in the additive case. In consequence, responding to X (proportional to the product $X_X V_{X\text{-Out}}$) is stronger, and blocking is weaker in the nonadditive case than in the additive case.

Maximality training preceding blocking

In Beckers et al.'s (2005, Experiment 1) maximality study, one maximal and one submaximal group were used. The maximal group received CX−/CX+/CX++ alternated trials (prior training), followed by alternated A++/Z− trials (elemental training), and finally by AX++/KL++/Z− alternated trials (compound training). Whereas CX denotes the context alone, Z, A, X, K and L denote neutral stimuli, and symbols −, + and ++ indicate no outcome, or the same outcome with different intensities. Responses to X and K (and L) were then compared. Submaximal groups received A+/Z− trials during elemental training, and AX+/KL+/Z− alternated trials during compound training. Although blocking was present in both groups, it was stronger in the submaximal than in the maximal case.

Computer simulations (results not shown here) demonstrate that the attentional-configural form of the SLG model also reproduces Beckers et al.'s (2005) results showing that blocking is stronger when the Outcome is submaximal. As mentioned in Chapters 2 and 3, the model incorporates an associative network controlled by a "constrained" competitive rule. According to this rule, CS-Outcome and CS-CS associations are limited to a maximum value, so that a CS of a given salience cannot fully predict the Outcome of maximal strength, and therefore block other CSs from becoming associated with that Outcome. Because in the submaximal case the Outcome is relatively weak, even with a limited CS-Outcome association, the CS *can* fully predict the Outcome strength and block the formation of other associations. Notice that in the maximal case, the model predicts that blocking will be present if the intensity of the blocking CS is increased.

Inferential reasoning

Some tasks involving relationships between temporally noncontiguous events (such as sensory preconditioning and transitivity) require the combination of multiple independent temporally contiguous pieces of information through inference, and can also be solved by the associative system (shown in Figures 16.3 and 2.1) in the attentional-configural form of the SLG model (Buhusi and Schmajuk; 1996).

Like the SLG model described in Chapter 2, the attentional-configural form of the model is also capable of generating inferences by chaining associations. Figure 16.3 shows a neural architecture that is able to combine CS-CS and CS-US associations stored in a recurrent network (Kohonen, 1977). The outputs of the system are the aggregate predictions of the CSs or the US (p_1, p_2, p_3 and p_{US}). Aggregate predictions p_i represent the expected magnitude of a CS or the US based upon all the CSs with a trace active at a certain time. As in the case

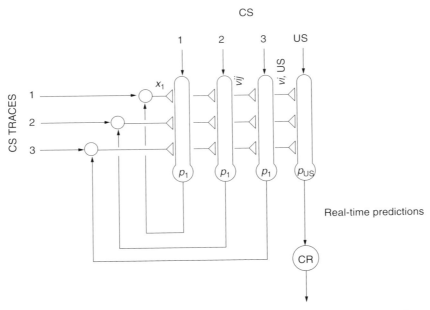

Figure 16.3 Inference generation. Associative system in the attentional–configural network (see Figure 2.1) that stores CS–CS and CS–US associations. $V_{i,j}$: CS_i–CS_j associations. $V_{i,US}$: CS_i–US associations. Aggregate predictions of CS_j, p_j, are fedback into the representation of CS_j. The real-time prediction of the US, p_{US}, controls the generation of the conditioned response (CR). Arrows: Fixed excitatory connections. Open triangles: Variable excitatory connections.

of the spatial cognitive map, predictions are inferred by reinjecting the predictions p_1 and p_2, into the traces of their corresponding CS_i. As in the case of spatial maps (see Schmajuk & Thieme, 1992), because activation spreads decrementally, the output of cell p_{US} decreases as the time interval (measured as the number of interposed CSs) beween CS_i and the US increases. Due to its recurrent property, the network is able to describe sensory preconditioning, second-order conditioning and transitivity. As mentioned in Chapter 2, whereas the activation of trace of CS_i by the stimulus is experienced as a perception, activation of the trace by p_i is experienced as an "image" of CS_i.

Sensory preconditioning

In sensory preconditioning (Brogden, 1939), two conditioned stimuli (CSs), CS_A and CS_B, are paired together in the absence of the US. In a second phase, CS_A is paired with the US. Finally, when CS_B is presented alone, it generates a conditioned response (CR). In second-order conditioning (Pavlov, 1927), a CS, CS_A, is paired with the US. In a second phase, CS_A and CS_B are paired together in the absence of the US. Finally, when CS_B is presented alone, it generates a CR (Figure 16.4, upper left panel).

Computer simulations, shown in Figure 16.4 (lower left panel), demonstrate that the attentional–configural network (Buhusi and Schmajuk, 1996) successfully describes sensory preconditioning, transitivity and symmetry. In terms of the network, sensory preconditioning is described as follows. During CS_A–CS_B trials $V_{A,B}$ association increases and during CS_B reinforced trials the $V_{B,US}$ association grows. When CS_A (never presented with the US before) is presented by itself on test trial, it activates the prediction of CS_B, B_B, which in turn activates prediction of the US.

Transitivity

In a transitivity task (Bunsey & Eichenbaum, 1996), when CS_A is presented CS_B is reinforced, and when CS_X is presented CS_Y is reinforced. In addition, when CS_B is presented CS_C is reinforced, and when CS_Y is presented CS_Z is reinforced. When CS_A is presented in a test trial, animals chose stimulus CS_C over CS_Z, but CS_Z over CS_C when CS_X is presented (Figure 16.4, upper right panel)

Computer simulations, shown in Figure 16.4 (lower right panel), demonstrate that the associative system attentional–configural network (Buhusi and Schmajuk, 1996) successfully describes transitivity and symmetry. The model describes transitivity as follows. During CS_A–CS_B trials, $V_{A,B}$ associations increase, during CS_B–CS_C trials, $V_{B,C}$ associations increase, and $V_{C,US}$ associations grow during CS_C reinforced trials. When CS_A (never presented with the US before) is presented by itself on test trial, it activates the prediction of CS_B, B_B, which in turn activates prediction of CS_C, B_C, which in turn activates the $V_{C,US}$ association, thereby generating a prediction of the US. Based on this prediction, C will be chosen over Z when A is presented. Symmetry (the ability to associate paired elements presented in the reverse of training order) is simply the result of the simultaneous increase in $V_{B,A}$ and $V_{A,B}$ associations.

Notice that the same basic mechanism used to describe sensory preconditioning and transitivity has been used, as indicated in Chapter 2, to describe latent learning and detour tasks in complex mazes in terms of $Place_i$–$Place_j$ associations (Schmajuk & Thieme, 1992), as well as problem solving (Schmajuk & Thieme, 1992). Furthermore, under the same assumptions made in Chapters 6 and 10 regarding the effect of hippocampal lesions (i.e. CS_i–CS_j associations are not formed in the cortex), the model describes the deleterious effects of these lesions on sensory preconditioning (Port & Patterson., 1984), transitivity (Bunsey & Eichenbaum, 1996), spatial navigation (Schmajuk, Thieme & Blair, 1993), and problem solving (Xu & Corkin, 2001).

412   Attentional, associative, configural and timing mechanisms

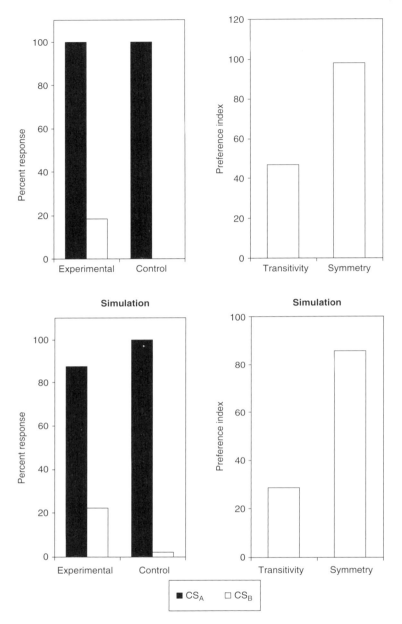

Figure 16.4 Sensory preconditioning and transitive inference. *Left upper panel*: Sensory Preconditioning. Data from Brogden (1939). Percentage of conditioned responses evoked by CS_A and CS_B after nonreinforced CS_A–CS_B trials followed by reinforced CS_A trials (experimental group), or reinforced CS_A trials (control group). *Right upper panel*: Transitivity and Symmetry. Data from Bunsey and Eichenbaum (1996). Preference index calculated as $(X - Y)/(X + Y)$ where X is the time in the transitive (or symmetry) choice and Y is the time in the alternate choice.

Summary

This chapter shows that the configural–attentional form of the SLG model is able to describe several results in causal learning, including additivity training before and after forward blocking, additivity training before backward blocking and maximality. In addition, computer simulations show that the model approximates (a) the facilitatory effect of subtractivity pre-training results on backward blocking (Mitchell *et al.*, 2005), and (b) higher-order retrospective revaluation (De Houwer & Beckers, 2002). Interestingly, whereas the propositional approach predicts no blocking following subadditive pre- and post-training, the SLG model can account for the reduced blocking observed in those cases (see Beckers *et al.*, 2005, pages 241 and 246). Furthermore, these finely graded results are also present in the model description of latent inhibition, in which weaker conditioned responding is observed (see Chapter 5).

Beckers *et al.* (2006, page 93) suggested that human reasoning need not be based on syllogistic logic (e.g. if submaximal and additive premises are true, then the rating of the blocked stimulus as cause for the outcome should be relatively small blocking). Instead, other forms of "controlled analytical processing might be involved." In line with this concept, the attentional–configural SLG model suggests that people "reason" that in some circumstances more attention should be paid to the blocked stimulus, whose image is activated through a CS_i–CS_j association, thereby decreasing blocking.

Caption for Fig. 16.4 (Cont.)

Left bottom panel: Sensory Preconditioning. Simulated peak CR amplitude evoked by CS_A and CS_B after 120 nonreinforced CS_A–CS_B trials followed by 120 reinforced CS_A trials (experimental group), or 120 context-alone trials followed by 120 reinforced CS_A trials (control group). *Right bottom panel*: Transitivity and Symmetry. Simulated reference indexes $(X - Y)/(X + Y)$ for transitivity and symmetry tests after 150 trials of alternated reinforced CS_A–CS_B and CS_X–CS_Y trials and nonreinforced CS_A–CS_Y and CS_X–CS_B trials, followed by 150 trials of alternated reinforced CS_B–CS_C and CS_Y–CS_Z trials and nonreinforced CS_Y–CS_C and CS_B–CS_Z trials. For the transitivity test, X was computed as the average of the peak CR amplitudes evoked by CS_C following CS_A presentations and by CS_Z following CS_X presentations, Y was computed as the average of the peak CR amplitudes evoked by CS_Z following CS_A presentations and by CS_C following CS_X presentations. For the symmetry test, X was computed as the average of the peak CR amplitudes evoked by CS_B following CS_C presentations, and by CS_Y following CS_Z presentations, Y was computed as the average of the peak CR amplitudes evoked by CS_Y following CS_C presentations and by CS_B following CS_Z presentations. Simulated results were obtained with the attentional–configural model (Buhusi and Schmajuk, 1996).

Furthermore, those CS_i-CS_j associations in the model's associative system also participate in other cognitive functions such as sensory preconditioning and transitivity, as well as spatial navigation, problem solving (e.g. Tower of Hanoi problem, see Schmajuk & Thieme, 1992) and creativity (see Chapter 7).

Part V CONCLUSION: MECHANISMS OF CLASSICAL CONDITIONING

17
Conclusion: mechanisms of classical conditioning

This book analyzes attentional, associative, configural and timing mechanisms that are embedded in different computational, neural network theories of classical conditioning.

Part I serves as an introduction to the problem and possible solutions. Chapter 1 briefly introduces a number of well-known conditioning paradigms and describes both theories and computational models that have been proposed to describe and explain them.

Part II concentrates on attentional and associative mechanisms. Chapters 2, 3, 5, 7, 8 and 9 describe an attentional–associative model and apply it to the description of excitatory and inhibitory conditioning, compound conditioning, latent inhibition, overshadowing, blocking, backward blocking, recovery from overshadowing, recovery from blocking, recovery from backward blocking, extinction and creative processes. Chapters 4, 6, 7 and 10 describe the neurobiological mechanisms involved in fear conditioning, latent inhibition, creative processes and extinction.

Part III is devoted to configural mechanisms. Chapters 11 and 12 describe a configural model and apply it to the description of occasion setting. The model provides a very precise description of under what conditions a CS functions as a simple CS or an occasion setter, and how these different functions are mechanistically achieved. In the model, direct and indirect CS–US associations compete, through the same rule that governs blocking, overshadowing and conditioned inhibition, to decide the role of the CS as a simple CS or as an occasion setter. Chapter 13 analyzes the neurobiological bases of occasion setting.

Part IV deals with different combinations of attentional, associative, configural and timing mechanisms. Chapters 14, 15 and 16 present two "mixed" neural networks and apply them to the description of timing in simple conditioning and occasion setting, extinction cues, causal learning and inferential reasoning.

In addition, in some cases we describe how different brain areas implement the mechanisms at work in the neural network models, including how brain activity might code some of the variables in the models, how those variables might be computed in a biologically plausible fashion, and how brain lesions and drug administration affect the computation of those variables and affect behavior.

Although in most cases the models are able to describe existing data, in some cases they "implicitly" predict those results, and in some cases they are capable of explicitly predicting the experimental results. Examples of predictions include Huff *et al.*'s (2007]) report on the effect of a time interval between acquisition and extinction on renewal, and Pothuisen *et al.*'s (2006) report on the preservation of latent inhibition in a conditioned taste-aversion paradigm following lesions of the shell of the accumbens.

Mechanisms needed to describe classical conditioning

The book seems to support the notion that more than one mechanism is needed to account for the reported results on classical conditioning. It is apparent that the phenomenon cannot be reduced to one simple formula, like that of the kinetic theory of gasses. Conditioning can be described by a rather complex combination of different mechanisms. This combination might include not only the attentional, associative, configural and timing mechanisms analyzed in this book, but also additional attentional processes that reflect the predictive quality of a CS, as suggested by Mackintosh (1975; see also Schmajuk & Moore, 1989) and Grossberg (1975; see also Schmajuk & DiCarlo, 1991b). Such integration of multiple attentional systems is present in the models offered by Buhusi and Schmajuk (1996) and LePelley (2004).

This view is supported by the fact that, in the relatively short history of behavioral modeling, simple and elegant models such as Rescorla–Wagner's (1972) delta rule, Pearce and Hall's (1980) attentional rule, and Miller and Schachtman's (1985) first comparator hypothesis failed to explain important aspects of conditioning. Instead, the field has moved to increasingly complex models. It is clear that the complexity of these models puts them beyond the ability of our intuitive thinking, and makes computer simulations irreplaceable.

The fact that multiple mechanisms are needed, and that those mechanisms are implemented by neural networks built by assembling "neural elements" of relatively weak computational power, implies that the number of parameters in the model will be relatively large. For instance, according to the SLG model, the CR strength is determined by the intensity of the internal representation of the CS activating its associative connection with the US. In turn, this internal representation is computed as the activation of an "attentional" synapse by the combination of (a) the short-term trace of the CS; and (b) that CS prediction by other CSs (attentional mechanism), while the X_{CS}–US association is controlled by an error term of the US (associative mechanism). In addition, these two relatively simple mechanisms need to be implemented by neural mechanisms that carry out sub-functions, such as maintaining a short-term memory of the CS, conveying the value of the prediction of the CS from a CS–CS memory storage to the input of the system, averaging the value of the actual US strength and other information processing actions.

Therefore, even if parsimony indicates that a small number of model parameters should be preferred, it is likely that this number will not be insignificant when a neural network approach is used. Even a simple function such as a hyperbola, which can be mathematically defined with one parameter, needs two or three parameters to be implemented by a neural network. However, the number of parameters should not be a cause of concern given the complexity of the behavior at hand, the moment-to-moment detail with which the model describes the behavior and the intricacy of the biological organ that regulates the behavior.

Combining mechanisms

The book suggests that the different mechanisms required for a full description of the data can be analyzed separately and then combined into an integrated model. The method is reminiscent of the Wright brothers' approach to airplane design, which consisted of the independent development and testing of the individual components of the plane before assembling them together into a flying machine (Padfield & Lawrence, 2003). Interestingly, that approach was based on a study of bird flight by George Cayley (1773–1857), who realized that the lift function and the thrust function of bird wings were separate and distinct, and could be imitated by separate systems on a fixed-wing craft. Imitation of each separate, relatively simple function permits the explanation of how each function is achieved. It is important to point out that some additions are sometimes necessary (e.g. a different type of attentional mechanism) when given mechanisms (e.g. configuration) are integrated into a general model (see Chapter 15; Buhusi & Schmajuk, 1996).

Emergent properties

Even if the study of different separate, relatively simple mechanisms permits the explanation of how each of those mechanisms achieves its function, some properties become apparent only when the different parts are integrated into one single behaving system. As indicated in Chapter 2, the SLG model was initially developed to address a few properties of classical conditioning (sensory preconditioning, latent inhibition, competition between CSs). However, it proved capable of correctly describing many experimental results for which it was not specifically designed: those results are "emergent properties" of the model. A few examples of emergent properties include the facilitation of fear acquisition by CS preexposure (Chapter 5) and spontaneous recovery (Chapter 9). Similarly, the SD model, initially developed to address negative and positive patterning, was later shown to be capable of describing response form in occasion setting (Chapter 12).

Evaluation of the models: imitating and explaining the experimental data

In order to permit the comparison of experimental data and simulated results, many figures in the book show data and simulations next to each other (e.g. Figure 8.1 for recovery from overshadowing). In many, but not all cases, it is difficult to distinguish the experimental data from the computer simulated results. In those cases, as suggested by Church (1990), the model has passed a Turing (1950) test by correctly "imitating" the behavior under study.

Notice that imitation refers to both the model's predictions *and* postdictions. We refer to "explicit correct predictions" as those cases in which the model has correctly described the experimental results before the experiment was carried out. "Implicit correct predictions" are those cases of experimental results, reported after the model was first offered, that the unmodified model correctly describes. Implicit predictions are still predictions, because the model, even if capable of correctly describing the data, had not been explicitly applied to the experiment of interest. Finally, "correct postdictions" are those cases of experimental results, reported before the model was offered, that the model correctly describes. In this case, even if the data was not used to develop the model, it was already available at the time when the model was designed.

In order to quantify the quality of a model's imitation, we have used different quantitative methods. These methods include (a) correlations (e.g. Schmajuk *et al.*, 1998), (b) χ^2 (Schmajuk *et al.*, 2001), and (c) analysis of variance using the actual variance of the experimental subjects (Schmajuk & Larrauri, 2006). Of these alternative methods, we prefer to use correlations because, although they disregard the importance of the variance in the data, they indicate when to reject the null hypothesis (that simulated values and experimental data are not correlated). Instead, χ^2 indicates when to accept the null hypothesis (that simulated values and experimental data are equivalent). Finally, the analysis of variance used in Schmajuk and Larrauri (2006) is rather convoluted and requires knowledge of the values of the variance of the data, which is not always reported in the experimental studies.

In addition to imitating the experimental results (e.g. Figure 9.6 for an approximate "imitation" of spontaneous recovery), models can also explain those results, i.e. indicate how those results are reached by showing why and how they occurred. In the book, explanations are provided by the figures illustrating the value of the variables in the model during the course of a given experiment (e.g. Figure 9.1 for an explanation of spontaneous recovery). These figures explain *why* a behavior occurs by showing its causes. For example, the cause of spontaneous recovery is the increased attention to the still excitatory CS before attention to the inhibitory context increases (see Figure 9.1). These figures also show *how* a behavior occurs by describing the series of causal processes that lead to the occurrence of that behavior. For example, spontaneous recovery is present when (a) Novelty' and attention to the CS and the CX decrease during extinction, and (b) Novelty' and attention to the CS and CX increase again when the CS is presented after being absent for a period of time (see Figure 9.1).

Importantly, having a causal explanation for a given behavior might facilitate its control. This might be particularly important in certain conditions (e.g. post-traumatic stress disorder), in which "spontaneous recovery" is equivalent to a relapse that should be avoided.

The neurophysiology of behavior

It has been suggested (Hebb, 1949; Gray; 1975; McNaughton, 2004) that in order to relate brain and behavior, one should first develop a "conceptual nervous system" to handle behavioral data, and second, find out whether brain structures and neural elements carry out the operations described by the conceptual system. Following that suggestion, this book (a) describes how different neural networks describe classical conditioning; (b) identifies brain structures, brain activity and neurotransmitters (among many structures, activities and neurotransmitters) that seemed to carry out the information processing described by the network when performing a given behavior; and (c) based on this neural network–brain mapping, describes the effect of brain lesions, drug administration, as well as neural activity and neurotransmitter release in several brain areas during different classical conditioning paradigms. When model variables used to explain the behavioral results can be related to brain neural activity (e.g. Chapters 4, 6 and 13), we can say that the model provides a neurophysiological explanation for the behavioral results.

An interesting illustration of how the models address the relationship between brain and behavior is the case of the neurophysiology of latent inhibition (Chapter 6). In this instance, our approach was able to clarify the apparently conflicting results regarding the effect of hippocampal lesions on latent inhibition. Chapter 6 suggests that the notion that latent inhibition (a label that encompasses many alternative experimental protocols) is, or is not affected by hippocampal lesions, should be replaced by a detailed analysis of how Novelty', the key variable that controls latent inhibition in the SLG model, varies as a function of the experimental procedure and brain manipulation.

Indeed, the fact that the variables introduced to describe behavior found later a correlate in the brain speaks in favor to this "reverse engineering" approach. The success of the approach suggests that neural networks can provide the link between the different levels of description proposed by Marr (1982): computational (the computation of attention, association and configuration), representation (representations of inputs and outputs), algorithmic (what are the algorithms underlying the different mechanisms?) and biological (the neural implementation of those algorithms).

Evaluation of the data

In addition to the question of the evaluation of the model addressed above, the issue of the robustness of the data was an important concern during our modeling efforts. For example, in Chapter 12 we refer to the contradictory results for the transfer to a trained and extinguished CS in different contexts following serial feature-positive discrimination (see Figure 12.24). Similarly, in Chapter 8 we mentioned contradictory results (Figures 8.5 and 8.7) of the effect the extinction of the blocker on the responding to the blocked stimulus. Because the model explains away these contradictions in terms of differences in experimental parameters or procedures used by different scientists, parametric studies are needed to test those explanations. In addition to testing the model, these parametric studies would serve to determine the range in which some reported results are valid and can be replicated.

Future challenges

In most cases, the models analyzed in this book proved to be proficient at describing a large part of the existing data. We also referred to data that the models do not seem able to describe, they might describe if improved, or that they have not fully addressed so far. The paradigms that need further study include: the effect of massed vs. spaced trials (Chapter 3), the interaction between noncontingent training and extinction (Chapter 3), conditioned inhibition as a "slave process" (Chapter 3), learned irrelevance (Chapter 5), the conflicting results for the overshadowing–LI interactions (Chapter 6), potentiation (Chapter 6), the participation of configural cues in generalization and discrimination (Chapter 11) and causal learning (Chapter 16). Other experimental discoveries regarding causal learning (e.g. Blaisdell *et al.*, 2006) present challenges to be addressed.

Future work should also focus on the fact that the design of the models presented in this book neither takes into account any specific preparation (e.g. rabbit's eyeblink conditioning, rat's conditional emotional response, taste aversion, human ratings); nor the different experimental values (e.g. duration of the CS, salience of the CS, the duration and strength of the US, context salience, intertrial interval, trials to criterion) used in the experiments run with those preparations. Therefore, our models are "generic" models of classical conditioning. We expect that future models will (a) adopt parameters appropriate for specific preparations, and (b) use simulation values (e.g. stimulus duration and salience, trials to criterion) that are scaled to those used in the corresponding experiments. The resulting models will provide more accurate descriptions of the data.

Advantages of neural network theorizing in psychology

In the Conclusion of *Animal Learning and Cognition: A Neural Network Approach*, I analyzed the advantages of neural network theorizing in psychology. As I indicated there, the approach (a) integrates and condenses large amounts of data, (b) provides accurate real-time descriptions of behavior and neural activity, (c) permits the comparison of computer simulations of real-time paradigms with experimental data, (d) permits continuous variables to precisely describe complex designs, (e) describes the implementation of simple computational mechanisms in terms of complex networks built with simple neural elements, (f) describes the emergence of system properties that go beyond the individual mechanisms in which they are based, and (g) permits the integration of behavioral and neurophysiological descriptions. All those features seem well illustrated in the different chapters of this book.

Summary

In sum, in this book we have shown that (a) *multiple mechanisms* are needed to describe classical conditioning; (b) that *computer simulations* are needed to describe how those mechanisms work separately and together;

(c) *emergent properties* appear when different mechanisms are integrated into one single system; (d) once behavior is captured by the model, observation of the *model variables* serve to identify *brain structures, brain activity* and *neurotransmitters* that might carry out the information processing that takes place when performing a given behavior; (e) the brain-mapped models describe the effect of *brain lesions, drug administration and neural activity*; (f) the models can both *imitate and explain the experimental data*; (g) it is suggested that *parametric experimental studies are needed* to determine the range in which the reported results can be replicated; and (h) future work should consider using model parameters appropriate for specific preparations, as well as simulation values (e.g. stimulus duration and salience, trials to criterion) that reflect the actual values used in the corresponding experiments.

References

Amabile, T. M. (1983). The social psychology of creativity: a componential conceptualization. *Journal of Personality and Social Psychology*, **45**, 357–376.

Amsel, A. (1958). The role of frustrative nonreward in noncontinuous reward situations. *Psychological Bulletin*, **55**, 102–119.

Andreasen, N. J. & Powers, P. S. (1974). Overinclusive thinking in mania and schizophrenia. *British Journal of Psychiatry*, **125**, 452–456.

Andreasen, N. J. & Powers, P. S. (1975). Creativity and psychosis. An examination of conceptual style. *Archives of General Psychiatry*, **32**, 70–73.

Ashby, F. G., Isen, A. M. & Turken, A. U. (1999). A neuropsychological theory of positive affect and its influence on cognition. *Psychological Review*, **106**, 529–550.

Atre-Vaidya, N., Taylor, M. A., Seidenberg, M., Reed, R., Perrine, A. & Glick-Oberwise, F. (1998). Cognitive deficits, psychopathology, and psychosocial functioning in bipolar mood disorder. *Neuropsychiatry, Neuropsychology, & Behavioral Neurology*, **11**, 120–126.

Aylward, E., Walker, E. & Bettes, B. (1984). Intelligence in schizophrenia: meta-analysis of the research. *Schizophrenia Bulletin*, **10**, 430–459.

Ayres, J. J. B., Albert, M. & Bombace, J. C. (1987). Extending conditioned stimuli before versus after unconditioned stimuli: implication for real-time models of conditioning. *Journal of Experimental Psychology: Animal Behavior Processes*, **13**, 168–181.

Ayres, J. J. B., Philbin, D., Cassidy, S. & Belling, L. (1992). Some parameters of latent inhibition. *Learning and Motivation*, **23**, 269–287.

Baker, A. G. (1974). Conditioned inhibition is not the symmetrical opposite of conditioned excitation: a test of the Rescorla-Wagner model. *Learning and Motivation*, **5**, 396–379.

Baker, A. G. & Mercier, P. (1982). Extinction of the context and latent inhibition. *Learning and Motivation*, **13**, 391–416.

Baker, A. G., Haskins, C. E. & Hall, G. (1990). Stimulus generalization decrement in latent inhibition to a compound following exposure to the elements of the compound. *Animal Learning & Behavior*, **18**, 162–170.

Baker, A.G., Mercier, P., Gabel, J. & Baker, P.A. (1981). Contextual conditioning and the US preexposure effect in conditioned fear. *Journal of Experimental Psychology: Animal Behavior Processes*, **7**, 109–128.

Barela, P. (1999). Theoretical mechanisms underlying the trial-spacing effect in Pavlovian fear conditioning. *Journal of Experimental Psychology: Animal Behavior Processes*, **25**, 177–195.

Barnet, R.C., Grahame, N.J. & Miller, R.R. (1993). Temporal encoding as a determinant of blocking. *Journal of Experimental Psychology: Animal Behavior Processes*, **19**, 327–341.

Baruch I., Hemsley D. & Gray J.A. (1988). Differential performance of acute and chronic schizophrenics in a latent inhibition task. *Journal of Nervous and Mental Diseases*, **176**, 598–606.

Beckers, T., De Houwer, J., Pineno, O. & Miller, R.R. (2005). Outcome additivity and outcome maximality influence cue competition in human causal learning. *Journal of Experimental Psychology: Learning, Memory, and Cognition*, **31**, 238–249.

Beckers, T., Miller, R.R., De Houwer, J. & Urushihara, K. (2006). Reasoning rats: forward blocking in Pavlovian animal conditioning is sensitive to constraints of causal inference. *Journal of Experimental Psychology: General*, **135**, 92–102.

Bennett, C.H., Maldonado, A. & Mackintosh, N.J. (1995). Learned irrelevance is not the sum of exposure to CS and US. *The Quarterly Journal of Experimental Psychology B: Comparative and Physiological Psychology*, **48B**, 117–128.

Berger, T.W. & Orr, W.B. (1983). Hippocampectomy selectively disrupts discrimination reversal conditioning of the rabbit nictitating membrane response. *Behavioral Brain Research*, **8**, 49–68.

Berger, T.W. & Thompson, R.F. (1978a). Neuronal plasticity in the limbic system during classical conditioning of the rabbit nictitating membrane response. I. The hippocampus. *Brain Research*, **145**, 323–346.

Berger, T.W. & Thompson, R.F. (1978b). Neuronal plasticity in the limbic system during classical conditioning of the rabbit nictitating membrane response. II. Septum and mammillary bodies. *Brain Research*, **156**, 293–314.

Berger, T.W. & Thompson, R.F. (1982). Hippocampal cellular plasticity during extinction of classically conditioned nictitating membrane behavior. *Behavioral Brain Research*, **4**, 63–76.

Berger, T.W., Clark, G.A. & Thompson, R.F. (1980). Learning dependent neuronal responses recorded from limbic system brain structures during classical conditioning. *Physiological Psychology*, **8**, 155–167.

Berger, T.W., Rinaldi, P.C., Weisz, D.J. & Thompson, R.F. (1983). Single unit analysis of different hippocampal cell types during classical conditioning of rabbit nictitating membrane response. *Journal of Neurophysiology*, **50**, 1197–1219.

Berthier, N.E. & Moore, J.W. (1990). Activity of deep cerebellar nuclear cells during classical conditioning of nictitating membrane extension in rabbits. *Experimental Brain Research*, **83**, 44–54.

Best, M.R., Gemberling, G.A. & Johnson, P.E. (1979). Disrupting the conditioned stimulus preexposure effect in flavor aversion learning: effects of interoceptive distractor manipulations. *Journal of Experimental Psychology: Animal Behavior Processes*, **5**, 321-334.

Blaisdell, A. P., Denniston, J. C. & Miller, R. R. (1998). Temporal encoding as a determinant of overshadowing. *Journal of Experimental Psychology: Animal Behavior Processes*, **24**, 72-83.

Blaisdell, A. P., Gunther, L. & Miller, R. (1999). Recovery from blocking achieved by extinguishing the blocking CS. *Animal Learning & Behavior*, **27**, 63-76.

Blaisdell, A., Bristol, A., Gunther, L. & Miller, R. (1998). Overshadowing and latent inhibition counteract each other: further support for the comparator hypothesis. *Journal of Experimental Psychology: Animal Behavior Processes*, **24**, 335-251.

Blaisdell, A. P., Sawa, K., Leising, K. J. & Waldmann, M. R. (2006). Causal reasoning in rats. *Science*, **311**, 1020-1022.

Blough, D. S. (1975). Steady state data and a quatitative model of operant generalization and discrimination. *Journal of Experimental Psychology: Animal Bheavior Processes*, **104**, 3-21.

Boden, M. A. (1999). Computer models of creativity. In *Handbook of Creativity*, ed. R. J. Stenberg. Cambridge: Cambridge University Press.

Bonardi, C. & Hall, G. (1994a). Occasion-setting training renders stimuli more similar: acquired equivalence between the targets of feature positive discriminations. *Quarterly Journal of Experimental Psychology*, **47B**, 63-82.

Bonardi, C. & Hall, G. (1994b). A search for blocking of occasion setting using a nonexplicit training procedure. *Learning and Motivation*, **25**, 105-125.

Bonardi, C., Hall, G. & Ong, S. Y. (2005). Analysis of the learned irrelevance effect in appetitive Pavlovian conditioning. *The Quarterly Journal of Experimental Psychology B: Comparative and Physiological Psychology*, **58B**, 141-162.

Bottjer, S. W. (1982). Conditioned approach and withdrawal behavior in pigeons: effects of a novel extraneous stimulus during acquisition and extinction. *Learning and Motivation*, **13**, 44-67.

Boughner, R. L. & Papini, M. R. (2006). Appetitive latent inhibition in rats: preexposure performance does not predict conditioned performance. *Behavioural Processes*, **72**, 42-51.

Bouton, M. E. (1984). Differential control by context in the inflation and reinstatement paradigms. *Journal of Experimental Psychology: Animal Behavior Processes*, **10**, 56-74.

Bouton, M. E. (1986). Slow reacquisition following the extinction of conditioned suppression. *Learning and Motivation*, **17**, 1-15.

Bouton, M. E. (1993). Context, time, and memory retrieval in the interference paradigms of Pavlovian learning. *Psychological Bulletin*, **114**, 80-99.

Bouton. M. E. (1994). Conditioning, remembering, and forgetting. *Journal of Experimental Psychology: Animal Behavior Processes*, **20**, 219-231.

Bouton, M. E. (2002). Context, ambiguity, and unlearning: sources of relapse after behavioral extinction. *Biological Psychiatry*, **52**, 976-986.

Bouton, M. E. & Bolles, R. C. (1979a). Role of conditioned contextual stimuli in reinstatement of extinguished fear. *Journal of Experimental Psychology: Animal Behavior Processes*, **5**, 368-378.

Bouton, M. E. & Bolles, R. C. (1979b). Contextual control of the extinction of conditioned fear. *Learning and Motivation*, **10**, 445–466.

Bouton, M. E. & King, D. A. (1983). Contextual control of the extinction of conditioned fear: tests for the associative value of the context. *Journal of Experimental Psychology: Animal Behavior Processes*, **9**, 248–265.

Bouton, M. E. & King, D. A. (1986). Effect of context on performance to conditioned stimuli with mixed histories of reinforcement and nonreinforcement. *Journal of Experimental Psychology: Animal Behavior Processes*, **12**, 4–15.

Bouton, M. E. & Nelson, J. B. (1994). Context-specificity of target versus feature inhibition in a feature-negative discrimination. *Journal of Experimental Psychology: Animal Behavior Processes*, **20**, 51–65.

Bouton, M. E. & Swartzentruber, D. (1986). Analysis of the associative and occasion-setting properties of contexts participating in a Pavlovian discrimination. *Journal of Experimental Psychology: Animal Behavioral Processes*, **12**, 333–350.

Bouton, M. E. & Swartzentruber, D. (1989). Slow reacquisition following extinction: context, encoding, and retrieval mechanisms. *Journal of Experimental Psychology: Animal Behavior Processes*, **15**, 43–53.

Bouton, M. E., Woods, A. M. & Pineño, O. (2004). Occasional reinforced trials during extinction can slow the rate of rapid reacquisition. *Learning and Motivation*, **35**, 371–390.

Bouton, M. E., Rosengard, C., Achenbach, G. G., Peck, C. A. et al. (1993). Effects of contextual conditioning and unconditioned stimulus presentation on performance in appetitive conditioning. *The Quarterly Journal of Experimental Psychology B: Comparative and Physiological Psychology*, **46B**, 63–95.

Bower, G. H. & Hilgard, E. R. (1981). *Theories of Learning*. Englewood Cliffs, NJ: Prentice-Hall.

Brandon, S. E. & Wagner, A. R. (1998). Occasion setting: influences of conditioned emotional responses and configural cues. In *Occasion Setting: Associative Learning and Cognition in Animals*. ed. N. A. Schmajuk & P. C. Holland. Washington, DC: American Psychological Association, pp. 343–382; p. xxi.

Brandon, S. E. & Wagner, A. R. (1991). Modulation of a discrete Pavlovian conditioned reflex by a putative emotive Pavlovian conditioned stimulus. *Journal of Experimental Psychology: Animal Behavior Processes*, **17**, 299–311.

Brandon, S. E., Vogel, E. H. & Wagner, A. R. (2000). A componential view of configural cues in generalization and discrimination in Pavlovian conditioning. *Behavioural Brain Research*, **110**, 67–72.

Brandon, S. E., Vogel, E. H. & Wagner, A. R. (2002). Computational theories of classical conditioning. In *A Neuroscientist's Guide to Classical Conditioning*, ed. J. W. Moore. New York: Springer, pp. 232–310.

Breese, C. R., Hampson, R. E. & Deadwyler, S. A. (1989). Hippocampal place cells: stereotypy and plasticity. *Journal of Neuroscience*, **9**, 1097–1111.

Brodal, P. (1992). *The central nervous system. Structure and function*. Oxford: Oxford University Press.

Brogden, W. J. (1939). Sensory pre conditioning. *Journal of Experimental Psychology*, **25**, 323–332.

Brooks D. C. & Bouton M. E. (1993). A retrieval cue for extinction attenuates spontaneous recovery. *Journal of Experimental Psychology: Animal Behavior Processes*, **19**, 77–89.

Brooks, D. C. & Bouton, M. E. (1994). A retrieval cue for extinction attenuates response recovery (renewal) caused by a return to the conditioning context. *Journal of Experimental Psychology: Animal Behavior Processes*, **20**, 366–379.

Brooks, D. C. & Bowker, J. L. (2001). Further evidence that conditioned inhibition is not the mechanism of an extinction cue's effect: a reinforced cue prevents spontaneous recovery. *Animal Learning & Behavior*, **29**, 381–388.

Buchanan, S. L. & Powell, D. A. (1982). Cingulate cortex: its role in Pavlovian conditioning. *Journal of Comparative and Physiological Psychology*, **96**, 755–774.

Buchel, C., Morris, J., Dolan, R. J. & Friston, K. J. (1998). Brain systems mediating aversive conditioning: an event-related fMRI study. *Neuron*, **20**, 947–957.

Buhusi, C. V. & Schmajuk, N. A. (1996). Attention, configuration, and hippocampal function. *Hippocampus*, **6**, 621–642.

Buhusi, C. V. & Schmajuk, N. A. (1999). Timing in simple conditioning and occasion setting: a neural network approach. *Behavioral Processes*, **45**, 33–57.

Buhusi, C. V., Schmajuk N. A. & Dunn, L. (1999). Haloperidol administration at preexposure may impair latent inhibition. *Abstracts of the Society for Neuroscience*, Miami Beach, FL.

Buhusi, C. V., Gray, J. A. & Schmajuk, N. A. (1998). The perplexing effects of hippocampal lesions on latent inhibition: a neural network solution. *Behavioral Neuroscience*, **112**, 316–351.

Bunsey, M. & Eichenbaum, H. (1996). Conservation of hippocampal memory function in rats and humans. *Nature*, **379**, 255–257.

Burch, G. S. J., Hemsley, D. R., Pavelis, C. & Corr, P. J. (2006). Personality, creativity and latent inhibition. *European Journal of Personality*, **20**, 107–122.

Burkhardt, P. E., & Ayres, J. J. (1978). CS and US duration effects in one-trial simultaneous fear conditioning as assessed by conditioned suppression of licking in rats. *Animal Learning & Behavior*, **6**, 225–230.

Burstein, K. R. (1967). Spontaneous recovery: a (Hullian) non-inhibition interpretation. *Psychonomic Science*, **7**, 389–390.

Bush, R. R. & Mosteller, F. (1955). *Stochastic Models for Learning*. New York: Wiley.

Cain, C. K., Blouin, A. M. & Barad, M. (2003). Temporally massed CS presentations generate more fear extinction than spaced presentations. *Journal of Experimental Psychology: Animal Behavior Processes*, **29**, 323–333.

Campbell, D. T. (1960). Blind variation and selective retention in creative thought as in other knowledge processes. *Psychological Review*, **67**, 380–400.

Campolattaro, M. M. & Freeman, J. H. (2006a). Perirhinal cortex lesions impair simultaneous but not serial feature-positive discrimination learning. *Behavioral Neuroscience*, **120**, 970–975.

Campolattaro, M. M., & Freeman, J. H. (2006b). Perirhinal cortex lesions impair feature-negative discrimination. *Neurobiology of Learning & Memory*, **86**, 205–213.

Capaldi, E.J. (1967). A sequential theory of instrumental training. In *The Psychology of Learning and Motivation*, ed. K.W. Spence & J.T. Spence. New York: Academic, pp. 67–156.

Capaldi, E.J. (1971). Memory and learning: A sequential viewpoint. In *Animal Memory*, eds. W.K. Honig & P.H.R. James. New York: Academic, pp. 111–154.

Carelli, R.M. & Deadwyler, S.A. (1994). A comparison of nucleus accumbens neuronal firing patterns during cocaine self-administration and water reinforcement in rats. *Journal of Neuroscience*, **14**, 7735–7746.

Carr, A.F. (1974). Latent inhibition and overshadowing in conditioned emotional response conditioning with rats. *Journal of Comparative and Physiological Psychology*, **86**, 718–723.

Carson, S.H., Peterson, J.B. & Higgins, D.M. (2003). Decreased latent inhibition is associated with increased creative achievement in high-functioning individuals. *Journal of Personality and Social Psychology*, **85**, 499–506.

Cassaday, H.J., Mitchell, S.N., Williams, J.H. & Gray, J.A. (1993). 5,7-Dihydroxytryptamine lesions in the fornix-fimbria attenuate latent inhibition. Behavioral *Neural Biology*, **59**, 194–207.

Chan, K-H, Jarrard, L.E. & Davidson, T.L. (2003). The effects of selective ibotenate lesions of the hippocamapus on conditioned inhibition and extinction. *Cognitive Affective Behavioral Neuroscience*, **3**, 111–119.

Channell, S. & Hall, G. (1981). Facilitation and retardation of discrimination learning after exposure to the stimuli. *Journal of Experimental Psychology: Animal Behavior Processes*, **7**, 437–446.

Chapman, G. (1991). Trial order affects cue interaction in contingency judgement. *Journal of Experimental Psychology: Learning, Memory & Cognition*, **17**, 837–854.

Chorazyna, H. (1962). Some properties of conditioned inhibition. *Acta Biologiae Experimentalis*, **22**, 5-13.

Church, R.M. (1990). A Turing test for computational and associative theories of learning. *Current Directions in Psychological Sciences*, **10**, 132–136.

Church, R.M. & Broadbent, H.A. (1990). A connectionist model of timing. In *Quantitative Models of Behavior: Neural Networks and Conditioning*, ed. M.L. Commons, S. Grossberg & J.E.R. Staddon. Hillsdale, NJ: Lawrence Erlbaum Associates, pp. 225–240.

Colwill, R.M. & Rescorla, R.A. (1990). Evidence for the hierarchical structure of instrumental learning. *Animal Learning and Behavior*, **18**, 71–82.

Corcoran, K.A. & Maren, S. (2004). Factors regulating the effects of hippocampal inactivation on renewal of conditional fear after extinction. *Learning & Memory*, **11**, 598–603.

Coutureau, E., Galani, R., Gosselin, O., Majchrzak, M. & Di Scala, G. (1999). Entorhinal but not hippocampal or subicular lesions disrupt latent inhibition in rats. *Neurobiology of Learning and Memory* **72**, 143–157.

Crowell, C.R. & Anderson, D.C. (1972). Variations in intensity, interstimulus interval, and interval between preconditioning CS exposures and conditioning with rats. *Journal of Comparative and Physiological Psychology*, **79**, 291–298.

Csernansky, J. G. & Bardgett, M. E. (1998). Limbic cortical neuronal damage and the pathophysiology of schizophrenia. *Schizophrenia Bulletin*, **24**, 231–248.

Cunningham, C. L. (1979). Alcohol as a cue for extinction: state dependency produced by conditioned inhibition. *Animal Learning and Behavior*, **7**, 45–52.

Cunningham, C. L. (1981). Association between the elements of a bivalent compound stimulus. *Journal of Experimental Psychology: Animal Behavior Processes*, **7**, 425–436.

D'Esposito, M., Postle, B. R. & Rypma, B. (2000). Prefontal cortical contributions to working memory: evidence from event-related fMRI studies. *Experimental Brain Research*, **133**, 3–11.

Daly, H. B. & Daly, J. T. (1982). A mathematical model of reward and aversive nonreward: its application in over 30 appetitive learning situations. *Journal of Experimental Psychology: General*, **111**, 441–480.

Daum, I., Channon, S., Polkey, C. E. & Gray, J. A. (1991). Classical conditioning after temporal lobe lesions in man: impairment in conditional discrimination. *Behavioral Neuroscience*, **105**, 396–408.

Davidson, T. L. (1993). The nature and function of interoceptive signals to feed: toward an integration of physiological and learning perspectives. *Psychological Review*, **100**, 640–657.

Davidson, T. L. & Jarrard, L. E. (1989). Retention of concurrent conditional discriminations in rats with ibotenate lesions of hippocampus. *Psychobiology*, **17**, 49–60.

Davidson, T. L. & Jarrard, L. E. (1993). A role for the hippocampus in the utilization of hunger signals. *Behavioral and Neural Biology*, **59**, 167–171.

Davidson, T. L. & Rescorla, R. A. (1986). Transfer of facilitation in the rat. *Animal Learning and Behavior*, **14**, 380–386.

Davidson, T. L., McKernan, M. G. & Jarrard, L. E. (1993). Hippocampal lesions do not impair negative patterning: a challenge to configural association theory. *Behavioral Neuroscience*, **107**, 227–234.

De Houwer, J. & Beckers, T. (2002). Higher-order retrospective revaluation in human causal learning. *The Quarterly Journal of Experimental Psychology*, **55B**, 137–151.

De la Casa, L. & Lubow, R. (2000). Super-latent inhibition with delayed conditioned taste aversion testing. *Animal Learning & Behavior*, **28**, 389–399.

De la Casa, L. & Lubow, R. (2002). An empirical analysis of the super-latent inhibition effect. *Animal Learning & Behavior*, **30**, 112–120.

De la Casa, L. G. & Lubow, R. E. (1995). Latent inhibition in conditioned taste aversion: the roles of stimulus frequency and duration, and amount of fluid ingested during preexposure. *Neurobiology of Learning and Memory*, **64**, 125–132.

Delamater, A. R. (1997). Selective reinstatement of stimulus-outcome associations. *Animal Learning & Behavior*, **25**, 400–412.

Delamater, A. R. (2004). Experimental extinction in Pavlovian conditioning: behavioural and neuroscience perspectives. *Quarterly Journal of Experimental Psychology*, **57B**, 97–132.

Denniston, J. C., Chang, R. C. & Miller, R. R. (2003). Massive extinction treatment attenuates the renewal effect. *Learning and Motivation*, **34**, 68–86.

Denniston, J. C., Savastano, H. & Miller R. R. (2001). The extended comparator hypothesis: learning by contiguity, responding by relative strength. In *Handbook of Contemporary Learning*, ed. R. R. Mowrer & S. B. Klein. Mahwah, NJ: Lawrence Erlbaum Associates, pp. 65–117.

Desmond, J. E. & Moore, J. W. (1982). A brain stem region essential for classical conditioned but not unconditioned nictitating membrane response. *Physiology and Behavior*, **28**, 1029–1033.

Desmond, J. E. & Moore, J. W. (1988). Adaptive timing in neural networks: the conditioned response. *Biological Cybernetics*, **58**, 405–415.

Desmond, J. E. & Moore, J. W. (1991). Single-unit activity in red nucleus during the classically conditioned rabbit nictitating membrane response. *Neuroscience Research*, **10**, 260–279.

Devenport, L. D. (1998). Spontaneous recovery without interference: why remembering is adaptive. *Animal Learning and Behavior*, **26**, 172–181.

DeVietti, T. L. & Barrett, O. V. (1986). Latent inhibition: no effect of intertrial interval of the preexposure trials. *Bulletin of the Psychonomic Society*, **24**, 453–455.

Devinsky, O., Morrell, M. J. & Vogt, B. A. (1995). Contributions of anterior cingulate cortex to behavior. *Brain*, **118**, 279–306.

Dickey, C. C., McCarley, R. W., Xu, M. L., Seidman, L. J., Voglmaier, M. M., Niznikiewicz, M. A., Connor, E. & Shenton, M. E. (2007). MRI abnormalities of the hippocampus and cavum septi pellucidi in females with schizotypal personality disorder. *Schizophrenia Research*, **89**, 49–58.

Dickinson, A. (1980). *Contemporary Animal Learning Theory*. Cambridge: Cambridge University Press.

Dickinson, A. & Burke, J. (1996). Within-compound associations mediate the retrospective revaluation of causality judgments. *Quarterly Journal of Experimental Psychology*, **49B**, 60–80.

Dickinson, A., Hall, G.,& Mackintosh, N. J. (1976). Surprise and the attenuation of blocking. *Journal of Experimental Psychology: Animal Behavior Processes*, **2**, 313–322.

Dickinson, A. & Mackintosh, N. J. (1978). Classical conditioning in animals. *Annual Review of Psychology*, **29**, 587–612.

Dietrich, A. (2004). The cognitive neuroscience of creativity. *Psychonomic Bulletin and Review*, **11**, 1011–1026.

DiMattia, B. D. & Kesner, R. P. (1988). Spatial cognitive maps: differential role of parietal cortex and hippocampal formation. *Behavioral Neuroscience*, **102**, 471–480.

Duda, R. O. & Hart, P. E. (1973). *Pattern Classification and Scene Analysis*. New York: Wiley.

Dunsmoor, J. E., Bandettini, P. A. & Knight, D. C. (2007). Impact of continuous versus intermittent CS-UCS pairing on human brain activation during Pavlovian fear conditioning. *Behavioral Neuroscience*, **121**, 635–642.

Durlach, P. J. & Rescorla, R. A. (1980). Potentiation rather than overshadowing in flavor-a version learning: an analysis in terms of within-compound associations. *Journal of Experimental Psychology: Animal Behavior Processes*, **6**, 175–187.

Dykes, M. & McGhie, A. (1976). A comparative study of attentional strategies of schizophrenic and highly creative normal subjects. *British Journal of Psychiatry*, **128**, 50–56.

Eysenck, H. J. (1995). *Genius: The Natural History of Creativity*. New York: Cambridge University Press.

Falls, W. A., Miserendino M. J. & Davis M. (1992). Extinction of fear-potentiated startle: blockade by infusion of an NMDA antagonist into the amygdala. *Journal of Neuroscience*, **12**, 854–863.

Fanselow, M. S. (1990). Factors governing one-trial contextual conditioning. *Animal Learning & Behavior*, **18**, 264–270.

Feldon, J., Shofel, A. & Weiner, I. (1991). Latent inhibition is unaffected by direct dopamine agonists. *Pharmacology, Biochemistry, and Behavior*, **38**, 309–314.

Fiorillo, C. D., Tobler, P. N. & Schultz, W. (2003). Discrete coding of reward probability and uncertainty by dopamine neurons. *Science*, **299**, 1898–1902.

Flaherty, A. W. (2005). Frontotemporal and dopaminergic control of idea generation and creative drive. *Journal of Comparative Neurology*, **493**, 147–153.

Fletcher, P. C., Anderson, J. M., Shanks, D. R., Honey, R., Carpenter, T. A., Donovan, T., Papdakis, N. & Bullmore, E. T. (2001). Responses of the human frontal cortex to surprising events are predicted by formal associative learning theory. *Nature Neuroscience*, **4**, 1043–1048.

Frey, P. W. & Sears, R. J. (1978). Model of conditioning incorporating the Rescorla–Wagner associative axiom, a dynamic attention process, and a catastrophe rule. *Psychological Review*, **85**, 321–340.

Frohardt R. J., Guarraci F. A. & Bouton M. E. (2000). The effects of neurotoxic hippocampal lesions on two effects of context after fear extinction. *Behavioral Neuroscience*, **114**, 227–240.

Fuster, J. M. (1973). Unit activity in prefrontal cortex during delayed-response performance: neuronal correlates of transient memory. *Journal of Neurophysiology*, **36**, 61–78.

Gabora, L. (2002). Cognitive mechanisms underlying the creative process. In *Proceedings of the Fourth International Conference on Creativity and Cognition, October 13–16*, ed. T. Hewett & T. Kavanagh. Loughborough University, UK, pp. 126–133.

Gal, G. & Weiner, I. (1998). Latent inhibition is disrupted by shell lesion but restored by an addition of core lesion. *European Journal of Neuroscience*, **10**, Suppl. 10, 118.04.

Galef, B. G., Jr. & Osborne, B. (1978). Novel taste facilitation of the association of visual cues with toxicosis in rats. *Journal of Comparative and Physiological Psychology*, **92**, 907–916.

Gallagher, P. C. & Holland, P. C. (1992). Preserved configural learning and spatial learning impairments in rats with hippocampal damage. *Hippocampus*, **2**, 81–88.

Gallagher, P. C., Graham, P. W. & Holland, P. C. (1990). The amygdala central nucleus and appetitive Pavlovian conditioning: lesions impair one class of conditioned behavior. *The Journal of Neuroscience*, **10**, 1906–1911.

Gallistel, C. R. & Gibbon, J. (2000). Time, rate, and conditioning. *Psychological Review*, **107**, 289–344.

Gallo, M. & Candido, A. (1995). Dorsal hippocampal lesions impair blocking but not latent inhibition of taste aversion learning in rats. *Behavioral Neuroscience*, **109**, 413–425.

Garcia-Gutierrez, A. & Rosas, J. M. (2003). Context change as the mechanism of reinstatement in causal learning. *Journal of Experimental Psychology: Animal Behavior Processes*, **29**, 292–310.

Garrud, P., Rawlins, J. N. P., Mackintosh, N. J., Goodal, G., Cotton, M. M. & Feldon, J. (1984). Successful overshadowing and blocking in hippocampectomized rats. *Behavioural Brain Research*, **12**, 39–53.

Gelperin, A. (1986). Complex associative learning in small neural networks. *Trends in Neuro Sciences*, **9**, 323–328.

Gelperin, A., Hopfield, J. J. & Tank, D. W. (1985). The logic of Limax learning. In *Model Neural Networks and Behavior*, ed. A. Selverston. New York: Plenum, pp. 237–261.

Getzels, J. & Csikzentmihalyi, M. (1976). *The Creative Vision*. New York: Wiley.

Gewirtz, J. C., Falls, W. A. & Davis, M. (1997). Normal conditioned inhibition and extinction of freezing and fear-potentiated startle following electrolytic lesions of medical prefrontal cortex in rats. *Behavioral Neuroscience*, **111**, 712–726.

Gibbon, J. (1977). Scalar expectancy and Weber's Law in animal timing. *Psychological Review*, **84**, 279–325.

Gibbon, J. & Balsam, P. (1981). Spreading association in time. In *Autoshaping and conditioning theory, eds.* C. M. Locurto, H. S. Terrace & J. Gibbon. New York: Academic, pp. 219–253.

Gibbons, H. & Rammsayer, T. H. (1999). Differential effects of personality traits related to the P-ImpUSS dimension on latent inhibition in healthy female subjects. *Personality and Individual Differences*, **27**, 1157–1166.

Gleitman, H. (1971). Forgetting of long-term memories in animals. In *Animal Memory*, ed. W. K. Honig & P. H. R. James. New York: Academic, pp. 1–44.

Gluck, M. A. & Myers, C. E. (1993). Hippocampal mediation of stimulus representation: a computational theory. *Hippocampus*, **3**, 491–516.

Gluck, M. A. & Myers, C. E. (1994). Context, conditioning, and hippocampal representation in animal learning. *Behavioral Neuroscience*, **108**, 835–847.

Goddard, M. J. & Jenkins, H. M. (1988). Blocking of a CS US association by a US US association. *Journal of Experimental Psychology: Animal Behavior Processes*, **14**, 177–186.

Gonzalez, F., Quinn, J. J. & Fanselow, M. S. (2003). Differential effects of adding and removing components of a context on the generalization of conditional freezing. *Journal of Experimental Psychology: Animal Behavior Processes*, **29**, 78–83.

Good M. & Honey R. C. (1991). Conditioning and contextual retrieval in hippocampal rats. *Behavioral Neuroscience*, **105**, 499–509.

Good, M. & Honey, R. (1993). Selective hippocampus lesions abolish contextual specificity of latent inhibition and conditioning. *Behavioral Neuroscience*, **107**, 23–33.

Goodwin, F. K., Murphy, D. L., Brodie, H. K. & Bunney, W. E. (1970). L-dopa, catecholamines, and behavior: a clinical and biochemical study in depressed patients. *Biological Psychiatry*, **2**, 341–366.

Gordon, W. C. & Weaver, M. S. (1989). Cue-induced transfer of CS preexposure effects across contexts. *Animal Learning and Behavior*, **17**, 409–417.

Gormezano, I. & Moore, J. W. (1969). Classical conditioning. In *Learning Processes*, ed. M. H. Marx. New York: Macmillan.

Gough, H. G. (1979). A creative personality scale for the Adjective Check List. *Journal of Personality and Social Psychology*, **37**, 1398–1405.

Gould, T. J., Collins, A. C. & Wehner, J. M. (2001). Nicotine enhances latent inhibition and ameliorates ethanol-induced deficits in latent inhibition. *Nicotine & Tobacco Research*, **3**, 17–24.

Grahame, N. J., Barnet, R., Gunther, L. & Miller, R. (1994). Latent inhibition as a performance deficit resulting from CS-context associations. *Animal Learning and Behavior*, **22**, 395–408.

Grahame, N. J., Hallam, S. C., Geier, L. & Miller, R. R. (1990). Context as an occasion setter following either CS Acquisition and extinction or CS acquisition alone. *Learning and Motivation*, **21**, 237–265.

Gray, J. A. (1971). *The Psychology of Fear and Stress*. London: Weidenfeld and Nicholson.

Gray, J. A. (1975). *Elements of a two-process theory of learning*. London: Academic.

Gray, J. A., Buhusi, C. V. & Schmajuk, N. A. (1997). The transition from automatic to controlled processing. *Neural Networks*, **10**, 1257–1268.

Gray, J. A., Feldon, J., Rawlins, J. N. P., Hemsley, D. R. & Smith, A. D. (1991). The neuropsychology of schizophrenia. *Behavioral and Brain Sciences*, **14**, 20–84.

Gray, J. A., Mitchell, S. N., Joseph, M. H., Grigoryan, G. A., Date, S. & Hedges, H. (1994). Neurochemical mechanisms mediating the behavioral and cognitive effects of nicotine. *Drug Development Research*, **31**, 3–17.

Gray, N. S., Hemsley, D. R. & Gray, J. A. (1992). Abolition of latent inhibition in acute, but not chronic schizophrenics. *Neurology, Psychiatry and Brain Research*, **1**, 83–89.

Groenewegen, H. J., Vermeulen-Van der Zee, E., te Kortschot, A. & Witter, M. P. (1987). Organization of the projections from the subiculum to the ventral striatum in the rat. A study using anterograde transport of Phaseolus vulgaris leucoagglutinin. *Neuroscience*, **23**, 103–120.

Grossberg, S. (1975). A neural model of attention, reinforcement, and discrimination learning. *International Review of Neurobiology*, **18**, 263–327.

Grossberg, S. & Merrill, J. W. L. (1992). A neural network model of adaptively timed reinforcement learning and hippocampal dynamics. *Cognitive Brain Research*, **1**, 3–38.

Grossberg, S. & Schmajuk, N. A. (1989). Neural dynamics of adaptive timing and temporal discrimination during associative learning. *Neural Networks*, **2**, 79–102.

Groves, P. M. & Thompson, R. F. (1970). Habituation: a dual-process theory. *Psychological Review*, **77**, 419–550.

Guez, D. & Miller, R. R. (2008). Blocking and pseudoblocking: the reply of Rattus norvegicus to Apis mellifera. *The Quarterly Journal of Experimental Psychology (Colchester)*, **61**, 1186–98.

Guilford, J. P. (1950). Creativity. *American Psychologist*, **5**, 444–454.
Hall, G. (1991). *Perceptual and Associative Learning*. Oxford: Clarendon.
Hall, G. (1996). Learning about associatively activated stimulus representations: implications for acquired equivalence and perceptual learning. *Animal Learning & Behavior*, **24**, 233–255.
Hall, G. & Channel, S. (1985a). Latent inhibition and conditioning after preexposure to the training context. *Learning and Motivation*, **16**, 381–481.
Hall, G. & Channel, S. (1985b). Differential effects of contextual change on latent inhibition and on the habituation of an orienting response. *Journal of Experimental Psychology: Animal Behavior Processes*, **11**, 470–481.
Hall, G. & Minor, H. (1984). A search for context-stimulus associations in latent inhibition. *The Quarterly Journal of Experimental Psychology*, **36B**, 145–169.
Hall, G. & Pearce, J. M. (1979). Latent inhibition of a CS during CS-US pairings. *Journal of Experimental Psychology: Animal Behavior Processes*, **5**, 31–42.
Hall, G. & Pearce, J. M. (1982). Restoring the associability of a preexposed CS by a surprising event. *The Quarterly Journal of Experimental Psychology*, **34B**, 127–140.
Hall, G. & Schachtman, T. R. (1987). Differential effects of a retention interval on latent inhibition and the habituation of an orienting response. *Animal Learning & Behavior*, **15**, 76–82.
Hamann, S. B. & Squire, L. R. (1995). On the acquisition of new declarative knowledge in amnesia. *Behavioral Neuroscience*, **109**, 1027–1044.
Hampson, S. E. (1990). *Connectionistic Problem Solving: Computational Aspects of Biological Learning*. Cambridge, MA: Birkhauser.
Han, J-S., Gallahger, M. & Holland, P. (1995). Hippocampal lesions disrupt decrements but not increments in conditioned stimulus processing. *Journal of Neuroscience*, **11**, 7323–7329.
Harris, J. A. (2006). Elemental representations of stimuli in associative learning. *Psychological Review*, **113**, 584–605.
Harris, J. A. & Westbrook, R. F. (1998). Evidence that GABA transmission mediates context-specific extinction of learned fear. *Psychopharmacology*, **140**, 105–115.
Harris, J. A., Jones, M. L., Bailey, G. K. & Westbrook, R. F. (2000). Contextual control over conditioned responding in an extinction paradigm. *Journal of Experimental Psychology: Animal Behavior Processes*, **26**, 174–185.
Hebb, D. O. (1949). *The Organization of Behavior: A Neuropsychological Theory*. New York: Wiley/Interscience.
Heith, C. D. & Rescorla, R. A. (1973). Simultaneous and backward fear conditioning in the rat. *Journal of Comparative and Physiological Psychology*, **8**, 434–43.
Higgins, J., Mednick, S. A. & Thompson, R. E. (1966). Acquistion and retention of remote associates in process-reactive schizophrenia. *Journal of Nervous and Mental Disease*, **142**, 418–423.
Hinzman, D. L. (1991). Why are formal models useful in psychology? In *Relating Theory and Data: Essays on Human Memory in Honor of Bennet B. Murdock*, ed. W. E. Hockley & S. Lewandosky. Hillsdale, NJ: Lawrence Erlbaum Associates.
Hirsh, R. (1974). The hippocampus and contextual retrieval of information from memory: a theory. *Behavioral Biology*, **12**, 421–444.

Hirsh, R., Leber, B. & Gillman, K. (1978). Fornix fibers and motivational states as controllers of behavior: a study stimulated by the contextual retrieval theory. *Behavioral Biology*, **22**, 463–478.

Hobin, J. A., Ji, J. & Maren, S. (2005). Ventral hippocampal muscimol disrupts context-specific fear memory retrieval after extinction in rats. *Hippocampus*, **16**, 174–182.

Holland, P. C. (1977). Conditioned stimulus as a determinant of the form of the Pavlovian conditioned response. *Journal of Experimental Psychology: Animal Behavior Processes*, **3**, 77–104.

Holland, P. C. (1983). Occasion-setting in Pavlovian feature positive discriminations. In *Quantitative Analyses of Behavior: Discrimination Processes*, ed. M. L. Commons, R. J. Herrnstein & A. R. Wagner. New York: Ballinger, vol. 4, pp. 183–206.

Holland, P. C. (1986a). Temporal determinants of occasion setting in feature positive discriminations. *Animal Learning & Behavior*, **14**, 111–120.

Holland, P. C. (1986b). Transfer after serial feature positive discrimination training. *Learning and Motivation*, **17**, 243–268.

Holland, P. C. (1989a). Acquisition and transfer of conditional discrimination performance. *Journal of Experimental Psychology: Animal Behavior Processes*, **15**, 154–165.

Holland, P. C. (1989b). Feature extinction enhances transfer of occasion setting. *Animal Learning & Behavior*, **17**, 269–279.

Holland, P. C. (1989c). Occasion setting with simultaneous compounds in rats. *Journal of Experimental Psychology: Animal Behavior Processes*, **15**, 183–193.

Holland, P. C. (1989d). Transfer of negative occasion setting and conditioned inhibition across conditioned and unconditioned stimuli. *Journal of Experimental Psychology: Animal Behavior Processes*, **15**, 311–328.

Holland, P. C. (1990). Event representation in Pavlovian conditioning: image and action. *Cognition*, **37**, 105–131.

Holland, P. C. (1991). Learning, thirst, and drinking. In *Thirst: Physiological and Psychological Aspects, eds.* D. J. Ramsey & D. Booth. New York: Springer, pp. 279–295.

Holland, P. C. (1992). Occasion setting in Pavlovian conditioning. *The Psychology of Learning and Motivation*, **28**, 69–125.

Holland. P. C. (1998). Temporal control in Pavlovian occasion setting. *Behavioral Processes*, **44**, 225–236.

Holland, P. C. (1999). Overshadowing and blocking as acquisition deficits: no recovery after extinction of overshadowing or blocking cues. *The Quarterly Journal of Experimental Psychology*, **52B**, 307–333.

Holland, P. C. (2000). Trial and intertrial durations in appetitive conditioning in rats. *Animal Learning and Behaviour*, **28**, 121–135.

Holland, P. C. & Forbes, D. R. (1980). Effects of compound or element preexposure on compound flavor aversion conditioning. *Animal Learning & Behavior*, **8**, 199–203.

Holland, E. C. & Haas, M. L. (1993). The effects of target salience in operant feature positive discriminations. *Learning and Motivation*, **24**, 119–140.

Holland, E. C. & Lamarre, J. (1984). Transfer of inhibition after serial and simultaneous feature negative discrimination training. *Learning and Motivation*, **15**, 219–243.

Holland, P. C., Hamlin, P. A. & Parsons, J. P. (1997). Temporal specificity in serial feature positive discrimination learning. *Journal of Experimental Psychology: Animal Behavior Processes*, **23**, 95–109.

Holland, P. C., Lamoureux, J. A., Han, J-S. & Gallagher, M. (1999). Hippocampal lesions interfere with Pavlovian negative occasion setting. *Hippocampus*, **9**, 143–157.

Honey, R. & Good, M. (1993). Selective hippocampal lesions abolish the contextual specificity of latent inhibition and conditioning. *Behavioral Neuroscience*, **107**, 23–33.

Honey, R. C. & Hall, G. (1988). Overshadowing and blocking procedures in latent inhibition. *The Quarterly Journal of Experimental Psychology*, **49B**, 163–186.

Honey, R. C. & Hall, G. (1989). Attenuation of latent inhibition after compound pre-exposure: associative and perceptual explanations. *The Quarterly Journal of Experimental Psychology B: Comparative and Physiological Psychology*, **41**, 355–368.

Horvitz, J. C. (2000). Mesolimbocortical and nigrostriatal dopamine responses to salient non-reward events. *Neuroscience*, **96**, 651–656.

Huff, N. C., Alba Hernandez, J., Blanding, N. & LaBar, K. S. (2007). Effects of immediate versus delayed extinction on the renewal of conditioned fear in humans. *Society for Neuroscience Abstracts*.

Hull, C. L. (1943). *Principles of Behavior*. New York: Appleton–Century–Crofts.

Ishii, K., Haga, Y. & Hishimura, Y. (1999). Distractor effect on latent inhibition of conditioned flavor aversion in rats. *Japanese Psychological Research*, **41**, 229–238.

Jarrard, L. E. & Davidson, T. L. (1991). On the hippocampus and learned conditional responding: effects of aspiration versus ibotenate lesions. *Hippocampus*, **1**, 107–117.

Jenkins, J. J. & Sainsbury, R. (1969). The development of stimulus control through differential reinforcement. In *Fundamental Issues in Associative Learning*, eds. N. J. Mackintosh & W. K. Honig. Halifax, Nova Scotia: Dalhousie University Press, pp. 123–161.

Ji, J. & Maren, S. (2005). Electrolytic lesions of the dorsal hippocampus disrupt renewal of conditional fear after extinction. *Learning and Memory*, **12**, 270–276.

Johnson, D. M., Baker, J. D. & Azorlosa, J. L. (2000). Acquisition, extinction, and reinstatement of Pavlovian fear conditioning: the roles of the NMDA receptor and nitric oxide. *Brain Research*, **857**, 66–70.

Jones, S. H., Gray, J. A. & Hemsley, D. R. (1990). The Kamin blocking effect, incidental learning and psychoticism. *British Journal of Psychology*, **81**, 95–109.

Jones, S. H., Gray, J. A. & Hemsley, D. R. (1992). Loss of the Kamin blocking effect in acute but not chronic schizophrenics. *Biological Psychiatry*, **32**, 739–755.

Joseph, M. H., Peters, S. L. & Gray, J. A. (1993). Nicotine blocks latent inhibition in rats: evidence for a critical role of increased functional activity of dopamine in the mesolimbic system at conditioning rather than pre-exposure. *Psychopharmacology*, **110**, 187–192.

Kamil, A. C. (1969) Some parameters of the second-order conditioning of fear in rats. *Journal of Comparative and Physiological Psychology*, **67**, 364–369.

Kamin, L. (1968). Attention-like processes in classical conditioning. In *Miami Symposium on the Prediction of Behavior: Aversive Stimulation*, ed. D. J. Ramsey & D. Booth. Coral Gables, FL: University of Chicago Press, pp. 9–32.

Kamin, L. (1969a). Predictability, surprise, attention, and conditioning. In *Punishment and aversive behavior*, ed. B. A. Campbell & R. M. Church. New York: Appleton–Century–Crofts, pp. 279–296.

Kamin, L. J. (1969b). Selective association and conditioning. In *Fundamental Issues in Associative Learning*, ed. N. J. Mackintosh & W. K. Honig. Proceedings of a symposium held at Dalhousie University, Halifax, June 1968. Halifax: Dalhousie University Press, pp. 42–64.

Kamprath, K. & Wotjak, C. T. (2004). Nonassociative learning processes determine expression and extinction of conditioned fear in mice. *Learning & Memory*, **11**, 770–786.

Kasprow, W., Schachtman, T. & Miller, R. (1987). The comparator hypothesis of conditioned response generation: Manifest conditioned excitation and inhibition as a function of relative excitatory associative strengths of CS and conditioning context at the time of testing. *Journal of Experimental Psychology: Animal Behavioral Processes*, **13**, 395–406.

Kasprow, W., Catterson, D., Schatchman, T. & Miller, R. (1984). Attenuation of latent inhibition by postacquisition reminder. *Quarterly Journal of Experimental Psychology*, **36B**, 53–63.

Kaufman, M. A. & Bolles, R. C. (1981). A nonassociative aspect of overshadowing. *Bulletin of the Psychonomic Society*, **18**, 318–320.

Kaye, H. & Pearce, J. M. (1984). The strength of the orienting response during Pavlovian conditioning. *Journal of Experimental Psychology: Animal Behavioral Processes*, **10**, 90–109.

Kehoe, E. J. (1986). Summation and configuration in conditioning of the rabbit's nictitating membrane response to compound stimuli. *Journal of Experimental Psychology: Animal Behavior Processes*, **12**, 186–195.

Kehoe, E. J. (1988). A layered network model of associative learning: learning to learn and configuration. *Psychological Bulletin*, **95**, 411–422.

Kehoe, E. J., Weidemann, G. & Dartnall, S. (2004). Apparatus exposure produces profound declines in conditioned nictitating-membrane responses to discrete conditioned stimuli by the rabbit (Oryctolagus cuniculus). *Journal of Experimental Psychology: Animal Behavior Processes*, **30**, 259–270.

Kiernan, M. J. & Westbrook, R. F. (1993). Effects of exposure to a to-be-shocked environment upon the rat's freezing response: evidence for facilitation, latent inhibition, and perceptual learning. *Quarterly Journal of Experimental Psychology*, **46B**, 271–288.

Killcross, A. S., Dickinson, A. & Robbins, T. W. (1994a). Amphetamine-induced disruptions of latent inhibition are reinforcer mediated – implications for animal models of schizophrenic attentional dysfunction. *Psychopharmacology*, **115**, 185–195.

Killcross, A. S., Dickinson, A. & Robbins, T. W. (1994b). Effects of the neuroleptic α-flupenthixol on latent inhibition in aversely- and appetitively-motivated paradigms: evidence for dopamine-reinforcer interactions. *Psychopharmacology*, **115**, 196–205.

Killcross, A. S., Stanhope, K. J., Dourish, C. T. & Piras, G. (1997). WAY100635 and latent inhibition in the rat: selective effects at preexposure. *Behavioural Brain Research*, **88**, 51–57.

Killcross, S. (2001). Loss of latent inhibition in conditioned taste aversion following exposure to a novel flavour before test. *Quarterly Journal of Experimental Psychology*, **54B**, 271–288.

Killcross, S. & Balleine, B. (1996). Role of primary motivation in stimulus preexposure effects. *Journal of Experimental Psychology: Animal Behavior Processes* **96**, 32–42.

Killcross, S., Robbins, T. W. & Everitt, B. J. (1997). Different types of fear conditioned behaviour mediated by separate nuclei within amygdala. *Nature*, **388**, 377–380.

Killeen, P. R. & Fetterman, J. G. (1988). A behavioral theory of timing. *Psychological Review*, **95**, 274–295.

Kishimoto, H., Yamada, K., Iseki, E., Kosaka, K. & Okoshi, T. (1998). Brain imaging of affective disorders and schizophrenia. *Psychiatry and Clinical Neuroscience*, **52**, Suppl, S212-S214.

Knight, D. C., Nguyen, H. T. & Bandettini, P. A. (2005). The role of the human amygdala in the production of conditioned fear responses. *Neuroimage*, **26**, 1193–2000.

Knudsen, E. I. & Brainard, M. S. (1995). Creating a unified representation of visual and auditory space in the brain. *Annual Review of Neuroscience*, **18**, 19–43.

Kohonen, T. (1977). *Associative Memory. A System Theoretical Approach*. New York: Springer.

Konorski, J. (1948). *Conditioned Reflexes and Neuron Organization*. New York: Cambridge University Press.

Konorski, J. (1967). *Integrative Activity of the Brain*. Chicago: University of Chicago Press.

Krabbendam, L., Arts, B., van Os, J. & Aleman, A. (2005). Cognitive functioning in patients with schizophrenia and bipolar disorder: a quantitative review. *Schizophrenia Research*, **80**, 137–149.

Kraemer, P. J., Randall, C. K. & Carbary, T. J. (1991). Release from latent inhibition with delayed testing. *Animal Learning & Behavior*, **19**, 139–145.

Kruschke, J. K. (2001). Toward a unified model of attention in associative learning. *Journal of Mathematical Psychology*, **45**, 812–863.

Kruschke, J. K. & Blair, N. J. (2000). Blocking and backward blocking involve learned inattention. *Psychonomic Bulletin & Review*, **7**, 636–645.

LaBar, K. S. & Phelps, E. A. (2005). Reinstatement of conditioned fear in humans is context dependent and impaired in amnesia. *Behavioral Neuroscience*, **119**, 677–686.

Lamarre, J. & Holland, P. C. (1987). Acquisition and transfer of serial feature negative discriminations. *Learning and Motivation*, **18**, 319–342.

Lamoureux, J. A., Buhusi, C. V. & Schmajuk, N. A. (1998). A real time theory of Pavlovian conditioning: simple stimuli and occasion setters. In: *Occasion Setting: Associative Learning and Cognition in Animals*, ed. N. A. Schmajuk, & P. C. Holland. American Psychological Association, Washington, D.C., pp. 383–424.

Lantz, A. E. (1973). Effect of number of trials, interstimulus interval, and dishabituation on subsequent conditioning in a CER paradigm. *Animal Learning & Behavior*, **1**, 273–277.

Le Pelley, M. E. (2004). The role of associative history in models of associative learning: a selective review and a hybrid model. *The Quarterly Journal of Experimental Psychology*, **57B**, 193–243.

LeDoux, J. E. (1992). Emotion and the amygdala. In *The Amygdala: Neurobiological Aspects of Emotion, Memory, and Mental Dysfunction*, ed. J. Aggleton. New York: Wiley-Liss.

LeDoux, J. E. (2000). Emotion circuits in the brain. *Annual Review of Neuroscience*, **23**, 155–184.

Legault, M. & Wise, R. A. (2001). Novelty-evoked elevations of nucleus accumbens dopamine: dependence on impulse flow from the ventral subiculum and glutamatergic neurotransmission in the ventral tegmental area. *European Journal of Neuroscience*, **13**, 819–828.

Lencz, T., Smith, C. W., McLaughlin, D., Auther, A., Nakayama, E., Hovey, L. & Cornblatt, B. A. (2006). Generalized and specific neurocognitive deficits in prodromal schizophrenia. *Biological Psychiatry*, **59**, 863–871.

Lipska, B. K., Jaskiw, G. E., Chrapusta, S., Karoum, F. & Weinberger, D. R. (1992). Ibotenic acid lesion of the ventral hippocampus differentially affects dopamine and its metabolites in the nucleus accumbens and prefrontal cortex in the rat. *Brain Research*, **585**, 1–6.

Lipska, B. K., Jaskiw, G. E., Karoum, F., Phillips, I., Kleinman, J. E. & Weinberger, D. R. (1991). Dorsal hippocampal lesion does not affect dopaminergic indices in the basal ganglia. *Pharmacology, Biochemistry and Behavior*, **40**, 181–184.

Livesey, E. J. & Boakes, R. A. (2004). Outcome additivity, elemental processing and blocking in human causality judgments. *Quarterly Journal of Experimental Psychology B*, **57**, 361–379.

Loechner, K. J. & Weisz, D. J. (1987). Hippocampectomy and feature-positive discrimination. *Behavioural Brain Research*, **26**, 63–73.

LoLordo, V. M. (1979). Selective associations. In *Mechanisms of Learning and Motivation*, ed. A. Dickinson & R. A. Boakes. Hillsdale, NJ: Lawrence Erlbaum Associates, pp. 367–398.

Lovibond, P. E., Been, S. L., Mitchell, C. J., Bouton, M. E. & Frohardt, R. (2003). Forward and backward blocking of causal judgment is enhanced by additivity of effect magnitude. *Memory & Cognition*, **31**, 133–42.

Lovibond, P. F., Preston, G. C. & Mackintosh, N. J. (1984). Context specificity of conditioning, extinction, and latent inhibition. *Journal of Experimental Pyschology: Animal Behavior Processes*, **10**, 360–375.

Lubow, R. E. (1989). *Latent Inhibition and Conditioned Attention Theory*. Cambridge: Cambridge University Press.

Lubow, R. E. & Moore, A. U. (1959). Latent inhibition: the effect of nonreinforced preexposure to the conditional stimulus. *Journal of Comparative and Physiological Psychology*, **52**, 415–419.

Lubow, R. E., Weiner, I. & Schnur, P. (1981). Conditioned attention theory. In *The Psychology of Learning and Motivation*, ed. G. H. Bower. New York: Academic, vol. 15, pp. 1–49.

Lubow, R. E., Weiner, I., Schlossberg, A. & Baruch, I. (1987). Latent inhibition and schizophrenia. *Bulletin of the Psychonomic Society*, **25**, 464–467.

Lysle, D. T. & Fowler, H. (1985). Inhibition as a "slave" process: deactivation of conditioned inhibition through extinction of conditioned excitation. *Journal of Experimental Psychology: Animal Behavior Processes*, **11**, 71–94.

Machado, A. (1997). Learning the temporal dynamics of behavior. *Psychological Review*, **104**(2), 241–265.

Mackintosh, N. J. (1973). Stimulus selection: learning to ignore stimuli that predict no change in reinforcement. In *Constraints of Learning*, ed. R. A. Hinde & J. S. Hinde. London: Academic.

Mackintosh, N. J. (1975). A theory of attention: variations in the associability of stimuli with reinforcement. *Psychological Review*, **82**, 276–298.

Mackintosh, N. J. (1983). *Conditioning and Associative Learning*. Oxford: Clarendon Press.

Mackintosh, N. J. & Turner, C. (1971). Blocking as a function of novelty of CS and predictability of UCS. *The Quarterly Journal of Experimental Psychology*, **23**, 359–366.

Maes, J. H. R. & Vossen, J. M. H. (1996). Differential inhibition using contextual stimuli. *Behavioural Processes*, **37**, 167–184.

Maren, S. & Chang, C. (2006). Recent fear is resistant to extinction. *Proceedings of the National Academy of Sciences of the United States of America*, **103**, 18020–18025.

Markman, A. (1989). LMS rules and the inverse base-rate effect: comment on Gluck and Bower (1988). *Journal of Experimental Psychology: General*, **118**, 417–421.

Marlin, N. A. (1982). Within-compound associations between the context and the conditioned stimulus. *Learning and Motivation*, **13**, 526–541.

Marr, D. (1982). *Vision: A Computational Investigation into the Human Representation and Processing of Visual Information*. San Francisco: W. H. Freeman.

Martindale, C. (1995). Creativity and connectionism. In *The Creative Cognition Approach*, ed. S. M. Smith, T. B. Ward & R. A. Finke. Cambridge, MA: MIT Press.

Martindale, C. (1999). Biological bases of creativity. In *Handbook of Creativity*, ed. R. J. Sternberg. New York: Cambridge University Press.

Martindale, C. & Hasenfus, N. (1978). EEG differences as a function of creativity, stage of the creative process, and effort to be original. *Biological Psychology*, **6**, 157–167.

Matzel, L., Brown, A. & Miller, R. (1987). Associative effects of US preexposure: modulation of conditioned responding by an excitatory training context. *Journal of Experimental Psychology: Animal Behavior Processes*, **13**, 65–72.

Matzel, L., Schachtman, T. & Miller R. (1985). Recovery of an overshadowed association achieved by extinction of the overshadowing stimulus. *Learning and Motivation*, **16**, 398–412.

McLaren, I. P. L., Kaye, H. & Mackintosh, N. J. (1989). An associative theory of the representation of stimuli: applications to perceptual learning and latent inhibition. In *Parallel Distributed Processing: Implications for Psychology and Neurobiology*, ed. R. G. M. Morris. Oxford: Clarendon.

McCloskey, M. & Cohen, N. J. (1989). Catastrophic interference in connectionist networks: The sequential learning problem. In *The Psychology of Learning and Motivation*, ed. G. H. Bower. New York: Academic, vol. 24, pp. 109–165.

McCormick, D. A., Steinmetz, J. E. & Thompson, R. F. (1985). Lesions of the inferior olivary complex cause extinction of the classically conditioned eyeblink response. *Brain Research*, **359**, 120–130.

McKinzie, D. L. & Spear, N. E. (1995). Ontogenetic differences in conditioning to context and CS as a function of context saliency and CS-US interval. *Animal Learning & Behavior*, **23**, 304–313.

McLaren, I. P. L. & Mackintosh, N. J. (2000). An elemental model of associative learning: latent inhibition and perceptual learning. *Animal Learning and Behavior*, **28**, 211–246.

McLaren, I. P. L., Kaye, H. & Mackintosh, N. J. (1989). An associative theory of the representation of stimuli: applications to perceptual learning and latent inhibition. In *Parallel Distributed Processing: Implications for Psychology and Neurobiology*, ed. R. G. M. Morris. Oxford: Clarendon.

McNaughton, N. (2004). The conceptual nervous system of J. A. Gray: anxiety and neuroticism. *Neuroscience and Biobehavioral Reviews*, **28**, 227–228.

McPhee, J. E., Rauhut, A. S. & Ayres, J. J. B. (2001). Evidence for learning deficit versus performance deficit theories of latent inhibition in Pavlovian fear conditioning. *Learning and Motivation*, **32**, 274–305.

Meck, W. H. & Church, J. (1982). Abstraction of temporal attributes. *Journal of Experimental Psychology: Animal Behavior Processes*, **10**, 1–29.

Meck, W. H. & Church, J. (1987). Cholinergic modulation of the content of temporal memory. *Behavioral Neuroscience*, **101**, 457–464.

Mednick, S. A. (1962). The associative basis of the creative process. *Psychological Review*, **69**, 220–232.

Meeter, M. Myers, C. E. & Gluck, M. A. (2005). Integrating incremental learning and episodic memory models of the hippocampal region. *Psychological Review*, **112**, 560–585.

Melchers, K. G., Wolff, S. & Lachnit, H. (2006). Extinction of conditioned inhibition through nonreinforced presentation of the inhibitor. *Psychonomic Bulletin & Review*, **13**, 662–667.

Merten, T. (1992). Word association and schizophrenia: an empirical study. *Nervenarzt*, **63**, 401–408.

Micco, D. J., & Schwartz, M. (1972). Effects of hippocampal lesions upon the developments of Pavlovian internal inhibition in rats. *Journal of Comparative and Physiological Psychology*, **76**, 371–377.

Milad, M. R., Orr, S. P., Pitman, R. K. & Rauch, S. L. (2005). Context modulation of memory for fear extinction in humans. *Psychophysiology*, **42**, 456–464.

Miller, R. & Matute, H. (1996). Biological significance in forward and backward blocking: resolution of a discrepancy between animal conditioning and human causal judgement. *Journal of Experimental Psychology: General*, **125**, 370–386.

Miller, R., Hallam, S. & Grahame, N. (1990). Inflation of comparator stimuli following CS training. *Animal Learning & Behavior*, **18**, 434–443.

Miller, R. R. & Matzel, L. (1988). The comparator hypothesis: a response rule for the expression of associations. In *The Psychology of Learning and Motivation*, ed. G. H. Bower. Orlando, FL: Academic, vol. 2, pp. 51–92.

Miller, R. R. & Schachtman, T. (1985). Conditioning context as an associative baseline: implications for response generation and the nature of conditioned inhibition. In *Information Processing in Animals: Conditioned Inhibition*, ed. R. R. Miller & N. E. Spear. Hillsdale, NJ: Lawrence Erlbaum Associates, pp. 51–88.

Miller R. R., Schachtman, T. & Matzel, L. (1988). *Failure to attenuate blocking by posttraining extinction of a blocking stimulus*. Unpublished raw data. Cited in Blaisdell *et al*. (1999).

Milner, B. R. (1966). Amnesia following operation on temporal lobes. In *Amnesia*, ed. C. W. N. Whitty & O. L. Zangwill. London: Butterworths.

Miserendino, M. J. D., Sananes, C. B., Melia, K. R. & Davis, M. (1990). Blocking of acquisition but not expression of conditioned fear-potentiated startle by NMDA antagonists in the amygdala. *Nature*, **345**, 716–718.

Mitchell, C. J., Lovibond, P. F. & Condoleon, M. (2005). Evidence for deductive reasoning in blocking of causal judgments. *Learning and Motivation*, **36**, 77–87.

Moody, E. W., Sunsay, C. & Bouton, M. E. (2006) Priming and trial spacing in extinction: effects on extinction performance, spontaneous recovery, and reinstatement in appetitive conditioning. *The Quarterly Journal of Experimental Psychology*, **59**, 809–829.

Moore, J. W., & Choi, J. S. (1998). Conditioned stimuli are occasion setters. In *Occasion Setting: Associative Learning and Cognition in Animals*, eds. N. A. Schmajuk & P. C. Holland. *American Psychological Association*, Washington, DC, pp. 279–318.

Moore, J. W. &. Schmajuk, N. A. (2008). Kamin blocking. *Scholarpedia* **3**, 3542.

Moore, J. W. & Stickney, K. J. (1980). Formation of attentional-associative networks in real time: role of the hippocampus and implications for conditioning. *Physiological Psychology*, **8**, 207–217.

Moore, J. W., Yeo, C. H., Oakley, D. A. & Russell, I. S. (1980). Conditioned inhibition of the nictitating membrane response in decorticate rabbits. *Behavioural Brain Research*, **1**, 397–409.

Morell, J. R. & Holland, E. C. (1993). Summation and transfer of negative occasion setting. *Animal Learning and Behavior*, **21**, 145–153.

Morris, R. G. M., Garrud, P., Rawlins, J. N. P. & O'Keefe, J. (1982). Place navigation impaired in rats with hippocampal lesions. *Nature*, **297**, 681–683.

Morris, R. G. M., Schenk, F., Tweedie, F. & Jarrard, L. E. (1990). Ibotenate lesions of hippocampus and/or subiculum: dissociating components of allocentric spatial learning. *European Journal of Neuroscience*, **2**, 1016–1028.

Morrow, B. A., Elsworth, J. D., Rasmusson, A. M. & Roth, R. H. (1999). The role of mesoprefrontal dopamine neurons in the acquisition and expression of conditioned fear in the rat. *Neuroscience*, **92**, 553–564.

Myers, C. E., Gluck, M. A. & Granger, R. (1995). Dissociation of hippocampal and entorhinal function in asociative learning: a computational approach. *Psychobiology*, **23**, 116–138.

Myers, K. M., Ressler, K. J. & Davis, M. (2006). Different mechanisms of fear extinction dependent on length of time since fear acquisition. *Learning and Memory*, **13**, 216–223.

Nagaishi, T. & Nakajima, S. (2008). Further evidence for the summation of latent inhibition and overshadowing in rats' conditioned taste aversion. *Learning and Motivation*, **39**, 221–242.

Nakajima, S. & Nagaishi, T. (2005). Summation of latent inhibition and overshadowing in a generalized bait shyness paradigm of rats. *Behavioural Processes*, **69**, 369–377.

Napier, R. M., Macrae, M. & Kehoe, E. J. (1992). Rapid reacquisition in conditioning of the rabbit's nictitating membrane response. *Journal of Experimental Psychology: Animal Behavior Processes*, **18**, 182–192.

Nicholson, D. A. & Freeman, J. H. Jr. (2002). Medial dorsal thalamic lesions impair blocking and latent inhibition of the conditioned eyeblink response in rats. *Behavioral Neuroscience*, **116**, 276–285.

Oades, R. D. & Halliday, G. M. (1987). Ventral tegmental (A10) system: neurobiology. 1. Anatomy and connectivity. *Brain Research*, **434**, 117–165.

Oakley, D. A. & Russell, I. S. (1972). Neocortical lesions and Pavlovian conditioning. *Physiology and Behavior*, **8**, 915–926.

Oakley, D. A. & Russell, I. S. (1973). Differential and reversal conditioning in partially neodecorticate rabbits. *Physiology and Behavior*, **13**, 221–230.

Oakley, D. A. & Russell, I. S. (1975). Role of cortex in Pavlovian discrimination learning. *Physiology and Behavior*, **15**, 315–321.

O'Donnell, P. & Grace, A. A. (1998). Dysfunctions in multiple interrelated systems as the neurobiological bases of schizophrenic symptom clusters. *Schizophrenia Bulletin*, **24**, 267–283.

O'Keefe, J. & Nadel, L. (1978). *The Hippocampus as a Cognitive Map*. Oxford: Clarendon.

Orr, W. B. & Berger, T. W. (1985). Hippocampectomy disrupts the topography of conditioned nictitating membrane responses during reversal learning. *Journal of Comparative and Physiological Psychology*, **99**, 35–45.

Oswald, C. J. P., Yee, B. K., Rawlins, J. N. P., Bannerman, D. B., Good, M. & Honey, R. C. (2002). The influence of selective lesions to components of the hippocampal system on the orienting response, habituation and latent inhibition. *European Journal of Neuroscience*, **15**, 1983–1990.

Otto, T. & Poon, P. J. (2006). Dorsal hippocampal contributions to unimodal contextual conditioning. *Journal of Neuroscience*, **26**: 6603–6609.

Packard, M. G. & McGaugh, J. L. (1992). Double dissociation of fornix and caudate nucleus lesions on acquisition of two water maze tasks: further evidence for multiple memory systems. *Behavioral Neuroscience*, **106**, 439–446.

Padfield, P. D. & Lawrence, B. (2003). The birth of flight control: an engineering analysis of the Wright brothers' 1902 glider. *The Aeronautical Journal*, December, 697–718.

Partridge, D. & Rowe, J. (2002). Creativity: A computational modeling approach. In *Creativity, Cognition, and Knowledge: An Interaction*, ed. T. Dartnall. Westport, CT: Praeger-Greenwood.

Pavlov, I. P. (1927). *Conditioned Reflexes*. London: Oxford University Press.

Pearce, J. M. (1987). A model for stimulus generalization in Pavlovian conditioning. *Psychological Review*, **94**, 61–75.

Pearce, J. M. (1994). Similarity and discrimination: a selective review and a connectionist model. *Psychological Review*, **101**, 587–607.

Pearce, J. M. & Hall, G. (1979). Loss of associability by a compound stimulus comprising excitatory and inhibitory elements. *Journal of Experimental Psychology: Animal Behavior Processes*, **5**, 19–30.

Pearce, J. M. & Hall, G. (1980). A model for Pavlovian conditioning: variations in the effectiveness of conditioned but not unconditioned stimuli. *Psychological Review*, **87**, 332–352.

Pearce J. M., Kaye, H. & Hall, G. (1983). Predictive accuracy and stimulus associability: development of a model of Pavlovian learning. In *Quantitative Analyses of Behaviour*, ed. M. Commons, R. J. Herrnstein & A. R. Wagner. Cambridge, MA: Ballinger, vol. 3, pp. 241–255.

Peet, M. & Peters, S. (1995). Drug-induced mania. *Drug Safety*, **12**, 146–153.

Penick, S. & Solomon, P. R. (1991). Hippocampus, context, and conditioning. *Behavioral Neuroscience*, **105**, 611–617.

Perry, B., Luchins, D. & Schmajuk, N. A. (1993). *Hippocampal lesions and dopamine receptor density*. Annual Meeting of the American Psychiatric Association, San Francisco, May 1993.

Pineño, O., Urushihara, K. & Miller, R. R. (2005). Spontaneous recovery from forward and backward blocking. *Journal of Experimental Psychology: Animal Behaviour Processes*, **31**, 172–183.

Pineño, O., Urushihara, K., Stout, S., Fuss, J. & Miller, R. (2006). When more is less: extending training of the blocking association following compound training attenuates the blocking effect. *Learning and Behavior*, **34**, 21–36.

Port, R. L. & Patterson, M. M. (1984). Fimbrial lesions and sensory preconditioning. *Behavioral Neuroscience*, **98**, 584–589.

Port, R. L., Beggs, A. L. & Patterson, M. M. (1987). Hippocampal substrate of sensory associations. *Physiology & Behavior*, **39**, 643–647.

Pothuizen, H. H. J., Jongen-Rêlo, A. L., Feldon, J. & Yee, B. K. (2006). Latent inhibition of conditioned taste aversion is not disrupted, but can be enhanced, by selective nucleus accumbens shell lesions in rats. *Neuroscience*, **137**, 1119–1130.

Poulos, A. M., Pakaprot, N., Mahdi, B., Kehoe, E. J. & Thompson, R. F. (2006). Decremental effects of context exposure following delay eyeblink conditioning in rabbits. *Behavioral Neuroscience*, **120**, 730–734

Prados, J. (2000). Effects of varying the amount of preexposure to spatial cues on a subsequent navigation task. *Quarterly Journal of Experimental Psychology*, **53B**, 139–148.

Puga, F., Barrett, D. W., Bastida, C. C. & Gonzalez-Lima, F. (2007). Functional networks underlying latent inhibition learning in the mouse brain. *NeuroImage*, **38**, 171–183.

Purves, D., Bonardi, C. & Hall, G. (1995). Enhancement of latent inhibition in rats with electrolytic lesions of the hipopcampus. *Behavioral Neuroscience*, **109**, 366–370.

Rauhut, A. S., Thomas, B. L. & Ayres, J. J. B. (2001). Treatments that weaken Pavlovian conditioned fear and thwart its renewal in rats: implications for treating human phobias. *Journal of Experimental Psychology: Animal Behavior Processes*, **27**, 99–114.

Rauhut, A. S., McPhee, J. E., DiPietro, N. T. & Ayres, J. J. B. (2000). Conditioned inhibition training of the competing cue after compound conditioning does not reduce cue. *Animal Learning & Behavior*, **28**, 92–108.

Redhead, E. S. & Pearce, J. M. (1995a). Similarity and discrimination learning. *The Quaterly Journal of Experimental Psychology*, **48B**, 46–66.

Redhead, E. S. & Pearce, J. M. (1995b). Stimulus salience and negative patterning. *The Quaterly Journal of Experimental Psychology*, **48B**, 67–83.

Reed, P. (1991). Blocking latent inhibition. *Bulletin of the Psychonomic Society*, **29**, 292–294.

Reilly, S., Harley, C. & Revusky, S. (1993). Ibotenate lesions of the hippocampus enhance latent inhibition in conditioned taste aversion and inrease resistance to extinction in conditioned taste preference. *Behavioral Neuroscience*, **107**, 966–1004.

Reiss, S. & Wagner, A. R. (1972). CS Habituation produces a "latent inhibition effect" but no active "conditioned inhibition". *Learning and Motivation*, **3**, 237–245.

Rescorla, R. A. (1971b). Variation in the effectiveness of reinforcement and nonreinforcement following prior inhibitory conditioning. *Learning and Motivation*, **2**, 113–123.

Rescorla, R. A. (1973). Evidence for a unique-cue account of configural conditioning. *Journal of Comparative and Physiological Psychology*, **85**, 331–338.

Rescorla, R. A. (1974). Effect of inflation of the unconditioned stimulus value following conditioning. *Journal of Comparative and Physiological Psychology*, **86**, 101–106.

Rescorla, R. A. (1975). Pavlovian excitatory and inhibitory conditioning. In *Handbook of Learning and Cognitive Processes*, ed. W. K. Estes. Hillsdale, NJ: Lawrence Erlbaum Associates, vol. 2, pp. 7–35.

Rescorla, R. A. (1976). Pavlovian excitatory and inhibitory conditioning. In *Handbook of Learning and Cognitive Processes*, ed. W. K. Estes. Hillsdale, NJ: Lawrence Erlbaum Associates, vol. 2, pp. 7–35.

Rescorla, R. A. (1979). Conditioned inhibition and extinction. In *Mechanisms of Learning and Motivation: A Memorial Volume for Jerzy Konorski*, ed. A. Dickinson & R. A. Boakes. Hillsdale, NJ: Lawrence Erlbaum Associates, pp. 83–110.

Rescorla, R. A. (1982). Some consequences of associations between the excitor and the inhibitor in a conditioned inhibition paradigm. *Journal of Experimental Psychology: Animal Behavior Processes*, **8**, 288–298.

Rescorla, R. A. (1984). Associations between Pavlovian CSs and context. *Journal of Experimental Psychology: Animal Behavior Processes*, **10**, 195–204.

Rescorla, R. A. (1985). Conditioned inhibition and facilitation. In *Information Processing in Animals: Conditioned Inhibition*, ed. R. R. Miller & N. E. Spear. Hillsdale, NJ: Lawrence Erlbaum Associates, pp. 299–326.

Rescorla, R. A. (1986). Facilitation and excitation. *Journal of Experimental Psychology: Animal Behavior Processes*, **12**, 325–332.

Rescorla, R. A. (1988). Facilitation based on inhibition. *Animal Learning and Behavior*, **16**, 169–176.

Rescorla, R. A. (1989). Simultaneous and sequential conditioned inhibition in autoshaping. *The Quarterly Journal of Experimental Psychology B: Comparative and Physiological Psychology*, **41**, 275–286.

Rescorla, R. A. (1992). Hierarchical associative relations in Pavlovian conditioning and instrumental training. *Current Directions in Psychological Science*, **1**, 66–70.

Rescorla, R. A. (2000). Associative changes in excitors and inhibitors differ when they are conditioned in compound. *Journal of Experimental Psychology: Animal Behavior Processes*, **26**, 428–438.

Rescorla, R. A. (2001). Unequal associative changes when excitors and neural stimuli are conditioned in compound. *The Quarterly Journal of Experimental Psychology B: Comparative and Physiological Psychology*, **54B**, 53–68.

Rescorla, R. A. (2002). Effect of following an excitatory-inhibitory compound with an intermediate reinforcer. *Journal of Experimental Psychology: Animal Behavior Processes*, **28**, 163–174.

Rescorla, R. A. (2003). Protection from extinction. *Learning and Behavior*, **31**, 124–132.

Rescorla, R. A. (2004a). Spontaneous recovery. *Learning and Memory*, **11**, 501–509.

Rescorla, R. A. (2004b). Spontaneous recovery varies inversely with the training-extinction interval. *Learning and Behavior*, **32**, 401–408.

Rescorla, R. A., & Cunningham, C. L. (1977). The erasure of reinstated fear. *Animal Learning and Behavior*, **5**, 386–394.

Rescorla, R. A. & Cunningham, C. L. (1978). Recovery of the US representation over time during extinction. *Learning and Motivation*, **9**, 373–391.

Rescorla, R. A. & Durlach, P. J. (1981). Within-event learning in Pavlovian conditioning. In *Information Processing in Animals: Memory Mechanisms*, ed. N. E. Spear & R. R. Miller. Hillsdale, NJ: Lawrence Erlbaum Associates, pp. 81–112.

Rescorla, R. A. & Heith, C. D. (1975). Reinstatement of fear to an extinguished conditioned stimulus. *Journal of Experimental Psychology: Animal Behavior Processes*, **1**, 88–96.

Rescorla, R. A. & Wagner, A. (1972). A theory of Pavlovian conditioning: variations in the effectiveness of reinforcement and non-reinforcement. In *Classical conditioning II: Current Research and Theory*, ed. A. H. Black & W. F. Prokasy. New York: Appleton–Century–Crofts, pp. 64–99.

Revusky, S. (1971). The role of interference in association over delay. In *Animal Memory*, eds. W. K. Honig & P. H. R. James. New York: Academic, pp. 155–213.

Richards, R. W. & Sargent, D. M. (1983). The order of presentation of conditioned stimuli during extinction. *Animal Learning and Behavior*, **11**, 229–236.

Ricker, S. T. & Bouton, M. E. (1996). Reacquisition following extinction in appetitive conditioning. *Animal Learning & Behavior*, **24**, 423-436.

Rickert, E. J., Bent, T. L., Lane, P. & French, J. (1978). Hippocampectomy and the attenuation of blocking. *Behavioral Biology*, **22**, 147-160.

Rickert, E. J., Lorden, J. F., Dawson, R., Smyly, E. & Callahan, M. F. (1979). Stimulus processing and stimulus selection in rats with hippocampal lesions. *Behavioral and Neural Biology*, **27**, 454-465.

Robbins, S. J. (1990). Mechanisms underlying spontaneous recovery in autoshaping. *Journal of Experimental Psychology: Animal Behavior Processes*, **16**, 235-249.

Roberts, W. A., Cheng, K. & Cohen, J. S. (1989). Timing light and tone signals in pigeons. *Journal of Experimental Psychology: Animal Behavior Processes*, **15**, 23-25.

Rochford, J., Sen, A. P. & Quirion, R. (1996). Effect of nicotine and nicotinic receptor agonists on latent inhibition in the rat. *The Journal of Pharmacology and Experimental Therapeutics*, **277**, 1267-1275.

Rosas, J. M. & Bouton, M. E. (1996). Spontaneous recovery after extinction of a conditioned taste aversion. *Animal Learning and Behavior*, **24**, 341-348.

Rosenfield, M. E. & Moore, J. W. (1995). Connection to cerebellar cortex (Larsell's HVI) in the rabbit: A WGA-HRP study with implication for classical eyeblink conditioning. *Behavioral Neuroscience*, **109**, 1106-1118.

Ross, R. T. (1983). Relationships between the determinants of performance in serial feature positive discriminations. *Journal of Experimental Psychology: Animal Behavior Processes*, **9**, 349-373.

Ross, R. T. & Holland, P. C. (1981). Conditioning of simultaneous and serial feature positive discriminations. *Animal Learning & Behavior*, **9**, 293-303.

Ross, R. T., Orr, W. B., Holland, P. C. & Berger, T. W. (1984). Hippocampectomy disrupts acquisition and retention of learned conditional responding. *Behavioral Neuroscience*, **98**, 211-225.

Rudell, A. P., Fox, S. E. & Ranck, J. B. (1980). Hippocampal excitability phase-lock to theta rhythm in walking rats. *Experimental Neurology*, **68**, 87-96.

Rudy, J. W. & Sutherland, R. J. (1989). The hippocampal formation is necessary for rats to learn and remember configural discriminations. *Behavioral Brain Research*, **34**, 97-109.

Rudy, J. W. & Sutherland, R. J. (1995). Configural association theory and the hippocampal formation: an appraisal and reconfiguration. *Hippocampus*, **5**, 375-389.

Rudy, J. W., Krauter, E. E. & Gaffuri, A. (1976). Attenuation of the latent inhibition effect by prior exposure to another stimulus. *Journal of Experimental Psychology: Animal Behavior Processes*, **2**, 235-247.

Rudy, J. W., Rosenberg, L. & Sandell, J. H. (1977). Disruption of taste familiarity effect by novel exteroceptive stimulation. *Journal of Experimental Psychology: Animal Behavior Processes*, **88**, 665-669.

Rumelhart, D. E., Hinton, G. E. & Williams, G. E. (1986). Learning internal representations by error propagation. In *Parallel Distributed Processing: Explorations in the Microstructure of Cognition*, eds. D. E. Rumelhart & J. L. McClelland. Foundations. Cambridge, MA: Bradford Books, MIT Press, vol. 1.

Ruob, C., Weiner, I. & Feldon, J. (1998). Haloperidol-induced potentiation of latent inhibition: interaction with parameters of conditioning. *Behavioral Pharmacology*, **9**, 245–253.

Ruob, C., Elsner, J., Weiner, I. & Feldon, J. (1997). Amphetamine-induced disruption and haloperidol-induced potentiation of latent inhibition depend on the nature of the stimulus. *Behavioural Brain Research*, **88**, 35–41.

Russell, W. A. & Jenkins, J. J. (1954). *The complete Minnesota norms for responses to 100 words from the Kent–Rosanoff word association test.* Studies on the Role of Language in Behavior, Technical Report No. 11, University of Minnesota.

Santosa, C. M., Strong, C. M., Nowakowska, C., Wang, P. W., Rennicke, C. M. & Ketter, T. A. (2006). Enhanced creativity in bipolar disorder patients: a controlled study. *Journal of Affective Disorders*, Electronic publication.

Sass, L. A. (2000–2001). Schizophrenia, modernism, and the "creative imagination": on creativity and psychopathology. *Creativity Research Journal*, **13**, 55–74.

Saul'skaya N. B. & Gorbachevskaya A. I. (1998). Conditioned reflex release of dopamine in the nucleus accumbens after disruption of the hippocampal formation in rats. *Neuroscience and Behavioral Physiology*, **28**, 380–385.

Saunders, R. C., Kolachana, B. S., Bachevalier, J. & Weinberger D. R. (1998). Neonatal lesions of the medial temporal lobe disrupt prefrontal cortical regulation of striatal dopamine. *Nature*, **393**, 169–171.

Savastano, H. I., Arcediano, F., Stout, S. C. & Miller, R. R. (2003). Interaction between preexposure and overshadowing: further analysis of the extended comparator hypothesis. *The Quarterly Journal of Experimental Psychology*, **56B**, 371–395.

Schmajuk, N. A. (1987). SEAS: A dual memory architecture for computational cognitive mapping. *Proceedings of the Ninth Annual Conference of the Cognitive Science Society*, Hillsdale, NJ: Lawrence Erlbaum Associates, pp. 644–654.

Schmajuk, N. A. (1990). Role of the hippocampus in temporal and spatial navigation: an adaptive neural network. *Behavioral Brain Research*, **39**, 205–229.

Schmajuk, N. A. (1997). *Animal learning and cognition: a neural network approach.* New York: Cambridge University Press.

Schmajuk, N. A. (2001) Hippocampal dysfunction in schizophrenia. *Hippocampus*, **11**, 599–613.

Schmajuk, N. A. (2002) *Latent inhibition and its neural substrates.* Norwell, MA: Kluwer Academic.

Schmajuk, N. A. (2005). Brain–behaviour relationships in latent inhibition: a computational model. *Neuroscience and Biobehavioral Reviews*, **29**, 1001–1020.

Schmajuk, N. A. (2008a). Classical conditioning. *Scholarpedia*, **3**, 2316.

Schmajuk, N. A. (2008b). Computational models of classical conditioning. *Scholarpedia*, **3**, 1664.

Schmajuk, N. A. & Blair, H. T. (1993). Stimulus configuration, spatial learning, and hippocampal function. *Behavioural Brain Research*, **59**, 103–117.

Schmajuk, N. A. & Blair, H. T. (1995). Time, space, and the hippocampus. In *Neurobehavioral Plasticity: Learning, Development, and Response to Brain Insult*, eds. N. E. Spear, L. P. Spear & M. Woodruff. Hillsdale, NJ: Lawrence Erlbaum Associates.

Schmajuk, N. A. & Buhusi, C. V. (1997). Occasion setting, stimulus configuration, and the hippocampus: a neural network approach. *Behavioral Neuroscience*, **111**, 235–258.

Schmajuk, N. A. & DiCarlo, J. J. (1989). A neural network approach to hippocampal function in classical conditioning. *Behavioral Neuroscience*, **105**, 82–110.

Schmajuk, N. A. & DiCarlo, J. J. (1991a). Neural dynamics of hippocampal modulation of classical conditioning. In *Neural Network Models of Conditioning and Action*, ed. M. Commons, S. Grossberg & J. E. R. Staddon. Hillsdale, NJ: Lawrence Erlbaum Associates, pp. 149–180.

Schmajuk, N. A. & DiCarlo, J. J. (1991b). A neural network approach to hippocampal function in classical conditioning. *Behavioral Neuroscience*, **105**, 82–110.

Schmajuk, N. A. & DiCarlo, J. J. (1992). Stimulus configuration, classical conditioning, and the hippocampus. *Psychological Review*, **99**, 268–305.

Schmajuk, N. A. & Holland, P. C. (1995). *Multiple response systems in classical conditioning*. Proceedings of the World Congress on Neural Networks, Washington, DC, vol. 1, pp. 700–703.

Schmajuk, N. A. & Kutlu, G. M. (2009). The computational nature of associative learning. *Behavioral Brain Science*, **32**, 223–224.

Schmajuk, N. A. & Larrauri, J. A. (2006). Experimental challenges to theories of classical conditioning: application of an attentional model of storage and retrieval. *Journal of Experimental Psychology: Animal Behavior Processes*, **32**, 1–20.

Schmajuk, N. A. & Larrauri, J. A. (2008). Associative models describe both causal learning and conditioning. *Behavioral Processes*, **77**, 443–445.

Schmajuk, N. A. & Moore, J. W. (1985). Real-time attentional models for classical conditioning and the hippocampus. *Physiological Psychology*, **13**, 278–290.

Schmajuk, N. A. & Moore, J. W. (1988). The hippocampus and the classically conditioned nictitating membrane response: a real-time attentional-associative model. *Psychobiology*, **16**, 20–35.

Schmajuk, N. A. & Moore, J. W. (1989). Effects of hippocampal manipulations on the classically conditioned nictitating membrane response: simulations by an attentional associative model. *Behavioral Brain Research*, **32**, 173–189.

Schmajuk, N. A. & Thieme, A. D. (1992). Purposive behavior and cognitive mapping: an adaptive neural network. *Biological Cybernetics*, **67**, 165–174.

Schmajuk, N. A. & Tyberg, M. (1991). The hippocampal lesion animal model of schizophrenia. In *Animal Models in Psychiatry*, ed. A. Boulton, G. Baker & M. T. Martin Iverson. Clifton, NJ: Humana Press.

Schmajuk, N. A., Aziz, D. R. & Bates, M. J. B. (2009). Attentional-associative interactions in creativity. *Creativity Research Journal*, **21**, 92–103.

Schmajuk, N. A., Buhusi, C. V. & Gray, J. A. (1998). The pharmacology of latent inhibition: a neural network approach. *Behavioural Pharmacology*, **9**, 711–730.

Schmajuk, N. A., Christiansen, B. A. & Cox, L. (2000). Haloperidol reinstates latent inhibition impaired by hippocampal lesions: data and theory. *Behavioral Neuroscience*, **114**, 659–670.

Schmajuk, N. A., Cox, L. & Gray, J. A. (2001). Nucleus accumbens, entorhinal cortex and latent inhibition: a neural network model. *Behavioral Brain Research*, **118**, 123–141.

Schmajuk, N. A., Gray, J. A. & Larrauri, J. A. (2005). A pre-clinical study showing how dopaminergic drugs administered during pre-exposure can impair or facilitate latent inhibition. *Psychopharmacology*, **177**, 272–279.

Schmajuk, N. A., Lam, P. & Christiansen, B. A. (1994). Hippocampectomy disrupts latent inhibition of the rat eyeblink conditioning. *Physiology and Behavior*, **55**, 597–601.

Schmajuk, N. A., Lam, P. & Christiansen, B. A. (1994). Hippocampectomy disrupts latent inhibition of the rat eyeblink conditioning. *Physiology and Behavior*, **55**, 597–601.

Schmajuk, N. A., Lam, Y. & Gray, J. A. (1996). Latent inhibition: a neural network approach. *Journal of Experimental Psychology: Animal Behavior Processes*, **22**, 321–349.

Schmajuk, N. A., Lamoureux, J. A. & Holland, P. C. (1998). Occasion setting and stimulus configuration: a neural network approach. *Psychological Review*, **105**, 3–32.

Schmajuk, N. A., Spear, N. E. & Isaacson, R. L. (1983). Absence of overshadowing in rats with hippocampal lesions. *Physiological Psychology*, **11**, 59–62.

Schmajuk. N. A., Thieme, A. D. & Blair, H. T. (1993). Maps, routes, and the hippocampus: a neural network approach. *Hippocampus*, **3**, 387–400.

Schmajuk, N. A., Larrauri, J. A., De la Casa, L. G. & Levin, E. D. (2009). Attenuation of auditory startle and prepulse inhibition by unexpected changes in ambient illumination through dopaminergic mechanisms. *Behavioural Brain Research*, **197**, 251–261.

Schneider, W. & Shiffrin, R. M. (1977). Controlled and automatic human information processing: detection, search and attention. *Psychological Review*, **84**, 1–66.

Schnur, P. & Lubow, R. E. (1976). Latent inhibition: the effects of ITI and CS intensity during preexposure. *Learning and Motivation*, **7**, 540–550.

Schrag, A & Trimble, M. (2001). Poetic talent unmasked by treatment of Parkinson's disease. *Movement Disorders*, **16**, 1175–1176.

Schreurs, B. G. & Westbrook, R. F. (1982). The effect of changes in the CS–US interval during compound conditioning upon an otherwise blocked element. *Quarterly Journal of Experimental Psychology*, **34B**, 19–30.

Schultz, W. (1998). Predictive reward signal of dopamine neurons. *Journal of Neurophysiology*, **80**, 1–27.

Schultz, W. & Dickinson, A. (2000). Neuronal coding of prediction errors. *Annual Review of Neuroscience*, **23**, 473–500.

Sears, L. L. & Steinmetz, J. E. (1990) Acquisition of classically conditioned-related activity in the hippocampus is affected by lesion of the cerebelllar interpositus nucleus. *Behavioral Neuroscience*, **104**, 681–692.

Seidman, L. J. (1983). Schizophrenia and brain dysfunction: an integration of recent neurodiagnostic findings. *Psychological Bulletin*, **94**, 195–238.

Seillier, A., Dieu, Y., Herbeaux, K., Di Scala, G., Will, B. & Majchrzak, M. (2007). Evidence for a critical role of entorhinal cortex at pre-exposure for latent inhibition disruption in rats. *Hippocampus*, **17**, 220–226.

Shanks, D. R. (1985). Forward and backward blocking in human contingency judgment. *Quarterly Journal of Experimental Psychology*, **37B**, 1–21.

Shepard, R. N. (1991). Integrality vs. separability of stimulus dimensions. In *The Perception of Structure*, ed. G. R. Lockhead & J. R. Pomerantz. Washington, DC: American Psychological Association, pp. 53–77.

Sherman, J. E. & Maier, S. F. (1978). The decrement in conditioned fear with increased trials of simultaneous conditioning is not specific to the simultaneous procedure. *Learning and Motivation*, **9**, 31–53.

Shevill, I. & Hall, G. (2004). Retrospective revaluation effects in the conditioned suppression procedure. *The Quarterly Journal of Experimental Psychology B: Comparative and Physiological Psychology*, **57B**, 331–347.

Shimamura, A. P. & Squire, L. R. (1984). Paired-associate learning and priming effects in amnesia: a neuropsychological study. *Journal of Experimental Psychology: General*, **113**, 556–570.

Shohamy, D., Allen, M. T. & Gluck, M. A. (2000). Dissociating entorhinal and hippocampal involvement in latent inhibition. *Behavioral Neuroscience*, **114**, 867–874.

Sidman, M. (1986). Functional analysis of emergent verbal classes. In *Analysis and Integration of Behavioral Units*, ed. T. Thompson & M. D. Zeiler. Hillsdale, NJ: Lawrence Erlbaum Associates, pp. 213–235.

Siegel, S. & Domjan, M. (1971). Backward conditioning as an inhibitory procedure. *Learning and Motivation*, **2**, 1–11.

Silbersweig, D. A., Stern, E., Frith, C., Cahill, C., Holmes, A., Grootoonk, S., Seaward, J., McKenna, P., Chua, S. E., Schnorr, L., Jones, T. & Frackowiak, R. S. J. (1995). A functional neuroanatomy of hallucinations in schizophrenia. *Nature*, **378**, 176–179.

Skinner, B. E. (1938). *The Behavior of Organisms: An Experimental Analysis*. Englewood Cliffs, NJ: Prentice-Hall.

Skinner, B. F. (1950). Are theories of learning necessary? *Psychological Review*, **57**, 193–216.

Smith, M. (1968). CS–US interval and US intensity in classical conditioning of the rabbit's nictitating membrane response. *Journal of Comparative and Physiological Psychology*, **66**, 679–687.

Snyder, S. H. (1980). *Biological Aspects of Mental Disorder*. New York: Oxford.

Sokolov, E. N. (1960). Neuronal models and the orienting reflex. In *The Central Nervous System and Behavior*, ed. M. A. B. Brazier. New York: Macy Foundation.

Sokolov, Y. N. (1963). *Perception and the Conditioned Reflex*. Oxford: Pergamon.

Solomon, P. R. (1977). Role of the hippocampus in blocking and conditioned inhibition of rabbit's nictitating membrane response. *Journal of Comparative and Physiological Psychology*, **91**, 407–417.

Solomon, P. R., Brennan, G. & Moore, J. W. (1974). Latent inhibition of the rabbit's nictitating membrane response as a function of CS intensity. *Bulletin of the Psychonomic Society*, **4**, 445–448.

Solomon, P. R., Crider, A., Winkelman, J. W., Turi, A., Kamer, R. M. & Kaplan, L. J. (1981). Disrupted latent inhibition in the rat with chronic amphetmaine or haloperidol-induced supersensitivity: relationship to schizophrenic attention disorder. *Biological Psychiatry*, **16**, 519–537.

Soltysik, S. (1985). Protection from extinction: new data and a hypothesis of several varieties of conditioned inhibition. In *Information Processing in Animals: Conditioned Inhibition*, ed. R. R. Miller & N. E. Spear. Hillsdale, NJ: Lawrence Erlbaum Associates.

Spear, N. E. (1971). Forgetting as retrieval failure. In *Animal Memory*, ed. W. K. Honig & P. H. R. James. New York: Academic, pp. 45–109.

Spear, N. E. (1981). Extending the domain of memory retrieval. In *Information Processing in Animals: Memory, Mechanisms*, ed. R. R. Miller & N. E. Spear. Hillsdale, NJ: Lawrence Erlbaum Associates, pp. 341–378.

Spear, N. E., Miller, J. S. & Jagielo, J. A. (1990). Animal memory and learning. *Annual Review of Psychology*, **41**, 169–211.

Spence, K. W. & Norris, E. B. (1950). Eyelid conditioning as a function of the inter-trial interval. *Journal of Experimental Psychology*, **40**, 716–720.

Squire, L. R., Shimamura, A. P. & Amaral, D. G. (1989). Memory and the hippocampus. In *Neural Models of Plasticity*, ed. J. H. Byrne & W. O. Berry. San Diego, CA: Academic, pp. 208–239.

Staddon, J. E. R. & Higa, J. (1999). Time and memory: towards a pacemaker free theory of interval timing. *Journal of the Experimental Analysis of Behavior*, **71**, 215–251.

Sternberg, R. J. & Lubart, T. I. (1995). An investment perspective on creative insight. In *The Nature of Insight*, eds. R. J. Sternberg & J. E. Davidson. Cambridge, MA: MIT Press.

Stout, S. C. & Miller, R. R. (2007). Sometimes-competing retrieval (SOCR): a formalization of the comparator hypothesis. *Psychological Review*, **114**, 759–783.

Strasser, H. C., Lilyestrom, J., Ashby, E. R., Honeycutt, N. A., Schretlen, D. J., Pulver, A. E., Hopkins, R. O., Depaolo, J. R., Potash, J. B., Schweitzer, B., Yates, K. O., Kurian, E., Barta, P. E. & Pearlson, G. D. (2005). Hippocampal and ventricular volumes in psychotic and nonpsychotic bipolar patients compared with schizophrenia patients and community control subjects: a pilot study. *Biological Psychiatry*, **57**, 633–639.

Suiter, R. D. & LoLordo, V. M. (1971). Blocking of inhibitory Pavlovian conditioning in the conditioned emotional response procedure. *Journal of Comparative and Physiological Psychology*, **76**, 137–144.

Sutherland, N. S., Mackintosh, N. J. & Mackintosh, J. (1963). Simultaneous discrimination training of Octopus and transfer of discrimination along a continuum. *Journal of Comparative and Physiological Psychology*, **56**, 150–156.

Sutherland, R. J. & Rudy, J. W. (1989). Configural association theory: the role of the hippocampal formation in learning, memory, and amnesia. *Psychobiology*, **17**, 129–144.

Sutton, R. S. & Barto, A. G. (1981). Toward a modern theory of adaptive networks: expectation and prediction. *Psychological Review*, **88**, 135–170.

Sutton, R. S. & Barto, A. G. (1990). Time derivative models of Pavlovian reinforcement. In: *Learning and Computational Neuroscience: Foundations of Adaptive Networks,* eds. M. Gabriel & J. Moore. Cambridge, MA: MIT Press, pp. 497–537.

Swartzentruber, D. (1995). Modulatory mechanisms in Pavlovian conditioning. *Animal Learning and Behavior*, **23**, 123–143.

Swerdlow, N. R. & Koob, G. F. (1987). Dopamine, schizophrenia, mania, and depression: toward a unified hypothesis of cortico-striato-pallido-thalamic function. *Behavioral and Brain Sciences*, **10**, 197–245.

Talk, A. C., Gandhi, C. C. & Matzel, L. D. (2002) Hippocampal function during behaviourally silent associative learning: dissociation of memory storage and expression. *Hippocampus*, **12**, 648–656.

Tamai, N. & Nakajima, S. (2000). Renewal of formerly conditioned fear in rats after extensive extinction training. *International Journal of Comparative Psychology*, **13**, 137–146.

Tassoni, C. (1995). The least mean squares network with information coding: a model of cue learning. *Journal of Experimental Psychology: Learning, Memory & Cognition*, **21**, 193–204.

Taylor, K. M., Joseph, V. T., Balsam, P. D. & Bitterman, M. E. (2008). Target-absent controls in blocking experiments with rats. *Learning & Behavior*, **36**, 145–148.

Terrace, H. S. & McGonigle, B. (1994). Memory and representation of serial order by children, monkeys, and pigeons. *Current Directions in Psychological Science*, **3**, 80–185.

Testa, T. J. & Ternes, J. W. (1977). Specificity of conditioning mechanisms in the modification of food preferences. In *Learning Mechanisms in Food Selection*, eds. L. M. Barker, M. R. Best & M. Domjan. Waco, TX: Baylor University Press, pp. 229–253.

Thomas, B. L. & Ayres, J. J. B. (2004). Use of the ABA fear renewal paradigm to assess the effects of extinction with co-present fear inhibitors or excitors: implications for theories of extinction and for treating human fears and phobias. *Learning and Motivation*, **35**, 22–52.

Thomas, B. L. & Papini, M. R. (2001). Adrenalectomy eliminates the extinction spike in autoshaping with rats. *Physiology and Behavior*, **72**, 543–547.

Thomas, B. L., Larsen, N. & Ayres, J. J. B. (2003). Role of context similarity in ABA, ABC, and AAB renewal paradigms: implications for theories of renewal and for treating human phobias. *Learning and Motivation*, **34**, 410–436.

Thompson, R. E. (1986). The neurobiology of learning and memory. *Science*, **233**, 941–947.

Tolman, E. C. (1932). *Purposive Behavior in Animals and Men*. New York, NY: Irvington.

Torrance, E. P. (1968). Examples and rationales of test tasks for assessing creative abilities. *Journal of Creative Behavior*, **2**, 165–178.

Totterdell, S. & Meredith, G. E. (1997). Topographical organization of projections from the entorhinal cortex to the striatum of the rat. *Neuroscience*, **78**, 715–729.

Trobalon, J. B., Chamizo, V. D. & Mackintosh, N. J. (1992). Role of context in perceptual learning in maze discriminations. *The Quarterly Journal of Experimental Psychology*, **44B**, 57–73.

Turing, A. M. (1950). Computing machinery and intelligence. *Mind*, **59**, 433–460.

Urcelay, G. P. & Miller, R. R. (2008). Counteraction between two kinds of conditioned inhibition training. *Psychonomic Bulletin & Review*, **15**, 103–107.

Van Hamme, L. & Wasserman, E. (1994). Cue competition in causality judgments: the role of nonpresentation of compound stimulus elements. *Learning and Motivation*, **25**, 127–151.

Vansteenwegen, D., Hermans, D., Vervliet, B., Francken, G., Beckers, T., Baeyens, F. & Helen, P. (2005). Return of fear in a human differential conditioning paradigm caused by a return to the original acquisition context. *Behaviour Research and Therapy*, **43**, 323–336.

Vervliet, B., Vansteenwegen, D., Baeyens, F., Hermans, D. & Helen, P. (2005). Return of fear in a human differential conditioning paradigm caused by a stimulus change after extinction. *Behaviour Research and Therapy*, **43**, 357–371.

Wagner, A. R. (1976). Priming in STM: an information-processing mechanism for self-generated or retrieval-generated depression in performance. In *Habituation: Perspectives from Child Development, Animal Behavior, and Neurophysiology*, ed. T. J. Tighe & R. N. Leaton. Hillsdale, NJ: Lawrence Erlbaum Associates, pp. 95–128.

Wagner, A. R. (1978). Expectancies and the priming of STM. In *Cognitive Processes in Animal Behavior*, eds. S. H. Hulse, H. Fowler & W. K. Honig. Hillsdale, NJ: Lawrence Erlbaum Associates, pp. 177–209.

Wagner, A. R. (1979). Habituation and memory. In *Mechanisms of Learning and Motivation*, eds. A. Dickinson & R. A. Boakes. Hillsdale, NJ: Lawrence Erlbaum Associates.

Wagner, A. R. (1981). SOP: A model of automatic memory processing in animal behavior. In *Information Processing in Animals: Memory Mechanisms*, eds. N. E. Spear & R. R. Miller. Hillsdale, NJ: Lawrence Erlbaum Associates, pp. 5–47.

Wagner, A. R. (1992). Some complexities anticipated by AESOP and other dual-representation theories. Paper abstracted in H. Kimmel (Chair), Symposium on Pavlovian Conditioning with Complex Stimuli, XXV International Congress of Psychology. *International Journal of Psychology*, 101–102.

Wagner, A. R. & Brandon, S. E. (1989). Evolution of a structured connectionist model of Pavlovian conditioning (AESOP). In *Contemporary Learning Theories: Pavlovian Conditioning and the Status of Traditional Learning Theory*, ed. S. B. Klein & R. R. Mowrer. Lawrence Erlbaum Associates, Hillsdale, NJ, pp. 149–189.

Wagner, A. R., Logan, F. A., Haberlandt, K. & Price, T. (1968). Stimulus selection in animal discrimination learning. *Journal of Experimental Psychology*, **76**, 171–180.

Wallach, M. A. (1970). Creativity. In *Carmichael's Manual of Child Psychology*, 4th edn., ed. P. H. Mussen. New York: Wiley, pp. 1211–1272.

Wallach, M. A. & Wing, C. W. (1969). *The Talented Student*. New York: Holt, Rinehart & Winston.

Wallas, G. (1926). *The Art of Thought*. London: Watts.

Warburton, E. C., Mitchell, S. N. & Joseph M. H. (1996). Calcium dependent dopamine release following a second amphetamine challenge: relation to the disruption of latent inhibition. *Behavioral Pharmacology*, **7**, 119–129.

Ward, W. C. (1969). Creativity and environmental cues in nursery school children. *Developmental Psychology*, **1**, 543–547.

Ward-Robinson, J. & Hall, G. (1996). Backward sensory preconditioning. *Journal of Experimental Psychology: Animal Behavior Processes*, **22**, 395–404.

Ward-Robinson, J., Coutureau, E., Good, M., Honey, R. C., Killcross, A. S. & Oswald, C. J. (2001). Excitotoxic lesions of the hippocampus leave sensory preconditioning intact: implications for models of hippocampal function. *Behavioral Neuroscience*, **115**, 1357–1362.

Weinberger, N. M. (1995). Dynamic regulation of receptive fields and maps in the adult cortex. *Annual Review of Neuroscience*, **18**, 129–158.

Weiner, I. (1990). Neural substrates of latent inhibition: the switching model. *Psychological Bulletin*, **108**, 442–461.

Weiner, I. (2003). The "two-headed" latent inhibition model of schizophrenia: modeling positive and negative symptoms and their treatment. *Psychopharmacology*, **169**, 257–297.

Weiner, I. & Feldon, J. (1997). The switching model of latent inhibition: an update of neural substrates. Behavioral *Brain Research*, **88**, 11–25.

Weiner, I., Lubow, R. E. & Feldon, J. (1984). Abolition of the expression but not the acquisition of latent inhibition by chronic amphetamine in rats. *Psychopharmacology*, **83**, 191–199.

Weiner, I., Lubow, R. E. & Feldon, J. (1988). Disruption of latent inhibition by acute administration of low doses of amphetamine. *Pharmacology, Biochemistry, and Behavior*, **30**, 871–878.

Weiner, I., Gal, G., Rawlins, J. N. P. & Feldon, J. (1996). Differential involvement of the shell and core subterritories of the nucleus accumbens in latent inhibition and amphetamine-induced activity. *Behavioural Brain research*, **81**, 123–133.

Weisberg, R. (1986) *Creativity: Genius and Other Myths*. New York: W. H. Freeman/ Times Books/ Henry Holt & Co.

Weiss, C., Kroforst-Colllins, M. A. & Disterhoft, J. F. (1996). *Activity of hippocampal pyramidal neurons during trace eyeblink conditioning*. Hippocampus, 6, 192–209.

Werbos, P. (1974). *Beyond regression: New tools for prediction and analysis in the behavioral sciences*. Doctoral dissertation, Harvard University, Cambridge, MA.

Werbos, P. (1987). Building and understanding adaptive systems: a statistical/ numerical approach to factory automation and brain research. *IEEE Transactions SMC*, March/April, 1987.

Wertheimer, M. (1959). *Productive Thinking*. Oxford: Harper.

Westbrook, R. F., Bond, N. W. & Feyer, A-M. (1981). Short-and long-term decrements in toxicosis-induced odor-aversion learning: the role of duration of exposure to an odor. *Journal of Experimental Psychology: Animal Behavior Processes*, **7**, 362–381.

Westbrook, R. F., Good, A. J. & Kiernan, M. J. (1997). Microinjection of morphine into the nucleus accumbens impairs contextual learning in rats. *Behavioral Neuroscience*, **111**, 996–1013.

Westbrook, R. F., Iordanova, M., McNally, G., Richardson, R. & Harris, J. A. (2002). Reinstatement of fear to an extinguished conditioned stimulus: two roles for context. *Journal of Experimental Psychology: Animal Behavior Processes*, **28**, 97–110.

Wheeler, D. S., Stout, S. C. & Miller, R. R. (2004). Interaction of retention interval with CS-preexposure and extinction treatments: symmetry with respect to primacy. *Learning & Behavior*, **32**, 335–347.

Whishaw, I. Q. & Tomie, J. (1991). Acquisition and retention by hippocampal rats of simple, conditional and configural tasks using tactile and olfactory cues: implications for hippocampal function. *Behavioral Neuroscience*, **105**, 787–797.

Wickelgren, W. A. (1979). Chunking and consolidation: a theoretical synthesis of semantic networks, configuring in conditioning, S-R versus cognitive learning, normal forgetting, the amnesic syndrome, and the hippocampal arousal system. *Psychological Review*, **86**, 44–60.

Wickens, C., Tuber, D. S. & Wickens, D. D. (1983). Memory for the conditioned response: the proactive effect of preexposure to potential conditioning stimuli and context change. *Journal of Experimental Psychology: General*, **112**, 41–57.

Widrow, B. & Hoff, M. E. (1960). Adaptive switching circuits. *1960 IRE WESCON Convention Record*, pp. 96–104.

Williams, B. (1996). Evidence that blocking is due to associative deficit: blocking history affects the degree of subsequent associative competition. *Psychonomic Bulletin & Review*, **3**, 71–74.

Williams, G. V., Rolls, E. T., Leonard, C. M. & Stern, C. (1993). Neuronal responses in the ventral striatum of the behaving macaque. *Behavioral Brain Research*, **55**, 243–252.

Wilson, A., Brooks, D. C. & Bouton, M. E. (1995). The role of the rat hippocampal system in several effects of context in extinction. *Behavioral Neuroscience*, **109**, 828–836.

Wilson, P. N. & Pearce, J. M. (1989). A role for stimulus generalization in conditional discrimination learning. *Quarterly Journal of Experimental Psychology*, **41B**, 243–273.

Wilson, P. N. & Pearce, J. M. (1990). Selective transfer of responding in conditional discriminations. *Quarterly Journal of Experimental Psychology*, **42B**, 41–58.

Wilson, P. N., Boumphrey, P. & Pearce, J. M. (1992). Restoration of the orienting response to a light by a change in its predictive accuracy. *The Quarterly Journal of Experimental Psychology*, **44B**, 17–36.

Winocur. G., Rawlins, J. N. P. & Gray, J. A. (1987). The hippocampus and conditioning to contextual cues. *Behavioral Neuroscience*, **101**, 617–625.

Winston, P. H. (1977). *Artificial Intelligence*. Reading, MA: Addison-Wesley.

Witcher, E. S. & Ayres, J. J. B. (1984). A test of two methods for extinguishing Pavlovian conditioned inhibition. *Animal Learning and Behavior*, **12**, 149–156.

Woodbury, C. B. (1943). The learning of stimulus patterns by dogs. *Journal of Comparative Psychology*, **35**, 29–40.

Xu, Y. & Corkin, S. (2001). H. M. revisits the Tower of Hanoi puzzle. *Neuropsychology*, **15**, 69–79.

Yadav, R. N., Kumar, N., Kalra, P. K. & John, J. (2006). Learning with generalized-mean neuron model. *Neurocomputing: An International Journal*, **69**, 2026–2032.

Yee, B. K., Feldon, J. & Rawlins, J. N. P. (1995). Latent inhibition in rats is abolished by NMDA-induced neuronal loss in the retrohippocampal region, but this lesion effect can be prevented by systemic haloperidol treatment. *Behavioral Neuroscience*, **109**, 227–240.

Yeo, C. H., Hardiman, M. J., Moore, J. W. & Rusell, I. S. (1984). Trace conditioning of the nictitating membrane response in decorticate rabbits. *Behavioural Brain Research*, **11**, 85–88.

Young, M. E. & Wasserman, E. A. (2002). Limited attention and cue order consistency affect predictive learning: a test of similarity measures. *Journal of Experimental Psychology: Learning, Memory, and Cognition*, **28**, 484–496.

Zackheim, J., Myers, C. & Gluck, M. (1998). A temporally sensitive recurrent neural network model of occasion setting. In *Occasion Setting: Associative Learning and Cognition in Animals*, ed. N. A. Schmajuk & P. C. Holland. Washington, DC, US: American Psychological Association, pp. 319–342, p. xxi.

Zimmer-Hart, C. L. & Rescorla, R. A. (1974). Extinction of Pavlovian conditioned inhibition. *Journal of Comparative and Physiological Psychology*, **86**, 837–845.

Author Index

Achenbach, G.G., 232
Alba Hernandez, J, 228, 418
Aleman, A., 135
Allen, M.T., 105
Amabile, T.M., 123
Amaral, D.G., 324
Amsel, A., 233
Anderson, D. C., 72, 73
Anderson, J.M., 133
Andreasen, N.J., 129, 130, 139
Arcediano, F., 149
Arts, B., 135
Ashby, E.R., 135
Atre-Vaidya, N., 139
Auther, A., 135
Aylward, E., 139
Ayres, J.J.B., 35, 72, 73, 91, 153, 195, 205
Aziz, D.R., 121
Azorlosa, J.L., 241

Bachevalier, J., 116
Bailey, G.K., 194, 195, 200, 221, 222
Baker, A.G., 28, 31, 80, 257
Baker, J.D., 241
Baker, P.A., 31
Balleine, B., 83, 85
Balsam, P.D., 180

Bandettini, P.A., 57, 59, 61–63, 98
Bannerman, D.B., 95
Barad, M., 173
Bardgett, M.E., 116
Barela, P., 171
Barnet, R.C., 85, 141, 366, 368, 371, 372, 374–376, 378, 380, 381, 391, 392, 394
Barrett, O. V., 72
Barta, P.E., 135
Barto, A.G., 10, 252, 367, 391
Baruch I., 117, 139
Bastida, C.C., 93
Bates, M.J.B., 121
Beckers, T., 30, 166, 411, 413, 414, 416, 417, 421
Beggs, A.L., 239
Belling, L., 72
Bent, T.L., 354
Berger, T.W., 99, 324, 326, 333
Berthier, N.E., 326
Best, M.R., 76
Bettes, B., 139
Bitterman, M.E., 145
Blair, H.T., 98, 99, 259, 323, 326, 328, 346, 351, 360, 419
Blair, N. J., 53, 164

Blaisdell, A.P., 11, 12, 15, 53, 141, 145, 149, 150, 152, 154, 155, 157, 158, 165, 166, 168, 431
Blanding, N., 228, 418
Blough, D.S., 30, 39
Blouin, A.M., 173
Boakes, R.A., 404, 406
Boden, M. A., 123, 135
Bolles, R.C., 141, 177, 195, 205–207, 211, 213, 214, 218, 219, 226, 232, 234, 236, 246
Bombace, J.C., 88
Bonardi, C., 93, 106, 317
Bond, N.W., 71
Bottjer, S.W., 185, 227
Boughner, R.L., 233
Boumphrey, P., 80, 81, 92
Bouton, M. E., 89, 169, 170, 173, 175–178, 181, 186, 188, 189, 193–195, 200–202, 204–207, 211, 213–223, 226, 228–234, 236, 243, 246, 251, 284, 285, 288, 299, 313, 314, 319, 320, 340, 343, 359, 400–402, 404, 406–408

Bower, G.H., 8
Bowker, J.L., 190, 191, 203, 401, 402, 407, 408
Brainard, M.S., 325
Brandon, S.E., 10, 30, 39, 54, 55, 179, 180, 259, 313, 314, 316, 317, 319
Breese, C.R., 356
Broadbent, H.A., 367, 390
Brodal, P., 326
Brodie, H.K., 135
Brogden, W.J., 24, 35, 418
Brooks D.C., 190, 191, 202–204, 233, 234, 400–402, 404, 406–408
Brown, A., 16
Buchanan, S.L., 333
Buchel, C., 64
Buhusi, C. V., 25, 33, 65, 103, 106, 120, 124, 138, 237, 239, 247, 354, 365, 389, 393, 399, 401, 409, 427
Bullmore, E.T., 133
Bunney, W.E., 135
Bunsey, M., 419
Burch, G.S.J., 139
Burke, J., 11, 14, 15, 145, 149, 159, 164, 166, 180
Burkhardt, P.E., 34
Burstein, K.R., 194
Bush, R. R., 10

Cahill, C., 116
Cain, C.K., 173, 175, 226
Callahan, M.F., 354
Campbell, D.T., 123
Campolattaro, M.M., 346
Candido, A., 106, 354
Capaldi, E.J., 170, 194, 223, 233, 235
Carbary, T.J., 89
Carelli, R.M., 100
Carpenter, T.A., 133
Carr, A.F., 69, 71

Carson, S.H., 123, 124, 127–130, 138
Cassaday, H.J., 106
Cassidy, S., 72
Catterson, D., 89
Chamizo, V.D., 81
Chan, K-H., 238
Chang, C., 184, 185
Chang, R.C., 197, 198, 200, 201, 223, 228
Channel, S., 80, 83
Channon, S., 336
Chapman, G., 11
Cheng, K., 366
Choi, J. S., 367, 368, 390–392
Chorazyna, H., 175, 194
Chrapusta, S., 116
Christiansen, B.A., 119, 138, 354
Chua, S.E., 116
Church, J., 366, 367, 390
Church, R.M., 367, 390, 428
Clark, G.A., 99
Cohen, J.S., 366
Cohen, N.J., 253
Collins, A.C., 103
Colwill, R.M., 314
Condoleon, M., 407, 413
Connor, E., 135
Corcoran, K.A., 243
Corkin, S., 419
Cornblatt, B.A., 135
Corr, P.J., 135
Cotton, M.M., 354
Coutureau, E., 99, 112
Cox, L., 33, 97, 119
Crider, A., 99
Crowell, C. R., 72, 73
Csernansky, J.G., 116
Csikzentmihalyi, M., 134
Cunningham, C. L., 176, 180, 211–213, 219, 226, 235, 246

Daly, H.B., 233
Daly, J.T., 233
Dartnall, S., 178
Date, S., 105
Daum, I., 336, 353
Davidson, T.L., 243, 266, 295, 320, 337, 339, 341, 346, 351–354, 360, 361
Davis M., 184
Dawson, R., 354
De Houwer, J., 30, 162, 403, 405–409, 413
De la Casa, L.G., 65, 133, 141, 168
Deadwyler, S.A., 100, 356
Delamater, A.R., 209, 218, 227, 246
Denniston, J.C., 8, 15, 16, 143, 159, 201, 202, 204, 205, 227, 232, 235, 236
Depaolo, J.R., 135
Desmond, J.E., 326, 368, 391, 392
D'Esposito, M., 62
Devenport, L.D., 169, 195
DeVietti, T. L., 72
Devinsky, O., 64
Di Scala, G., 95, 108, 109, 118
DiCarlo, J. J., 8, 12, 13, 30, 39, 89, 90, 223, 251–253, 255, 259, 308, 323, 324, 326, 328, 346, 356, 358, 359, 367, 369–371, 381, 382, 390, 392–394
Dickey, C.C., 139
Dickinson, A., 7, 9, 11, 14, 15, 54, 89, 102, 105, 145, 149, 159, 164, 166, 180
Dietrich, A., 137
Dieu, Y., 108, 109, 118
DiMattia, B.D., 352, 353
DiPietro, N.T., 153
Disterhoft, J.F., 324
Dolan, R.J., 62

Domjan, M., 35
Donovan, T., 133
Dourish, C.T., 101, 102
Duda, R.O., 11
Dunsmoor, J.E., 59–61, 63–65, 102
Durlach, P.J., 30, 36
Dykes, M., 139

Eichenbaum, H., 419
Elsner, J., 102
Elsworth, J.D., 241
Everitt, B.J., 101
Eysenck, H.J., 123, 124, 129, 133

Falls, W.A., 246
Fanselow, M.S., 71, 72, 170
Feldon, J., 97, 102, 103, 106, 114, 116, 136, 138, 354
Fetterman, J.G., 367, 390
Feyer, A-M., 71
Fiorillo, C.D., 54, 65, 102
Flaherty, A.W., 138, 139
Fletcher, P.C., 137
Forbes, D.R., 80
Fowler, H., 37, 57, 181
Frackowiak, R.S.J., 116
Francken, G., 239
Freeman, J.H., 346
Freeman, J.H. Jr., 136
French, J., 354
Friston, K.J., 62
Frith, C., 116
Frohardt R.J., 239, 240, 242, 243, 246
Fuss, J., 149
Fuster, J.M., 64

Gabel, J., 31
Gabora, L., 136
Gaffuri, A., 80
Gal, G., 114, 116
Galani, R., 95, 108
Galef, B. G., Jr., 36, 57

Gallagher, M., 106, 107
Gallagher, P.C., 326, 353, 354, 360
Gallistel, C.R., 170, 233
Gallo, M., 106, 354
Gandhi, C.C., 234
Garcia-Gutierrez, A., 209, 236
Garrud, P., 352, 354
Geier, L., 16, 200, 222, 225
Gelperin, A., 9
Gemberling, G.A., 76
Getzels, J., 134
Gewirtz, J.C., 246
Gibbon, J., 170, 180, 233, 367, 390
Gibbons, H., 133
Gillman, K., 341
Gleitman, H., 169, 194, 235
Glick-Oberwise, F., 135
Gluck, M.A., 313, 319, 358–360
Goddard, M.J., 366, 375
Gonzalez, F., 170
Gonzalez-Lima, F., 93
Good, M., 26, 106, 109, 115, 341, 343, 346, 352–354, 361
Goodal, G., 354
Goodwin, F.K., 139
Gorbachevskaya A.I., 117
Gordon, W.C., 90
Gormezano, I., 35
Gosselin, O., 95, 108
Gough, H.G., 129
Gould, T.J., 103–105
Grace, A.A., 116
Graham, P.W., 326
Grahame, N.J., 12, 16, 85, 141, 168, 194, 199, 204, 226, 229, 366, 368, 371, 394
Granger, R., 360
Gray J.A., 8, 9, 21, 24, 25, 33, 39, 59, 65, 97, 100–103,
105, 106, 116, 117, 133, 139, 336, 343, 429
Gray N.S., 117
Grigoryan, G.A., 105
Groenewegen, H.J., 99
Grootoonk, S., 116
Grossberg, S., 7, 12, 21, 22, 90, 365, 367–369, 372, 390–394
Groves, P.M., 75
Guarraci, F.A., 234–238, 241
Guez, D., 149
Guilford, J. P., 123
Gunther, L., 53, 85, 141, 150

Haas, M.L., 263
Haberlandt, K., 35, 56, 137
Haga, Y., 161
Hall, G., 7, 13, 17, 21, 24–26, 26, 30, 53, 64, 65, 69, 70, 76, 77, 80, 83, 89, 90, 93, 95, 106, 111, 149, 150, 163, 164, 169, 173, 174, 180, 223, 225, 263, 317, 400, 426
Hallam, S.C., 16
Halliday, G.M., 101
Hamann, S.B., 324
Hamlin, P.A., 366, 368, 371, 394
Hampson, R.E., 356
Hampson, S. E., 134
Han, J-S., 106–109
Hardiman, M.J., 354
Harley, C., 106
Harris, J.A., 8, 17, 55, 90, 198, 199, 204, 225, 226, 246, 247, 253
Hart, P.E., 11
Hasenfus, N., 135
Haskins, C.E., 80
Hebb, D.O., 7, 59, 429
Hedges, H., 105
Heith, C.D., 35

Helen, P., 182, 239
Hemsley, D.R., 102, 116, 117, 133
Herbeaux, K., 108, 109, 118
Hermans, D., 182, 239
Higa, J., 361, 384
Higgins, D.M., 123
Higgins, J., 125
Hilgard, E.R., 8
Hinton, G.E., 8, 10, 371
Hinzman, D.L., 9
Hirsh, R., 341, 356
Hishimura, Y., 161
Hobin, J.A., 243
Hoff, M.E., 7, 10
Holland, P. C., 35, 80, 106, 107, 150, 152–155, 159, 165, 166, 168, 177, 180, 219, 234, 251, 252, 260, 264–267, 269, 270, 274, 275, 277, 278, 281, 282, 288, 290, 292, 293, 295, 300, 303, 305–307, 313–318, 320, 323, 326, 328, 333, 337, 341, 346, 347, 353, 354, 359, 360, 366, 368, 371, 381–384, 387, 389, 393, 394
Holmes, A., 116
Honey, R. C., 26, 80, 95, 106, 109, 115, 341, 343, 346, 352–354, 361
Honeycutt, N.A., 135
Hopfield, J.J., 9
Hopkins, R.O., 135
Horvitz, J.C., 65
Hovey, L., 135
Huff, N.C., 232, 426
Hull, C.L., 10, 22, 135, 169, 194, 235

Iordanova, M., 205–208, 214, 215, 222, 241
Isaacson, R.L., 354
Iseki, E., 135

Isen, A.M., 134
Ishii, K., 165

Jagielo, J.A., 89
Jarrard, L.E., 97, 243, 295, 337, 339, 341, 346, 351–354, 360, 361
Jaskiw, G.E., 116, 117
Jenkins, H.M., 366, 375
Jenkins, J.J., 125
Ji, J., 243
John, J., 247
Johnson, D.M., 246
Johnson, P.E., 76
Jones, S.H., 120, 133
Jones, S.H., 116, 129
Jones, T., 116
Jongen-Relo, A.L., 110, 235
Joseph, M.H., 103–105
Joseph, V.T., 145

Kalra, P.K., 247
Kamer, R.M., 99
Kamil, A.C., 35
Kamin, L.J., 36, 133, 149, 366, 375
Kamprath, K., 230
Kaplan, L.J., 99
Karoum, F., 116
Kasprow, W., 16, 85, 89
Kaufman, M.A., 141
Kaye, H., 82, 83, 91
Kehoe, E.J., 173, 178, 180, 223, 233, 252, 253, 255, 256, 314, 318
Kesner, R.P., 352, 353
Ketter, T.A., 135
Kiernan, M.J., 71
Killcross, A.S., 83, 85, 101, 102, 105, 106
Killeen, P.R., 367
King, D. A., 175, 176, 207, 216–219, 226, 228–230, 236, 246, 400
Kishimoto, H., 139

Kleinman, J.E., 113
Knight, D.C., 63
Knudsen, E.I., 325
Kohonen, T., 27, 417
Kolachana, B.S., 116
Konorski, J., 24, 169, 180, 194, 235, 400
Koob, G.F., 136
Kosaka, K., 135
Krabbendam, L., 139
Kraemer, P.J., 85, 89, 92
Krauter, E.E., 80
Kroforst-Colllins, M.A., 324
Kruschke, J. K., 10, 32, 53, 164
Kumar, N., 247
Kurian, E., 135
Kutlu, G.M., 36

LaBar, K. S., 239, 240, 244, 246
Lachnit, H., 37
Lam, Y., 8, 9, 21, 39, 119, 138, 354
Lamarre, J., 259, 299, 310
Lamoureux, J.A., 234, 251, 252, 323, 389, 393
Lane, P., 354
Lantz, A.E., 71–73, 76, 77, 95
Larrauri, J.A., 21, 32, 33, 65, 67, 101, 127, 194, 428, 429
Larsen, N., 189, 191, 192, 194, 197, 200, 201, 220–223, 232
Lawrence, B., 427
Le Pelley, M.E., 8, 17, 30, 39, 55, 90, 426
Leber, B., 341
LeDoux, J. E., 63, 326
Legault, M., 65
Leising, K.J., 423
Lencz, T., 139
Leonard, C.M., 63
Levin, E.D., 65
Lilyestrom, J., 135

Lipska, B.K., 116, 117
Livesey, E.J., 404, 406
Loechner, K.J., 329, 346, 347, 351, 352, 360, 361
Logan, F.A., 35, 56, 137
LoLordo, V.M., 149, 277
Lorden, J.F., 354
Lovibond, P.F., 80, 198, 199–201, 204, 226, 229, 411
Lubart, T.I., 123
Lubow, R. E., 25, 68, 72, 85, 88, 90, 103, 117, 133, 139, 141, 168
Luchins, D., 113
Lysle, D.T., 37, 57, 181

Machado, A., 367, 390
Mackintosh, J., 347
Mackintosh, N.J., 7, 13, 17, 32, 53, 64, 80, 81, 89–91, 149, 164, 169, 347, 354, 384
Macrae, M., 173
Maes, J.H.R., 176, 177
Mahdi, B., 174–176, 220, 221
Maier, S. F., 31, 53, 64
Majchrzak, M., 95, 108, 109, 118
Maren, S., 184, 185, 243
Markman, A., 11
Marlin, N.A., 31
Marr, D., 430
Martindale, C., 123, 135, 136, 138
Matute, H., 160, 165
Matzel, L., 8, 11, 15, 16, 141–143, 150, 152, 165, 168, 180, 239
McCarley, R.W., 135
McCloskey, M., 253
McCormick, D.A., 359
McGaugh, J.L., 326
McGhie, A., 139

McGonigle, B., 387
McKenna, P., 116
McKernan, M.G., 354
McKinzie, D.L., 171
McLaren, I.P.L., 32, 91
McLaughlin, D., 135
McNally, G., 205–207, 214, 215, 221, 222, 241
McNaughton, N., 59, 429
McPhee, J.E., 153
Meck, W.H., 366, 367, 390
Mednick, S.A., 123–126, 131, 132, 134
Melchers, K.G., 37, 57, 181
Melia, K.R., 241
Mercier, P., 80
Meredith, G.E., 99
Merrill, J.W.L., 367, 368, 372, 390–392
Merten, T., 130
Micco, D.J., 354
Milad, M. R., 244
Miller, J.S., 89
Miller, R.R., 8, 15, 16, 37, 53, 85, 89, 92, 141, 143, 149, 150, 155, 159, 160, 165, 180, 235, 366, 368, 371, 394, 426
Milner, B. R., 324
Minor, H., 80
Miserendino M.J., 246
Mitchell, C.J., 421
Mitchell, S.N., 103, 105
Moody, E.W., 173, 175, 226
Moore, A.U., 68
Moore, J.W., 9, 13, 21, 35, 72, 89, 91, 149, 326, 354, 367, 368, 390–393
Morell, J.R., 308
Morrell, M.J., 64
Morris, J., 62
Morris, R.G.M., 352–354
Morrow, B.A., 246
Mosteller, F., 10
Murphy, D.L., 135

Myers, C.E., 313, 319, 358–360
Myers, K.M., 184, 185

Nadel, L., 243
Nagaishi, T., 165
Nakajima, S., 165, 201, 204
Nakayama, E., 135
Napier, R.M., 173, 220, 222
Nelson, J. B., 251, 314
Nguyen. H.T., 61
Nicholson, D.A., 136
Niznikiewicz, M.A., 135
Norris, E. B., 35
Nowakowska, C., 135

O'Donnell, P., 116
O'Keefe, J., 243, 352
Oades, R.D., 101
Oakley, D.A., 333, 353, 354
Okoshi, T., 135
Orr, S. P., 244
Orr, W.B., 333
Osborne, B., 36, 57
Oswald, C.J.P., 99
Otto, T., 243

Packard, M.G., 326
Padfield, P.D., 427
Pakaprot, N., 174–176, 220, 221
Papdakis, N., 133
Papini, M.R., 56, 173, 233
Parsons, J.P., 366, 368, 371, 394
Partridge, D., 123, 135
Patterson, M. M., 239, 419
Pavelis, C., 135
Pavlov, I.P., 24, 31, 32, 36, 53, 64, 141, 169, 182, 185, 192, 194, 205, 227, 234, 235, 418
Pearce, J.M., 7, 10, 13, 17, 21, 24, 25, 26, 53, 64, 65, 69, 70, 76, 77, 82, 83, 89,

90, 93, 95, 111, 150, 164, 169, 223, 234, 252, 256, 259, 263, 313, 318, 347, 400, 426
Pearlson, G.D., 135
Peet, M., 138
Penick, S., 32
Perrine, A., 135
Perry, B., 113
Peters, S.L., 103, 138
Peterson, J.B., 123
Phelps, E. A., 239, 240, 244, 246
Philbin, D., 72
Phillips, I., 113
Pineno, O., 152
Piras, G., 101, 102
Pineño, O., 141, 160-165, 223
Pitman, R. K., 244
Polkey, C.E., 336
Poon, P. J., 243
Port, R. L., 239, 419
Postle, B.R., 62
Potash, J.B., 135
Pothuizen, H.H.J., 114, 240
Poulos, A.M., 178-180, 225
Powell, D.A., 333
Powers, P.S., 129, 130, 139
Prados J., 71
Preston, G.C., 80
Price, T., 35, 56, 137
Puga, F., 97
Pulver, A.E., 135
Purves, D., 106

Quinn, J.J., 170
Quirion, R., 103

Rammsayer, T.H., 133
Randall, C.K., 89
Rasmusson, A.M., 241
Rauch, S. L., 244
Rauhut, A.S., 153, 201
Rawlins, J.N.P., 102, 114, 116, 138, 343, 352, 354
Redhead, E.S., 347
Reed, P., 80
Reed, R., 135
Reilly, S., 106, 109-111
Reiss, S., 68
Rennicke, C.M., 135
Rescorla, R.A., 7, 8, 10, 11, 21, 27-30, 35-39, 41-57, 64, 68, 91, 144, 150, 157, 159, 164, 166, 169, 170, 175, 177, 178, 181, 182, 184, 185, 192, 194, 195, 205, 211-213, 219, 226, 227, 233, 235, 236, 246, 251-253, 256, 257, 263, 265, 266, 276, 295, 308, 310-313, 316, 317, 320, 337, 366, 367, 371, 390, 396, 401, 411, 414, 426
Ressler, K.J., 184
Revusky, S., 89, 92, 106
Richards, R.W., 175, 178, 179, 180, 225
Richardson, R., 205, 206, 207, 208, 214, 215, 221, 222, 241
Ricker, S.T., 194, 220, 221, 223, 226, 228, 232, 243
Rickert, E.J., 354
Rinaldi, P.C., 99, 324
Robbins, S.J., 169, 173, 182, 186-189, 192-195, 227, 231, 234
Robbins, T.W., 101, 102, 105, 194
Roberts, W.A., 366
Rochford, J., 103, 105
Rolls, E.T., 65
Rosas, J.M., 188, 189, 193, 209, 228, 236
Rosenberg, L., 76
Rosenfield, M.E., 326
Ross, R. T., 264, 266, 267, 277, 278, 313, 333, 337, 346, 347, 351, 352, 361
Roth, R.H., 241
Rowe, J., 123, 135
Rudy, J. W., 53, 76, 80, 314, 354, 357, 358
Rumelhart, D.E., 8, 10, 253, 255, 256, 371
Ruob, C., 102, 103
Russell, I.S., 333, 353, 354
Russell, W.A., 125
Rypma, B., 62

Sainsbury, R., 257
Sananes, C.B., 241
Sandell, J. H., 76
Santosa, C.M., 139
Sargent, D.M., 175, 178, 179, 180, 225
Sass, L.A, 139
Saul'skaya N.B., 117
Saunders, R.C., 116
Savastano, H., 143, 149, 235
Sawa, K., 423
Schachtman, T.R., 8, 15, 16, 83, 89, 92, 141, 143, 150, 155, 159, 174, 426
Schenk, F., 353
Schlossberg, A., 117
Schmajuk, N. A., 3, 8, 9, 12, 13, 21, 25, 30, 32, 33, 35-37, 39, 57, 65, 67, 80, 81, 83, 89, 90-92, 97-99, 101-103, 106, 113, 115-120, 124, 125, 127, 134-136, 138, 139, 149, 194, 223, 234, 237, 251-253, 255, 259, 308, 323, 324, 326, 328, 337, 346, 351, 354, 356, 358-360, 365, 367-372, 368, 369, 370, 371, 372, 381, 389-394, 399, 401, 409, 418, 419, 422, 427-429, 428, 429

Schneider, W., 25
Schnorr, L., 116
Schnur, P., 72, 89, 90
Schrag, A., 138
Schretlen, D.J., 135
Schreurs, B.G., 366, 375
Schultz, W., 54, 65, 102
Schwartz, M., 354
Schweitzer, B., 135
Sears, L.L., 8, 89, 91, 324
Seaward, J., 116
Seidenberg, M., 135
Seidman, L.J., 117
Seillier, A., 112, 122
Sen, A.P., 103
Shanks, D.R., 133, 160
Shenton, M.E., 135
Shepard, R.N., 347, 361
Sherman, J. E., 31, 53, 64
Shevill, I., 30, 173, 180, 225
Shiffrin, R.M., 25
Shimamura, A.P., 139, 240, 324
Shofel, A., 106
Shohamy, D., 109
Sidman, M., 230, 314
Siegel, S., 35
Silbersweig, D.A., 120
Skinner, B.F., 194, 251
Smith, A.D., 102, 116
Smith, C.W., 135
Smith, M., 35, 366, 368, 371, 372, 394
Smyly, E., 354
Snyder, S.H., 116
Sokolov, E.N., 24, 27
Sokolov, Y.N., 194
Solomon, P.R., 32, 72, 103, 354
Soltysik, S., 175, 194
Spear, N.E., 89, 169, 171, 194, 235, 354
Spence, K. W., 35
Squire, L.R., 139, 240, 324
Staddon, J.E.R., 361, 384

Stanhope, K.J., 101, 102
Steinmetz, J.E., 324, 359
Stern, C., 65
Stern, E., 120
Sternberg, R. J., 123
Stickney, K.J., 13, 89, 367, 390, 391, 393
Stout, S.C., 8, 15, 149, 235
Strasser, H.C., 139
Strong, C.M., 135
Suiter, R.D., 149
Sunsay, C., 173
Sutherland, N.S., 347
Sutherland, R.J., 314, 354, 357, 358
Sutton, R.S., 10, 252, 367, 391
Swartzentruber, D., 175-178, 181, 186, 194, 200, 201, 220-222, 226, 229-231, 234, 236, 251, 284, 285, 288, 313, 319, 340, 343, 359
Swerdlow, N.R., 136

Talk, A.C., 252
Tamai, N., 197, 200
Tank, D.W., 9
Tassoni, C., 11
Taylor, K.M., 145
Taylor, M.A., 135
te Kortschot, A, 95
Ternes, J.W., 86, 90
Terrace, H.S., 387
Testa, T.J., 86, 90
Thieme, A.D., 30, 134, 135, 326, 418, 419, 422
Thomas, B.L., 56, 173, 194-196, 198, 201, 204, 205, 225-227, 236
Thompson, R.E., 125
Thompson, R.F., 75, 99, 324, 326, 359
Tobler, P.N., 65, 102
Tolman, E.C., 8, 9, 134
Tomie, J., 354

Torrance, E.P., 129
Totterdell, S., 99
Trimble, M., 138
Trobalon, J.B., 81
Tuber, D.S., 80
Turi, A., 99
Turing, A.M., 428
Turken, A.U., 134
Turner, C., 52
Tweedie, F., 353
Tyberg, M., 120

Urcelay, G.P., 37, 150
Urushihara, K., 141

Van Hamme, L., 11, 12, 144, 149, 159, 166, 180
Van Os, J., 135
Vansteenwegen, D., 244
Vermeulen Van der Zee, E., 95
Vervliet, B., 186
Vogel, E.H., 10, 30, 39, 259, 314
Voglmaier, M.M., 135
Vogt, B.A., 64
Vossen, J.M.H., 176, 177

Wagner, A.R., 7, 8, 10, 11, 14, 21, 27-30, 32, 36, 37-39, 54, 57, 64, 68, 80, 89, 91, 92, 95, 141, 144, 145, 150, 157, 159, 164, 166, 169, 170, 177-180, 194, 195, 205, 233, 234, 252, 253, 256, 259, 263, 310, 311, 313, 314, 316, 317, 319, 366, 367, 371, 390, 401, 411, 414, 426
Waldmann, M.R., 423
Walker, E., 139
Wallach, M.A., 133
Wallas, G., 136
Wang, P.W., 135
Warburton, E.C., 103
Ward, W.C., 129

Ward-Robinson, J., 163, 239
Wasserman, E., 11, 12, 144, 149, 159, 166, 180, 412
Weaver, M.S., 90
Wehner, J.M., 103
Weidemann, G., 178
Weinberger, D.R., 116
Weinberger, N.M., 325
Weiner, I., 89, 90, 97, 102, 103, 106, 112, 114, 116, 117, 136
Weisberg, R., 135, 136
Weiss, C., 324
Weisz, D.J., 99, 324, 329, 346, 347, 351, 352, 360, 361
Werbos, P., 255, 371
Wertheimer, M., 123
Westbrook, R.F., 71, 73, 116, 209, 210, 211, 218, 219, 226, 246, 247, 366, 375

Wheeler, D.S., 85
Whishaw, I.Q., 354
Wickelgren, W.A., 356, 357
Wickens, C., 80
Wickens, D.D., 80
Widrow, B., 7, 10
Will, B., 108, 109, 118
Williams, G.E., 8, 10, 65, 371
Williams, G.V., 63
Williams, J.H., 102
Wilson, A., 242
Wilson, P.N., 83, 95, 234, 252, 313
Wing, C.W., 133
Winkelman, J.W., 99
Winocur, G., 343
Winston, P.H., 134
Wise, R.A., 65
Witcher, E.S., 195

Witnauer, J.E., 150
Witter, M.P., 95
Wolff, S., 37, 181
Woodbury, C. B., 253
Woods, A.M., 223
Wotjak, C.T., 230

Xu, M.L., 135
Xu, Y., 419

Yadav, R. N., 253
Yamada, K., 135
Yates, K.O., 135
Yee, B.K., 138
Yeo, C.H., 354
Young, M.E., 412

Zackheim, J., 319
Zimmer Hart, C.L., 28, 36, 195, 257, 396

Subject Index

Absolute value 24, 26, 88
Abstract thinking 133
Accumulator process 361, 384
Acquisition 3, 4, 11, 14, 27, 31, 33, 61, 68, 71, 79, 80, 87, 88, 120, 139, 141, 149, 151, 154, 158, 159, 165, 168, 169, 171, 173–175, 178, 180, 181, 185, 186, 190–195, 197, 200, 201, 206, 209–219, 223–225, 227–229, 231–234, 238, 239, 241, 252, 253, 262, 263, 265, 266, 268, 273, 276, 279, 291, 293, 294, 297, 299–302, 305, 309, 310, 319, 321, 326, 327, 330–332, 337, 339, 344–347, 349, 350, 392, 399, 401, 420
    deficit 18, 68
    extinction interval 180, 181, 188, 191, 223, 232
    extinction series of delay conditioning 13, 253
    of delay and trace conditioning 13, 253
Action system 130, 131
Activity 7, 10, 12, 13, 17, 27, 53, 57–63, 95, 96, 98, 113, 132, 133, 135, 247–249, 252, 254, 317, 319, 329, 349, 361, 389, 390
    neural 57, 63, 96, 117, 317, 319, 349, 350, 354, 421, 423, 424
Additive 7, 405–408, 413
Additivity 162, 403, 404, 405, 406, 407, 408, 413
    training following blocking 406
    training preceding backward blocking 408, 413
    training preceding blocking 405
AESOP model 313
Aggregate prediction 7, 13, 26, 27, 31, 94, 95, 103, 110, 246, 251, 252, 254, 317, 320, 324, 329, 337, 339, 344, 347, 353, 390, 409
Amnesia 234, 317, 349, 352
    anterograde 317
    limited retrograde 317
Amnesic patients 235, 239, 240, 241, 242
Amygdala 57, 58, 60–63, 96, 97, 117, 133, 319, 320
Analysis of variance 420, 421
Anterior
    cingulate 57, 58, 60–63, 134
    insula 57
Appetitive 12, 32, 64, 149, 150, 163, 164, 168, 289, 302, 311
    responding 64, 163
Approach 8, 13, 29, 40, 46, 49, 53, 64, 86, 93, 119, 120, 130, 132, 136, 163, 165, 181, 190, 247, 306, 309, 319, 324, 351, 384, 406, 413, 419, 422, 423
Architecture 9, 12, 249, 317, 353, 409
Artificial intelligence 119
Aspiration lesions 94, 113, 115, 134

Subject Index

Associability 7, 13, 17, 21, 52, 86, 87, 88, 89, 90
Associative 5, 7–10, 17–19, 21, 27, 30, 33, 35–38, 43, 46, 47, 49–52, 54, 55, 61–63, 86, 88, 119, 121, 122, 125, 130, 132, 135, 155, 161, 162, 165, 166, 169, 177, 188, 192, 203, 214, 218–220, 224, 230, 233, 245–247, 258, 308, 315, 317, 319, 349, 351, 357, 359–366, 370, 372–376, 378, 381, 383–389, 392, 394, 402–404, 409, 411, 414, 417, 419
Assumptions 93, 121, 161, 230, 232, 234, 235, 238, 307, 308, 309, 310, 311, 312, 313, 314, 316, 317, 339, 351, 353, 387, 403, 404, 411
Asymptotic learning 27
Attentional 26, 38, 52, 54, 98, 101, 102, 115, 117, 132, 137, 149, 160, 200, 201, 217, 218, 220, 230, 231, 392, 394, 407, 419
  associability 17
  –associative model 8, 21, 38, 63, 88, 161, 162, 219, 394, 402, 417
  buffer 18
  component 117
  –configural model 347, 392, 394, 396, 397, 399, 401–404, 406–411, 413
  control 21, 33, 53, 55
  decrement 36
  decrements during blocking 17
  deficits 135

mechanisms 19, 21, 39, 43, 44, 47, 49–54, 151, 162, 166, 169, 174, 177, 182, 190, 194, 201, 214, 220, 222, 224, 233, 357, 392, 403, 417, 419
  memory 24, 31, 120, 123
  model 347
  network 12
  processes 98, 136, 188, 233, 393
  rule 24, 88, 418
  solutions 230
  theories 7, 8, 13, 86, 87
Attention-modulated representation 24
Attenuated 77, 82, 90, 145, 149, 188, 200, 214, 223, 228, 395, 401
Attenuation 15, 16, 77, 82, 92, 102, 139, 145, 151, 160, 161, 197, 201, 228, 397
Auditory stimuli 258, 271
Automatic processing 25
Autoshaping 32, 40, 43, 46, 55, 105, 178, 270, 271, 302, 311, 335, 337, 354
Aversive 12, 32, 57, 62, 138, 146, 163, 168, 235

Backpropagate 249
  See Backpropagation
Backpropagation 8, 250, 312, 332, 349, 351
Backward blocking (BB) 4, 11, 12, 15, 16, 35, 52, 63, 137, 155, 156, 158, 159, 161, 162, 404, 406, 408, 413, 417
  recovery 5, 35, 63, 137, 155, 156, 158, 161, 162, 417
Backward conditioning 34

Bar pressing 64
Behavioral 10, 57, 90, 93, 103, 116, 245, 316, 332, 418, 421, 423
  inhibition 163
  level 9
Best predictor 13, 86
Biologically plausible 250, 418
Blocker 5, 15, 137, 145, 146, 149, 151, 154, 155, 161, 162, 406, 422
Blocking 4, 6, 9, 11, 13, 14, 15, 16, 17, 18, 22, 24, 27, 30, 35, 44, 52, 53, 88, 116, 129, 137, 145–149, 151, 154, 155, 161, 162, 246, 253, 282, 314, 347, 360, 362, 365, 367–370, 372–375, 385, 386, 388, 392, 403, 404, 406–409, 413, 417
Brain 93, 101, 317, 349, 422
  abnormalities 115
  activity 58, 61, 418, 421, 424
  areas 10, 63, 97, 132, 135, 418, 421
  –behavior relationships 57, 93, 234, 421, 422
  circuitry 62, 93, 95, 117, 132, 249, 320, 353
  imaging studies 62
  lesions 320, 349, 354, 418, 421, 424
  –mapped model 317, 424
  mapping 421
  neurotransmitters 87, 120, 421, 424
  regions 57, 93, 95, 120, 162, 235, 317, 339
  responses 57
  structures 53, 57, 87, 421, 424

470 Subject Index

CA1 317
 hippocampal fields 249
 lesions 234
 region 95, 102
CA3
 hippocampal fields 249
 region 95, 102
Catastrophic interference 247
Caudate 319, 320, 353
Causal learning 205, 403, 413, 418, 423
Central nucleus (CN) 319
Chaining 30, 34, 40, 206, 215, 387, 409
Chi Square 420
Chunking 349
Cingulate gyri 116
CL *See* Cortical lesions
Classical Conditioning (CC) 3, 7-11, 18, 21, 30, 32, 38, 62, 63, 86, 88, 98, 119, 136, 137, 165, 201, 205, 245-247, 253, 256, 302, 304, 307, 308, 313, 316-319, 337, 339, 349, 353, 354, 378, 384, 415, 417, 418, 420, 421, 423
 Data 3
Cognitive 129, 130, 133
 deficits 129, 135
 functions 414
 impairment 134, 135
 inhibition *See* Latent inhibition
 map 9, 30, 130, 410
 mapping 9, 30
 performance 135
 processes 133
 tasks 134
 theory 119
Combination 21, 25, 93, 119, 120, 122, 125, 166, 215, 220, 230, 239, 242, 247, 250, 253, 282, 289, 296, 306-308, 311, 312, 366, 409, 417, 418
 of CSs 21, 22, 34, 77, 89, 206, 207, 252, 419
 of multiple conditioning events 6
Combinational model 131
 neural network 131
 of creativity 131
Combined configural/ elementary accounts 307
Common 60, 119, 125, 127, 167, 173, 190, 340, 344, 367, 384, 404
 elements 122, 125, 340, 344, 345
 error term 27, 37, 38, 51
Comparator 8, 15, 16, 119, 139, 145, 155, 176
Comparator hypothesis 8, 15, 16, 89, 146, 151, 154, 155, 160, 162, 231, 418
 extended 6, 17, 145, 155, 231
 limited 151
Competing memories theory 165, 231
Competition 13, 18, 21, 33, 34, 144, 154-156, 192, 246, 308, 309, 316, 317, 324, 325, 326, 329, 337, 339, 344, 345, 353, 362-364, 366, 373-375, 386, 403, 420
 associative 359, 362, 364, 366, 372, 386, 388
 temporal 362, 364, 388
 to predict the duration of the US 372, 374
 to predict the intensity of the US 372
 to predict the moment of US presentation 369
 to predict the time and duration of the US 373
Competitive 21, 137, 145, 246, 253, 256, 282, 301, 409
 mechanism 35, 52, 188
Compound 35-52, 54-56, 77, 82, 92, 247, 253, 257-260, 262, 264-268, 271, 273, 276, 277, 282, 284, 286, 287, 289, 296, 297, 299, 302, 307, 308, 309, 311, 312, 322, 326, 340, 350, 353, 404-409
 conditioning 4, 9, 34-36, 35, 36, 39, 44, 54, 56, 149, 246, 249, 359-362, 384, 385, 417
 CSs 10, 258
 recovery 5
Computational 94, 378, 422, 423
 level 10
 models 3, 9, 10, 388, 417
Computer simulations 32-36, 39-41, 43, 46, 49, 51, 52, 54, 55, 57, 59-62, 64, 80, 90, 96-98, 101, 104, 105, 106, 108, 110, 115, 116, 120, 121, 123, 124, 125, 127, 129, 145, 146, 148, 151, 156, 159-161, 163, 166-169, 171, 174-178, 180, 181, 183-185, 188, 190, 192, 197, 199, 203, 206, 208, 210, 214, 215, 219, 220, 226, 227, 229, 230, 234-236, 238, 239, 253,

257, 260–262, 265–267, 269, 272, 273, 276, 279, 286, 290, 291, 294, 301, 306, 314, 315, 321, 322, 325, 326, 328, 329, 337, 353, 362, 365, 366, 369, 370, 373–378, 383, 388, 389, 391, 399, 401, 404, 408, 409, 411, 413, 418, 420, 423, 424
Conceptual nervous system 57, 93, 234, 421
Conditional 90, 231, 349, 423
  -contextual
    discriminations 321, 333, 335, 337, 339, 344–347, 354
  discriminations 257, 301, 326, 329, 331, 335, 339, 345, 346, 349, 353
Conditioned 80, 87, 90, 145, 155, 163, 180, 182, 185, 191, 194–196, 207, 208, 216, 218, 222, 253, 254, 258, 279, 332, 378, 389
  attention theory (CAT) 87
  emotional response (CER) 64, 110, 150, 175, 191
  excitation 11, 36, 258
  inattention 87
  inhibition 3, 4, 11, 14, 16, 17, 22, 27, 34, 35, 36, 44, 46, 52, 82, 172, 180, 200, 226, 246, 251, 253, 258, 271, 282, 302, 347, 392, 393, 399, 417, 423
  inhibitor 36
  orienting response 319, 320

reinforcement learning 12, 87
response (CR) 3, 7, 10, 21, 32, 37, 89, 105, 163, 165, 229, 231, 245, 259, 287, 310, 312, 319, 322, 352, 360, 377, 390, 410, 413
stimulus (CS) 9, 37, 51, 141, 158, 165, 208, 210, 218, 245, 246, 248, 254, 364, 365, 370, 410
taste aversion 105, 110, 112, 118, 168, 235, 418
Conditioning 3–6, 10, 13–15, 18, 21, 22, 25, 28, 29, 30, 34, 40, 46, 49, 51, 54, 58, 64, 66, 68, 69, 77, 80, 82, 86–90, 95, 96, 98–102, 104, 105, 107–110, 112, 113, 115, 117, 118, 123, 137, 138, 141, 142, 144, 153, 165, 166, 174, 289, 384, 385
  differential
    See Differential conditioning
  excitatory *See* Excitatory conditioning
  inhibitory *See* Inhibitory conditioning
  second-order *See* Second-order conditioning
Configural 246, 350, 361, 394
  accounts 307, 309
  approach 309
  associations 256, 267, 278, 287, 293, 300, 304, 305, 310, 312, 313, 314, 316, 351, 352

cue 253, 331, 423
discrimination 94
element 394
hidden units 255, 301, 308, 309, 311, 312
learning 310, 349, 351
mechanisms 166, 233, 243, 393, 404, 417
model 245, 246, 253, 307, 309, 359, 417
processes 246, 306, 307, 309, 310, 351
representations 256, 308, 339
stimulus 226, 230, 233, 241, 246, 249, 250, 255, 256, 304, 317–320, 329, 332, 333, 349, 350, 394, 396, 397, 404
system 308, 339, 350, 351
theory 253, 311, 339
-timing model 359, 362, 365, 387, 389
units 306, 309, 312, 351
Configuration 95, 195, 230, 247, 250, 308, 309, 312, 313, 316, 335, 337, 339, 352, 353, 359, 384, 392, 393, 394, 403, 404, 419, 422
  of stimuli 316, 317
  role of 230
Conflicting data 108, 118, 289, 329
Connection strengths 246
Conscious 25, 133
Consciousness 133
Constrained error term 38
Context 4, 5, 8, 11, 14, 15, 18, 21, 26, 27, 29, 30, 55, 65, 77, 78, 80, 82,

86–90, 92, 95, 101, 103, 105, 108, 111, 137, 138, 140, 145–149, 151, 154, 157, 159, 160, 166–168, 171–178, 180–182, 184, 191–210, 212–216, 221–233, 235, 237–239, 241, 242, 248, 251, 275, 278–281, 289–294, 297, 299, 306, 308, 313, 314, 332, 333, 335, 337, 340, 344–346, 347, 352, 354, 364, 389, 390, 392, 393, 394, 400, 402, 409, 422
-dependent
    representations 313
-independent latent
    inhibition 111, 112
preexposure 4, 77
presentation 15
salience 60, 167, 172, 192, 194, 200, 210, 311, 423
-specific latent
    inhibition 26, 87, 105, 108
Contextual 173, 238, 293, 347, 349
    associations 167–172, 203, 226
    changes 80, 112, 117, 180
    conditioning 167, 282
    cues 166, 169, 172, 175–177, 194, 219, 226, 227, 242, 289, 308, 331, 337, 349, 394
    discrimination 172, 227, 230, 231, 251, 257, 278, 279, 281, 282, 306, 313, 314, 333, 335, 337, 340, 345, 346, 352, 354, 402

effects 117, 392
exposure 92
inhibition 171, 173, 233, 393
representations 167
stimuli 168, 173, 404
system 349
Contiguity 86, 119, 122, 309
Contingency 3, 4, 11
Contradictory results 422
Controlled processing 25
Correct postdictions 420
Correlation 138, 141, 148, 151, 154, 157, 163, 420
Cortical 320, 349, 351
    activity 132
    areas 132, 134, 234
    arousal 131
    associations 94, 95, 103
    circuits 249, 351, 353
    hidden units 317, 351, 352
    inputs 320
    learning 318
    lesions (CLs) 320, 321, 326, 329, 337, 339, 340, 344, 346, 347, 353
    networks 351
    neurons 349
    regions 351, 352
    system 320
Counteract 12, 14, 16, 113, 137, 141, 165, 205, 212, 215, 269, 329
Counteraction 36, 161, 162
Counterconditioning 259, 270, 271, 282, 284, 286
CR *See* Conditioned response (CR)
    generation 4, 7, 8, 22, 113, 223, 237, 252, 301

    topography 95, 362, 365, 366, 385, 388
CR Timing 362
Creative achievement
    questionnaire (CAQ) 123, 125, 127
Creative personality scale (CPS) 125
Creative thinking 119
Creativity 119, 120, 124–136, 414
    exploratory model 131
CS *See* Conditioned stimulus (CS)
CS associations 8, 9, 13, 21, 22, 26, 27, 30, 33, 35, 80, 88, 97, 101, 162, 234, 307, 316, 409
    duration 34, 51, 68, 70, 92, 99, 101, 102, 118, 361, 387, 394, 423
    intensity 69, 70, 167
    salience 10, 59, 88, 108, 161, 167, 168, 423
-specific CR 3
–US associations 8–11, 14–17, 21, 22, 25, 27–29, 31–36, 38, 54, 56, 58, 61, 71, 80, 86–90, 97, 101, 103, 113, 120, 144, 146, 154, 155, 161, 165, 166, 168, 169, 171, 174, 176, 177, 179–182, 184, 188, 189, 191, 192, 197, 198, 200, 201, 203, 205, 206, 210–212, 214–219, 221–225, 228, 230–232, 237, 238, 246, 256, 282,

287, 293, 314, 386, 393, 397, 409, 417

DA 63, 93, 94, 97–99, 101, 109, 112, 113, 115–118, 120, 132–135
   receptor antagonists 98, 117
Darwinian theory 119
Decay 12, 23, 24, 26, 122, 169, 248, 261, 324, 389
Decision processes 8
Declarative 9
Deficits 89, 116, 123, 129, 134, 135, 149, 151, 235, 241, 302, 347
Delay conditioning 13, 34, 156, 311, 347
Delta rule 7, 246, 249, 250, 252, 253, 350, 418
Dentate gyrus 95, 102, 249
Differential conditioning 4
Direct association 173, 230, 248, 250, 252, 255, 257, 263, 267, 269, 275, 278, 281, 282, 293, 294, 301, 306, 313, 314, 321, 324, 329, 352, 366, 390
Discrete cues 392
Discrimination 6, 30, 77, 78, 172, 227, 230, 231, 247, 249, 251, 253, 257, 259–263, 265–273, 276–282, 284, 286–289, 293–297, 299–302, 305, 306, 308, 310–315, 319, 321, 322, 324–326, 328, 329, 331–333, 335, 337, 339, 340, 344, 345, 346, 347, 349, 352, 353, 354, 360, 362, 365, 375–378, 380, 381, 383, 386, 387, 388, 392, 402, 422, 423

acquisition 13, 253, 301, 302, 347, 392
reversal 13, 253, 347
serial 281, 282, 284, 289, 314, 321, 329
simultaneous 281, 282, 287, 299, 314, 321
Discriminative stimulus 245, 344, 378
Dishabituation 80
Disinhibition 5, 168, 177, 178, 181, 182, 189, 215, 223, 231, 233, 394
   external *See* External disinhibition
Distal landmarks 253, 345, 355
Divergent thinking 120, 122–126, 129, 130, 135, 136
Dopamine 53, 120, 136, 241, 242
   and creativity 134
Dopaminergic (DA) 24, 63, 96, 98, 116, 117, 135
   system 93
Dorsal hippocampus 113, 238, 317
Dorsolateral prefrontal cortex (dlPFC) 57, 59, 61, 98, 133
Drinking rate 163
Drive representation 12, 13, 87
Drug administration 108, 116, 118, 418, 421, 424

Effectiveness
   of the CS 7, 8, 86, 88, 307
   of the US 7, 8, 46, 86, 88, 286, 307
Efference copy 252, 317

Electrolytic lesions 102, 238, 241
Elevation scores 32, 168, 229
Emergent properties 22, 26, 32, 71, 253, 419, 420, 424
Empirical data 52, 98, 257, 260
Entorhinal cortex (EC) 94, 95, 102, 108, 118, 134, 249, 317
   effects of reversible inactivation 108
Environmental stimuli 24, 26
Equation 10, 11, 17, 21, 23–32, 34, 35, 36, 38, 39, 40, 42–46, 49, 51–54, 64, 80, 96, 101, 103, 110, 115, 145, 163, 234, 236, 239, 246, 248–252, 255, 261, 265, 272, 320, 321, 324, 329, 352, 365, 366, 389, 408
Error signals 53, 317, 319, 353
   for cortical hidden units (EH) 320, 321, 339, 352
   for output units (EO) 246–254, 256, 318,
Error-correction rule 38, 39, 247
Evaluation of the models 420, 422
Event 7, 9, 17, 26, 27, 89, 116, 299, 310, 314, 389, 409
Excitatory 5, 12, 14, 15, 17, 25, 27–31, 34, 36–38, 40–50, 54–56, 61, 68, 88, 89, 96, 117, 131, 165, 169, 173, 174, 176, 177, 178, 182, 185, 188,

191, 194, 196, 198, 200, 201, 203, 205, 206, 209–212, 214, 215, 219, 222, 224–228, 231–233, 237–239, 241, 242, 246, 247, 250–254, 256, 258, 259, 267–270, 277, 278, 284, 286, 289, 296, 297, 301, 302, 305, 306, 310, 311, 324, 329, 333, 339, 393, 397, 421
    association 6, 14, 61, 88, 165, 174, 196, 198, 200, 203, 205, 209, 214, 237, 238, 241, 254, 258, 259, 267, 270, 278, 284, 286, 293, 296, 297, 306, 311, 324, 329, 339, 344
    conditioning 3, 17, 65, 66, 145, 417
    -inhibitory compound 38, 40, 46, 49
    -neutral compound 44
Excitotoxic hippocampal lesions 93, 95, 108, 134, 234, 238, 239, 242
Exclusive-or 8, 247
Expectancies 8, 9, 58, 61
Experimental 40, 52, 82, 96, 107, 115, 126, 136, 138, 141, 147, 154, 160, 161, 163, 167, 174, 178, 233, 239, 306, 307, 312, 346, 378, 420, 422, 423, 424
    animals 369, 370, 372, 373, 374
    context 383
    data 10, 40, 43, 57, 68, 80, 82, 99, 102, 103, 105, 108, 110, 113, 116, 123–125, 127, 129, 138, 141, 146, 148, 149, 151, 154, 157, 163, 172, 173, 175, 181, 185, 188, 191, 192, 196, 208, 219, 235, 239, 251, 264, 267, 272, 278, 282, 284, 286, 287, 288, 294, 297, 302, 311, 315, 326, 328, 337, 340, 347, 366, 404–406, 420, 423
    design 118, 192, 331
    group 52, 123, 146, 160, 174, 185, 197, 225, 226, 233, 372, 374, 408
    procedure 38, 108, 118, 168, 368, 378, 422
    results 9, 38, 46, 49, 52, 56, 65, 77, 110, 113, 115, 120, 138, 141, 146, 149, 161, 166, 169, 184, 189, 191, 197, 199, 200, 210, 225, 229, 231–233, 241, 242, 260, 268, 271, 275, 282, 284, 286, 287, 289, 297, 299, 302, 314, 322, 326, 331, 335, 339, 340, 353, 376, 392–394, 396, 399, 403, 418, 420, 421
Explaining experimental data 18, 420, 424
Explicit correct predictions 420
Exposure 4, 24, 77, 82, 90, 92, 108, 116, 118, 146, 149, 174–177, 192, 197, 203, 205, 208, 214, 215, 222, 224, 235, 239, 241, 278, 333

Extensive conditioning 31, 52
External 116, 173
    disinhibition 5, 16, 168, 177, 178, 181, 182, 189, 223, 231, 233, 394
    inhibtion 32
Extinction 9, 11, 13, 14, 15, 22, 27–30, 33, 35, 36, 49, 53, 55, 56, 63, 82, 89, 120, 131, 137–141, 146–149, 151–155, 160–162, 165–169, 171–219, 221–235, 237–239, 241, 242, 251, 253, 258–260, 264, 282, 284, 286, 289–294, 297, 299, 305, 309, 331, 347, 349, 352, 381, 383, 386, 390, 392–397, 399–401, 417, 418, 421, 422, 423
    burst 55, 169, 233
    cues (ECs) 168, 186, 190, 198–200, 229, 230, 233, 392, 394, 397, 399, 400, 418
    mediated *See* Mediated extinction
    of compounds 49
    of conditioned inhibition 28, 35, 36, 56, 177, 191, 390
    of context 14–16, 82, 89, 155, 202, 203, 205, 293
    of reinstatement 215, 224
    -test intervals 178, 180
Eyeblink conditioning 32

Facilitation 68, 71, 100, 102, 176
    of extinction 176, 177, 229

of fear acquisition 420
of LI 98, 99, 101, 105, 108, 110, 112, 117, 118, 234, 235
Facilitator 101, 245, 307
Fear 61, 71, 180, 182, 191–193, 235, 238, 239, 420
  conditioning 4, 57, 58, 61–63, 95, 417
Feature 6, 17, 30, 32, 33, 167, 173, 194, 226, 246, 251, 252, 257, 260, 261, 267–272, 275, 278, 282, 287, 289, 293, 294, 296, 297, 299, 306, 308, 309, 312–314, 322, 324–326, 328, 332, 340, 344, 347, 354, 360, 361, 364, 375, 376, 378, 380, 381, 383, 386, 387, 391, 392, 404, 423
  -negative discrimination 257–259, 275, 278, 282, 284, 286, 287, 297
  -positive discrimination 253, 257–259, 261, 265, 271, 278, 282, 284, 287, 289, 296, 301, 302, 306, 309, 312, 325, 326, 360, 383
  pretraining 300, 309
  -target interval 360, 376, 377, 382, 384, 386
  -target similarity 299, 310
Feedback 12, 21, 22, 31, 120, 389
Fimbrial lesions 234
fMRI 57
Formal models 9, 131
Forward blocking 16, 158, 160, 413

recovery 5, 11, 15, 35, 63, 137, 146, 151, 154, 155, 161, 162, 417
Freezing 32, 33, 64, 68, 71, 166, 168, 180, 206
Frequency 3, 121, 165

Generalization 3, 11, 60, 119, 122, 125, 129, 136, 159, 160, 167, 200, 205, 250, 253, 256, 289, 290, 294, 296, 297, 300, 301, 304, 307, 310, 312, 362, 404, 423
  decrement 166, 190, 229, 231
Generalized 13, 86, 166, 184
  characteristics of contexts 166
  delta rule 8
Generation of an excitatory CS 28
Generic models 423
Gestalt theory 119
Goal 30, 80, 130, 131, 132, 133
  -seeking mechanism 130, 411

Habituation 72, 103, 190, 230, 234, 239, 319
Hallucinations 113, 115, 116, 134
Haloperidol (HAL) 94, 98, 99, 108, 110, 111, 114–118, 135
Head jerk behavior 254, 258, 260, 271
Hidden units 18, 246, 247, 249–254, 256, 257, 262, 263, 265–270, 274, 275, 277, 278, 280, 281, 284, 286, 289, 291, 292, 294, 296, 297, 301, 305, 307, 308, 311, 312, 314, 317, 319, 321, 324–326, 329, 332, 333, 337, 345–347, 351, 352, 359, 361–365, 378, 380, 381, 386–388, 390
Hidden-unit 248, 249, 250, 251, 252, 256, 294, 297, 305, 310, 311, 315, 317, 320, 321, 329, 332, 333, 337, 345, 346, 349, 363, 365, 376, 388–391, 394
  layer 8, 246, 247, 249, 250, 254, 255, 267, 269, 275, 278, 281, 293, 308, 324, 329, 332, 337, 339
  output associations 252, 320
Hippocampal 93–95, 102, 103, 108, 111–118, 120, 132, 134, 135, 234, 235, 237–242, 249, 316–318, 320, 321, 323, 327, 330, 336, 337, 340, 347–353, 411, 421
  and schizophrenia 112
  dysfunction 112, 116, 118, 134, 241
  formation 94, 102, 103, 112, 116, 316, 350, 351
  involvement in latent inhibition 102, 117
  pyramidal cells 349
  region 351
  regions 94
Hippocampus 93, 94, 103, 112, 116, 117, 234, 235, 316–319, 349, 350, 351, 353
  and creativity 134
  hypoxic damage 239, 242

Hippocampus (cont.)
   ibotenic acid lesions
      103, 105, 113, 316,
      339
   kainic acid lesions 234,
      320
   of the formation lesions
      (HFLs) See Lesions
      of the hippocampal
      formation (HFL)
   proper (HP) 94, 95, 102,
      235, 321
   proper lesions (HPLs)
      See Lesions of the
      hippocampus
      proper (HPL)
   selective lesions 316,
      321
History of reinforcement
   185, 194, 214, 224
Home cage 77, 84, 159, 166,
   167, 169, 172, 174, 175,
   177-180, 184, 185, 188,
   192, 196, 203, 206, 210,
   223, 395
Human ratings See Rating
Hybrid model 8, 16, 17, 87

Image 410, 413
Impairment of LI 98, 101,
   103, 110, 112, 115, 117,
   118
Implicit correct predictions
   420
Inattention 25, 87, 160
Incentive motivation 12,
   13, 87
Inference 8, 9, 30, 409
   generation 8
Inferential 403, 409, 418
   explanation 403
   process 403
   process view 403
   reasoning 403, 409, 418

Inferior olive 352
Inflation 29, 209-212, 214,
   222, 224
   and reinstatement 209
Infusions 10, 238
Inhibitory 4, 12, 15, 17, 18,
   27, 28, 29, 30, 31, 36,
   37, 38, 40, 41, 43-46, 50,
   54-56, 84, 89, 96, 102,
   131, 164, 165, 169, 171,
   173, 174, 175, 177, 191,
   193, 194, 196, 200, 201,
   212, 214, 215, 219, 221,
   222, 224-227, 230-232,
   237, 241, 247, 250-253,
   259, 267, 269, 274,
   275, 277, 278, 280, 289,
   292-294, 293, 294, 297,
   301, 302, 305, 310, 311,
   329, 335, 337, 345, 393,
   397, 399, 401, 402
   associations 6, 14, 18, 27,
      29, 36, 48, 56, 165,
      166, 169, 171-174,
      177, 180-182, 188,
      189, 191, 193,
      197, 198, 200, 201,
      203-206, 209-212,
      214-217, 219,
      221-226, 228,
      230-232, 237, 238,
      241, 242, 246, 247,
      250, 254, 256, 258,
      259, 267, 269, 270,
      278, 284, 289, 292,
      293, 294, 296, 297,
      305, 310, 311, 324,
      329, 333, 337, 344,
      383, 393, 394, 397,
      402
   conditioning 3-5, 34, 35,
      36, 65, 66, 145, 347,
      392, 417
   context 25, 29, 173, 177,

178, 182, 184, 185,
191, 224, 225, 227,
293, 397, 421
memory 191
occasion setter 256
Initial associations 36, 50,
   51
Input 8, 12, 21, 24, 25, 27,
   54, 88, 113, 115, 116,
   131, 132, 230, 235,
   246, 247, 249-256, 274,
   278, 281, 305, 308, 315,
   319, 320, 329, 332, 333,
   337, 339, 340, 349, 351,
   353, 393, 400, 402, 419,
   422
   -hidden associations
      251, 274, 278, 281,
      329
   -output associations
      250, 251, 320
   units 247, 248, 250, 254
Intelligence 119, 135
Intensity 7, 27, 31, 64, 98,
   101, 108, 121, 123, 125,
   147, 150, 167, 248, 257,
   263, 264, 281, 314, 317,
   321, 322, 354, 359, 364,
   365, 366, 372, 373, 375,
   386, 389, 390, 391, 394,
   406, 409, 418
Interfere 86, 89, 184, 301,
   339
Interference 89, 90, 160, 205
Intermediate reinforcement
   46, 47
Internal contexts 80
Internal model of the
   environment 27
Interstimulus interval (ISI)
   6, 11, 34, 281, 301, 321,
   360, 361, 365, 366, 373,
   374, 375, 384
   effects 6

Intertrial interval (ITI) 6, 11, 25, 28, 29, 39, 60, 69, 70, 92, 121, 123, 125, 128, 150, 151, 154, 167, 170–172, 174, 178, 181, 182, 184, 185, 187, 188, 191, 192, 195–197, 199, 202, 203, 206, 208, 210, 213, 214, 216, 223, 239, 248, 394, 396, 399, 400, 423
Investment theory 119

Kent-Rosanoff Word Association Test 121
Kinetic theory of gasses 418
Knowing
   how 9
   that 9

Latent Inhibition (LI) 4, 11, 13, 14, 17, 18, 21, 24–26, 52, 63–65, 72, 86, 87, 92, 93, 95, 98, 108–110, 112, 116–120, 123, 125, 129, 132, 134–137, 141, 144, 161, 162, 190, 218, 235, 239, 242, 347, 392, 413, 417, 420, 422
   and schizophrenia 108, 112, 118
   facilitation of
      See Facilitation of latent inhibition (LI)
   impairment of
      See Impairment of latent Inhibition (LI)
   preservation of
      See Preservation of latent Inhibition (LI)
   recovery 5
Lateral nucleus (LN) 319
Lateral septum 349
Layer 246–250, 252, 254, 307, 308, 351
Learned irrelevance 4, 17, 90, 92, 423
Learning 3, 7, 8, 10, 12, 14, 17, 18, 30, 33, 37, 57, 61, 62, 86, 87, 92, 94, 110, 122, 159, 160, 219, 235, 246, 251, 253, 254, 257, 268, 270, 282, 284, 286, 296, 302, 306, 308–310, 312, 317, 318, 321, 337, 345, 346, 347, 349, 350, 351, 354, 355, 359, 360, 361, 363, 365, 369, 375, 376, 384, 385, 387, 392, 411
Learning theories 3, 7, 360
Lesions 10, 93, 94, 95, 102–108, 110–113, 115–118, 120, 134, 135, 234–242, 316, 317, 319, 320, 321, 323, 327, 330, 331, 334, 336, 337, 339, 340, 347–349, 351–354, 418, 421, 422, 424
   of the hippocampal formation (HFLs) 94, 95, 102, 103, 110, 112, 115–117, 316, 320, 322, 324–329, 331, 332, 335, 337, 339, 340, 344, 345, 347, 349, 351, 353
   of the hippocampus proper (HPLs) 93, 102, 103, 115, 234
Licking period 369, 391
Licking time 32, 33, 163
Limitations of the model 229
Limited-capacity STM 12
Limiting term for $V_{CS,US}$ associations 29
Linear combination 10, 246, 247, 253

Liquid consumed 163, 185
Long-term memory (LTM) 12, 133
   access 133

Manipulations 71, 77, 82, 93, 145, 166, 225, 241, 282, 309, 310, 317, 422
Mapping 9, 30, 93, 108, 249, 320, 353, 421
Massed trials 34
Maximal 61, 96, 98, 162, 253, 393, 397, 402, 403, 409
Maximality 162, 403, 404, 409, 413
   effects 30, 404
   training 409
Medial geniculate body (MGB) 319
Medial septal/Medial septum 95, 249, 319, 349
Mediated 95, 98, 134, 169, 176, 215, 221, 291–294
   acquisition 30, 215
   extinction 16, 30, 169, 176
Mediation 119, 122
Memory systems 349, 350
Microfeatures 17
Mismatch 27, 92, 349
Model
   connectionist 307
   variables 421, 424
Models
   AESOP
      See AESOP model
   attentional–associative
      See Attentional associative model
   attentional–configural
      See Attentional configural model
   brain-mapped See Brain-mapped model

Models (cont.)
   combinational
      See Combinational model
   computational
      See Computational models
   configural See Configural model
   configural-timing
      See Configural timing model
   creativity exploratory
      See Creativity exploratory model
   creativity-combinational
      See Combinational model of creativity
   formal See Formal models
   generalizational decrement 166, 190, 229, 231
   generic See Generic models
   hybrid
      See Hybrid model
   internal environmental
      See Internal model of the environment
   Pearce-Hall See Pearce-Hall model
   replaced elements
      See Replaced elements model
   Rescorla-Wagner (RW)
      See Rescorla-Wagner (1972) (RW) model
   Schmajuk-DiCarlo (SD)
      See Schmajuk-DiCarlo (SD) model
   Schmajuk-Lam-Gray (SLG)
      See Schmajuk-Lam-Gray (SLG) model
   Schmajuk-Lamoureux-Holland (SLH)
      See Schmajuk-Lamoureux-Holland (SLH) model
   spectral timing
      See Spectral timing model
   Wagner's sometimes opponent process (SOP) See Wagner's sometimes opponent process (SOP) theory
Modulate/Modulation 8, 24, 51, 62, 88, 95, 101, 117, 132, 155, 169, 173, 175, 245, 259, 260, 264, 284, 286, 306, 307, 310, 319, 353
Modulatory 260, 306, 307, 308, 309, 310, 311
   accounts 307, 309
   mechanisms 308
Moment-to-moment 21, 33, 246, 385, 419
Motivational 307, 314, 335, 344–347, 354
   effects 63, 80
   states 80, 82, 349
Multiple
   CSs 360, 385, 402
   mechanisms 418, 423
   response systems 254, 256, 304, 339, 353
   traces 359, 362, 363, 376, 387, 388
Muscimol 238

Negative 6, 8, 11, 17, 18, 24, 25, 28, 29, 31, 36, 38, 39, 46, 53, 64, 166, 173, 198, 200, 217, 230, 232, 247, 257, 274, 308, 339, 340, 347, 353, 372, 381, 390, 392, 393, 394, 420
   patterning 6, 8, 11, 18, 247, 253, 308, 340, 347, 353, 392, 394
   spatial and temporal occasion setter 166, 173, 230
Neocortex (NCX) 94, 95, 317
Neural
   activity See Activity neural
   elements 10, 57, 254, 317, 418, 421, 423
   network 7, 10, 21, 32, 86, 130, 131, 245, 254, 256, 302, 316, 317, 359, 363, 392, 403, 417–419, 421, 422, 423
Neuroanatomical level 10
Neurobiological principles 250
Neurobiology 57, 93, 132, 234, 316
   of conditioning 57
   of creativity 132
   of extinction 234
   of latent inhibition 93
   of occasion setting 316
Neuroimaging 57
Neurophysiological 10, 30, 53, 93, 116, 162, 241, 249, 314, 316–318, 352, 384, 421, 423
   basis of error-correcting mechanisms 53
   level 10
Neurotoxic hippocampal lesions 234, 235, 237, 238, 240

Subject Index

Neurotransmitters 30, 101, 134, 421
Nicotine 98, 99, 101, 102, 117, 134
  administration 33, 99–102, 118
  cholinergic reseptors 99
Nictitating membrane (NM) 95, 167, 175, 319, 339
  response (NMR) 365, 366, 368, 391
Node 12, 23–25, 27, 30–32, 93, 117, 131, 132, 248, 250, 317, 320, 353, 404
Nonlinear 32, 57, 60, 61, 246, 252, 281, 314, 322, 326, 333, 340, 344–346
  classical conditioning paradigms 247
  combination 5, 11
Nonreinforced
  3–6, 37, 39, 44, 45, 49–51, 53–55, 80, 87, 138, 165, 169, 172–174, 176, 177, 184, 187, 201, 204, 205, 208, 214, 219, 221–223, 225, 227–229, 251, 257, 260, 261, 271, 273, 276, 278, 282, 284, 286, 291, 292, 311, 329, 331–333, 335, 337, 340, 344, 345, 353–355
  compound 37, 40, 43, 46, 49, 50
Nonselective 316, 320, 321, 339, 349
  aspiration lesions 316, 339
  hippocampal lesions 316, 320, 321
Normalized value 24, 26, 113
Novel 5, 10, 90, 125, 131,
160, 174, 177, 239, 347, 387
  cue 225, 226, 227, 228
  event 73
  stimuli 5, 32, 63, 136, 163, 178, 181, 182, 223, 225–228, 233, 408
Novelty 21, 24–27, 31, 33, 36, 39, 42–44, 52, 53, 60–63, 66, 71, 74, 77, 80, 82, 84, 88, 92, 96, 98, 101, 102, 105, 108–111, 113, 115, 117, 120, 121, 123, 125, 129, 132–134, 135, 138, 144, 146, 148, 149, 151, 158, 159, 162, 169–171, 178–182, 184, 190, 192, 194, 195, 198, 203, 205, 206, 209, 219, 223, 225, 227–229, 233, 237, 241, 242, 347, 394, 404, 408, 421, 422
  level 26, 60, 61, 171, 231
Nucleus accumbens (NAC) 93, 95, 96, 101, 109, 112, 113, 116, 117, 118, 120, 132–136
  core 96, 109, 110, 113, 116–118
  shell 95, 96, 110, 112, 113, 116–118, 235
Number of CS preexposures 68, 70

Occasion setter 6, 173, 200, 230, 245, 246, 252, 253, 256, 257, 260, 267, 270, 278, 281, 282, 294, 299, 305–308, 311, 313, 314, 321, 322, 324, 333, 335, 337, 339, 340, 349, 352, 353, 359, 360–363, 365,
375, 376, 378, 381, 385–388, 393, 417
Occasion setting 6, 173, 230, 232, 245, 246, 254, 257–260, 265, 270, 282, 286, 287, 294, 299, 301, 302, 305–314, 316, 329, 331, 332, 333, 339, 352, 353, 359–363, 376, 378, 381, 383, 384, 386–388, 392, 393, 402, 417, 418, 420
  effects of extinction and counterconditioning 282
Odor cues 191, 192
Omission 63, 92, 105, 107, 145, 166
  of an expected event 73
Operant conditioning 245, 360, 361, 378, 384
Ordinal positions 387, 388
Orienting response (OR) 24, 26, 32, 33, 63, 72, 79, 80, 92, 95, 108, 117, 163, 164, 170, 190, 229, 319
Oscillator 361, 384
Outcome 13, 62, 158, 162, 226, 232, 254, 301, 306, 309, 320, 385, 403–409, 413
Output 7, 12, 13, 21, 23, 24, 30–32, 88, 96, 117, 118, 120, 132, 246–256, 274, 278, 281, 297, 302, 305, 307, 308, 310, 311, 314, 317, 319, 320, 329, 332, 349, 351, 352, 353, 359, 361, 409, 410, 422
  units 246–250, 252–254, 256, 305, 361
Overexpectation 5, 17
Overinclusion 120, 123, 125, 129, 136

Overinclusive thinking 125
Overshadowing 5, 11-18, 22, 24, 27, 35, 66, 77, 88, 137, 138-141, 144, 145, 149, 151, 154, 155, 161, 162, 246, 253, 282, 314, 347, 392, 417, 423
   recovery 35, 63, 137, 161, 162, 417, 420
Overtraining reversal effect 17

Paired associates 121
Paradigms 3, 8-11, 24, 34, 35, 56, 64, 110, 118, 119, 159, 161, 166, 191, 219, 220, 226, 241, 245, 247, 253, 256, 302, 319-320, 339, 347, 354, 360, 362, 365, 368, 369, 372, 374, 375, 384, 385, 388, 389, 391, 417, 418, 421, 422, 423
Parameter 11, 33, 40, 46, 68, 99, 123, 144, 151, 154, 167, 185, 190, 197, 199, 229, 239, 247, 248, 309, 314, 354, 361, 390, 391, 399, 400, 406, 418, 419, 422-424
   of preexposure 68, 190
   values 33, 40, 123, 260, 314, 315, 320, 321, 365
Parametric studies 299, 422
Parsimony 419
Partial reinforcement 3, 11, 29, 34, 46, 47, 49, 96, 98, 212-214, 222, 224, 229, 325
   and reinstatement 212
   extinction effect (PREE) 5, 33, 229

of compound 37, 47, 49
Passage of time 82, 231
Path 30, 130, 131, 165, 185, 189
Pavlovian 11, 263, 302
   conditioning 3, 10, 11, 245, 359, 360-362, 384, 388
Pearce-Hall model 13, 17, 74, 80, 90, 92, 219
Pecking 32, 178
Perception 116, 133, 410
Perceptual learning 77, 78, 88
Performance 8, 10, 31, 37, 43, 44, 51, 55, 77, 89, 92, 123, 129, 130, 134, 135, 136, 151, 155, 254, 257, 259, 267, 268, 299, 310, 324, 346
   deficit 18, 151
   theories 8
Place 11, 29, 55, 108, 118, 166, 167, 173, 180, 192, 193, 194, 195, 197, 199, 210, 212, 226, 235, 239, 253, 309, 344, 345, 353, 376, 392, 394, 400, 424
   learning 94, 253, 345, 346, 347, 349, 355
   -place associations 30, 411
Poor predictor 13, 17, 86, 87
Positive patterning (PP) 5, 253, 301, 347, 392, 420
Postconditioning manipulations 82, 92
Posterior intralaminar nucleus (PIN) 319
Postsynaptic activity 7
Postsynaptic neural population 7
Potentiation 4, 35, 56, 423
Predicted outcomes 404

Predictions 7-11, 23, 26, 29, 30, 31, 36, 39, 49, 51, 53, 54, 55, 57, 60, 61, 86, 87, 94, 95, 111, 112, 116-118, 120, 121, 129, 130, 132, 135, 146, 159, 160, 163, 176, 190, 200, 219, 225, 227, 228, 232, 233, 235, 238, 242, 246, 262, 266, 267, 268, 269, 273-275, 277, 278, 280, 281, 291, 292, 311, 312, 317, 324, 337, 346, 347, 363-366, 370, 372, 373, 378, 381, 385, 388, 390, 404, 410, 411, 418-420
   of the model 54, 226, 227
Preexposure 4, 5, 11, 13, 15, 18, 21, 65, 66, 68-71, 73, 77-80, 82, 86-90, 99-102, 105, 107, 108, 110-113, 115, 123, 141, 142, 144, 145, 399, 420
   effects 4, 16
   procedure 65
Prefrontal cortex (PFC) 96, 98, 133, 241
Preresponse (PRE) 96, 97
Preservation of LI 102, 105, 108, 110, 112, 117, 118, 418
Presubiculum 95, 102
Presynaptic activity 7
Presynaptic neural population 7
Problem solving 130
   Tower of Hanoi 130, 414
Procedural 9
Properties 18, 21, 22, 24, 25, 28, 29, 32, 35, 56, 64, 90, 92, 154, 161, 166, 168, 173, 177, 179, 190,

214, 224, 230, 233, 245, 246, 282, 287, 294, 311, 313, 324, 326, 352, 353, 359, 360, 362, 363, 365, 374–378, 381, 383, 384, 385, 386, 388, 392, 394, 402, 410, 420, 423
Psychometric theory 119
Putamen 116, 320
Pyramidal activity 349

Quantitative descriptions 9

Rabbit's eyeblink conditioning 423
Range 8, 33, 51, 119, 158, 168, 229, 306, 384, 392, 422, 424
Rate 10, 12, 17, 18, 23, 24, 26, 31, 33, 52, 57, 58, 60, 63, 68, 88, 90, 105, 128, 166, 182, 203, 218, 219, 227–231, 248, 302, 312, 351, 389, 390, 399
Rating 58, 404, 406, 408, 413
Reacquisition 5, 168, 216–219, 222, 228, 229, 231, 232, 238
    fast 5, 166, 215, 216, 217, 218, 219, 231, 238
    following extinction 215
    slow 5, 190, 216–219, 224, 230, 233, 238, 394
Reactive inhibition 165, 190, 231
Real-time 21, 88, 92, 155, 246–248, 247, 248, 256, 261, 272, 282, 302, 308, 310, 312, 313, 324, 328, 361–363, 384, 387, 388, 423

Rear(ing) behavior 150, 254, 256, 258, 260, 271, 304, 319, 339
Reasoning process 8
Recovery
    from backward blocking
        See Backward blocking recovery
    from forward blocking
        See Forward blocking recovery
    from latent inhibition
        See Latent inhibition recovery
    from overshadowing
        See Overshadowing recovery
Recurrent autoassociative network 9, 409
Reinforced presentations 6, 36, 37, 40, 43, 44, 54, 138, 141, 146, 149, 152, 153, 176, 177, 182, 183, 184, 195, 212, 213, 216, 229, 259, 284, 286, 335, 337
Reinforcement 5, 10, 12, 13, 29, 37, 39, 47, 49, 50, 51, 53, 59, 61–63, 87, 96, 98, 151, 152, 166, 173, 182, 214, 224, 229, 239, 241, 245, 359, 361, 362, 363, 370, 372, 373, 374, 375, 388
Reinstatement 5, 29, 30, 166, 168, 173, 180, 201–215, 222–224, 228, 230–232, 234, 235, 237–242, 393
    elimination 207
    forms 201
Relative validity 4, 35, 56, 137
Remote associate test (RAT) 120, 129

Renewal 5, 16, 33, 166, 168, 180, 191–201, 206, 207, 212, 221, 223, 224, 228–233, 237, 238, 239, 242, 293, 393, 394, 400–402, 418
    by massive extinction 197, 201, 228, 233, 394
    by pretreatment of the testing context 194, 225
Replaced elements model 253, 308, 313
Representation 7, 12, 14–16, 27, 29–32, 36, 38, 39, 57, 60–63, 86, 88, 89, 95, 97, 102, 120, 122, 130, 139, 141, 145, 154, 155, 160, 162, 165, 169, 176, 177, 179, 180, 182, 184, 188, 190, 192, 194, 198, 200, 203, 205, 215, 225, 226, 227, 231, 247, 258, 270, 307, 308, 318, 350, 351, 362, 363, 364, 365, 385, 387, 388, 394, 397, 399, 408, 418, 419, 422
Rescorla–Wagner (RW) model 8, 10, 11, 27–29, 88, 145, 154, 155, 162, 174, 191, 247, 304, 305, 365, 403, 406, 418
    Van Hamme and Wasserman version 11, 12, 140, 145, 155, 162, 176
Response 3, 5–7, 10, 11, 14, 24, 25, 32, 52, 53, 57, 61, 62, 64, 87, 96, 105, 110, 115, 121, 123, 126–130, 132, 137, 138, 146, 149, 150, 153, 155, 156, 163, 165, 175, 178, 184, 191,

206, 214, 239, 245–247, 252, 254–256, 258, 260, 264, 267, 269, 272, 275, 278, 282, 293, 294, 302, 304–307, 310, 312, 313, 314, 319, 322, 329, 332, 333, 340, 352–354, 360, 365, 366, 370, 377, 378, 381, 390, 391, 393, 402, 404, 410, 420, 423, 439, 451, 453
  form 258, 260, 264, 267, 269, 272, 278, 282, 293, 294, 305, 312–314, 329, 353, 420
  selection 130, 132
  system 254–256, 267, 304, 307, 339
Retardation 21, 36, 56, 144, 172, 173, 177, 200, 226, 227, 230, 301, 337, 393, 397, 399, 401, 402
  of conditioning 36, 56, 65
  test 172, 173, 200, 226, 227, 230, 393, 397, 399, 401, 402
Retrieval 21, 31–33, 68, 86, 88, 89, 120, 149, 161, 190, 349
Robust 33, 51, 161, 168, 177, 219, 315
Role of a CS 282, 314, 363, 375, 417

Salience 10, 17, 35, 36, 38–40, 39, 40, 54, 60, 87, 108, 157, 161, 167, 168, 169, 172, 174, 176, 178, 184, 188, 192, 194, 196, 200, 203, 206, 208, 210, 214, 263, 266, 268, 269, 270, 271, 289, 290, 294,
297, 299, 307, 311, 312, 331, 340, 396, 400, 402, 409, 423, 424
  associability 17, 87
Schizophrenia 93, 112, 113, 116, 118, 134, 135
Schmajuk–DiCarlo (SD) model 8, 18, 38, 172, 219, 245, 246, 247–250, 253–257, 260, 261, 265, 267, 272, 276, 282, 284, 286, 287, 293, 294, 297, 299–306, 308–314, 316–321, 324, 332, 339, 347, 349, 350, 352–354, 353, 354, 359, 361, 363–365, 375, 376, 384, 386, 387, 388, 392–394, 402, 403, 420
Schmajuk–Lam–Gray (SLG) model 21–23, 25, 32–36, 38, 39, 41, 51, 52, 54, 56, 64, 65, 66, 68, 71, 74, 77, 78, 80, 82, 86, 89–93, 98, 99, 101–103, 108, 109, 111, 114, 116–134, 136, 137, 145, 149, 151, 154–156, 158–162, 165–177, 166, 177, 188, 190, 200, 229, 230, 233, 234, 238, 241, 242, 246, 316, 347, 387, 392–397, 399–409, 413, 418, 420, 422
Schmajuk–Lamoureux–Holland (SLH) model 172, 230, 245, 246, 254, 255, 256, 272, 276, 282, 284, 286, 287, 289, 293, 294, 297, 299–302, 304–306, 309–314, 316, 317, 320, 324, 329, 332, 339, 347, 349, 350, 351, 353, 354, 357, 359, 361,
362, 392, 393, 403
Second-order conditioning 7, 9, 22, 24, 30, 34, 368, 369, 374, 385, 386, 410
Selective hippocampal lesions See Hippocampus selective lesions
Sensitization 72, 230
Sensory 6, 8, 9, 12, 13, 16, 96, 254, 287, 288, 296, 313, 319, 320, 351
  preconditioning 6, 8, 9, 16, 22, 24, 30, 34, 410
  representation 12, 13, 87, 313, 351
Separable error term 17
Sequences 132, 294, 361, 384, 387
Serial
  compound conditioning 9
  feature-negative discrimination 6, 249, 251, 257, 275, 276, 278, 282, 286, 287, 297, 299, 305, 306, 308, 313
  feature-positive discrimination 6, 249, 251, 257, 271–273, 278, 282, 285, 286, 287, 289, 293–295, 299, 300, 301, 305, 306, 308, 310, 311, 313, 314, 321, 326, 328, 329, 331, 332, 339, 340, 344, 345, 346, 347, 354, 360, 365, 375–378, 381, 383, 386, 392, 422
  order 387

Series 36, 49, 50, 56, 219, 233, 392, 421
  of reinforced and nonreinforced presentations 49
  of reinforced presentations 49
Serotonin 99, 101
Sham lesions (SLs) 104,105, 106, 107, 114, 115, 235, 236, 323, 327, 330-332, 334-337
Short-term memory (STM) 13, 21, 248, 250, 331
  trace 22, 23, 62, 96, 247, 248, 261, 324, 361
Sigmoid 12, 13, 32, 61, 250, 252, 389, 390, 404
Simple
  conditioning 34, 42, 56, 217, 245, 246, 257-260, 282, 286, 287, 301, 306, 315, 316, 359, 360, 361-363, 365, 381, 384-386, 388, 418
  CSs 252, 381
  discriminations 319, 321, 322, 326, 328, 331, 335, 337, 339, 340, 344, 345, 354
Simulation *See* Computer simulations
Simultaneous 34
  discrimination with a strong feature and a weak target 257
  discrimination with a weak feature and a strong target 257
  excitatory and inhibitory associations 393, 397
  feature-negative discrimination 6, 257, 268-271, 278, 284, 286-288, 305, 339
  feature-positive discrimination 6, 257, 260-263, 265-267, 278, 282, 284, 286, 287, 299-301, 305, 306, 308, 314, 321, 322, 324-326, 339, 340, 344-347, 354, 381, 392
Single response system 247, 252, 256, 267, 275
Skin conductance response (SCR) 57-62, 235, 239, 240
Slave process 35, 423
Social-psychological theory 119
Sometimes opponent process (SOP) theory 8, 14, 21, 32, 38, 53, 62, 88, 89, 92, 141, 155, 160, 162, 308
  Dickinson and Burke version 11, 14, 15, 141, 145, 155, 160, 162, 176
Spaced trials 34, 422
Spatial navigation 411, 414
Specific 3, 11, 26, 33, 40, 51, 54, 65, 87, 102, 166, 181, 193, 195, 201, 205, 221, 254, 260, 307, 310, 317, 360, 364, 365, 372, 384, 386-388
  characteristics of contexts 166
  preparations 423, 424
Spectral timing hypothesis 363, 385
Spectral timing model 361
Spontaneous recovery 5, 16, 25, 33, 63, 156, 160, 166, 167, 168, 177-194, 199, 205, 223, 224, 227, 229, 230-233, 237, 238, 241, 242, 393-395, 399-402, 420, 421
State 14, 15, 80, 82, 88, 130, 131, 141, 145, 162
Stimulus configuration 249, 256, 304, 305, 316, 317, 326, 332, 339, 353
STM *See* Short term memory
Storage 12, 31, 32, 68, 86, 88, 89, 134, 151, 318, 350, 419
Strategic planning 133
Strength 5, 7, 8, 10, 12, 15-18, 24, 32, 37, 46, 49, 52-54, 57, 58, 60, 61, 64, 68, 77, 88, 119, 120, 121, 127-132, 135, 138, 155, 162, 163, 165, 168, 171, 175-177, 190, 229, 246, 247, 249, 257, 281, 286, 299, 308, 311, 312, 321, 351, 360, 365, 370, 372-375, 385, 409, 418, 423
  of LI 68, 77, 190
Subadditive 404, 405, 406, 408, 413
Subicular excitotoxic lesions 95, 108
Subiculum (SUB) 95, 102, 134
Submaximal 162, 409, 413
Successive 6, 49
Summation 171-174, 177, 182, 196, 197, 200, 220, 222, 225, 226, 230-233, 246, 253, 258
  tests 171, 225, 226

Superconditioning/
    Supernormal
    conditioning 5, 17, 29,
    46, 48
Super-latent inhibition
    (super-LI) 82, 84, 92, 137
Suppression 100, 109, 110,
    279
  paradigm 32
  ratios 33, 64, 168, 191,
    195, 203, 229, 235,
    236
Surprising event 73
Symbols 22, 405, 409
Symmetry 411–413
  Synaptic weight 24, 27

Target 6, 32, 123, 137, 141,
    161, 166, 169, 173, 177,
    181, 182, 184, 200, 217,
    222, 223, 226, 227, 228,
    249, 251, 252, 257, 259,
    260, 261, 263, 267–272,
    275, 278, 281, 284, 287,
    289, 293, 294, 297, 299,
    306, 310–312, 314, 321,
    322, 324–326, 328, 329,
    331, 332, 337, 340,
    344–347, 354, 360, 376,
    378, 381, 386, 391
Taste aversion 35, 95, 161,
    184, 423
Teaching signal for cortical
    hidden units 352
Temporal 6, 375, 376, 381,
    389
  competition in simple
    conditioning 366
  information 360, 375
  parameters 11, 247,
    361
  properties 6, 361–363,
    365, 375, 384, 386,
    388, 389

Temporal specificity
    361–363, 375, 376, 378,
    383, 384, 386–388
  of an occasion setter
    365, 376
  of blocking 6, 366, 369,
    385, 389
  of occasion setting 6,
    376, 378, 388
  of serial FP
    discrimination 362,
    375, 376, 378, 381,
    383, 386, 388
  of the action of an
    occasion setter 375,
    385
Test 40, 41, 44–46, 48–50,
    53, 55, 105, 107, 121,
    124–127, 138, 141, 148,
    149, 151, 154, 156, 157,
    171, 172, 176, 177–185,
    188–196, 200, 203, 204,
    210, 212, 223–227, 233,
    239, 257, 260, 268, 272,
    275, 289, 291, 292, 293,
    365, 367, 369, 381, 395,
    397, 399, 400, 411, 420,
    422
Tetrodotoxin (TTX) 108, 109
Thalamus (THAL) 62, 96, 116,
    133, 319
Theories/Theorizing 3, 7–10,
    18, 86, 88, 89, 92, 101,
    102, 119, 120, 137, 161,
    165, 173, 176, 190, 229,
    231, 257, 306, 309, 311,
    349, 353, 360, 361, 384,
    403, 404
  computational 307, 313,
    351, 417
  neural network 417
Thirst 80, 82, 163
Time interval 188, 223, 251,
    306, 307, 363, 410

  between acquisition and
    extinction 228, 418
Time to complete a number
    of licks 64, 163, 168
Timing 310, 357, 359–366,
    372, 376, 383, 384, 385,
    387–389, 417, 418
  and occasion setting 359
  model 359, 362, 365,
    387, 389
Timing mechanism 357, 361,
    362
  clock-based 384
  clock-free 384
Tonic 25, 337
  contextual stimuli 166,
    235
Total
  CS-preexposure time
    68, 70
  novelty 24, 26
Traces 13, 22, 34, 60–62, 96,
    115, 120, 159, 161, 170,
    188, 200, 247–251, 253,
    261, 272, 278, 311, 314,
    324, 329, 331, 344, 347,
    354, 359, 361–366,
    372–376, 378, 383–390,
    394, 397, 409, 410, 419
Training 4, 6, 13–16, 32, 37,
    39, 41–43, 50, 51, 55, 62,
    77, 78, 82, 84, 89, 115,
    121, 125, 128, 138, 139,
    145, 148, 149, 151, 152,
    155, 167, 171, 174, 180,
    188, 196, 201, 207, 210,
    235, 239, 250, 257–260,
    262, 266, 268, 272, 275,
    276, 278, 279, 282, 284,
    286–290, 294, 297, 299,
    301, 302, 305, 309,
    310–312, 324, 331, 332,
    337, 345, 347, 351, 352,
    360, 369, 376–378, 383,

404–409, 411, 413, 423
cage 156, 181, 223, 394
context 4, 15, 232, 293, 396
Transfer 17, 191-193, 225, 258-260, 264, 287-291, 293-299, 305, 308, 310, 312-314, 329-333, 342, 352, 422
  effects 258, 260, 287
Transitivity 409-414
Trial
  order 16
  -to-trial 246
Trials to criterion 124, 340, 423, 424
Two-layer networks 246

Unblocking 4, 11, 149, 368, 372, 373, 385, 386
Uncertainty 53, 59, 62
Unconditioned stimulus (US) 4–18, 21, 22, 24, 25-38, 40-42, 44, 46, 49-52, 54, 56-66, 68, 71, 74, 80, 82, 84, 86-90, 92, 94-99, 101-103, 105, 108, 110, 113, 117, 118, 120, 123, 125, 138-141, 144-147, 149, 150, 154-163, 165-169, 171-232, 235, 237-242, 245-263, 265-275, 277, 278, 280-282, 284, 286, 287, 289, 291-294, 296, 297, 299-302, 304-315, 317, 319-322, 324-326, 328, 329, 331-333, 337, 339, 340, 344, 345, 347, 349-353, 359-367, 369, 370, 372-376, 378, 380, 381, 383-388, 390, 391, 393-395, 397, 399, 400, 402, 403, 406, 409, 410, 411, 417, 419, 423
  duration
    34, 40, 125, 171, 248, 361, 386
  preexposure 4, 90
  preexposure effect 11
  strength 11, 17, 38, 46, 167, 169, 419
Unique-stimulus hypothesis 312

$V_{CS, CX}$ 94, 95, 102, 103, 112, 236
$V_{CS, US}$ 7, 22, 23, 68, 82, 103, 140, 143, 148, 152, 397, 399
Ventral 62, 94, 96, 113, 132, 133, 134, 238, 319
  hippocampus /
    hippocampal 113, 238
  pallidum (VP) 96, 117, 132
  tegmental area (VTA) 62, 132, 133
VH 248, 249, 250-252, 254, 318, 320, 321, 332, 363, 364, 365, 389, 390, 391
Visual stimuli 104, 123, 239, 254, 258, 271
VN 248, 249, 251-256, 262, 267, 269, 274, 277, 280, 317-321, 363-365, 390
VS 248, 250-256, 262, 267, 268, 273, 277, 280, 315, 317, 318, 320

Water licking 64, 150
Weak
  computational power 418
  conditioning and spontaneous recovery 184
Weber law 362
Within
  -category transfer 294, 314
  -compound association 11, 160, 162
Word pairs 235
  Working memory 62, 63, 133

RECEIVED
FEB 1 4 2011
GUELPH HUMBER LIBRARY
205 Humber College Blvd
Toronto, ON   M9W 5L7